ROUTLEDGE HANDBOOK OF SOUTHEAST ASIAN DEVELOPMENT

Southeast Asia is one of the most diverse regions in the world – hosting a wide range of languages, ethnicities, religions, economies, ecosystems and political systems. Amidst this diversity, however, has been a common desire to develop. This provides a uniting theme across landscapes of difference.

This Handbook traces the uneven experiences that have accompanied development in Southeast Asia. The region is often considered to be a development success story; however, it is increasingly recognized that growth underpinning this development has been accompanied by patterns of inequality, violence, environmental degradation and cultural loss. In 30 chapters, written by established and emerging experts of the region, the Handbook examines development encounters through four thematic sections:

- Approaching Southeast Asian development,
- Institutions and economies of development,
- People and development and
- Environment and development.

The authors draw from national or sub-national case studies to consider regional scale processes of development – tracing the uneven distribution of costs, risks and benefits. Core themes include the ongoing neoliberalization of development, issues of social and environmental justice, and questions of agency and empowerment.

This important reference work provides rich insights into the diverse impacts of current patterns of development and in doing so raises questions and challenges for realizing more equitable alternatives. It will be of value to students and scholars of Asian Studies, Development Studies, Human Geography, Political Ecology and Asian Politics.

Andrew McGregor is Associate Professor in the Department of Geography and Planning at Macquarie University, Australia. He is a human geographer with interests in political ecology, critical development studies and climate mitigation strategies in Indonesia, Timor-Leste and Australia. He is author of *Southeast Asian Development* (Routledge, 2008).

Lisa Law is Associate Professor at James Cook University, Cairns, Australia. She is an urban social geographer with interests in the politics of urban spaces in Southeast Asia and tropical Australia. She is currently Editor in Chief of *Asia Pacific Viewpoint*.

Fiona Miller is a Senior Lecturer in the Department of Geography and Planning at Macquarie University, Australia. She is a human geographer with an interest in political ecology, social vulnerability, society-water relations and climate change adaptation in Vietnam, Cambodia and Australia. She is currently Southeast Asian Editor of *Asia Pacific Viewpoint*.

ROUTLEDGE HANDBOOK OF SOUTHEAST ASIAN DEVELOPMENT

*Edited by Andrew McGregor,
Lisa Law and Fiona Miller*

LONDON AND NEW YORK

First published 2018
by Routledge
2 Park Square, Milton Park, Abingdon, Oxon OX14 4RN

and by Routledge
711 Third Avenue, New York, NY 10017

Routledge is an imprint of the Taylor & Francis Group, an informa business

© 2018 selection and editorial matter, Andrew McGregor, Lisa Law and Fiona Miller; individual chapters, the contributors

The right of Andrew McGregor, Lisa Law and Fiona Miller to be identified as the authors of the editorial material, and of the authors for their individual chapters, has been asserted in accordance with sections 77 and 78 of the Copyright, Designs and Patents Act 1988.

All rights reserved. No part of this book may be reprinted or reproduced or utilised in any form or by any electronic, mechanical, or other means, now known or hereafter invented, including photocopying and recording, or in any information storage or retrieval system, without permission in writing from the publishers.

Trademark notice: Product or corporate names may be trademarks or registered trademarks, and are used only for identification and explanation without intent to infringe.

British Library Cataloguing-in-Publication Data
A catalogue record for this book is available from the British Library

Library of Congress Cataloging-in-Publication Data
Names: McGregor, Andrew, 1971– editor. |
Law, Lisa, 1967– editor. | Miller, Fiona, 1971– editor.
Title: Routledge handbook of Southeast Asian development / edited by
Andrew McGregor, Lisa Law and Fiona Miller.
Description: New York : Routledge, 2018. |
Includes bibliographical references and index.
Identifiers: LCCN 2017025616 | ISBN 9781138848535 (hbk) |
ISBN 9781315726106 (ebk)
Subjects: LCSH: Economic development—Southeast Asia. |
Southeast Asia—Economic conditions. | Southeast Asia—Economic conditions—Regional disparities. | Southeast Asia—Economic policy.
Classification: LCC HC441 .R678 2018 | DDC 338.959—dc23
LC record available at https://lccn.loc.gov/2017025616

ISBN: 978-1-138-84853-5 (hbk)
ISBN: 978-1-315-72610-6 (ebk)

Typeset in Bembo
by Apex CoVantage, LLC

Printed and bound by CPI Group (UK) Ltd, Croydon, CR0 4YY

To our wonderful children Finley, Jarvis, Madeleine, Dylan and Audrey.

CONTENTS

List of figures xi
List of tables xii
Contributors xiv
Acknowledgments xxi

PART 1
Approaching Southeast Asian development 1

1 Approaching Southeast Asian development 3
  Andrew McGregor, Lisa Law and Fiona Miller

2 What is development in Southeast Asia and who benefits? Progress, power and prosperity 14
  Katharine McKinnon

3 Neoliberalism in Southeast Asia 27
  Simon Springer

4 Aggregate trends, particular stories: tracking and explaining evolving rural livelihoods in Southeast Asia 39
  Jonathan Rigg and Albert Salamanca

5 'Nature' embodied, transformed and eradicated in Southeast Asian development 53
  Victor R. Savage

PART 2
Development institutions and economies in Southeast Asia: introduction   65
Andrew McGregor, Lisa Law and Fiona Miller

6   Neoliberalism and multilateral development organizations in Southeast Asia   69
      Toby Carroll

7   The International Labour Organization as a development actor in Southeast Asia   85
      Michele Ford, Michael Gillan and Htwe Htwe Thein

8   Justice processes and discourses of post-conflict reconciliation in Southeast Asia: the experiences of Cambodia and Timor-Leste   96
      Rachel Hughes

9   Civil society participation in the reformed ASEAN: reconfiguring development   109
      Kelly Gerard

10   Industrial economies on the edge of Southeast Asian metropoles: from gated to resilient economies   120
      Delik Hudalah and Adiwan Aritenang

11   Community economies in Southeast Asia: a hidden economic geography   131
      Katherine Gibson, Ann Hill and Lisa Law

12   Implications of non-OECD aid in Southeast Asia: the Chinese example   142
      May Tan-Mullins

13   'Timeless Charm'? Tourism and development in Southeast Asia   153
      John Connell

PART 3
People and development: introduction   169
Lisa Law, Fiona Miller and Andrew McGregor

14   Family, migration and the gender politics of care   173
      Brenda S.A. Yeoh and Shirlena Huang

15	Healthcare entitlements for citizens and trans-border mobile peoples in Southeast Asia *Meghann Ormond, Chan Chee Khoon and Sharuna Verghis*	186
16	Migration, development and remittances *Philip Kelly*	198
17	Children, youth and development in Southeast Asia *Harriot Beazley and Jessica Ball*	211
18	Ethnic minorities, indigenous groups and development tensions *Sarah Turner*	224
19	Globalization, regional integration and disability inclusion: insights from rural Cambodia *Alexandra Gartrell and Panharath Hak*	238
20	Religion and development in Southeast Asia *Orlando Woods*	250
21	A feminist political ecology prism on development and change in Southeast Asia *Bernadette P. Resurrección and Ha Nguyen*	261
22	Rethinking rural spaces: decropping the Southeast Asian countryside *Tubtim Tubtim and Philip Hirsch*	271

PART 4
Environment and development: introduction
Fiona Miller, Andrew McGregor and Lisa Law   281

23	Material, discursive and cultural framings of water in Southeast Asian development *Fiona Miller*	285
24	Agriculture and land in Southeast Asia *Yayoi Fujita Lagerqvist and John Connell*	300
25	Labor, social sustainability and the underlying vulnerabilities of work in Southeast Asia's seafood value chains *Simon R. Bush, Melissa J. Marschke and Ben Belton*	316

26 Oil palm cultivation as a development vehicle: exploring the
    trade-offs for smallholders in East Malaysia   330
    Fadzilah Majid Cooke, Adnan A. Hezri, Reza Azmi, Ryan Morent Mukit,
    Paul D. Jensen and Pauline Deutz

27 Disasters and development in Southeast Asia: toward equitable
    resilience and sustainability   342
    Frank Thomalla, Michael Boyland and Emma Calgaro

28 Upscaled climate change mitigation efforts: the role of regional
    cooperation in Southeast Asia   362
    Noim Uddin and Johan Nylander

29 Can Payments for Ecosystem Services (PES) contribute to sustainable
    development in Southeast Asia?   376
    Andreas Neef and Chapika Sangkapitux

30 Forest-led development? A more-than-human approach to forests in
    Southeast Asian development   392
    Andrew McGregor and Amanda Thomas

Index   *408*

FIGURES

1.1	Map of Southeast Asia	5
4.1	Gender and generational work in two northeastern Thai villages, 1982 and 2008	43
10.1	Map of industrial concentration in Jakarta Metropolitan Area	127
23.1	Total renewable water resources per capita (m^3/inhab/year) 1992–2014	290
23.2	Total annual freshwater withdrawals (m^3) as a percentage of total (2013)	291
23.3	Annual freshwater withdrawals (m^3) as a percentage of total (2013) for agriculture, industry and domestic	291
24.1	Land under cereal crop production for selected Southeast Asian countries	303
24.2	Yield of cereal crop production (kg/ha) for selected Southeast Asian countries	304
24.3	Share of agriculture, percentage of GDP for selected Southeast Asian countries	307
25.1	Capture fisheries and aquaculture output in Southeast Asia, 1950–2013	317
25.2	Social issue areas in the seafood employment chain	319
27.1	Mortality risk distribution of selected hydro-meteorological hazards (tropical cyclone, flood, rain-triggered landslide) in Southeast Asia	343
27.2	Total number of disasters [(i) hydro- meteor- and climate-logical disasters, and (ii) geophysical disasters] that have occurred in Southeast Asia, 1970–2015	344
27.3	The total economic impact (damages) of disasters in Southeast Asia, 1970–2015	347
27.4	Trends showing urban slum population numbers and the proportion of urban populations living in slums in Southeast Asia, 1990–2010	349
29.1	Map of Mae Sa watershed, Chiang Mai Province, Northern Thailand	385
29.2	Proposed PES Model for Mae Sa watershed	386
30.1	Regional extent of tropical forest in Southeast Asia (including Papua New Guinea) derived from Spot Vegetation 1 km data 2000	393
30.2	Change in forested area by country and year	395
30.3	Regional pattern of main areas and causes of forest change in Southeast Asia	396
30.4	Value of forest products exports (in millions USD at 2011 prices and exchange rates)	397
30.5	Contribution of the forestry sector to total GDP	397

TABLES

1.1	Southeast Asian development indicators	8
4.1	The Southeast Asian countryside: the big picture (1960–2012)	40
4.2	Asia's greying farmers	44
4.3	Trends in average farm size, Indonesia, Philippines and Thailand (ha)	45
4.4	Rural, urban and national poverty rates for selected countries of developing Asia, earliest and latest years	47
6.1	Three phases of neoliberal development policy	73
6.2	ADB 'sovereign' approvals (including loans, grants and official co-financing and technical assistance) for Southeast Asian developing member countries	75
6.3	World Bank (International Development Association and International Bank for Reconstruction and Development) commitments for Southeast Asian developing member countries	75
6.4	Total historical cumulative MDO allocations to Southeast Asian countries (as of 2014) in millions of dollars	75
10.1	Total number of SMIs in Jakarta Metropolitan Area	124
10.2	Total employment in SMIs in Jakarta Metropolitan Area	124
10.3	The concentration of SMIs in Jakarta Metropolitan Area	126
10.4	Location Quotient of SMI employment clusters in Jakarta Metropolitan Area	128
12.1	Geographical distribution of China's foreign assistance funds, 2010–2012	144
12.2	Timeline of Chinese aid with particular reference to Southeast Asia	145
12.3	Similarities and differences between OECD and Chinese aid	148
13.1	Regional sources of visitors to ASEAN	155
13.2	Tourist arrivals by country in Southeast Asia	155
16.1	Migration and remittances in Southeast Asia	201
17.1	Positional ranking based on children's well-being	213
18.1	Ethnic minorities and indigenous peoples in Southeast Asia	225
23.1	Multiple dimensions of water and development	286
23.2	Irrigation water withdrawals	293
23.3	Percentage of Southeast Asian population with access to improved drinking water sources	295
23.4	Basic water, sanitation and health indicators	295

24.1	Land reforms in selected Southeast Asian countries	302
24.2	Yield of cereal crop in selected Southeast Asian countries (Unit: kg/ha)	303
24.3	Rural population, percentage of total population	306
25.1	Timeline of key media, NGO and academic coverage	318
25.2	Employment in capture fisheries in selected Southeast Asian countries	320
27.1	The 2016 World Risk Index for Southeast Asian nations	345
27.2	Major disasters affecting Southeast Asia in terms of loss of life and economic damage, 1970–2016	346
28.1	Regional cooperation in the Southeast Asian region	369
29.1	Major prerequisites for PES schemes in the Southeast Asian context	379

CONTRIBUTORS

Adiwan Aritenang is an Assistant Professor in the Urban and Regional Program in Institute Technology Bandung (ITB), Indonesia. He was a Postdoctoral Fellow in the Institute of Southeast Asian Studies (ISEAS) in Singapore. His research interests include urban and regional economics, decentralisation, supra-regional integration and creative cities.

Reza Azmi holds a doctorate in Plant Ecology and Systematics and is the Founder and Executive Director of Wild Asia, a Malaysia-based environmental and social enterprise focused on helping oil palm smallholders to self-organize, overcome their own challenges and be supported by industry through certification.

Jessica Ball is a Professor in Child and Youth Care at the University of Victoria, Canada. She is active in research, teaching and consulting throughout Southeast Asia with a focus on policies and practices that manufacture the marginalization of children and families, particularly indigenous and minoritized populations and economic and forced migrants.

Harriot Beazley is a children's geographer and community development practitioner, focusing on rights-based, child-focused participatory research in the Southeast Asia region (especially Indonesia). Harriot is Program Leader (International Development) and Senior Lecturer (Human Geography) at the University of the Sunshine Coast, Australia. She is Commissioning Editor for the journal *Children's Geographies*.

Ben Belton is an Assistant Professor of International Development in the Department of Agricultural, Food and Resource Economics, Michigan State University, USA. His research focuses on the political economy of aquaculture and capture fisheries development in South and Southeast Asia, agricultural value chains, and food and nutrition security.

Michael Boyland is a Research Associate with Stockholm Environment Institute, Bangkok. Michael is a researcher for the SEI Asia Research Cluster on Reducing Disaster Risk and the global SEI Initiative on Transforming Development and Disaster Risk. He holds an MA in Disasters, Adaptation and Development from King's College London.

Contributors

Simon R. Bush is Professor and Chair of the Environmental Policy Group at Wageningen University, The Netherlands. Simon's research focuses on the interaction between public and private governance arrangements within the sustainable seafood movement.

Emma Calgaro is a Research Fellow with the Hazards Research Group, School of Geosciences, University of Sydney, Australia. Emma has ten years' experience in Disaster Risk Reduction (DRR), vulnerability and climate change research with a strong regional focus on Southeast Asia, Australia and the South Pacific. Her research focuses on understanding the complex set of contextual and cultural factors that impede and/or improve resilience and vulnerability levels to risk.

Toby Carroll is Associate Professor in the Department of Asian and International Studies at City University of Hong Kong. His research concentrates on the political economy of development and development policy, with a particular geographical focus on Asia.

Chan Chee Khoon is a Health Policy Analyst at the University of Malaya. He graduated from Harvard University with a doctorate in epidemiology and has served on the editorial advisory boards of *International Journal for Equity in Health*, *Global Health Promotion*, *Global Social Policy* and *Oxford Bibliographies in Public Health*.

John Connell is Professor of Geography in the School of Geosciences, University of Sydney, Australia. His research focuses on migration and remittances in the Pacific. He has written and edited more than 30 books, including *Tourism at the Grassroots: Villagers and Visitors in the Asia Pacific* (with B. Rugendyke) and *Islands at Risk: Environments, Economies and Contemporary Change*.

Pauline Deutz is a Senior Lecturer at the University of Hull, United Kingdom. She takes an interdisciplinary social science approach to, primarily, waste policy issues. She has numerous publications on industrial ecology, including the 2015 co-edited book *International Perspectives on Industrial Ecology*. Pauline is vice-president of the International Sustainable Development Research Society.

Michele Ford is Professor of Southeast Asian Studies, Australian Research Council Future Fellow and Director of the Sydney Southeast Asia Centre at the University of Sydney, Australia. Her research focuses on labor movements in Southeast Asia. Michele is the author of *Workers and Intellectuals: NGOs, Unions and the Indonesian Labour Movement* (NUS/Hawaii/KITLV 2009). She has also edited or co-edited several volumes including *Social Activism in Southeast Asia* (Routledge 2013) and *Beyond Oligarchy: Wealth, Power, and Contemporary Indonesian Politics* (Cornell SEAP 2014).

Alexandra Gartrell is a Lecturer in Human Geography, School of Social Sciences at Monash University, Australia. Her current research interests include gendered socio-spatial politics of disability and employment, sexual and reproductive health and rights of persons with disabilities, and disability inclusive disaster risk reduction.

Kelly Gerard is a Lecturer in Political Science and International Relations at the University of Western Australia. Her research interests span political economy and development and social movements in Southeast Asia. She is the author of *ASEAN's Engagement of Civil Society: Regulating Dissent* (Routledge 2014).

Contributors

Katherine Gibson is a Professorial Research Fellow in the Institute for Culture and Society at the Western Sydney University, Australia. She is an economic geographer with an international reputation for innovative research on economic transformation and over 30 years' experience of working with communities to build resilient economies.

Michael Gillan is a Senior Lecturer at the UWA Business School at the University of Western Australia. His current research focuses on Global Union Federations; employment relations in global production networks; labor movements and politics in India; and employment relations in Myanmar.

Panharath Hak is a graduate student at Monash University, Australia, majoring in environmental management and sustainability. He has bachelor degrees in environmental science and international relations from Cambodia and is now conducting a master's research project on the sustainability of Cambodia's community forest management under the UN's REDD+ framework.

Adnan A. Hezri is Adjunct Fellow in the Fenner School of Environment and Society, the Australian National University and a Member of the United Nations' International Resource Panel. His specialization is comparative public policy spanning areas such as green economy, natural resources governance and sustainable development strategy.

Ann Hill is an Assistant Professor in Education at the University of Canberra, Australia, currently teaching and conducting research in community development, community economies and global education. She has extensive research experience in Southeast Asia and is currently a researcher on Australian Research Council Discovery Project "Strengthening Economic Resilience in Monsoon Asia."

Philip Hirsch is Professor of Human Geography in the School of Geosciences at the University of Sydney, Australia. His research interests are in agrarian change, natural resource management and the politics of environment in Thailand and the wider Mekong region.

Shirlena Huang is Associate Professor of Geography and Vice-Dean (Graduate Studies) at the Faculty of Arts and Social Sciences, National University of Singapore. Her research mainly examines issues at the intersection of migration, gender and families (with a particular focus on care labor migration and transnational families within the Asia-Pacific region) as well as urbanization and heritage conservation (particularly in Singapore).

Delik Hudalah is Associate Professor of Urban and Regional Planning at the School of Architecture, Planning and Policy Development and senior researcher at the Research Center for Infrastructure and Regional Development, Bandung Institute of Technology, Indonesia. He focuses on the transformation of urban frontiers in emerging cities, metropoles and mega-regions in the context of Asian countries' transition to decentralization and democracy.

Rachel Hughes is an Australian Research Council Postdoctoral Research Fellow in the School of Geography at The University of Melbourne, Australia. She has wide-ranging interests in the geographies of law, geopolitics, public memory, and visual and material cultures. Her current project examines the relationship between the non-judicial legacies of the Khmer Rouge

Tribunal and wider social and political change. She is the author of a number of book chapters and journal articles on the memorialization of the Cambodian genocide, and a co-editor of the collection *Observant States: Geopolitics and Visual Culture* (2010).

Paul D. Jensen is a freelance corporate sustainability consultant and researcher at the University of Hull, United Kingdom, currently working on sustainable supply chains. He focuses on industrial ecology in addition to authoring business sustainability best practice reports for regional and international governmental organizations.

Philip Kelly is Professor of Geography at York University, Canada, and Director of the York Centre for Asian Research. His research has examined the global dimensions of Philippine development, immigrant labor market processes and intergenerational mobility in Canada, and transnational economic ties between Canada and the Philippines.

Yayoi Fujita Lagerqvist is a lecturer of Human Geography in the School of Geosciences at the University of Sydney, Australia. Her research interests are in natural resource management and rural livelihood in the mainland Southeast Asia.

Lisa Law is an Associate Professor in the Centre for Tropical Urban and Regional Planning at James Cook University in Cairns, Australia. She is an urban social geographer with interests in the politics of urban spaces in Southeast Asia and tropical Australia. She is Editor in Chief of *Asia Pacific Viewpoint*, a Wiley journal publishing articles in geography and allied disciplines about the region.

Fadzilah Majid Cooke is Professor of Sociology at Universiti Malaysia Sabah, working in political ecology, on local rights and access to natural resources as well as the socio-politico-technological context of conservation. She is a steering committee member of the International Science Council's Asia Pacific program on the Sustainability Initiative for the Marginal Seas of East and South Asia (SIMSEA).

Melissa J. Marschke is an Associate Professor in the School of International Development and Global Studies at the University of Ottawa, Canada. Melissa's research focuses on livelihoods, labor and governance issues in the seafood sector.

Andrew McGregor is Associate Professor in the Department of Geography and Planning at Macquarie University, Australia. He has conducted research in Indonesia, Timor-Leste and Myanmar and previously worked for UNICEF Australia. His current interests are on climate mitigation strategies, more-than-human geographies and critical development studies. He is author of *Southeast Asian Development* (Routledge 2008) and former Editor in Chief of *Asia Pacific Viewpoint*.

Katharine McKinnon is Senior Lecturer with the Community Planning and Development Program at La Trobe University, Australia. She is a social geographer whose work focuses on the politics of development, particularly in relation to professional practice, community economies and gender, and is informed by post-development theory and feminist economic geography.

Fiona Miller is a Senior Lecturer in the Department of Geography and Planning, Macquarie University, Sydney, Australia. She conducts research on the social and equity dimensions of environmental change in the Asia Pacific, with a particular interest in political ecology, social vulnerability, society-water relations and climate change adaptation.

Ryan Morent Mukit is a postgraduate student of Sociology and Anthropology at the Faculty of Humanities, Arts and Heritage (FKSW) at Universiti Malaysia Sabah and a research assistant at the same university. He is working on livelihoods and access to land for indigenous oil palm smallholders and its political economic context.

Andreas Neef is Professor and Director of the Development Studies program at the University of Auckland, New Zealand. His research focuses on natural resource governance, land grabbing, development-induced displacement, adaptation to climate change, and post-disaster response and recovery. He served as Scientific Advisor to the German Parliament on issues of global food security and on societal and political discourses on the economic valorization of biodiversity and ecosystem services.

Ha Nguyen is a Research Associate for Gender, Environment and Development research cluster at the Stockholm Environment Institute, Sweden. Her research interests include gender and empowerment in large-scale investments and rural livelihoods, inclusive and sustainable agriculture supply chains. She has researched women's land tenure rights in Vietnam and women's empowerment in home gardens in Cambodia.

Johan Nylander is Director and Principal Consultant of Climate Policy and Markets Advisory International AB, Sweden. Johan's research focuses on climate policy, sustainable development and climate change mitigation, particularly carbon market mechanisms.

Meghann Ormond is Assistant Professor in Cultural Geography at Wageningen University and Research, The Netherlands. With a background in health geography and migration studies, Meghann focuses on how shifting visions of citizenship and belonging transform social and economic development agendas and impact healthcare systems.

Bernadette P. Resurrección is Senior Research Fellow at the Stockholm Environment Institute (SEI), Sweden. For more than 15 years, she has researched gender, natural resource management, livelihoods, climate change adaptation, disasters and mobility in Vietnam, the Philippines, Thailand and Cambodia. Her current feminist political ecology research interests include gender professionals in techno-scientific, development and environment policy fields, as well as disaster and large-scale economic land concession displacements.

Jonathan Rigg is Professor of Geography at the National University of Singapore and Director of the Asia Research Institute. He is the author of *Challenging Southeast Asian Development: The Shadows of Success* (Routledge, 2016) and has been working on issues of agrarian change and rural livelihoods since the early 1980s.

Albert Salamanca is a Research Fellow at the Stockholm Environment Institute in Bangkok, Thailand, where he leads SEI's initiative on Transforming Development and Disaster Risk. He has undertaken fieldwork across mainland and insular Southeast Asia.

Chapika Sangkapitux is an honorary academic with the Development Studies program at the University of Auckland, New Zealand. Her research expertise is in the field of environmental and resource economics, particularly choice modelling, payments for environmental services and agricultural multifunctionality. In a recent study funded by the Thailand Research Fund, she compared the experience of various OECD countries in promoting multifunctional agriculture.

Victor R. Savage is currently a Visiting Senior Fellow at the S. Rajaratnam School of International Studies at the Nanyang Technological University, Singapore. He is also the Honorary Vice-President of the Commonwealth Geographical Bureau (CGB) (2016–2020). His major research interests are climate and environmental change, human-nature relationships and urban landscapes in Southeast Asia.

Simon Springer is an Associate Professor in the Department of Geography at the University of Victoria, Canada. His research agenda explores the political, social and geographical exclusions that neoliberalization has engendered, particularly in post-transitional Cambodia, where he emphasizes the spatialities of violence and power. Recent books include *The Discourse of Neoliberalism: An Anatomy of a Powerful Idea* (Rowman & Littlefield, 2016), and *Violent Neoliberalism: Development, Discourse and Dispossession in Cambodia* (Palgrave Macmillan, 2015).

May Tan-Mullins is Professor in International Relations, Dean of Graduate School, Director of Institute of Asia and Pacific Studies, University of Nottingham, China, and Series Editor of Palgrave Series in Asia and Pacific Studies. Her research interests are political ecology of rising China, environmental and energy justice, poverty alleviation and building resilience for the poorest and most vulnerable.

Htwe Htwe Thein is a Senior Lecturer in International Business at the School of Management, Curtin University, Australia, where she teaches international business, management in Asia, and culture and ethics in business. Her research focuses on economic development and international business investment in Myanmar and on institutional theory within the field of international business.

Frank Thomalla is a Senior Research Fellow at Stockholm Environment Institute, Bangkok, Thailand. He leads the SEI Asia Research Cluster on Reducing Disaster Risk and co-leads the global SEI Initiative on Transforming Development and Disaster Risk (TDDR). He is also a Member of the United Nations Office for Disaster Risk Reduction (UNISDR) Asia Science, Technology, and Academia Advisory Group (ASTAAG).

Amanda Thomas is a Lecturer in Environmental Studies at Victoria University of Wellington in Aotearoa New Zealand. She is a human geographer and researches nature society relations and decision-making about the environment.

Tubtim Tubtim is an independent researcher based in Sydney, Australia, and Chiang Mai, Thailand. Since the early 1990s she has worked on community-based natural resource management in the Mekong Region, and more recently she has written on social change in peri-urban Thailand.

Sarah Turner is a Professor in the Department of Geography, McGill University, Canada. Her research explores how ethnic minority communities in the Sino-Vietnamese borderlands create sustainable livelihoods in the face of state modernisation and 'development' plans. She co-authored *Frontier Livelihoods: Hmong in the Sino-Vietnamese borderlands* (2015, University of Washington Press), and edited *Red Stamps and Gold Stars: Fieldwork Dilemmas in Upland Socialist Asia* (2013, University of British Columbia Press).

Contributors

Noim Uddin is currently a Postdoctoral Associate at Stony Brook University, Long Island New York, USA. He is also a Senior Consultant with Climate Policy and Markets Advisory International AB and is an Honorary Associate with the Department of Geography and Planning of Macquarie University. Noim's research focuses on sustainable/low-carbon energy strategies, climate change policy and greenhouse gas risk management.

Sharuna Verghis is with the Jeffrey Cheah School of Medicine and Health Sciences, Monash University Malaysia and Health Equity Initiatives, Malaysia. She has worked extensively on health and mobility, nationally and internationally. Her professional interests include community-based health interventions, equity of access to healthcare and community-based participatory research.

Orlando Woods is an independent researcher, currently based in London, United Kingdom. He previously spent ten years in Singapore, where he completed a PhD in Geography at the National University of Singapore in 2012. In 2016, his book *Religion and Space: Competition, Conflict and Violence in the Contemporary World* (co-authored with Professor Lily Kong) was published by Bloomsbury.

Brenda S.A. Yeoh is Professor, Department of Geography, National University of Singapore. Her research interests include the politics of space in colonial and postcolonial cities as well as transnational migration in the Asian context. She co-authored *Return: Nationalizing Transnational Mobility in Asia* (Duke University Press, 2013) and *Transnational Labour Migration, Remittances and the Changing Family in Asia* (Palgrave Macmillan, 2015).

ACKNOWLEDGMENTS

Projects of this scale require a supportive environment to come to fruition, and we are especially grateful to our Editorial Board who helped steer us through the process. We were fortunate to have editorial guidance from eminent scholars whose work has transformed the way we think about Southeast Asian development, including Professors Jonathan Rigg (National University of Singapore), Katherine Gibson (Western Sydney University), Phillip Kelly (York University), Victor R. Savage (National University of Singapore), James Sidaway (National University of Singapore) and Sarah Turner (McGill University). We also acknowledge the financial support provided by the Department of Geography and Planning at Macquarie University and are particularly appreciative of the tireless efforts of Dr. Claire Colyer who kept us focused and organized throughout the many months it took to bring this collection to print.

PART 1

Approaching Southeast Asian development

1
APPROACHING SOUTHEAST ASIAN DEVELOPMENT

Andrew McGregor, Lisa Law and Fiona Miller

Introduction

Southeast Asia is typically presented as a development success story. Since the collapse of colonialism most countries have experienced significant improvements in health, education, incomes and opportunities, and boast swelling middle classes. The region has avoided inter-state conflict for an extended period under the auspices of the Association of Southeast Asian Nations (ASEAN), and while many political freedoms remain restricted, at a regional scale progress in countries like Myanmar and Indonesia suggest they have been improving. The infrastructure and facilities of cities like Singapore and Bangkok have made them globally significant finance and transportation hubs, and the region continues to attract high levels of foreign investment, bolstered by initiatives such as the recent formation of the ASEAN Economic Community. Improvements in agricultural production alongside enhanced mobility have diversified rural incomes and opportunities, while initiatives oriented at conserving the region's rich natural resources and biodiversity have proliferated. There are many challenges ahead, particularly in terms of positioning itself alongside the neighboring political economies of India and China; however, the future of the region is generally considered to be bright.

Such glossy regional interpretations provide a narrative that is attractive to many, particularly political and business elites within and outside the region. However it is only part of the story. As countless studies have shown, Southeast Asia is a region of immense diversity, not only in terms of society, culture, economy, environment and politics, but also in terms of its development experiences. Southeast Asia has indeed developed at an impressive pace over the last few decades, but, as is well recognized, development has been uneven and comes with its own set of challenges and costs. More critical accounts highlight the huge disparities in wealth and opportunity dividing rich and poor, the millions of people left behind by development – even in ostensibly middle income countries, the lack of security or services typifying sprawling informal urban settlements and impoverished rural villages, the harsh labor conditions sustained by foreign investment in export processing zones, widespread human rights abuses and abuse of power, and the ongoing degradation of the natural environment to fuel primary industries and rampant consumption. These stories are also true, providing a counterpoint to narratives of success.

Given such diversity the challenge of putting together this *Routledge Handbook of Southeast Asian Development* is a considerable one. We could side with either inflection to provide an update on development from those perspectives – reinforcing one set of stories over the other. We have chosen not to do that. Instead we have invited a range of outstanding regional scholars to each provide a chapter analyzing an aspect of development that reflects cutting edge scholarship on the topic. In particular we asked authors to move beyond mere description or critique to identify the processes of development and how more equitable, sustainable and empowering forms of development might be pursued. In this sense the Handbook provides a level of understanding that goes far beyond the statistical analyses that dominate development reports on the region. Such statistics are important but fall well short of capturing how and why development is occurring and what the intended and unintended impacts may be. Instead we have sought to provide a perspective on development in the region that goes beyond statistics and simplistic good/bad binaries from the multiple viewpoints of those who have spent their careers studying it.

In taking on the task of analyzing Southeast Asian development, two sets of issues immediately become apparent. First, what is Southeast Asia and how and why should we approach it as a region. Second, what is development and how should we approach it in the Southeast Asian context. In what follows we will build from previous scholarship on these topics to argue that a regional approach to development is important for understanding how and why development occurs in some places and not others. Our intention is not to smooth out the uneven experience of development across the region – a regional GDP does not feature! – but instead to highlight the interconnections that are bringing about diverse development geographies. The regional scale, existing between the nation-state and the global, is under-represented in academia and practice, and yet it reveals much about the nature of development and its variable impacts.

Southeast Asia as a region

The region examined in this collection incorporates what is sometimes known as mainland Southeast Asia – Myanmar, Thailand, Cambodia, the Lao People's Democratic Republic (Laos), the Socialist Republic of Vietnam, Malaysia and Singapore – as well as maritime or insular Southeast Asia comprised of the Philippines, Indonesia, Brunei Darussalam and Timor-Leste (see Figure 1.1). The grouping is driven by geography: the region is nestled between China and India to the north and northwest, the Pacific and Indian oceans to the east and west, and Papua New Guinea and Australia in the southeast. However, the borders of the region, or where the region ends, have been driven as much by colonialism, nationalism and geopolitics as any essential geographic feature. The indigenous people of Papua, for example, in Indonesia's easternmost province, have much more in common with their Melanesian cousins in Papua New Guinea on the eastern side of the island than people in Java, or broader Southeast Asia. Similarly ongoing unresolved tensions concerning the large gas deposits beneath the Timor Sea involving Australia, Indonesia and Timor-Leste, or in regards to the natural resources and geopolitically vital sea routes of the South China Sea involving claimants from Malaysia, Vietnam, the Philippines, Taiwan and China, prevent firm maritime boundaries from being drawn at all.

These lingering boundary disputes reflect a longer lineage of uncertainty regarding the very existence of an identifiable region. Such uncertainty is structured around a dialectic of unity and diversity. On the one hand the region defined as Southeast Asia is seen as a space of shared cultures; on the other the diversity of the region is readily apparent and gives it a distinctive quality. Unity is identified in social and cultural traits that are shared widely across the region, some of which are thought to have derived from long patterns of internal and external trade, and others from patterns of wet and dry rice cultivation linked with the tropical monsoon climate (Gillogy and Adams 2011, 5). Milton Osborne (2004), for example, argues that women and the nuclear family are generally more valued in the region than in neighboring states and much has been

Approaching Southeast Asian development

Figure 1.1 Map of Southeast Asia

made of a traditional mandala political structure, in which pre-colonial kingdoms set up tributary systems that had no set territorial boundaries but faded in influence with distance from the core. The selective appropriation of Indian and Chinese influences, evident in, for example, the absence of India's caste system, also suggest particular cultural norms and values are shared across the region. The extensive Chinese diaspora throughout the region is another common feature across many societies. In contrast diversity within the region is also very apparent. No other world region boasts the same degree of geographic, religious, linguistic, cultural, ethnic, economic, ecological and political difference.

Historically, the region has been framed in part by its position in relation to two larger neighbors. Indians referred to it as *Suwarnadwipa* (Goldland) and the Chinese *Nanyang* (South Seas). Arab traders knew it as *Jawa*, and the Europeans as *Further India*. Within the region empires rose and fell, such as the Angkor kingdom centered in current day Cambodia, Pagan in Myanmar, and Chinese vassal state of Srivijaya that controlled east-west trade to China from current day Indonesia and Malaysia. A consolidated regional power structure equivalent to India or China failed to form; instead existing divisions were accentuated during an extended colonial period when Portugal, Spain, Britain, France, the Netherlands and the United States established colonial boundaries that continue to mark the extent of state territories. However even then the concept of a distinct region had yet to develop, and it wasn't until the 1890s that German language scholarship first referred to the term Southeast Asia (*Südostasien*) in a purely geographical way (Reid

5

2015, 414). The term caught on and became more widely used, particularly during the Second World War and the subsequent Cold War when the region was of critical geopolitical importance. These external signifiers were formally internalized through the formation of ASEAN in 1967 when a Western-oriented alliance of Indonesia, Thailand, Malaysia, Singapore and the Philippines formed amidst the turmoil of the Vietnam/USA War. The collapse of the Soviet Union and the cessation for the Cold War eventually saw a broadening of ASEAN to include all states of the region with the exception of Timor-Leste – the region's newest country – which has applied for membership and is expected to be admitted soon.

Southeast Asian imaginaries now proliferate through maps, tourism, media and geopolitical strategies; however, it is unlikely that a strong Southeast Asian identity has swept through the diverse populations that make up the region. Different ways of imagining and dividing the region help illustrate this point. Timor-Leste, for example, despite sharing half of its island with the Indonesian province of West Timor, has observer status on the Pacific Island Forum – the main political grouping of Pacific Island countries – and has joined the Pacific Island Development Forum, forging links with similarly small island states. Other groupings such as East Asia, Asia Pacific, Pacific Rim, Indochina, Australasia, Oceania and Western Pacific, provide alternative ways of grouping and dividing the states of Southeast Asia. More challenging is Willem van Schendel's (2002) naming of Zomia to refer to the Tibeto-Burman language areas occupied by highland groups stretching from Vietnam, Laos, Thailand and Myanmar through Bangladesh, India, Nepal, Bhutan and China, occupying spaces conventionally divided between Southeast, East, South and, more recently, Central Asia. Concepts like Zomia call into question the self-evident nature of the regions that currently comprise the world in geographical maps, including Southeast Asia, and open possibilities for alternative research trajectories as evident in James Scott's (2009) subsequent anarchist history of the area.

Alternative regional imaginaries also highlight the problems of searching for particular traits at the regional scale – as presumably different traits would be found if different regional groupings, such as Zomia, were used. The once prominent Asian Values argument, for example, has faltered, in part, due to the sheer diversity of values inherent in Asian societies and the difficulties in even defining what or where Asia is. This does not mean that Malaysians and Indonesians don't share similar traits – clearly they do – but it is harder to identify the traits shared by middle class Chinese residents of Singapore, the Kachin people living in the mountains on the China-Myanmar border and the post-disaster rural fishing communities of Indonesia's Aceh. Similarly colonial empires have left cultural marks in the languages and institutions that link geographically diverse nations, such as Portugal, Timor-Leste and Angola, or Malaysia, Britain and India – creating imaginary post-colonial geographies that could equally be the focus of a book such as this.

Despite these possibilities it is the Southeast Asian regional identity that has stuck to become the dominant self-reinforcing geopolitical and cultural frame. Given its diversity and somewhat arbitrary boundaries and definition we approach the region not as a space of shared endemic traits but as a dynamic region that is continually forming and reforming in response to internal and external processes. We see value in Appadurai's (2000, 7) conception of process geographies, whereby attention is directed toward movement rather than stability, and regions are recast as "problematic heuristic devices for the study of global geographic and cultural processes." Our attention, in focusing on development, turns to what Anna Tsing refers to as the 'friction' of global encounters, how globalizing processes are engaged with, transformed and grounded in particular geographic spaces, often with unexpected outcomes. The study of such flows challenges homogenous and static images of regions, which are instead creatively likened to lattices, archipelagos, hollow rings and patchworks (Van Schendel 2002). Regions matter, not because of shared norms and values, although where they exist these are important, but because of the social, economic, political and biophysical interconnections that cross national boundaries, underpin regional formations and shape encounters with external and internal processes.

Development in Southeast Asia

We approach development in a similar way. Rather than focusing on a core indicator or trait, such as GDP, human rights or freedom, we borrow again from Appadurai (2000) to see development as comprising a set of flows characterized by what he calls relations of disjuncture. Development is far from a smooth and seamless project, instead the speed, impacts and forms of development differ spatially and temporally, between and within regions, states, sectors, cities, villages and households. There is no one development; instead there are a myriad of ideas and resources that have become associated with this powerful but slippery concept. Certainly, as post-development researchers have argued, development is about change and because of that it necessarily disrupts, and can destroy, what existed before. Disjunctures are created through the unevenness of these disruptions, whereby improved access to markets, technology, healthcare or education emerge unequally across time, space and society, reflecting the unpredictable friction of place-based encounters. Some benefit from development interventions while others are disadvantaged, or as Rigg (2015, 4) has observed more subtly "problems and tensions that have arisen from growth." For this reason we do not take a normative perspective on whether development is good or bad (contrast almost any pro-growth report from the World Bank with Wolfgang Sach's 2010 *Development Dictionary*), as its goodness or badness depends on time, space, perspective, culture, power relations and the materialities of particular initiatives. Perhaps most important is the capacity of those most affected to selectively engage with development and actively steer development processes toward desirable ends. A role for researchers is to highlight the injustices and inequalities that emerge, thereby making space for alternative approaches, when this is not the case.

In applying this lens to development in Southeast Asia we are interested in dynamism and diversity, seeking to understand how people and places are engaging with the globalizing forces of development. We do not pursue a regional economic development model (see Hill's 2014 review and dismissal of the idea), but we are interested in how the incorporation of countries into the region influences their development. Space and scale matter to development, and a regional optic can provide insights into processes that national, local and global analyses cannot. As James Sidaway (2013, 997) writes in relation to Area Studies, "It is imperative, however, to supplement historical and history with geographic and geographical, signifying spatial comparison, perspective and position." In focusing on the region the collection aims to understand how the flows and processes associated with development are burrowing across and through Southeast Asia and the diverse effects they are having in different spaces. As two of the editors argue elsewhere (Miller and McGregor forthcoming) some of the benefits of regional analyses include: enabling comparisons between places and the identification of shared experiences and trends; creating space for regional narratives and counter-narratives; highlighting the connectivities and influences of human and non-human regionally significant actors such as ASEAN, the Asian Development Bank or the Asian monsoon; and exposing intra-regional connectivities forming through increased mobility and uneven patterns of development (such as migration and remittance flows). The Vietnam/USA War and the Asian Economic Crisis of 1997–98 are just two events that emphasize the importance of regional analyses, with both having fundamental impacts on Southeast Asian development. Regional analyses, which are sensitive to shortcomings, difference, borderlands and minorities, can contribute to valuable knowledges and dialogue oriented toward improved or alternative development approaches.

Sensitivity to difference within regional analysis is important. National-scale differences are apparent in Table 1.1, which provides a snapshot of how individual countries within Southeast

Table 1.1 Southeast Asian development indicators

	Land area 2015*	Population 2015*	Human Development Index[1] 2015**	Ranking	Life expectancy 2014**	Mean years of schooling 2014**	GDP[2] 2015*	GDP per capita 2015*	GDP per capita 2015*	Income inequality 2005–2013** Gini coefficient[4] 2005–2013	Income poverty 2002–2012** % below $1.25 PPP/day	Infant mortality rate 2013** per 1,000 live births	Electrification in rural areas 2012** %	Share of women in parliament 2014** % seats held by women
	Sq. km	Million			years	years	US$ billions	US$	$US PPP[3]					
Brunei Darussalam	5,769	.4	31 (very high)	79	8.8		12.9	30,942	87,117	no data (n.d.)	n.d.	8.4	67	n.d.
Cambodia	181,035	15.4	143 (medium)	68		4.4	18.5	1,198	3,578	31.8	10	32.5	19	19
Indonesia	1,913,579	255.5	110 (medium)	69		7.6	857.6	3,357	11,108	38.1	16	29.3	93	17
Laos	236,800	6.9	141 (medium)	66		5	12.6	1,831	5,466	36.2	30	53.8	55	25
Malaysia	330,290	30.5	62 (high)	75		10	294.4	9,657	26,515	46.2	n.d.	7.2	100	14
Myanmar	676,577	52.5	148 (low)	66		4.1	65.4	1,246	5,275	n.d.	n.d.	39.8	31	5
Philippines	300,000	101.6	115 (medium)	68		8.9	289.5	2,850	7,241	43	19	23.5	82	27
Singapore	719	5.5	11 (very high)	83		10.6	291.9	52,744	85,021	n.d.	n.d.	2.2	99	25
Thailand	513,120	69	93 (high)	75		7.3	395.7	5,737	16,064	39.4	0.3	11.3	100	6
Timor-Leste	14,870 ***	1.2 ***	133 (medium)	68		4.4	1.4 ***	1,158 ***	2,399 ***	30.4	35	46.2	27	39
Vietnam	330,951	91.7	116 (medium)	76		7.5	193.4	2,109	6,083	35.6	2.4	19	98	24

* ASEANStats (2016) ASEANStats web portal (www.aseanstats.org downloaded 2 March 2017)
** United Nations Development Program (2015) Human Development Report 2015 Work for Human Development (http://hdr.undp.org/en/2015-report downloaded 2 March 2017)
*** The World Bank (2016) World Bank Open Data (http://databank.worldbank.org/data/home.aspx downloaded 2 March 2017)

1 The human development index ranks countries from 1–188 (highest to lowest) based on average achievement in three basic dimensions of human development – a long and healthy life, knowledge and decent standard of living.
2 Gross Domestic Product – sum of gross value of all resident producers in the economy (note: GDP often does not capture contributions from informal economies).
3 PPP – Purchasing power parity takes the cost of living into account by comparing how much it costs in local currencies to purchase particular goods.
4 The Gini coefficient calculates income inequality where perfect equality = 0 and absolute inequality (one person owns all) = 1.

Asia are currently faring according to some common development indicators. Clearly the encounters with development have been uneven with the city state of Singapore seemingly benefiting most from its small size and positioning as the service and financial hub of the region while Brunei's growth has been propelled by sales of oil and gas. Malaysia has benefited from its positioning between these successful neighbors and through growing industry and service sectors, but it struggles more than most to address forms of income inequality linked to ethnic diversity. Indonesia, Thailand, Vietnam and the Philippines are all considered to be developing reasonably well; however millions still live in poverty in these countries, particularly in Indonesia and the Philippines, both of which were severely affected by the Asian Economic Crisis and subsequent structural adjustment programs. Timor-Leste, Cambodia, Laos and Myanmar face continuing challenges with high rates of poverty, infant mortality and low rates of electrification and education, partly reflecting regional and national histories of conflict. In this volume we do not drill down into the national histories, plans, politics, economies and ecologies that have influenced these development trajectories; instead we focus on trends, connections, commonalities and differences across the region. In adopting a regional optic we hope to bring to light development encounters and challenges that may remain hidden or under-recognized within national and sub-national analyses.

While such goals are admirable we certainly do not claim to have fulfilled them in this Handbook; instead we see this work as making a contribution to an ongoing effort devoted to regional scale understandings of development (see for example Rigg 2015, 2012, 2003, Nevins and Peluso 2008; McGregor 2008; Hill 2008; Leinbach and Ulack 2000). One issue we have encountered is the limited number of researchers analyzing development processes at a regional scale. Most of us, editors included, tend to specialize in one or two sectors in one or two countries. Taking a step back from national or sub-national specialisms to think about development regionally is not an easy or straightforward process. The contributors to the volume have admirably risen to this task. In a similar vein while we are pleased that close to half the chapters involved an author from the region, we would have liked to have engaged more. In some cases potential collaborators were too busy, in others we could not identify an appropriate person to approach. The differing pressures on Southeast Asian researchers who are often engaged in much more policy-oriented work (with clear exceptions like Singapore), and those academic researchers from neoliberalizing Western institutions, whose work is quantified and assessed through academic publications and citations, can act against collaboration on projects like these. A final concern has been around the perennial question of just what is development – and what should be included and what should be left out. Reviewers can no doubt take us to task regarding the content of the Handbook and our decision to focus particularly upon institutions and economies, people and environment. Even within these sections there are clear gaps – indeed whole Handbooks could (and in the case of the environment – have!) been written on each of these themes. Nevertheless we are confident each chapter provides useful insights into Southeast Asian development which, when taken as a whole, provides a comprehensive introduction to regional development processes.

About this book

The Handbook is divided into four sections. Section One provides an introduction to some of the key themes and issues that recur throughout the volume. Katharine McKinnon provides an initial wide-ranging review that highlights the inseparability of development from politics. She focuses on Cold War politics and the birth of development alliances and industries in the region; the mobilization of development resources to pursue particular political goals that can dispossess minorities or suppress political rights; and the cultural and gender politics that influence who has access to development benefits. Her work ends on a hopeful note that emphasizes the agency of actors to engage with development opportunities in unexpected but beneficial ways, something that is echoed by many contributors to the Handbook. Simon Springer's subsequent review of neoliberalism in Southeast Asia takes a much darker view. Springer argues, as do many other contributors, that neoliberalism has come to dictate how development is understood and pursued within the region. He traces the roots of neoliberalism and its uptake in Southeast Asia, emphasizing how it has been used to legitimize and extend authoritarian structures rather than challenge them, as theorized by free market advocates. The uneven costs and benefits of existing neoliberal development models are subsequently explored throughout much of the collection.

Jonathan Rigg and Albert Salamanca provide an introduction to transformations in rural spaces and dwell on the surprising resilience of smallholders despite prevailing development trends favoring rural agglomeration. Like McKinnon they emphasize the agency of rural households to creatively diversify in an age of rapidly expanding mobility where migration and remittances challenge traditional conceptions of urban-rural divides. However they also express concern about development processes that lead to dispossession through state sanctioned land grabs or from the creeping uptake of market logics that fracture communities and result in a type of dispossession from below. They also worry about the likely impacts of climate change on some of the region's most vulnerable people, a theme also taken up by Victor R. Savage in his wide-ranging review of human-nature relationships in Southeast Asia. Savage sees a collision course between dominant linear conceptions of resource-based development that have roots in pre-colonial and colonial trade and the sustainability or even viability of the ecological systems on which we all depend. He contrasts the ecological blindness of capitalist approaches with the deep ecological knowledge of indigenous people, concluding with a plea for more biophilic approaches to development.

Section Two looks at the institutions and economies of development in Southeast Asia. These range from the large multilateral development institutions like the World Bank, Asian Development Bank and the International Monetary Fund that are the focus of Toby Carroll's chapter through to the community institutions focused on by Gibson, Hill and Law. Carroll's analysis backs up Springer's earlier argument that neoliberal rationalities now dominate development policy. This influence is subsequently explored by Ford, Gillan and Thein in terms of the changing role of the International Labour Organization in facilitating foreign investment, by Kelly Gerard in the instigation of the ASEAN Economic Community, and by Hudalah and Aritenang regarding shifting geographies of industrial investment. Rachel Hughes' study of international justice tribunals in Cambodia and Timor-Leste also argues that the limited framing of such institutions, intentionally or not, improves the climate for international investment. A second theme addressed in this section is the growing influence of China with May Tan-Mullins providing a unique insight into the opaque operations of Chinese aid. John Connell reflects on how Chinese tourists, now the most popular extra-regional source country for Southeast Asian tourism, along with a general maturing of the industry, are transforming tourism and development. Within each of the chapters there is a concern for justice and an interest in how the most marginalized are being, or likely to be, impacted by these changing development flows. Gerard sees some hope in the regionalization of civil society advocacy, Gibson, Hill and Law emphasize the strength and capacity of community institutions, and Hughes notes the national and community healing that flows from justice tribunals. The overall feeling, though, is that the dominant expressions of neoliberalism in Southeast Asia are embedding uneven patterns of development, creating considerable challenges for more just and equitable pathways.

Section Three continues the theme of regional integration and neoliberal economic reform, but explores these issues from the vantage point of the people whose everyday lives are most impacted by such development. Critical questions are asked about whether development pathways across the region shape more equitable, sustainable and empowering social worlds. The chapters in this section suggest that economic development, and the mobility it relies on, is an uneven and relational process. Brenda S.A. Yeoh and Shirlena Huang discuss the complex web of gendered mobilities that feed into child- and elder-care deficits in the region, for example, showing how families in receiving places like Singapore and sending places the Philippines are reworking ideas of the 'modern' family at both sites. While labor mobility can reduce poverty and provide much needed investment in places like the Philippines, Philip Kelly discusses the downsides (brain drain, exploitation and family separation) and raises questions about how migration contributes to genuine well-being in sending countries. Alexandra Gartrell and Panharath Hak remind us, however, that not everyone migrates; indeed, those with disabilities are most likely to be left behind in villages, to be uneducated, underemployed and in poverty. Orlando Woods similarly suggests articulations between religion and the modernizing impulses of development. Meghann Ormond, Chan Chee Khoon and Sharuna Verghis examine how mobility is transforming ideas about who is 'entitled to' and 'responsible for' health and social care in ASEAN, and they too find disparity: high income mobile individuals are more likely to have better healthcare than migrant workers. In their analysis of children, youth and development, Harriot Beazley and Jessica Ball remind us that even forced migrant children have visions about what development in the region should look like. Another major theme in this section is the powerful connections between social identities, livelihoods and environment in the region. Sarah Turner's chapter shows how processes such as agrarian transition, resettlement policies and environmental destruction have particular impacts on ethnic minority and indigenous groups. Tubtim Tubtim and Philip Hirsch explore what regional integration and neoliberal reform looks like from the vantage point of the 'village': perhaps the original site of development imaginations. Bernadette P. Resurrección and Ha Nguyen make the case for understanding this process through feminist political ecology.

In Section Four the authors position questions of values, rights, knowledge, equity and scale as central to their analysis of environment and development, providing a clear articulation of some of the challenges confronting development in contemporary Southeast Asia as well as some of the more hopeful strategies by which society-environment relations might be reconfigured in more sustainable and equitable ways. The chapter by Miller provides an overview of the changing position and value of water in the region, focusing on how changes in material, discursive and cultural relations with water have accompanied modernist development. A rise in competition over increasingly scarce resources, such as water, is a theme addressed throughout this section, and is explicitly addressed in the chapter by Fujita Lagerqvist and Connell with reference to land. They document the quite radical shifts in rural livelihoods in the region, accompanying the increased commercialization and intensification of agriculture, arguing that though the importance of agriculture to the economies of the region has declined, it remains an important part of rural people's lives and livelihoods. The following two chapters by Bush, Marschke and Belton and Majid Cooke et al. frame their analyses of environment-development issues much more from a livelihoods perspective, highlighting how the increased integration of people's livelihoods into global commodity chains present opportunities for income improvements but also the possibility of increased exploitation and insecurity. Bush, Marschke and Belton focus in particular on the social dimension of sustainability concerning the highly interconnected capture fisheries and aquaculture industries in the region, identifying particular governance challenges concerning labor conditions. The chapter by Majid Cooke et al. focuses on the very particular ways in which land rights, knowledge and power relations influence the extent to which smallholder palm oil farmers are able to navigate the opportunities and risks associated with this rapidly expanding cash crop.

As highlighted elsewhere in the collection, the wealth and development of the region has come at great cost to the environment. Yet, as shown in the two chapters on disasters and climate change, development itself is now seriously threatened by environmental variability and change. The chapter by Thomalla, Boyland and Calgaro demonstrates how as wealth in the region has grown, so too has the number of people, assets, infrastructure and services at risk of disaster. Southeast Asia is one of the most disaster-prone regions in the world, with the authors arguing the nature of development directly contributes to rising vulnerability. Disasters are likely to worsen with climate change, unless development can shift to more adaptive and low-carbon models. Uddin and Nylander's chapter offers a strong regional level analysis of efforts to transition toward low-carbon development through the upscaling of climate change mitigation efforts. Both chapters highlight the potential for more regionally coordinated responses to reduce vulnerability to disasters and improve climate change mitigation efforts.

The chapter by Neef and Sangkapitux considers the wider environmental governance context in which payment for ecosystem services is situated as a means of promoting conservation. They present analysis of a series of case studies of conservation efforts, concluding that highly mixed outcomes are apparent. The final chapter in our collection, by McGregor and Thomas, engages with more-than-human theories, opening up clear opportunities to rethink development by foreshadowing human and non-human entanglements. As such, the chapter critiques the hyper-separation of humans and nature that lies at the core of modernist development, widely seen as responsible for widespread environmental degradation.

As a collection, the book emphasizes the dynamism associated with the disjunctive flows of development across the region. Cities, villages, societies and environments are being transformed as people engage with the new opportunities and constraints of development in vastly different ways. As the regional imaginary continues to materialize through development initiatives and networks, national economies, societies and environments are becoming increasingly mobile and interconnected, meaning a change in one Southeast Asian location vibrates through the networks bound up with it. It is these interconnected processes of change that are shaping development in the region – they are diverse, unpredictable and uneven, however they are Southeast Asian.

References

Appadurai, A. (2000). Grassroots globalization and the research imagination. *Public Culture*, 12, pp. 1–19.
Gillogy, K. and Adams, K. (2011). Introduction: Southeast Asia and everyday life. In: K. Adams and K. Gillogy, eds., *Everyday life in Southeast Asia*. Bloomington and Indianapolis: Indiana University Press, 1–8.
Hill, H. (2008). *The economic development of Southeast Asia*. Cheltenham: Edward Elgar.
Hill, H. (2014). Is there a Southeast Asian development model? *Malaysian Journal of Economic Studies*, 51, pp. 89–111.
Leinbach, T. and Ulack, R. (2000). *Southeast Asia: Diversity and development*. Upper Saddle River, NJ: Prentice Hall.
McGregor, A. (2008). *Southeast Asian development*. London: Routledge.
Miller, F. and McGregor, A. (under review). Rescaling political ecology? Regional approaches to climate change in the Asia Pacific. *Progress in Human Geography*.
Nevins, J. and Peluso, N. (2008). *Taking Southeast Asia to market*. Ithaca, NY and London: Cornell University Press.
Osborne, M. (2004). *Southeast Asia: An introductory history*. Sydney: Allen and Unwin.

Reid, A. (2015). *A history of Southeast Asia: Critical crossroads*. Chichester: Wiley-Blackwell.
Rigg, J. (2003). *Southeast Asia: The human landscape of modernization and development*. London: Routledge.
Rigg, J. (2012). *Unplanned development: Tracking change in South East Asia*. London: Zed Books.
Rigg, J. (2015). *Challenging Southeast Asian development: The shadows of success*. London: Routledge.
Sachs, W. (2010). *The development dictionary: A guide to knowledge as power*. London and New York: Zed Books.
Scott, J. (2009). *The art of not being governed: An Anarchist history of upland Southeast Asia*. New Haven, CT and London: Yale University Press.
Sidaway, J. D. (2013). Geography, globalization, and the problematic of Area Studies. *Annals of the Association of American Geographers*, 103, pp. 984–1002.
Van Schendel, W. (2002). Geographies of knowing, geographies of ignorance: Jumping scale in Southeast Asia. *Environment and Planning D: Society and Space*, 20, pp. 647–668.

2
WHAT IS DEVELOPMENT IN SOUTHEAST ASIA AND WHO BENEFITS? PROGRESS, POWER AND PROSPERITY

Katharine McKinnon

Introduction

Development means different things to different people, but everywhere it is associated with ideas of progress and betterment. Development makes people more prosperous, delivers the advantages of modernity (clean water, healthcare, electricity) and, ideally, creates a just society of active and empowered citizens at the same time. In recent decades all of these hopeful visions have been strongly critiqued, and observers from all walks of life have argued that so often, and in so many unanticipated ways, development has failed: More people might have access to electricity and telecommunications, but at the same time the global divide between rich and poor has been growing larger; the world might be more interconnected, and in richer places people are living longer than ever before, yet in many other parts of the world people are dying from diseases long ago eradicated from the minority world.[1]

These trends are seen as symptomatic of the failures of the development era and of the noxious effect of the spread of global capitalism. On a global scale we continue to fall short of the targets set through, for example, the Millennium Development Goals (MDGs). The MDGs were established to target eight areas for global efforts to address global poverty, including, for example, eradicating extreme poverty and hunger, reducing child mortality, improving maternal health and achieving universal primary education for all children. These targets were set at the Millennium Summit of 2000, and 189 countries signed the agreement, committing themselves to meeting these goals by 2015. As UNICEF's 2015 *Progress for Children* report shows, while there have been some gains toward achieving the MDGs there remain stark inequalities in indicators of child and maternal health, and in some cases the inequalities are increasing. For example, the measure used to assess undernutrition in children, stunting, has significantly decreased in some regions. But at the same time, "more low-income countries show increasing stunting inequities than decreasing inequities" (UNICEF 2015, 10), reflecting a trend of increasing inequalities within nations. While there are successes at the regional level, a closer look shows that for the most vulnerable members of our global society things seem to be getting worse rather than better (Childs, 2015).

The development industry is rife with stories of things that went wrong and objectives not achieved or great intentions derailed. At the same time, however, in examples of objectives not achieved or intentions derailed there are often stories of something unexpected emerging, new possibilities being created and new opportunities emerging. While sharing disaster stories of the failures and challenges of development is a favorite pastime, even the most jaded and disappointed of development professionals will be able to cite examples of how development efforts can make a difference in the world. But to see this one has to look closely and be prepared that the successes of development may come in unexpected ways. Even in the success stories, unanticipated impacts are often the most important. In Southeast Asia there are many such examples in which unanticipated outcomes can have powerful effects. Or where the outcomes that are hoped for are achieved in surprising ways.

This chapter explores development in Southeast Asia through the lens of what development is supposed to achieve, giving consideration to concerns about failure, tracing the often convoluted pathways toward success and identifying who it is that benefits along the way. Development discourses identify a series of altruistic goals that national development programs and individual aid projects are supposed to achieve. These include: progress and modernization, justice and empowerment, greater equity and prosperity. These magical transformations are supposed to result from the dedicated efforts of development workers and community members, pushed forward by national development policies, backed in many cases by international funding and the support of global institutions like the UN. Yet, as this overview of Southeast Asian development will demonstrate, all such development efforts are also inevitably about politics and power, responding as much to the interests of local and global elites as to the felt needs of community members. I begin the chapter with a discussion of how this confluence of politics and development is evident in the roots of development in Southeast Asia. I discuss the origins of aid in the region and how foreign investment in development was tied in to the emergence of Southeast Asia as a geopolitical entity. Following this, I discuss several examples of Southeast Asian development in practice, showing how each one speaks to the altruistic goals of development more broadly. Each example shows how development efforts are bound up in a complex mix of altruism and politics that sometimes achieves what it is meant to, usually leads to something unexpected and often challenges our expectations in the process.

Southeast Asia as a region

Before we speak about what development is in Southeast Asia it is important to identify where and what Southeast Asia is. The boundaries of the region are imprecise and contested. There is an underlying logic to the existence of a geographical boundary to group the region together, but as with any geographic region there is also a great deal of diversity and contradiction. When examined closely the contradictions are enough to question whether the definition of the region makes sense at all. The term 'Southeast Asia' came into use in the mid-1800s, and according to historian Donald Emmerson (1984) the first scholarly usage of the name was inaugurated by British anthropologist J.R. Logan in the first of a series of articles in *The Journal of the Indian Archipelago and Eastern Asia*. But the name did not gain popularity until after WWII, which Emmerson attributes to the adoption of the name during wartime. War required "the demarcation of regional theaters, one of which was Admiral Lord Louis Mountbatten's South-East Asia Command (SEAC), created in 1943" (Emmerson 1984, 7). As van Schendel (2002) points out, it was in the post-war period that the current boundaries of Southeast Asia took shape. As colonial authorities withdrew from the region and domestic governments took over, the contemporary nations and states of Southeast Asia took shape. This process was determined by a range of factors. Pre-colonial governments did not simply re-instate their power over fixed territories. Instead the shifting territorial claims made by colonial authorities intermingled with complex emerging aspirations of nationhood, remnant pre-colonial state claims and an emerging post-war world order. In some regions the shape of national boundaries remained in flux until the end of the century: Timor-Leste only gained independence in 1999, and in West Papua there is an ongoing battle for independence from Indonesia. While cartographic borders have become fixed, the questions of what those borders mean, and who and what they contain have been much contested.

Officially, Southeast Asia is made up of Indonesia, Singapore, the Philippines, Timor-Leste and Brunei, with mainland Southeast Asia consisting of Cambodia, Laos, Myanmar, Thailand, Vietnam and Malaysia. The fact that this collection of nations is taken to constitute a singular 'Southeast Asia' is the result to large degree of the agreement taken to form an Association of Southeast Asian Nations (ASEAN) in 1967 and the shifting composition of signatories to that agreement. The ASEAN motto is 'One Vision, One Identity, One Community,' and indeed there are some characteristics that appear to be held in common across the region. The historic influence of cultural traditions and religious beliefs that originated in today's India and China are apparent across the region. Territories that became Burma, Thailand, Cambodia, Malaysia and Indonesia were in early days profoundly transformed by aspects of caste and state-craft associated with Hinduism and Buddhism. On the coastal strip that became Vietnam, Chinese influences predominated. The arrival of Islam from the west added another important dimension, especially the contemporary states of Indonesia, Malaysia and parts of the Philippines and Burma. Certain foods (such as rice), sports (*sepak takraw*) and clothing are similar across the region. Look more closely, however, and there are as many reasons for discounting these as shared characteristics. More than any shared characteristic Southeast Asia is a region of astounding diversity in cultures, languages, religions, economies and politics.

The fixity of political and cartographic designations of Southeast Asia bely the contradictory nature of remnant pre-colonial state power structures and more contemporary modes of nation-state imported from Europe (Thongchai 1994). In pre-colonial times state borders were not the pencil thin lines we are so used to seeing now. Instead the border was a more negotiated thing and the idea of who belonged to which state power was much less precise. The pre-colonial state operated on a principle not of hard and fast claims to strictly defined territory. Instead it was relationships that the pre-colonial ruler was after: relationships through which they could claim loyalty, call on contributions of labor or soldiers if necessary, and the payment of regular tithes to finance the state. State power sat at the center of a radiating zone of influence, reaching outwards through a chain of loyalty. Wolters (1999) visualized this system as a mandala, with Southeast Asia made up of many overlapping mandala states. This was a system in which small communities may have found themselves playing to the competing interests of two distinct state powers, rather than being bound exclusively to one. The territories of these mandala states were similarly imprecise compared to today's conceptualization of a state space. Within the sphere of influence of the mandala state could be pockets of non-state spaces (Scott 2009). In pockets of dense forest, on difficult to access mountain ridges, many communities thrived, existing beyond the direct influence of any state power. This is something unthinkable in today's world.

The arrival of colonial powers brought a very different mode of state power to the region. From the mid- to late 1800s European powers introduced a new cartographic imperative in which spheres of influence were to be supplanted by strict territorial claims. For example, in 1905 when the British and Siamese finally agreed where the border between Siam and Burma should be drawn, the cartographic line divided the people who had co-existed in that mountain region for generations, literally drawing some into Burma and others into Siam (modern-day Thailand). The legacy of this process continues to this day as communities may have been drawn into the state but do not necessarily identify with it and may not be recognized as full citizens. The borders may have been decided upon, but many cross-border populations, such as the ethnically diverse mountain peoples in mainland Southeast Asia (a region dubbed 'Zomia' by van Schendel 2002) remain in a quasi-stateless space.

These tangled and complex dynamics of territory and identity remain an issue in contemporary Thailand and are echoed across the region. While we may speak of the region as composed neatly of 11 countries, look closer and the logic is harder to maintain. In this diverse and difficult to define region development has been woven into the mix as Southeast Asian states of the modern era pursue progress and modernization.

Progress and modernization: the politics of development in Southeast Asia

In the post-war period the Southeast Asian region saw rapid change and political upheaval, characterized at this time by ideological conflict that fueled both state warfare and violent revolution. The Cold War defined postwar politics, in which conflict was as much ideological as territorial, and the competing superpowers of the USSR and USA sought to cement allegiances from countries around the world and demonstrate the superiority of their own economic and political structures: socialist government and centralized economies among the so-called 'communist bloc' and democracy and market capitalism among the allies of the United States. The industry of international development emerged in this context and was as much a part of Cold War politics as the arms race. While the Cold War never escalated into direct armed conflict between the two superpowers, Southeast Asia became the primary proxy site for violent confrontation.

Through the Korean War (1950–1953) and what many Vietnamese refer to as the 'American War' (1955–1975), the United States and allied forces from Australia, New Zealand, Thailand and elsewhere fought not just the troops of socialist aligned national forces, but sought to stall the potential spread of communism throughout Southeast Asia. As Laos, Vietnam and Cambodia took on socialist aligned systems, the United States and its allies focused their attention on strengthening alliances with other nations in the region. It was during this period that international aid first became important in the region. Strategic alliances such as the Southeast Asian Treaty Organization (SEATO) called for investment in domestic economies and aid assistance aimed at increasing prosperity. But the aims of SEATO were also explicitly anti-communist (see Anderson 2004). The organization was founded in the 1954 signing of the Southeast Asia Collective Defense Treaty in Manila. The signatories were the United States, Britain, France, Australia, New Zealand, the Philippines, Thailand and Pakistan. The Philippines and Thailand were the only Southeast Asian nations to join, with Pakistan signing on for "essentially anti-Indian reasons" (Emmerson 1984, 9). SEATO proceeded to provide funding for a range of initiatives to address social and economic conditions, with research, education and training given high priority (Franklin 2006). Under the rubric of international development, allied forces claimed to be bringing the advantages of modernized economies to the 'underdeveloped' nations of the region. As with much development of the era, the assumption was that the industrialized economies of the minority world provided a template for transformation that could be applied to the poorer, majority countries of the world, 'lifting' them from an unnecessary state of poverty.[2] Development supported by aid dollars and technical expertise and training from the West was meant to bring widespread prosperity and in doing so cement ideological and political allegiances.

It was not only the so-called allied forces that were engaged in such efforts. The USSR also put resources into the dual investments of aid and political allegiances. By 1956, for example, the USSR had committed US$500 million in aid expenditure to India, Afghanistan, Cambodia, Indonesia and Burma, placing itself as a direct competitor for the allegiance solicited through

SEATO (Wolf 1957). There is very little research to account for the impacts of Soviet aid and investment across the region, but a glimpse into the investments made in Vietnam (see Hoan 1991) suggests a significant and ongoing commitment that matched that of the allied powers, and was equally plagued by failure and inefficiency.

Politics and empowerment: development in northern Thailand

As a means for bringing greater prosperity, development was meant to primarily be of benefit to ordinary communities across Southeast Asia. But when understood as simultaneously an ideological and political enterprise the question of who were the main beneficiaries of development becomes less clear. I turn now to the example of post-war aid and development investment in Thailand. This case shows clearly how geopolitical interests, as much as a concern for underdevelopment and poverty, were (and, arguably, remain) the driving forces for international development efforts. Between 1957 and 1959 the level of aid money being fed into the Thai economy was so great that aid from the United States alone made up 2 percent of Thai GNP (Muscat 1990, 24) and this does not account for a considerable volume of military assistance. From 1951 to 1957, US$149 million in economic aid and US$222 million in military aid flowed from the United States to Thailand (Wyatt 1984). The investment in military hardware including aircraft, tanks, guns and credits along with the cost of training in elite military academies alone amounted to more than 30 percent of GNP (Wyatt 1984).[3] What this investment enabled was a significant US presence in Thailand. USAID projects were initiated across the country to build infrastructure and modernize Thai agricultural production. Military aid enabled the United States to work closely with their Thai counterparts to combat the apparent 'communist threat' and to utilize bases in the country from which to launch, among other things, secret bombing raids against Cambodian and Laotian targets.

Aid rhetoric has it that development assistance attends to the well-being and prosperity of the people, but there are numerous examples of how funding allocations seem to in fact prioritize strategic political aims. Foreign aid investment certainly played an important role in improving public health, restructuring Thai agriculture and establishing the nation as the 'rice bowl of Asia.' Yet these transformations came at a cost. Across Southeast Asia similar change has thrown up many new challenges as tensions emerge between the interests of the rising middle classes and the changing hopes and aspirations of rural farmers (see Riggs 2001). In northern Thailand, programs put in place to benefit highland minority groups highlight additional challenges, as the actions needed to benefit community members appeared to be at odds with the political concerns of the time. Numerous highland development programs were put in place from the 1970s, ostensibly to ensure that highland minority peoples could benefit from more stable incomes, the advantages of membership in the national community and increased health and well-being, as well as educational opportunities. After decades of community development, and general social and political change in Thailand, highland communities have indeed been transformed, but not always for the better. Almost every mountain village now has a school where children gain a primary school education. Many villages have electricity and mobile phone connection, and young people increasingly travel to the lowlands to work. Many households have modern amenities such as a gas cooker and electricity to run a refrigerator and TV. At the same time, however, many highlanders still have not been given formal recognition of their right to be in Thailand either through land tenure or citizenship. Without such formal recognition of their rights and standing in the nation highlanders are vulnerable to discriminatory treatment. As with many indigenous people around the world, the imposition of cultural and political change has gone hand in hand with rising rates of substance abuse signaling the negative impacts of cultural and political dispossession (see Kampe and McCaskill 1997; Venkateswar 2012). According to research conducted on the impacts of the Thai-German Highland Development Program, project interventions themselves may have contributed to escalating addiction rates (Chupinit and Gebert 1993). Across the board, the consequences of development programs have not been altogether positive.

While the outcomes for highland communities were mixed, the benefits that accrued to those implementing the programs were clear. As programs were put in place foreign aid workers gained employment and foreign governments gained a foothold in these key strategic territories in the Cold War. Thai employees gained incomes and were taught new skills and expertise. Piggy backing on overseas aid dollars the Thai state was also assisted to extend its authority into the borderlands, at last claiming full control over these former non-state spaces out to the edge of the cartographic borderline (see McKinnon 2005, 2011).

Part of the justification for foreign investment in aid by allied powers was, of course, that development would bring not only prosperity, but justice and empowerment. The record of highland development programs shows few gains in this regard. However empowerment did sometimes occur unexpectedly, often through interpersonal relationships and professional integrity rather than from explicit goals of development programs. For example, in Nan Province, in northeastern Thailand one project implemented in the late 1990s involved the introduction of 'new' fire management strategies in communities situated next to national park land.[4] Project staff were asked to address the risk that annual burn-off of rice fields posed to the forest and offer education to community members on how to avoid the risk. Over many months of meetings and discussion between project staff and village representatives, project staff began to understand that the villages *already* had fire management strategies in place. Project staff instead began to gently persuade Royal Forest Department personnel to give these existing rules formal recognition. The final product was a document entitled Village Watershed Network Rules and Regulations, which included detailed guidelines and an agreed system of fines that now apply at the District Government level. Without making it an explicit aim, what the project succeeded in doing was to reverse the norm in which regulations decided by the state are imposed on villages. In this case, a set of governing practices decided upon at the village level came to be adopted by a branch of the state, one especially powerful in the highlands of the north. Forest Department personnel, rather than villagers, were the party that needed education, but saying so directly would not have achieved results. To be successful in benefitting and empowering local communities, development practice needs sometimes (probably most of the time) to approach the process with subtlety, flexibility and political awareness.

Democracy and development: the case of the NICs

Another major rationale for aid investment during the Cold War, and one that remains relevant today, is the belief that development and democracy go hand in hand. In contemporary parlance, good governance, transparency and accountability are seen to be essential to development, and democratic state structures are assumed to contribute to greater prosperity. The logic for this is that under free and democratic systems corruption cannot flourish, thus the populace can be motivated to create economic opportunities knowing that they will be able to benefit from the results. Australian development aid priorities, for example, highlight the promotion of stable democratic government as a core concern alongside economic development (DFAT 2014, 8). But do economic growth and democracy go together? I turn now to the example of Singapore, and the way that national development plans have challenged the assumption that political freedoms and economic development go hand in hand.

In Singapore the label of Newly Industrialized Country (NIC) is used to reflect the exceptionally rapid industrial growth that has occurred there since the 1960s. Compared to neighboring economies, Singapore now boasts a high per capita income (76,860 PPP dollars, compared with 13,450 in Thailand and 2,890 in Cambodia, World Bank 2016). While development is meant to herald the arrival of a just society, the extraordinarily rapid transformation of the Singaporean economy suggests that justice and democracy are not requirements for development. As measured by GNP the NICs (including Singapore, Hong Kong, South Korea and Taiwan) have been incredibly successful examples of increasing prosperity through the introduction of modern economies and close integration with global capital. Singapore led the move to a 'developmental state model' that emphasized close regulation of the labor market, taxation and fiscal incentives, and state-owned enterprises. This model for a planned economy was heralded as a "model for less-developed countries aiming to become late industrializers" (Huff 1995). The living circumstances of very many ordinary families in the NICs has improved dramatically as a result of the measures introduced. Yet authoritarian rule and strict social controls exist alongside a highly prosperous and educated population. Singaporeans are subject to a strict regime of social control through major public institutions that stratify society, force people into wage labor and induce political loyalty (Tremewan 1996). Although Singapore is ostensibly a multiparty republic, the People's Action Party has won every election since self-government in 1959. At the same time, the benefits of prosperity are far from evenly shared. An underclass of imported wage labor supports many essential areas of daily life in household work, child and elder care, and menial labor. The low socio-economic status of migrant workers is the bedrock of domestic productivity, yet regulations that ensure a lower pay rate and poorer working conditions for migrants deliberately create and sustain a situation of inequity within these highly prosperous nations.

In this case successful development, as measured by high overall living standards and a very healthy GNP, does not necessarily equate to a greater level of justice or empowerment for all. At the lowest end of the income spectrum workers are getting low pay for providing essential services, and often under what many would regard as unfair conditions with unrestricted working hours. Even those at the high income end of the spectrum have a price to pay in terms of their own political and social freedoms.

Economic growth and prosperity: overseas contract workers in the Philippines

While GNP is relied upon by mainstream economists as a measure of a nation's level of prosperity, and by extension the nation's development status, the expectations of what prosperity looks like and how it should be achieved are contested. GNP provides an insight to a nation's financial wealth, but financial wealth does not necessarily reflect all definitions of prosperity or well-being.

The definition of household prosperity as something based in cash incomes and material possessions seems self-evident. After all, having enough money to pay for basic services such as electricity, water and healthcare, having enough food in the cupboards and the security of a roof over your head makes an enormous difference to households that are struggling to make ends meet. Likewise, being able to afford luxury items such as a refrigerator, washing machine, bicycle or motorbike can impact enormously on the capacity of individuals in the home to engage in productive work. Yet prosperity cannot only be measured by the money coming into a household or the assets that are owned. In many places people refuse to be categorized as poor although their cash incomes and their asset ownership may be low. For many it is access to productive land, and the food supplies that the land provides, that indicates prosperity. In addition, it is strong family and community networks that support individual households in hard times and allow for a culture of sufficiency to emerge. It is *sufficiency* of what is needed to live well, more than access to ready cash, that defines wealth and prosperity for many communities.

At the national level, sufficiency is not usually stated as the desired outcome of economic development.[5] Instead governments look to economic development to boost the nation's GNP with the prosperity of the nation measured only by the amount of money changing hands through business investments, employment in the formal workplace and the import/export of taxable goods. In the Philippines, for example, government investment is meant to boost economic growth through greater investment in export products:

> For those in the mainstream, development in the Philippines is seen to depend on getting into the globalised economy by promoting economic growth derived from exports.
>
> (Gibson-Graham 2005, 9)

What this means in the Philippines is that small-scale gardens that assure food security and a sufficient livelihood for households is seen as secondary to investment in the large-scale production of palm oil for export markets. In addition, the Philippines government is continuing to promote the benefits of overseas work, and the export of contract migrant labor is a major source of foreign capital flows as workers send remittance payments back to families in the Philippines.

The view that exports of goods and labor are the best way to promote prosperity in the Philippines is being challenged. While there is no doubt that both are a significant source of revenue nationally, it is more debatable whether the average household is benefitting as much as the nation's GNP. Community-based organizations are engaged in an effort to redirect remittance revenues toward enterprises that will benefit the wider community and create local livelihoods – promoting an ethic of community benefit and shared sufficiency. As Gibson-Graham (2005) demonstrates, one example of this is the work of Unlad Kabayan Migrant Services Foundation Inc. This community-based organization runs a range of programs that re-envision Overseas Contract Workers (OCWs), not as disempowered victims of global capitalism, but as "as investors in community-based enterprises and as contributors to different pathways for local development in their home communities" (Gibson-Graham 2005, 7). At the heart of Unlad Kabayan's multifaceted programs is:

> a refusal to see capitalist globalisation and global proletarianisation as the only development process that enrolls migrant workers; a refusal to accept the linearity and singularity of the mainstream development dream of capitalist industrialisation as the only way of increasing standards of living in the Philippines; and a refusal to accept that the local and international knowledges that OCWs possess might not be sufficient to the task of building different economic opportunities.
>
> (Gibson-Graham 2005, 7)

By harnessing OCW investment toward community enterprises that build on pre-existing skills and resources of regional communities, Unlad Kabayan is able to redirect remittances toward social enterprises that generate further incomes for local people. These enterprises, including processing local produce, ginger tea production and rice milling, build on and work through the diverse ways in which local communities already maintain livelihoods, including practices of gifting, sharing and offering reciprocal exchange. The focus is on enterprises that support community well-being, rather than enterprises that will lead to individual profit. While the efforts may be small scale they offer concrete examples of a different development pathway that will enable communities to prosper. Examples of similar local innovations exist across the region, and not always in formal ways such as the programs being run by Unlad Kabayan (see Gibson et al. this volume). Turner et al. (2015), for example, detail the everyday ways that Hmong in the Sino-Vietnamese borderlands are responding to the pressures of neoliberalism through "everyday politics and covert resistance to form culturally specific and locally adapted livelihood approaches" (p. 172).

Equality and economic empowerment: gender and small-scale agriculture in Eastern Indonesia

The shift in development discourse to an interest in economic empowerment provides one conduit for national and foreign development investment to refocus on community level prosperity. Community-based and community-driven economic development is being tried in various ways by larger development agencies, with positive results for community prosperity. One such example is the Value Chain Development approach, which seeks to work with small-scale producers and enterprises to help them maximize their access to markets (Saarelainen and Sievers 2011). An example is World Vision's Local Value Chain Development (LVCD) program, first piloted on the island of Flores in Nusa Tenggara Timur (NTT), eastern Indonesia. NTT is one of the disadvantaged and, materially speaking, poorest provinces in Indonesia (Brown 2009, 156). NTT is amongst the least urbanized regions in Indonesia with little transportation networks and over 80 percent of the population living in rural areas (Williams 2005, 402, see also Williams 2007).

World Vision's LVCD program worked with small-scale farmers in the region, seeking to improve household incomes by improving their ability to obtain good prices for their produce and get better access to markets. Farmers producing cashew nut, tamarind, candle nut and cacao were supported to create collective selling groups, in which key individuals would take on the role of 'market facilitator' maintaining direct contact with buyers, discussing ways to improve the quality of the product and negotiating a fair price rather than relying on middle men. The successes of the program saw the household incomes of collective members increase, and families reported a significant benefit in becoming less reliant on borrowing money to cover household expenses and gaining increased ability to pay for the education of their children (McKinnon et al. 2015). While the LVCD program in Flores addressed concerns with the disadvantaged position of eastern Indonesian farmers, what was forgotten in the early days of the pilot was how local and social cultural practices were shaping place-based dynamics of disadvantage, and in particular the gendered nature of local agricultural economies.

Local economies in Flores, indeed across Indonesia and Southeast Asia, are highly gendered. Women and men take on different and complementary roles in the family livelihood. In East Flores the tasks that are traditionally women's responsibility include childcare, household provisioning (including barter trade) and the processing of home-grown cotton into woven textiles for clothing and exchange (Graham 2008, 123). Additionally, a great deal of rural women's time is spent on gathering firewood and water, picking vegetables, husking rice and feeding pigs (Graham 2008, 123). Women generally work longer daily hours than men, and are engaged significantly more in subsistence production and household chores (Gondowarsito 2002, 252), while men take a greater role in public affairs and formal governance outside the household (de Jong 2000). Due to the cultural norm of men taking a greater role in public affairs, it was predominantly men who stepped forward to represent their families in the sellers collectives of the LVCD program and men who took part in meeting and votes that decided which community members would be put forward to manage the collective. As a result, the majority of community leaders in the collectives were men. Women, who manage the household budget and who work alongside men in the fields, often did not have their complementary knowledge and experience on the table.

The failure to recognize the gendered nature of local economies, and understand women and men's different roles in livelihood creation, continues to be a major issue in development practice (McKinnon et al. 2016). The tendency to forget women's productive contributions often means that development programs that may succeed in raising household incomes also have the effect of raising women's overall workloads. The Flores pilot was no exception to this. However, there was also a surprising impact on gender relations. Although the traditional gendered divisions of labor are still adhered to where possible, in these communities there is an exceptionally high number of female-headed households as many men become OCWs seeking work in Java, Kalimantan and further afield (in some villages the proportion of men working as OCWs is as high as 30 percent, Graham 2008). As women step in to fill the labor gap, taking on what was traditionally 'men's work' in the fields, they have also been stepping into new roles as the public representative of their households. The public meetings held as part of the management of collective selling sat outside traditional public forums of community governance and ceremonial leadership around *adat* (customary) practices. With encouragement from project staff, more women have been taking the opportunity to speak up about their perspectives and concerns. The collective selling groups also provided a space where women could step into leadership roles without challenging the patriarchal structuring of traditional public life. With time, increasing numbers of women took up the kind of leadership opportunities the LVCD program presented, and men have had to learn to think differently about what women can do:

> Before the men thought that women are not capable so that's why we put them behind, but after we realized that they have very good thinking about some things that's when we realized that they are more good with the details that sort of thing.
> (Male interview respondent, cited in McKinnon et al. 2015)

For the different contributions of women and men to be acknowledged and valued has required both women and men to change, and for their efforts to be supported. World Vision too has had to learn how to understand and engage with the gendered dynamics of local economies in subsequent iterations of the LVCD model, and to work with women and men to ensure that the benefits of improved incomes are felt equitably.

Conclusion

In Southeast Asia, development originated alongside a complex web of overlapping international and political interests. While in many cases life transformed for the better as development unfolded, it has seldom been the needs and aspirations of local communities that have determined the direction of aid investment or national development plans. The direction of aid tends to have been driven by the needs and interests of domestic and international elites, whether informed by Cold War geopolitical concerns, the desire to secure and maintain political power or the belief that European experiences of industrialization provided a good model for how the rest of the world should 'develop.' The example of aid investment in Thailand through the late 1900s is an especially clear case of geopolitical concerns driving development while the case of Singapore shows how models of 'success' do not necessarily serve the interests of the poor or create a more just or democratic society. The story is not always negative, however. Justice and empowerment, greater equity and prosperity have accompanied development efforts – but often in unexpected ways.

In this chapter I have discussed examples where development efforts have helped to empower communities. The story of forest fire management in Nan Province, northern Thailand, showed how this can happen through a practice of political subtlety combined with a commitment to a participatory ethic to shift the usual balance of power. Formal and informal community-based organizations are also working to shift the balance toward community, rather than fiscal, benefit. The example of reinvestment of OCW earnings in community enterprises in the Philippines demonstrates how places that seem to have been 'left behind' by economic growth are in fact working to create well-being and prosperity in accordance with a more community-minded vision of sufficiency. Finally, I discussed how even efforts to work closely with communities to build new economic opportunities can go awry when important dynamics of local social and cultural norms are not attended to, like the gendered nature of agricultural production in eastern Indonesia. With all the examples I have discussed there is inevitably a gap between what development is meant to achieve and the impacts that it actually has on communities in the region. In the Southeast Asian context this has not, however, stopped people from making the most of the opportunities that development aid has brought, as the women of Flores show by taking up the chance to become vocal advocates and leaders in their communities, challenging their men in the process. Development in Southeast Asia has always been about politics, but with the right support and collaboration it can also present real possibilities for positive change. If there is one lesson to be learned amongst the dazzling diversity of the region and stories of development challenges and inefficiencies, it is that local people, in all sorts of ways and in all sorts of places, find their own pathways to change, making the most of the opportunities at hand to enact their own pragmatic and often unorthodox solutions.

Notes

1 The term 'minority world,' in complement to 'majority world', is increasingly being used in place of terms such as 'First World' and 'Third World.' The language of First and Third worlds is associated with a view of development in which richer industrialized nations were seen to provide a model for how poorer nations should develop, and development itself was presented as a way for poorer nations to 'catch up' with the First World. The terms 'minority' and 'majority' recognize that those supposed 'models' for development are in the minority globally and encourages us to think more carefully about majority world perspectives and experience.
2 It is important to recognize that the assumptions of modernist development thinking as exemplified here, and the validity of the term 'underdeveloped,' have long been contested. See discussion in, for example, Escobar 1995, Sachs 1992.
3 The level of aid investment to Thailand during the Cold War is by no means uncommon. At present international aid investment into Laos and Cambodia is also a significant percentage of GDP, giving rise to considerable concern about the level of aid dependency and its potential to skew domestic economies (see Ear 2013 and Phraxayavong 2009).
4 This case is discussed in more depth in McKinnon 2011.
5 One exception to this may be Vanuatu, where the government has been working to put in place national level indicators for prosperity that acknowledge the important role played by subsistence production and extended family exchange networks in creating community well-being. See Regenvanu 2009.

References

Anderson, D. L. (2004). SEATO. In: J. Chambers, ed., *The Oxford companion to American military history*. Oxford: Oxford University Press. www.oxfordreference.com/view/10.1093/acref/9780195071986.001.0001/acref-9780195071986-e-0831?rskey=G69Tx0&result=829. Accessed 26 July 2017

Brown, I. (2009). *The territories of Indonesia*, ed. I. Brown. London and New York: Routledge.

Childs, A. (2015). How the Millenium Development Goals failed the world's poorest children. *The Conversation*. Available at: http://theconversation.com/how-the-millennium-development-goals-failed-the-worlds-poorest-children-44044. Posted 1 July 2015.

Chupinit, K and Gebert, R. (1993). *Drug Abuse in Pang Ma Pha Sub-district: Genesis and Current Situation*. Chiang Mai: Thai-German Highland Development Programme.

de Jong, W. (2000). Women's networks in cloth production and exchange in Flores. In: J. Koning, M. Nolten, J. Rodenburg and R. Saptari, ed., *Women and households in Indonesia: Cultural notions and social practices*. Surrey: Curzon Press, 264–280.

DFAT. (2014). *Australian aid: Promoting prosperity, reducing poverty, enhancing stability*. Barton ACT: Commonwealth of Australia, Department of Foreign Affairs and Trade.

Ear, S. (2013). *Aid dependence in Cambodia: How foreign assistance undermines democracy*. New York: Columbia University Press.

Emmerson, D. K. (1984). "Southeast Asia": What's in a name? *Journal of Southeast Asian Studies*, 15(1), pp. 1–21.

Franklin, J. K. (2006). *The hollow pact: Security and the Southeast Asia treaty organisation*, unpublished PhD dissertation, Texas Christian University.

Gibson-Graham, J. K. (2005). Surplus possibilities: Postdevelopment and community economies. *Singapore Journal of Tropical Geography*, 26(1), pp. 4–26.

Gondowarsito, R. (2002). Men, women and community development in East Nusa Tenggara. In: K. Robinson and S. Bessell, eds., *Women in Indonesia: Gender, equity and development*. Singapore: Institute of Southeast Asian Studies, 250–263.

Graham, P. (2008). Land, labour and liminality: Florenese women at home and abroad. In: P. Graham, ed., *Horizons of home: Nation gender and migrancy in island Southeast Asia*. Clayton: Monash University Press, 113–131.

Hoan, B. (1991). Soviet economic aid to Vietnam. *Contemporary Southeast Asia*, 12(4), pp. 360–376.

Huff, W. G. (1995). The developmental state, government, and Singapore's economic development since 1960. *World Development*, 23(8), pp. 1421–1438.

Kampe, K. and McCaskill, D. (ed.) (1997). *Development or domestication? Indigenous peoples of Southeast Asia*. Chiang Mai: Silkworm Books.

McKinnon, K. (2005). (Im)Mobilisation and hegemony: 'Hill Tribe' subjects and the 'Thai' state. *The Journal of Social and Cultural Geography*, 6(1), pp. 31–46.

McKinnon, K. (2011). *Development professionals in Northern Thailand: Hope, politics and power*. Singapore: ASAA Southeast Asia Publications Series, Singapore University Press in conjunction with University of Hawaii and NIAS.

McKinnon, K., Carnegie, M., Gibson, K. and Rowland, C. (2016). Generating a place-based language of gender equality in Pacific economies: Community conversations in the Solomon Islands and Fiji. *Gender Place and Culture*, 23(10), pp. 1376–1391.

McKinnon, K., Shamier, C. and Woodward, K. (2015). *The effects of economic development on women's empowerment and gender equality goals, East Flores, Indonesia*. Research Consultancy for World Vision, Australia, September.

Muscat, R. J. (1990). *Thailand and the United States: Development, security, and foreign aid*. New York: Columbia University Press.

Phraxayavong, V. (2009). *History of aid to Laos: Motivations and impacts*. Chiang Mai and Thailand: Mekong Press.

Regenvanu, R. (2009). *The traditional economy as a source of resilience in Melanesia*. Paper presented at the Lowy Institute Conference, The Pacific Islands and the World: The Global Economic Crisis, 3 August, Brisbane, Australia.

Riggs, J. (2001). *More than the soil: Rural change in Southeast Asia*. London and New York: Routledge.

Saarelainen, E. and Sievers, M. (2011). *Combining value chain development and local economic development*. ILO Value Chain Development Briefing Paper 1.

Sachs, W. (1992). *The Development Dictionary*. London: Zed Books.

Scott, J. (2009). *The art of not being governed: An anarchist history of upland Southeast Asia*. New Haven, CT and London: Yale University Press.

Thongchai, W. (1994). *Siam mapped: The history of the geo body of a nation*. Chiang Mai: University of Hawaii Press.

Tremewan, C. (1996). *The Political Economy of Social Control in Singapore*. London: Palgrave MacMillan.

Turner, S., Bonnin, C. and Michaud, J. (2015). *Frontier livelihoods: Hmong in the Sino-Vietnamese Borderlands.* Seattle and London: University of Washington Press.
UNICEF. (2015). Beyond the averages: learning from the MDGS. *Progress for Children.* New York: Division for Communication, UNICEF.
van Schendel, W. (2002). Geographies of knowing, geographies of ignorance: Jumping scale in Southeast Asia. *Environment and Planning D: Society and Space*, 20, pp. 647–668.
Venkateswar, S. (ed.) (2012). *The politics of indigeneity: Dialogues and reflections on indigenous activism.* London: Zed Books.
Williams, C. (2005). 'Knowing one's place': Gender, mobility and shifting subjectivity in Eastern Indonesia. *Global Networks*, 5(4), pp. 401–417.
Williams, C. (2007). *Maiden voyages: Eastern Indonesian women on the move.* Singapore: ISEAS Publishing.
Wolf, C. (1957). Soviet economic aid in Southeast Asia: Threat or windfall. *World Politics*, 10(1), pp. 91–101.
Wolters, O.W. (1999). *History, culture and region in Southeast Asian perspectives.* Rev. ed. Singapore: Institute of Southeast Asian Studies.
The World Bank Group. (2016). *GNI per capita PPP (current international $).* Available at: http://data.worldbank.org/indicator/NY.GNP.PCAP.PP.CD
Wyatt, D. (1984). *Thailand: A short history.* Chiang Mai: Silkworm Books.

Acknowledgments

This chapter is indebted to a great many research collaborators in Southeast Asia and elsewhere. In particular, I would like to acknowledge the support from The Australian National University for the fieldwork that informs the Thai case studies discussed here; World Vision Australia for their funding and support of the work in Eastern Indonesia, and in particular the contributions of co-researchers Clare Shamier and Kerry Woodward to that research. Thanks for the critical commentary from Andrew McGregor, Lisa Law and John McKinnon in revising this chapter.

3
NEOLIBERALISM IN SOUTHEAST ASIA

Simon Springer

Introduction

Neoliberalism is a slippery concept. The term is generally conceived as a new set of political, economic and social arrangements that place their emphasis on market relations, minimal states and individual responsibility. Neoliberalism has largely replaced earlier labels that referred to specific politicians and or political projects (Larner 2009), and has since become an identifier for a seemingly ubiquitous collection of market-oriented policies that are blamed for a wide range of social, political, ecological and economic ills. Critical scholars have taken up the mantle, examining the connections between neoliberalism and a vast collection of conceptual categories, including cities (Hackworth 2007; Leitner et al. 2007), citizenship (Ong 2007; Sparke 2006), development (Hart 2002; Power 2003), gender (Brown 2004; Oza 2006), homelessness (Klodawsky 2009; May et al. 2005), labor (Aguiar and Herod 2006; Peck 2002), migration (Lawson 1999; Mitchell 2004), nature (Bakker 2005; McCarthy and Prudham 2004), race (Haylett 2001; Goldberg 2009), sexualities (Oswin 2007; Richardson 2005) and violence (Springer 2008, 2012, 2015), which only begins to scratch the surface (Springer et al. 2016). Given the massive academic industry that has sprung up in response to neoliberalism, it is perhaps easy to forget that the phenomenon itself is quite recent, and its entry into the affairs of Southeast Asia even more so. The policies that are today considered standard practice within the contemporary neoliberal toolkit, namely privatization, deregulation and liberalization, probably seemed incomprehensible as recent as 60 years ago. In the aftermath of the Second World War the global north was thoroughly enamored with Keynesian economics, and owing to the ruination wrought by the Nazis, right wing ideologies had fallen completely out of favor. Within Southeast Asia there was a rise in nationalist sentiment in the wake of the Second World War, where attempts at decolonization resulted in protracted struggles like the war in Vietnam. This history makes the ascendancy of neoliberalism all the more surprising, and particularly so within Southeast Asia where a growing skepticism for 'Western' influence was emerging. Yet in the intervening years neoliberalism nonetheless rose to become the contemporary 'planetary vulgate' (Bourdieu and Wacquant 2001). A number of scholars have traced the unfolding of neoliberalism (Duménil and Lévy 2004; Harvey 2005; Peck 2010), where the common thread that ties these accounts together is the notion of a distinct historical lineage to the development of neoliberalism. In short, neoliberalism came from somewhere, and its trajectories were largely purposeful.

The roots of neoliberalism can be traced back to Europe and critiques of nineteenth-century laissez-faire (Peck 2008). Given the broader currents of Southeast Asian's history in relation to European colonialism, neoliberalism is positioned from its outset as a colonial appendage in the region. A key starting point for the idea is the 'Colloque Walter Lippmann' of 1938, where a group of 26 prominent liberal thinkers, including Friedrich Hayek, Michael Polanyi, Louis Rougier, Wilhelm Röpke and Alexander Rüstow, met in Paris to discuss Lippmann's (1937-2005) book *The Good Society*. Their goal was to reinvigorate classical liberalism and in particular its emphasis on the economic freedoms of the individual. The participants discussed potential names for their new philosophy of liberalism, eventually settling on 'neoliberalism,' giving the term both a birthday and an address (Mirowski and Plehwe 2009). The publication of Hayek's (1944/2001) *The Road to Serfdom* was one of the first products of this meeting, establishing him as a primary intellectual engineer of the 'neoliberal counter-revolution,' signaling a backlash against Keynesianism and its emphasis on low interest rates and government investment in infrastructure. Hayek had a profound influence over neoliberalism's various apostles, including the Chicago School of Economics' most (in)famous intellectual, Milton Friedman, and former Prime Minister of the United Kingdom, Margaret Thatcher. From there, the diffusion of neoliberalism may have seemed somewhat assured, as 'trickle-down economics' soon caught the attention of United States President Ronald Reagan, and much like Thatcher, neoliberalism became the defining feature of his time in office. Yet an initial explanation for the contemporary 'triumph' of neoliberalism is that the original group of neoliberals bought and paid for their own regressive 'Great Transformation' (Polanyi 1944).

Neoliberals understood that their ideas could, with time and relentless cultivation among political elites, have very material consequences (George 1999). The Colloque Walter Lippmann recognized that they would need to proselytize their ideas through formal channels, an idea that was given life when many of the original Paris participants reconvened in Switzerland for the founding of the Mont Pèlerin Society in 1947, the first of many neoliberal think tanks. From here, the group founded a platform for constructing an international network of institutes, foundations, research centers and journals to promote and advance neoliberal knowledge (Plehwe and Walpen 2006). The success of this campaign was not simply its growing virility among intellectuals, but rather its achievements hinged on the geographic dispersion of neoliberal discourse across multiple spaces of institutional engagement, including academia, business, politics and media. Getting political elites to buy into the idea was a key concern, and if David Harvey's (2005) theory of neoliberalism as a reconstitution of class power has merit, then this consideration represents a significant portion of neoliberalism's success story, including its penetration into Southeast Asian societies.

The rationale for this uptake in Southeast Asia is the primary contribution this chapter seeks to make. I begin by tracing some of the history of neoliberalism's diffusion into the region, highlighting the use of both 'shock' tactics and debt crisis as conduits to economic reform. In the following section I examine the limitations of democracy in the context of Southeast Asia's cultural proclivity for patronage and the ways in which this allows neoliberalism to be significantly manipulated into the service of elite interests. I then focus in on the authoritarian character that neoliberalism takes on within conditions of minimal accountability, a situation that is ripe for spurning social conflict. In the conclusion I emphasize the hybridity of neoliberalism and the limitations of adopting a regional perspective to a concept that continues to shift as it moves through different political, cultural, social and institutional fields.

A brief history of neoliberalism in Southeast Asia

In spite of all the early behind the scenes networking that eventually gave rise to neoliberalism, the economic theory remained in a state of virtual hibernation with respect to public policy for a number of years. It needed a flashpoint to ignite its slow burning intellectual fuse into a tangible political reality. The 1970s set neoliberalism aflame, as it was strategically poised to explode when a financial crisis hit. Between 1973 and 1979, world oil prices rose dramatically, setting off an economic recession on the global north, while the Soviet Block was derailed, careening into an economic tailspin that eventually led to its disappearance. For the global south a 'debt crisis' ensued, triggering a widespread condition of aid dependency that continues to this day. These disruptions to the global economy denoted the start of an economic paradigm shift away from Keynesianism and toward neoliberalism. Global north politicians, governments and citizens alike became increasingly disillusioned with the record of state involvement in social and economic life, leading to a growing acceptance of neoliberalism's primary proclamation: the most efficient economic regulator is to 'leave things to the market.' In Southeast Asia, governments had not yet been able to take advantage of neoliberal reform in the way the global north was able to, owing to its diffusion from Europe at a time when ideas didn't travel as quickly as they do today, however the so-called 'East Asian Miracle'[1] touted by the World Bank would soon change that. The Association of Southeast Asian Nations (ASEAN) first came together in 1967, anticipating that capitalism would triumph in the Cold War standoff, where Thailand, Indonesia, Malaysia, the Philippines and Singapore initially hedged their bets. Initially these countries were thought to have been following a 'developmental state' type of model, which seemed to diverge from the theories of neoliberalism. While being market-based, the state was nonetheless prepared to actively intervene in economic processes in order to produce favorable conditions for local economic growth. As has become clearer in recent years, this can fit within neoliberal theory, where apparent neoliberal state 'roll-back' always and inevitably comes with further state 'roll-out' (Peck and Tickell 2002). Indeed, this feature is now considered as one of the defining contradictions between neoliberalism-in-theory and 'actually existing neoliberalism' as a grounded empirical reality (Brenner and Theodore 2002). While the situation in Southeast Asia was at the time characterized by a significant degree of state-led development, the point is, this sort of orientation was easily accommodated within a neoliberal approach.

Among neoliberalism's defining, vanguard projects were Thatcherism in the UK and Reaganomics in the United States. Following this, more moderate forms of neoliberalism were unfurled in traditionally social-democratic countries like Canada, New Zealand, and Germany. But it was Chile's trauma that provided the lynchpin of turning this emergent economic theory into a grounded political development, and the country is widely recognized as the first state-level neoliberal experiment. In 1973, on the 'other 9-11,' the US government helped to orchestrate a coup that saw a despotic hand under Pinochet replace the country's elected socialist government (Challies and Murray 2008). This installation provided the model for what Naomi Klein (2007) refers to as the 'shock doctrine,' where collective crises or disasters, whether naturally occurring or manufactured, are used to push through neoliberal policies at precisely the moment when societies are too disoriented to organize meaningful contestation. Following the Chilean experiment, 'shock' tactics became the principle means of delivery in neoliberalism's selective exportation from the global north to the global south. Cambodia offers a case in point inasmuch as neoliberal reform was actually a mandated outcome of the United Nations peace agreement of the early 1990s following the Khmer Rouge atrocities and a decade of Vietnamese occupation (Springer 2010, 2015). When the Iron Curtain fell in 1989, the global political climate shifted and the aftermath of the Cambodian genocide could finally be addressed.

Cambodians only longed for peace, and the United Nations Transitional Authority in Cambodia (UNTAC) promised to give it to them, but in presiding over the country's transition, the population was not consulted or allowed to vote on what sort of economic policy would come into effect, and instead a particular version of free-market neoliberal economics was simply installed (Springer 2011).

Elsewhere in Southeast Asia the growing debt crisis opened a window of opportunity for neoliberalism as neocolonial relationships of aid dependency were fostered though the auspices of US-influenced multilateral agencies like the International Monetary Fund (IMF) and the World Bank. Loan dispersals and subsequent rescheduling hinged on conditionalities that subjected recipient countries to intervention, which reorganized their economies along neoliberal lines. Indonesia serves as the best example of the employment of this approach, where the country was compelled by the IMF to accept a US$43 billion bailout package to restore market confidence in the Indonesian rupiah, and in exchange the Indonesian government agreed to reduce energy and food subsidies, raise interest rates, and close 16 privately owned banks. The approach failed, as the bank closures triggered a run on other banks with people withdrawing their savings as the rupiah started to spiral downward leading to food hoarding (MacIntyre 1999). Nonetheless, neoliberal economics were packaged, marketed and sold to the global south as a series of nostrums that once implemented through the freeing of market forces, would supposedly lead to a prosperous future, where all of the world's peoples would come to live in a unified, harmonious 'global village.' Although neoliberalism's utopian promise was an empty one from its outset, powerful elites in countries across the region, and indeed the world, have only been too happy to oblige, as neoliberalism often opens up opportunities for well-connected government officials to informally control market and material rewards, thus allowing them to easily line their own pockets (Springer 2009). An extraordinary collection of regulatory reforms came with each consecutive wave of neoliberalism's dispersion. Beyond seeking to deregulate markets, advance 'free' trade and promote unobstructed capital mobility, neoliberalism typically includes the following finer points: it seeks to impede all forms of public expenditure and collective initiative through the imposition of user fees and the privatization of commonly held assets; to position individualism, competitiveness and economic self-sufficiency as incontestable virtues; to decrease or rescind all forms of social protections, welfare and transfer programs while promoting minimalist taxation and negligible business regulation; to control inflation even at the expense of full employment; and to actively push marginalized peoples into a flexible labor market regime of low-wage employment, where labor relations are unencumbered by unionization and collective bargaining (Peck 2001; Peck and Tickell 2002). These features have all become part of the Southeast Asian political economic landscape in recent years, where countries within the region have, to varying degrees, employed such policies. This ranges from the privatization of municipal water supplies in the Philippines, to union breaking tactics in Cambodia, to corporate tax holidays on light industry and electronics in Thailand, to the individualism of *Kiasu* culture in Singapore and its key characteristic of hyper-competitiveness. It should be emphasized that there is not a single model of neoliberalism in the region; instead each country has adopted neoliberal reforms in a piecemeal fashion that reflects unique political priorities and differing institutional matrixes.

Neoliberal patronage and the limits of democracy

For all the ideological purity of free market rhetoric, the actual practice of neoliberalism is inherently marked by contradictions and compromise (Peck 2010). This inconsistent character is precisely what made it so easy for the World Bank to shift its tone of espousing states like Singapore and Malaysia as models of neoliberal development prior to 1997, and then in the aftermath of the Asian Financial Crisis, lambaste them as 'developmental states' that were not following the assigned script. Although the underlying assumptions of neoliberalism and its naturalization of market relations remain largely constant, neoliberalism in its 'actually existing' circumstances (Brenner and Theodore 2002) has nonetheless varied greatly in terms of its dosages among regions, states, and cities, where and when it has been adopted. This includes the use of strong 'developmental' states to ram through markets when and where desired, but to intervene when deemed necessary. This situation really is no different than the supposed template state for neoliberalism, as the United States is in reality one of the most self-serving and protectionist economies in the world. Since elites have so much to gain by rigging the system, both in America and in Southeast Asia, neoliberalism has become an officially sanctioned racket, which promises improved lives for all, but in fact simply exacerbates inequalities by entrenching class difference (Gilbert 2014; Wade 2004). The various ways in which neoliberalism has been manipulated by local officials is part of its 'success' story. This chameleon-like pliability means that we must go beyond conceiving of 'neoliberalism in general' as a singular and fully realized policy regime, ideological form or regulatory framework, and work toward conceiving a plurality of hybridized or articulated neoliberalisms, with particular and unique characteristics arising from variable geohistorical outcomes. What constitutes 'actually existing neoliberalism' in any given Southeast Asian state is distinct unto itself, even if we can see general patterns that arise from similar and shared histories, cultures, and political structures. In particular, the deep roots of the patronage system have been a particularly influential articulation across the region, as they work so well with neoliberalism insofar as it allows local elites to co-opt, transform, and (re)articulate policies and reforms through a framework that enables favoritism in asset stripping public resources (Springer 2010). This is one of the key criticisms of the IMF's involvement in Indonesia's reform – its failure to address the patronage system. Suharto's hold on power prior to the financial crisis was maintained by offering powerful positions and/or monopolies to his family, friends, and enemies in exchange for financial and political support (Robison and Rosser 2003).

While the earliest, proto-historical roots of neoliberalism in Southeast Asia can be traced to the American interventions as part of its Cold War policies in the 1960s and 1970s, such as supporting the takeover of Indonesia by Suharto from Soekarno in 1965 and backing the overthrow of Philippine democracy by Ferdinand Marcos in 1972, neoliberalism as we currently know it only became entrenched in the region in the aftermath of the Asian financial crisis of 1997. In the wake of the Asian Crisis, the adoption of even greater neoliberal reforms in Southeast Asia has, by and large, simply meant that public monopoly became private monopoly while the authoritarian structure of the state has remained intact (Robison et al. 2005), albeit under new governments in many instances, notably Thailand, Indonesia, and Malaysia. This situation, of course, constitutes a paradox, as neoliberal ideology presents itself as being in favor of democracy (Harvey 2005). Yet in such transitional contexts, foreign and domestic pressures for the privatization of national economies and the opening of borders to trade and capital movements are far more prevalent than is support for democratization, accountability, and the economic assistance needed to ease the impacts of poverty on populations (Tetreault 2003). Neoliberals assuage their own misgivings by insisting that free markets constitute the necessary basis for freedoms in other arenas, enabling a system in which economic and social power is dispersed and able to accommodate numerous interests, and it is thus assumed as the most ideal way to encourage democratic reform (Robison 2002). Yet there is room for cynicism, as an alternative interpretation suggests that the best way to promote markets – particularly those that would be open to 'Western' capital – is to develop democracy, as this would, in theory, dislodge the politico-business oligarchies and patron favoritism advanced by an entrenched central authority.

Many Southeast Asian states are willing to play the lip-service game with regards to democracy, but when push comes to shove, they are only too happy to trample its norms. Actually existing neoliberalism in Southeast Asia, as in the United States and elsewhere, is not about a highroad approach to the morality of freedom, but is instead about getting the most for oneself through evoking such a feel-good rhetoric. Greed is its foundation, selfishness its core.

Democracy is easily one of the most abused words in the English language (Lummis 1996), so the (lack of) meaning of it in a neoliberal context is already moot. Indeed, it can be argued that 'democracy' offers no more of a solution to neoliberals than authoritarianism if it produces regimes where populist governments engage in policies of redistribution. Thus, from the outside, US democracy-promotion policies have sought to construct 'low-intensity democracies' across Southeast Asia, where elections do not subject rulers to pressures that might derail free market objectives (Robison 2004). Hence, for example, we see union-busting as a widespread practice in the region, as not only does it threaten the viability of locally produced goods on the global market, but so too do unions represent a mobilization of a politically active and democratically inclined population. In Stephen Gill's (1995) terms, 'disciplinary neoliberalism' is the Southeast Asian mainstay, as it ensures that through a variety of regulatory, surveillance and policing mechanisms, neoliberal reforms are instituted and 'locked in,' despite what the population might actually want. Insofar as democracy is concerned, neoliberalism represents an erosion of democratic control and accountability. Again, this is as true for Southeast Asia as it is for other parts of the world, as a neoliberal agenda employs a variety of legal and constitutional devices that insulate it from popular scrutiny and demands (Overbeek 2000). This is exactly what happened in Cambodia under the UNTAC in the early 1990s, when a 'liberal democracy' and hence a free market economy were enshrined in the country's new constitution with no prior public consultation (Springer 2010). At the same time, decision-makers who are privately motivated increasingly determine how public goods and services are distributed, or if they are to be sold off.

Accountability and authoritarianism under neoliberalism

The idea that neoliberalism is fundamentally unaccountable is further exemplified in the governance structures of the International Financial Institutions (IFIs). The IMF, for example, is completely unaccustomed to outside scrutiny and frequently responds to criticisms with defensive hubris (Bullard 2002). While democracy does not appear to have a place in their internal operations, outwardly this rhetoric is central in IFI discourse, touted as an imperative human value. This feature becomes most manifest in their criticisms of the failings of democracy in Southeast Asian countries where its reform policies have repeatedly resulted in social discontent and equally predictable authoritarian responses. This response, of course, begs the question, if the IFIs are themselves not democratic, accountable or transparent institutions, is their authority to make judgments on issues concerning democracy really tenable? Given the lack of introspection the IFIs display, it is not surprising that "the idea that authoritarian states could play a positive development role became attractive within the West at a time when growth rates lagged behind some Asian rates. . . . Within Western business circles many looked approvingly at the state's role in sweeping aside 'distributional coalitions' (labor, welfare, and environmental groups) and instituting low tax regimes" (Rodan et al. 2001, 14). Thus, although the relationship between neoliberalism and democracy is far from straight forward, neoliberal reform is rife with opportunities for political and economic elites with strong commercial interest to influence development away from democracy (Jonsson 2002). As demands for greater neoliberal reform come from the global north, in order for Southeast Asian states to preserve the privileges of their ruling classes they must confront two options: 1) the establishment of a new social compromise of their own (to align larger segments of the population with the prosperity of the wealthiest), a condition antithetical to neoliberalism; or 2) a shift toward an increasingly authoritarian regime, a position that neoliberalism can easily accommodate and one that seems to have been repeated across the region (Duménil and Lévy 2004) – Cambodia offers a particularly compelling case in point (Springer 2010). Political repression is, as such, a commonplace feature of Southeast Asian neoliberalism, where the persecution of activists and non-governmental organizations continues to play a leading role in the reform processes since at least the time of the Asian Crisis. While some might be keen to emphasize that neoliberalism has not been universally negative inasmuch as within existing authoritarian regimes there has been some opening of electoral reform, I am not convinced that this is enough. Within the context of Cambodia I have argued against the idea of democracy being reduced to mere electoralism, and have instead prioritized a more radical democratic basis located in a politics of resistance (Springer 2015). Despite the repression that continues to be meted out against protesters, there is some room for hope here as grassroots social movements have started to take hold in Southeast Asia, ranging from identity politics and belonging to place among indigenous minority groups in Upland Southeast Asia, to women's empowerment in response to forced evictions in Cambodia, to transnational labor activism, to the Red Shirt movement in Thailand (Parvanova and Pichler 2013).

That neoliberalism has coincided with a new and intensifying pattern of conflict, one that seems to be concerned with identity groups rather than the state is perhaps unsurprising given the ongoing exclusions of the poor and rising inequality (Rapley 2004). Contemporary political conflict is now predominantly *within* rather than between states (Demmers 2004; Desai 2006), which despite occasional territorial scuffles, such as the recent border conflict between Thailand and Cambodia over Preah Vihear, holds true in Southeast Asia. The bulk of conflict that does exist can be interpreted as a reflection of the geographic restructuring and uneven development that neoliberalism provokes (Harvey 2005). At the very least there is a new scalar logic, where particular cities like Phnom Penh and Manila have become the loci of development and investment, spurning mutually destructive place-making policies as cities strategically position themselves in favor of neoliberal policies. Meanwhile, peripheral areas are ignored, which makes individual countries prone to rural resentment and/or coveting of urban wealth. Marginalization of the rural frame is furthered as it becomes a site for extraction rather than investment, and differences can become magnified. Patterns of conflict can easily evolve among the dispossessed and downtrodden within the rural sphere in a competition for the scraps left at the neoliberal table among 'underdogs' (Uvin 2003), as 'top dogs' in the city meanwhile insulate themselves from reprisal through an ever-tightening security regime. This clampdown in Southeast Asia has come in the form the apparatus of the state, such as authoritarian showings of force in public space (Springer 2010), and private measures visible in the landscape, such as fenced properties patrolled by armed guards (Coleman 2004). Furthermore, any violence that bubbles up among underdogs engenders Orientalist discourses that problematically position 'local' cultures as being wholly responsible, erasing the contingency, fluidity and interconnectedness of the global political economy of violence. Such discourses license even more reform as neoliberalism is positioned as a 'civilizing' enterprise in the face of such purported 'savagery' (Springer 2015). The current implicit support for authoritarianism is not, of course, without historical precedent in Southeast Asia, but it is interesting to witness just how well neoliberalism is facilitated within this political landscape. We would do well to remember that following the Second World War and lodged within the prevailing rhetoric of modernization, the principal driver of American foreign policy was the threat to capitalism, not democracy (Rodan and Hewison 2004). Resistance to neoliberalism was to be expected as the state moved to ensure reforms, which required the threat of immediate state violence to dissenters and equally necessitated that local elites be brought on board. This exact scenario has played out repeatedly across Southeast Asia.

Convincing local elites that neoliberal reforms such as deregulation and privatization will offer them an opportunity for enrichment is not too hard of a sell, and indeed there is some indication that the final holdout in the region, Myanmar, is starting to move toward this precise understanding as it strengthens ties with the outside, and now thoroughly neoliberalized, world. Where elites have accepted this strategy, the tensions of neoliberal reform may be minimal and, in point of fact, even beneficial among the upper classes. Elsewhere I have argued that this is precisely the case in post-transitional Cambodia, where neoliberalization has become a useful part of the existing order (Springer 2010). While the potential for a small elite to consolidate their privilege is clearly not ideal for democracy, this has hardly been a stumbling block. In contrast to the experience of the upper class, though, neoliberal reform for the lower classes is much more problematic as any potential benefits are not as forthcoming. Inequality and poverty remain widespread in much of Southeast Asia, and the neoliberal fantasy of the 'trickle-down effect' has undoubtedly failed to bear fruit for the poor, particularly for those in rural areas. Consequently, neoliberalism in Southeast Asia may have actually played a key role in consolidating authoritarianism even more as those left behind in the wave of policy reforms so frequently come into conflict with those reaping neoliberalism's rewards (Canterbury 2005). Where and when those changes have been particularly rapid and a legitimizing discourse for neoliberalism has not already become widely circulated, the potential for conflict is even more pronounced. This is precisely why penetration of neoliberal ideas at an everyday level becomes a determinant of the degree of authoritarianism needed under neoliberalization (Springer 2015). Furthermore, Southeast Asia has been staged as a theater in the ongoing 'war on terror,' meaning that support for authoritarian regimes by the United States has been reinvigorated. This war of perpetuity has brought with it a revived nexus between police and military power, which has been exploited in the region. Preventive arrest and detention in concert with extensive powers of surveillance have generated panoptic conditions in many Southeast Asian states (Tyner 2006).

Conclusion

Across the region the general mood might be summed up by what is occurring in the Indonesian context where investor perspectives on the country are informed by an underlying sense of nostalgia for 'the good old days' of Soeharto when things were certain. In other words, investors are not primarily concerned with whether democracy is functioning or not, or even whether there is corruption or not. Instead, they simply require "a strong government, democratic or authoritarian, that provides predictability and keeps in check coalitions that might contest the terms under which they operate" (Robison 2002, 109). The global neoliberal marketplace is not necessarily hostile toward authoritarianism or supportive of democracy, but instead actually benefits from the presence of a strong state (Kinnvall 2002), which is precisely why neoliberalism can make such a good bedfellow for the 'developmental state'. Yet insofar as the 'strong state' thesis is concerned (Midgal 1988), neoliberalism requires precisely the opposite conditions to function, and not because of a mistaken belief that neoliberalization sees the state wither in the face of increased market power (Brenner and Theodore 2002; Peck 2010). Contra the common misconception that authoritarian regimes constitute 'strong states,' we should actually conceive of them as 'weak states' in the sense that they are unable to secure legitimacy among the population, requiring authoritarian regimes that resort to repressive measures (Jomo 2003). The lack of consultation with the general population over economic reforms and political decisions, coupled with reduced access to social provisions for the poor – who still continue to constitute an overwhelming majority in most Southeast Asian states – means that accountability is all but absent under neoliberalism. With little recourse for citizens and a system that is responsible only to shareholders, authoritarianism becomes the primary disposition of the Southeast Asian neoliberal state. In short, the desires of the capitalist class coincide with and come to dominate neoliberal policy orientations, making 'particularistic demands' the insignia of the neoliberalized state, which goes some way to explaining why it has had such a meteoric ascendency across Southeast Asia.

Neoliberalism as a conceptual category requires an appreciation for the complexity of exchanges between local, national, and regional forces operating within the global political economy. It is crucial to recognize and account for the traction of neoliberalization on its travels around the globe and to attend to how neoliberalism is always necessarily co-constituted with existing economic frameworks and political circumstances. Similarly, it is important to acknowledge that an inordinate focus on either external or internal phenomena to the exclusion of relational connections across space is insufficient in addressing the necessary features and significant articulations of neoliberalism as a series of 'glocal' processes. The abstraction of neoliberalism as a 'global' project, or even as something we can analyze at a regional level as I have attempted to do here, is necessarily contingent. In theorizing neoliberalization as an articulated, traveling, and hybridized phenomenon, my argument is not intended as plenary, and to be clear I am not suggesting that the substantive effects of neoliberalism are everywhere and always the same. Instead, I have attempted to draw out some of the articulations that neoliberalism has in the Southeast Asian context. As Southeast Asian states have become neoliberalized in their developmental agendas, planning agencies, decision-making powers and economic orientations, so too have they become increasingly integrated into transnational circuits of capital and expertise (Sneddon 2007). Patron politics, of course, predate the region's encounter with neoliberal ideas, but it is clear that such relations have since become inextricably bound-up in processes of neoliberalization. While Southeast Asia's particular versions of neoliberalism suggest that unique geohistorical power relations are at play, determining the precise and country-specific implications of neoliberalism requires a much more fine-grained analysis than I am able to offer here. I have only been able to speak to the broad regional patterns in neoliberalism's uptake, where the relationship between neoliberalism, patronage, and authoritarianism across Southeast Asia is rooted in similar cultural histories and political legacies. Importantly though, the degree to which a regional approach is meaningful is necessarily contradicted by neoliberalism's variegated stripes, as it is always context specific, protean, and checkered with contingencies and contradictions.

Note

1 This is a phrase employed by the World Bank, which initially touted the Four Asian Tigers of Hong Kong, Singapore, South Korea and Taiwan as models for development in the region.

References

Aguiar, L. L. M. and Herod, A. (eds.) (2006). *The dirty work of neoliberalism: Cleaners in the global economy*. Oxford: Blackwell.

Bakker, K. (2005). Neoliberalizing nature? Market environmentalism in water supply in England and Wales. *Annals of the Association of American Geographers*, 95(3), pp. 542–565.

Bourdieu, P. and Wacquant, L. (2001). NewLiberalSpeak: notes on the new planetary vulgate. *Radical Philosophy*, January/February, 105, pp. 2–5.

Brenner, N. and Theodore, N. (2002). Cities and the geographies of 'actually existing Neoliberalism'. *Antipode*, 34, pp. 349–379.

Brown, M. (2004). Between neoliberalism and cultural conservatism: Spatial divisions and multiplications of hospice labor in the United States. *Gender, Place and Culture: A Journal of Feminist Geography*, 11(1), pp. 67–82.

Bullard, N. (2002). Taming the IMF: How the Asian crisis cracked the Washington Concensus. In: P. P. Masina, ed., *Rethinking development in East Asia: From illusory miracle to economic crisis*. Richmond and Surrey: Curzon Press and The Nordic Institute of Asian Studies, 144–158.

Canterbury, D. C. (2005). *Neoliberal democratization and new authoritarianism*. Burlington and Vermont: Ashgate.

Challies, E. and Murray, W. (2008). Towards post-neoliberalism? The comparative politico-economic transition of New Zealand and Chile. *Asia Pacific Viewpoint*, 49(2), pp. 228–243.

Coleman, R. (2004). Images from a neoliberal city: The state, surveillance and social control. *Critical Criminology*, 12(1), pp. 21–42.

Demmers, J. (2004). Global neoliberalism and violent conflict: Some concluding thoughts. In: J. Demmers, A. E. Fernandez Jilberto and B. Hogenboom, eds., *Good governance in the era of global neoliberalism: Conflict and depolitization in Latin America, Eastern Europe, Asia, and Africa*. New York: Routledge, 331–341.

Desai, R. (2006). Neoliberalism and cultural nationalism: A danse macabre. In: D. Plehwe, B. Walpen and G. Neunhoffer, eds., *Neoliberal hegemony: A global critique*. New York: Routledge, 222–235.

Duménil, G. and Lévy, D. (2004). The nature and contradictions of neoliberalism. In: L. Panitch, C. Leys, A. Zuege and M. Konings, eds., *The globalization decade: A critical reader*. London: The Merlin Press, 245–274.

George, S. (1999). A short history of neo-liberalism: Twenty years of elite economics and emerging opportunities for structural change. *Economic Sovereignty in a Globalising World*, Conference, Bangkok, 24–26 March.

Gilbert, D. (2014). *The American class structure in an age of growing inequality*. London: Sage.

Gill, S. (1995). Globalisation, market civilization, and disciplinary neoliberalism. *Millennium: Journal of International Studies*, 24(3), pp. 399–423.

Goldberg, D. T. (2009). *The threat of race: Reflections on racial neoliberalism*. Malden: Blackwell.

Hackworth, J. (2007). *The neoliberal city: Governance, ideology, and development in American urbanism*. Ithaca, NY: Cornell University Press.

Hart, G. (2002). Geography and development: Development/s beyond neoliberalism? Power, culture, political economy. *Progress in Human Geography*, 26(6), pp. 812–822.

Harvey, D. (2005). *A brief history of neoliberalism*. New York: Oxford University Press.

Hayek, F. A. (1944/2001). *The road to serfdom*. Routledge Classics edition. New York: Routledge.

Haylett, C. (2001). Illegitimate subjects? Abject whites, neoliberal modernisation, and middle-class multiculturalism. *Environment and Planning D: Society and Space*, 19(3), pp. 351–370.

Jomo, K. S. (2003). Introduction: Southeast Asia's ersatz miracle. In: K. S. Jomo, ed., *Southeast Asian paper tigers: From miracle to debacle and beyond*. New York: Routledge Curzon, 1–18.

Jonsson, K. (2002). Globalization, authoritarian regimes and political change: Vietnam and Laos. In: C. Kinnvall and K. Jonsson, eds., *Globalization and democratization in Asia: The construction of identity*. New York: Routledge, 114–130.

Kinnvall, C. (2002). Analyzing the global-local nexus. In: C. Kinnvall and K. Jonsson, eds., *Globalization and democratization in Asia: The construction of identity*. New York: Routledge, 3–18.

Klein, N. (2007). *The shock doctrine: The rise of disaster capitalism*. Toronto: A.A. Knopf.

Klodawsky, F. (2009). Home spaces and rights to the city: Thinking social justice for chronically homeless women. *Urban Geography*, 30(6), pp. 591–610.

Larner, W. (2009). Neoliberalism. In: R. Kitchen and N. Thrift (eds., in-chief) *International encyclopedia of human geography*. Amsterdam: Elsevier, 374–378.

Lawson, V. (1999). Questions of migration and belonging: Understandings of migration under neoliberalism in Ecuador. *International Journal of Population Geography*, 4(5), pp. 261–276.

Leitner, H., Peck, J. and Sheppard, E. S. (eds.) (2007). *Contesting neoliberalism: Urban frontiers*. New York: Guilford Press.

Lippmann, W. (1937/2005). *The good society*. New Brunswick, NJ: Transaction Publishers.

Lummis, C. D. (1996). *Radical democracy*. Ithaca, NY: Cornell University Press.

MacIntyre, A. (1999). Political institutions and the economic crisis in Thailand and Indonesia. In: T. J. Pempel, ed., *The politics of the Asian economic crisis*. Ithaca, NY: Cornell University Press, 143–162.

May, J., Cloke, P. and Johnsen, S. (2005). Re-phasing neoliberalism: New labour and Britain's crisis of street homelessness. *Antipode*, 37(4), pp. 703–730.

McCarthy, J. and Prudham, S. (2004). Neoliberal nature and the nature of neoliberalism. *Geoforum*, 35(3), pp. 275–283.

Migdal, J. S. (1988). *Strong societies and weak states: State-society relations and state capabilities in the third world*. Princeton, NJ: Princeton University Press.

Mirowski, P. and Plehwe, D. (eds.) (2009). *The road from Mont Pelerin: The making of the neoliberal thought collective*. Cambridge, MA: Harvard University Press.

Mitchell, K. (2004). *Crossing the neoliberal line: Pacific Rim migration and the metropolis*. Philadelphia: Temple University Press.

Ong, A. (2007). Neoliberalism as a mobile technology. *Transactions of the Institute of British Geographers*, 32(1), pp. 3–8.

Oswin, N. (2007). Producing homonormativity in neoliberal South Africa: Recognition, redistribution, and the equality project. *Signs: Journal of Women in Culture and Society*, 32(3), pp. 649–669.

Overbeek, H. (2000). Transnational historical materialism: Theories of transnational class formation and world order. In: R. Palan, ed., *Global political economy: Contemporary theories*. New York: Routledge, 168–183.

Oza, R. (2006). *The making of neoliberal India: Nationalism, gender, and the paradoxes of globalization*. New York: Routledge.

Parvanova, D. and Pichler, M. (2013). Activism and social movements in South-East Asia. *ASEAS – Austrian Journal of South-East Asian Studies*, 6(1), pp. 1–6.

Peck, J. (2001). Neoliberalizing states: Thin policies/hard outcomes. *Progress in Human Geography*, 25, pp. 445–455.

Peck, J. (2002). Labor, zapped/growth, restored? Three moments of neoliberal restructuring in the American labor market. *Journal of Economic Geography*, 2(2), pp. 179–220.

Peck, J. (2008). Remaking laissez-faire. *Progress in Human Geography*, 32(1), pp. 3–43.

Peck, J. (2010). *Constructions of neoliberal reason*. Oxford: Oxford University Press.

Peck, J. and Tickell, A. (2002). Neoliberalizing space. *Antipode*, 34, pp. 380–404.

Plehwe, D. and Walpen, B. (2006). Between network and complex organization: The making of neoliberal knowledge and hegemony. In: D. Plehwe, B. Walpen and G. Neunhoffer, eds., *Neoliberal hegemony: A global critique*. New York: Routledge, 27–50.

Polanyi, K. (1944). *The great transformation: The political and economic origins of our rime*. Boston, MA: Beacon Press.

Power, M. (2003). *Rethinking development geographies*. New York: Routledge.

Rapley, J. (2004). *Globalization and inequality: Neoliberalism's downward spiral*. Boulder, CO: Lynne Rienner.

Richardson, D. (2005). Desiring sameness? The rise of a neoliberal politics of normalization. *Antipode*, 37(3), pp. 515–535.

Robison, R. (2002). What sort of democracy? Predatory and neo-liberal agendas in Indonesia. In: C. Kinnvall and K. Jonsson, eds., *Globalization and democratization in Asia: The construction of identity*. New York: Routledge, 92–113.

Robison, R. (2004). Neoliberalism and the future world: Markets and the end of politics. *Critical Asian Studies*, 36, pp. 405–423.

Robison, R., Rodan, G. and Hewison, K. (2005). Transplanting the neoliberal state in Southeast Asia. In: R. Boyd, ed., *Asian states: Beyond the developmental perspective*. New York: Routledge Curzon, 1–18.

Robison, R. and Rosser, A. (2003). Surviving the meltdown: Liberal reform and political oligarchy in Indonesia. In: R. Robison, M. Beeson, K. Jayasuriya and H-R. Kim, eds., *Politics and markets in the wake of the Asian crisis*. London: Routledge, 171–191.

Rodan, G. and Hewison, K. (2004). Closing the circle? Globalization, conflict, and political regimes. *Critical Asian Studies*, 36, pp. 383–404.

Rodan, G., Hewison, K. and Robison, R. (2001). Theorising South-East Asia's boom, bust, and recovery. In: G. Rodan, K. Hewison and R. Robison, eds., *The political economy of South-East Asia: Conflicts, crises and change*. New York: Oxford University Press, 1–41.

Sneddon, C. (2007). Nature's materiality and the circuitous paths of accumulation: Dispossession of freshwater fisheries in Cambodia. *Antipode*, 39, pp. 167–193.

Sparke, M. (2006). A neoliberal nexus: Economy, security and the biopolitics of citizenship on the border. *Political Geography*, 25(2), pp. 151–180.

Springer, S. (2008). The nonillusory effects of neoliberalisation: Linking geographies of poverty, inequality, and violence. *Geoforum*, 39(4), pp. 1520–1525.

Springer, S. (2009). The neoliberalzation of security and violence in Cambodia's transition. In: S. Peou, ed., *Human security in East Asia: Challenges for collaborative action*. New York: Routledge, 125–141.

Springer, S. (2010). *Cambodia's neoliberal order: Violence, authoritarianism, and the contestation of public space*. New York: Routledge.

Springer, S. (2011). Articulated neoliberalism: The specificity of patronage, kleptocracy, and violence in Cambodia's neoliberalization. *Environment and Planning A*, 43(11), pp. 2554–2570.

Springer, S. (2012). Neoliberalising violence: Of the exceptional and the exemplary in coalescing moments. *Area*, 44(2), pp. 136–143.

Springer, S. (2015). *Violent neoliberalism: Development, discourse and dispossession in Cambodia*. New York: Palgrave MacMillan.

Springer, S., Birch, K. and MacLeavy, J. (eds.) (2016). *The handbook of neoliberalism*. London: Routledge.

Tetreault, M. A. (2003). New odysseys in global political economy: Fundamentalist contention and economic conflict. In: M. A. Tetreault, R. A. Denemark, K. P. Thomas and K. Burch, eds., *Rethinking global political economy: Emerging issues, unfolding odysseys*. New York: Routledge, 3–20.

Tyner, J. A. (2006). *America's strategy in Southeast Asia: From the Cold War to the Terror War*. Lanham, MD: Rowman & Littlefield.

Uvin, P. (2003). Global dreams and local anger: From structural to acute violence in a globalizing world. In: M. A. Tetreault, R. A. Denemark, K. P. Thomas and K. Burch, eds., *Rethinking global political economy: Emerging issues, unfolding odysseys*. New York: Routledge, 147–161.

Wade, R. H. (2004). Is globalization reducing poverty and inequality? *World Development*, 32, pp. 567–589.

4
AGGREGATE TRENDS, PARTICULAR STORIES

Tracking and explaining evolving rural livelihoods in Southeast Asia

Jonathan Rigg and Albert Salamanca

Introduction

The challenge when it comes to interpreting changing rural livelihoods in Southeast Asia is to pinpoint what, exactly, is changing, why and with what consequences for human development. To be sure, much has changed in the region over the last half century. From subsistence and semi-subsistence modes of production, rural households today are thoroughly enmeshed in the market economy, not only producing agricultural commodities for sale but also engaging with a range of other commercial, often non-farm activities. These changes have, in turn, importantly shaped aspirations and cultural norms in the countryside.

This statement, though, highlights one of the key challenges with writing about 'rural' livelihoods: sustaining rural livelihoods and communities in the countryside depends increasingly on engagement with work in urban and non-farm contexts. Until the 1980s in most countries of the region, rural livelihoods were primarily located in the countryside and focused on farming. When scholars wrote about rural livelihoods the implication was that they were describing farm-based livelihoods and populations who lived in rural settlements – or 'villages.'

Today, and this applies not only to established middle-income economies like Malaysia and Thailand, but also to late developing countries such as Laos and Vietnam, the assumption that to understand rural livelihoods we need look little further than the fields, forests, rivers and lakes that surround villages is problematic, and increasingly so. To capture the essence of rural livelihoods, as this chapter will argue, we need to look far wider, both sectorally and spatially. Rural livelihoods have changed not just in terms of *what* they deliver, but also *how* and *where* rural livelihoods are sustained.

We approach this chapter in two ways. First of all, we set out some regional generalizations about the key changes to rural livelihoods. These, of course, do not have equal purchase in all countries but we maintain that they do have a level of regional resonance and are, therefore, useful generalizations. Second, we illustrate – and problematize – these generalizations with reference to country case studies, focusing on gender and generation, class and inequality, persistence and dispossession, poverty and well-being, and livelihoods and climate change. In this way, we aim to both outline the bigger story while also providing a degree of detail that enables these general propositions to have local-level traction.

Aggregate trends in the countryside

Aggregate trends in the countryside tell a familiar and expected story (Table 4.1). They track a relative decline in the proportion of the population defined as 'rural' as the countries of Southeast Asia become more urban, they show a continuing – perhaps even growing – concentration of a dwindling poor in rural areas (see Table 4.4), a decline in the importance of agriculture to national GDP, and a relative shrinking of the rural labor force. They reflect, therefore, quite fundamental transformations in the spatial distribution of populations, the sectoral distribution of economic activity and employment, and the location of poverty. All of these are important when it comes to understanding rural livelihoods.

Importantly, however, these data hide – or disguise – many of the important processes of transformation in the countryside that have implications for rural livelihoods. The data, and the

Table 4.1 The Southeast Asian countryside: the big picture (1960–2012)

	Selected indicators	1960	1970	1980	1990	2000	2010	2015
Cambodia	Rural population (%)	89.7	84	90.1	84.5	81.4	80.2	79.3
	Agriculture (% GDP)					37.8	36.0	28.2
	Agricultural employment (%)					73.7	54.2	51
Indonesia	Rural population (%)	85.4	82.9	77.9	69.4	58	50.1	46.3
	Agriculture (% GDP)	51.5	44.9	24.0	19.4	15.6	15.3	13.5
	Agricultural employment (%)			56.4	55.9	45.3	38.3	32.9
Laos	Rural population (%)	92.1	90.4	87.6	84.6	78	66.9	61.4
	Agriculture (% GDP)				61.2	45.2	32.7	27.4
	Agricultural employment (%)							
Malaysia	Rural population (%)	73.4	66.5	58	50.2	38	29.1	25.3
	Agriculture (% GDP)	34.3	29.4	22.6	15.2	8.6	10.4	8.5
	Agricultural employment (%)			37.2	26.0	18.4	13.3	12.5
Myanmar	Rural population (%)	80.8	77.2	76	75.4	73	68.6	65.9
	Agriculture (% GDP)		38.0	46.5	57.3	57.2	36.9	26.7
	Agricultural employment (%)			67.1	69.7			
Philippines	Rural population (%)	69.7	67	62.5	51.4	52	54.7	55.6
	Agriculture (% GDP)	26.9	29.5	25.1	21.9	14.0	12.3	10.3
	Agricultural employment (%)			51.8	45.2	37.1	33.2	29.1
Thailand	Rural population (%)	80.3	79.1	73.2	70.6	68.6	55.9	49.6
	Agriculture (% GDP)	36.4	25.9	23.2	12.5	9.0	12.4	9.1
	Agricultural employment (%)			70.8	64.0	48.8	38.2	32.3
Timor-Leste	Rural population (%)	89.9	87.1	83.5	79.2	75.7	70.5	67.2
	Agriculture (% GDP)					29.6	20.3	16.7
	Agricultural employment (%)						50.6	
Vietnam	Rural population (%)	85.3	81.7	80.8	79.7	75.6	69.6	66.4
	Agriculture (% GDP)				38.7	22.7	18.9	18.9
	Agricultural employment (%)					65.3	48.4	43.6

Source: World Development Indicators. Accessed 11 May 2017 (http://databank.worldbank.org/data/views/variableSelection/selectvariables.aspx?source=world-development-indicators#)

categorization of those data, have not kept up with the evolving nature of rural living (Rigg et al. 2012).

From the aggregate to the detailed: what's missing in the wider story

The first transition that needs to be questioned is the assumption that rural populations are leaving farming behind and taking up non-farm work. At an individual level this may, indeed, be the case. However if we assess rural livelihoods in household, rather than in individual terms, then the story is one of pluriactivity or occupational multiplicity, where households deftly combine farm and non-farm work. This has temporal and social permutations. In temporal terms, non-farm work is sometimes orchestrated around the seasonal demands of farming. In social terms, it has particular gender and generational characteristics. It also, thirdly, has a novel spatial signature in that rural livelihoods are often sustained through work in non-rural spaces.

The identification of these changes is not easy with the data available (such as those in Table 4.1). National and international agencies collect and record data with the assumption that individuals work in particular sectors, that they are fixed in space, and that the individual and not the household is the most appropriate unit of analysis. The reality of rural livelihoods in the second decade of the twenty-first century is that, across Southeast Asia, such livelihoods are multi-sited, encompass work in farming and non-farming, and that they are best viewed and understood in household terms.

These changes, then, raise some quite fundamental questions about how we view the rural household. Formerly, and classically, the rural household was a co-residential dwelling unit. The household, in other words, was spatially situated with its members living 'under one roof' or 'eating from one cooking pot,' and meeting their needs largely through work, mostly farming, in close proximity to the place of residence. This is not to say that rural populations were sedentary; historical work (e.g., Walker 1999a and 1999b on Laos) has shown that there was a degree of population mobility even before the modern era and that the sedentary peasant paradigm has long been empirically problematic (Rigg and Salamanca 2011). That said, while rural populations may not have been fixed in space, mobility was nonetheless limited.

Increasingly, Southeast Asia has become a region on the move where mobility and migration play a key role in defining rural livelihoods (Kelly 2011), with significant ramifications for the structure and operation of the household. Scholars have written quite extensively on the emergence of multi-sited (Rigg and Salamanca 2011) or shadow (Caces et al. 1985) households where household members spend much of their time residing and working in distant places. Taking the household as the co-residential dwelling unit would lead us to discount these individuals, and yet they self-evidently remain socially, emotionally and economically part of the household. This is perhaps clearest when children are 'left behind' to be raised either by one parent or, if both parents are absent, by their grandparents or another household member.

The key – although not the only – reason behind the emergence of such multi-sited households is because rural livelihoods can no longer be sustained at a socially acceptable level through farm-based work. In other words, the exigencies of delivering a sustainable rural livelihood have had knock-on effects for the operation of the household. The household has had to adapt to the demands placed on it by the changing signature of livelihoods in Southeast Asia; and this changing signature is importantly linked to *in situ* livelihood shortcomings.

There are two important aspects to this change, which are linked but not immutably. To begin with, rural household livelihoods are coming to depend on non-farm work – on work and income generated outside the agricultural sector. Second, livelihoods are becoming increasingly extra-local, sometimes requiring that individuals live for long periods away from 'home,' not infrequently in other countries.

Rural livelihoods, gender and generation

As noted above, this re-working of the rural household in the light of changing rural livelihoods has quite particular gender and generational permutations. In terms of generation, labor migration has been concentrated among the young. At the start of Southeast Asia's mobility revolution, it was mainly younger males (both married and single) who engaged in such work. In time, however, these younger men were joined by unmarried younger women. There was, in other words, a feminization of the migration stream (Piper 2008). In the 1970s, for example, women represented just 15 percent of the Filipino international migrant labor force; by 2010 they made up 55 percent of new hires (Cortes 2013). Such migrant work was – and remains – a component of rural livelihoods, with migrants regularly remitting funds to support their natal family.

The assumption in the literature from the 1980s and 1990s was that these young, unmarried migrant women would work away from home for a few years, and then return to their natal rural villages to marry and start a family. They were, as has been said, 'dutiful daughters' (Mills 1997, 1999) migrating for the sake of the family. More recent work, however, has shown how the dutiful daughters of the 1980s and 1990s have become the 'martyr mothers' (Parreñas 2005a and 2005b) of the twenty-first century as migration extends further through the life course. It has become, in effect, and in some areas, a permanent feature rather a temporary presence. Even so, and this is important for understanding rural livelihoods, the emotional and functional connection of migrants to their natal household in rural areas has often remained strong; return in later life is still the expectation (even if that is to retire rather than to work), and remittances continue to flow from migrants back to their rural homes.

But return for younger women, whether for work or for rest, is rarely easy. This is evident in Juliette Koning's account (2005) of returning female migrants from Rikmokèri, a rice-growing village in Central Java. These young women, having regularly remitted a portion of their income to support the natal household, return with new skills, habits, aspirations, preferences, beliefs, views and tastes. They may also physically look different in the ways they carry themselves and they talk, the foods that they consume and the clothes that they wear. For Koning, return is 'impossible' in the sense that these young women return transformed: values clash, intra-household relations become fraught, life course changes become era-defining, and new ideals and practices infiltrate so quickly that, for some, they simply cannot be accommodated:

> Gendered circular labour migration has led to changes in the socioeconomic and socio-cultural environments of family life and village life. Jakarta and urban values have become part and parcel of Rikmokèri life, co-conditioning it as such.
>
> (Koning 2005, 183)

The interplay of gender and generation is reflected in Figure 4.1, based on a longitudinal study of two rice-growing villages in Mahasarakham Province in northeast Thailand. The figures show that, in 1982, non-farm work was comparatively rare and was concentrated among men aged between 16 and 45; by 2008, non-farm work had become the principal occupation of most men as well as women aged between 16 and 45 and a not insignificant element among those aged between 46 and 60 (see Rigg et al. 2012 for further discussion).

This extension of migration further up the life course and, in some instances, to women as well as men has had consequences for those left behind and patterns of householding (Rigg

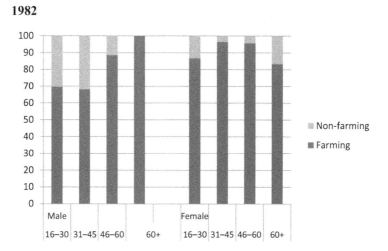

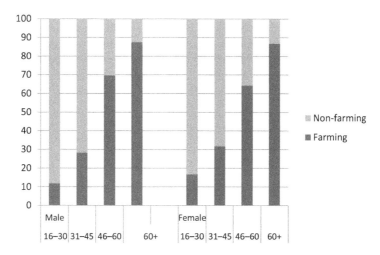

Figure 4.1 Gender and generational work in two northeastern Thai villages, 1982 and 2008

Source: Data collected by author, reported in Rigg et al. 2012

and Salamanca 2011; Douglass 2014; Huijsmans 2014; Locke et al. 2014). Most obviously, it has meant that caring responsibilities for children left behind have fallen to other household members, as noted. It has also led to a 'greying' of the countryside as the rural population ages and, therefore, to a 'geriatrification' of the farm labor force (Rigg et al. 2012) (Table 4.2). This would have happened in any case due to sharp falls in fertility in most countries of the region, but it has been accentuated by the selective out-migration of the young, and their continuing absence into middle age. This absence tends to be viewed in economic terms – it reflects the inability of farm-based livelihoods alone to deliver a reasonable livelihood – but it is also indicative of the sometimes low status of farming, particularly among the young (see Lahiri-Dutt et al. 2014).

Table 4.2 Asia's greying farmers

Country	Average age of farm labor force	Date of survey	Source
Thailand	55	2008	UNDP 2009: 71
Philippines	54	2012	Roque 2015. Accessed 19 July 2017.
Malaysia	53	2005	Shafiai and Moi 2015
Indonesia	80% >45 years old	2011	Sulthani 2011. Accessed 26 June 2016.

These demographic changes have implications for rural livelihoods because an aging farm labor force is not able to sustain farming without considerable adjustments. This may involve mechanization of production, the use of wage labor, or changes in land use and cropping patterns. In some instances, land is left idle or planted to tree crops due to the shortage of labor in rural areas. The fact that some rural households can entertain this option is because their livelihoods are sustained off the farm. As White (2012) writes, "Thinking about youth, farming and food raises fundamental questions about the future, both of rural young women and men, and of agriculture itself" (White 2012, 16). There is a three-fold challenge here; the first relates to the reproduction of the farm household; the second to generational succession or renewal; and the third to agricultural production and food security (see White 2015).

McKay's (2005) work in the community of Haliap in the foothills of Luzon's Ifugao Province in the Philippines reveals the degree to which livelihoods at home have been re-shaped by migration, leading to the emergence of what she terms 'remittance landscapes.' How those left behind manage their land in a situation where they are relatively cash rich (due to remittances) but labor poor (due to migrant absence) leads to important changes not only in the balance between different livelihood elements but also in the landscape itself:

> My respondents identified labour shortages as a key factor in the changing prevalence of crops on the local landscape. The most spectacular shift in the agricultural landscape has been the conversion of wet rice paddies into "gardens" of green beans.
>
> (McKay 2005, 92)

Rural livelihoods, class and inequality

Relatively understudied has been the way in which the issue of class intersects with changing rural livelihoods. This is partly because the 'story' is harder to demarcate and therefore to recount. There are thousands of poor rural households that have been displaced from the countryside. They have been marginalized by processes of market transition and have had little choice but to move to urban areas to eke out a living in the informal and semi-formal sectors of the economy. In some instances, they have lost their land (see below), or their work has vanished with mechanization or both. Their origins may be in the countryside – and they may continue to view themselves (and be viewed) as 'peasants' – but their displacement is permanent and their livelihoods have become urban out of necessity, rather than choice.

This pessimistic take on the effects of the advance of capitalism into Southeast Asia's rural spaces was very much part on the agrarian story as it was set out by radical scholars in the 1970s. Rex Mortimer (1973), for example, saw the Green Revolution in rice production in Indonesia, based on new varieties of rice and high applications of chemical inputs, leading to "further landlessness, unemployment and endemic rural poverty" (64). Rather later, James Scott (1985), in his highly influential work on agrarian change in Malaysia's Kedah State, wrote of the labor displacing effects of the combine harvester, a machine that was locally known as *mesin makan kerja*, the 'machine that eats work' (154). For him, "the 'facts' of the matter . . . are established with some assurance." These were the loss of harvest work and the income that comes from such work, and the loss of gleaning rights. These losses, of course, hit the poorest while the gains were reaped by the richest.

In fact, however, the rural livelihood story in Southeast Asia has not been one of absolute immiseration. Nor has there been a marked degree of dispossession in lowland, settled areas of the region. That said, the story is rather different in upland and forested areas, as we set out below. One of the real surprises with regard to rural livelihoods in Southeast Asia is the degree to which the rural smallholder appears to have survived through an extraordinary period of economic transformation. Rather than being levered out of the rural context by the inexorable forces of capitalist transformation, of which accumulation by dispossession is the most obvious (see below), smallholders on sub-livelihood farms have managed to survive through the creative combination of livelihood activities.

Land and rural livelihoods: the stubborn persistence of the smallholder

The assumption, and this continues to be reiterated in publications from many mainstream development institutions, is that, over time, smallholders will – indeed must – disappear if rural incomes and livelihoods are to be sustained. For the present, however, this has not occurred in Southeast Asia (or indeed in East and South Asia). The smallholder has proved to be remarkably persistent in the face of otherwise deep social and economic transformation (Table 4.3). Counter to the modernization theorists and their critics in the 1970s, landholdings have not been concentrated among a small number of large land owners; the smallholder remains very much a feature of the Southeast Asian rural landscape, leading some scholars (e.g., Hart 2006) to write of accumulation *without* dispossession in the countryside. The key, then, to understanding accumulation and capitalist transition in Southeast Asia is not through how producers (peasants) have been separated from their means of production (land), but how their *continuing* connections have permitted accumulation (Glassman 2006, 615). So, and to reiterate, rather than development transitions taking people out of rural areas so that their connections with the countryside are severed, it has been a case of rural livelihood diversification.

Table 4.3 Trends in average farm size, Indonesia, Philippines and Thailand (ha)

	Date	Average farm size
Indonesia	1960	1.0
	1973	1.1
	1993	0.9
	2003	0.8
Philippines	1960	3.6
	1971	3.6
	1991	2.2
	2002	2.0
Thailand	1960	3.5
	1980	3.7
	1993	3.4
	2003	3.2

Sources: Thapa 2009, Thapa and Gaiha 2011, Lipton 2010, Klatt 1972, and calculated by the author based on the World Census of Agriculture 2010 (www.fao.org/economic/ess/ess-wca/wca-2010/en/ [Accessed 26 June 2016])

On the face of it, this might seem to be welcome evidence of a moderation of capitalism's wealth-concentrating tendencies. However, for some mainstream scholars and institutions, the failure of Southeast Asia to follow the logic of the farm size transition (Hazell 2005, 94–95) is impeding rural progress, both nationally and at a household level. Indeed, it is seen as a threat both to national food security and to poverty amelioration in the countryside.

The World Bank's *Agriculture for Development* (2007) report, for example, argues that farm sizes are "becoming so minute that they can compromise survival if off-farm income opportunities are not available" (21). The answer is "the transfer of labor to the dynamic sectors of the economy" (World Bank 2007, 22), which would then permit the amalgamation of landholdings and the modernization of agriculture. For the World Bank and some economists, then, the persistence of the smallholder in Southeast (and East) Asia is a problem. Modernizing farming is assumed to be reliant on the re-organization of farms and, particularly, the amalgamation of smallholdings into larger, and tacitly more efficient, units of production. Otsuka et al. (2014) in a review of the state of Asian agriculture say that "this study strongly argues that unless drastic policy measures are taken to expand farm size, Asia as a whole is likely to lose comparative advantage in agriculture" (2014, 1; and see Otsuka 2012), with significant negative consequences.

Land and rural livelihoods: dispossession

While smallholders have proved to be surprisingly resilient as both a social formation and an economic unit of production, there are three important caveats. First, and as noted above, this has only been possible because of the way in which non-farm work has cross-subsidized farming, thus permitting what have often become sub-livelihood holdings, to survive. In a real sense, it is because some people reside and work off the farm and away from the village that others can continue to live in the village and farm. This includes, for instance, rural-urban migrants in Hanoi (Nguyen et al. 2012) and Lao cross-border migrants making their way to Thailand (Barney 2012). Second, there have been important implications arising from these new livelihood forms for the household, also as noted earlier. The smallholder household is structured and operates in new ways. The third caveat is that while the smallholder may not have been 'developed' out of existence in the manner anticipated by some earlier scholars, this is not to suggest either that smallholders have not been squeezed or that the region has been immune to the scourge of land grabbing and dispossession. The latter is a particularly significant feature of agrarian processes in upland areas of Southeast Asia where land concessions have been extended over millions of hectares of putatively 'empty' lands, with highly significant livelihood implications.

Of the countries of Southeast Asia, it is in Laos where debates over the developmental effects of large-scale land enclosure or land grabbing have been most vociferous. It has been suggested that between 2 and 3.5 million hectares of land or up to 15 percent of the country's total land area are under agri-business land concessions (see Kenney-Lazar 2012, 1023). This land has been 'grabbed' on the basis that it is 'empty,' 'undeveloped' or, at least, 'underused,' thus creating a frontier space – a fictive frontier – that capital can exploit through the transfer of land in the form of large-scale land concessions (Delang et al. 2013; Barney 2009). Upland peoples, mostly belonging to one of Laos' minorities and without secure tenure, have thus been dispossessed of their traditional lands and, as a result, their established livelihoods undermined. Having been squeezed off their land, these populations become available for work as wage labor. In Baird's (2011) study, for example, he describes how minority groups in Banchieng in southern Laos' Champasak Province have been displaced to make way for large-scale rubber concessions. These indigenous ethnic peoples – Jrou, Souay, Brou and Brao – are then hired back as casual wage labor on their former lands.

Tania Li (2002, 2007, 2014) provides a rather different insight into everyday processes of dispossession in the hills of Lauje, Sulawesi. Cocoa entered an agrarian context where land was a collective, lineage resource and farmed using techniques of shifting cultivation. The effect of cocoa, as a tree crop, was to propel processes of enclosure, privatization, commoditization and unequal accumulation. In a remarkably short space of time, a class of landless farmer emerged in the Lauje hills. Li writes:

> In these highlands ... capitalist relations emerged by stealth. No rapacious agribusiness corporations grabbed land from highlanders or obliged them to plant cacao. No government department evicted them. Nor was there a misguided development scheme that disrupted their old ways.
>
> (Li 2014, 9)

The important point, as Li makes clear, is that dispossession was in significant part enacted from below, by smallholders themselves.

Poverty and well-being in rural areas

Poverty reductions across Southeast Asia have been remarkable, reflecting the generally robust performance of the region's economies. A consistent feature of the geographical distribution of poverty has been that the poor have been – and remain – concentrated in rural areas. As Table 4.4 shows, based on national poverty lines, the incidence of poverty in rural areas, though it has declined, is between two and four times higher than in urban areas. Furthermore, for four of the five countries listed in the table, this rural-urban gap has widened so that, notwithstanding a declining poverty trend, poverty has become an increasingly rural problem. Finally, we can also say that most of these rural poor are smallholder households.

While the rural-urban poverty gap is tenacious, there is no question that material conditions in rural areas across the region have improved. Increases in agricultural production, not least linked to

Table 4.4 Rural, urban and national poverty rates for selected countries of developing Asia, earliest and latest years

	Percentage of population living in poverty				Rural poverty rate as ratio of urban poverty rate
	Rural	Urban	National	Date	
Cambodia	54.2	28.5	50.2	2004	1.9
	20.8	6.4	17.7	2012	3.3
Indonesia	19.8	13.6	17.6	1996	1.5
	14.2	8.3	11.3	2014	1.7
Malaysia	45.7	15.4	37.7	1976	3.0
	1.6	0.3	0.6	2014	5.3
Thailand	74.1	43.4	65.3	1988	1.7
	13.9	7.7	10.9	2013	1.8
Vietnam	26.9	6	20.7	2010	4.5
	18.6	3.8	13.5	2014	4.9

Notes: The poverty data are headcount poverty rates as percentage of urban, rural and national populations, based on national poverty lines. These data should not be used to compare levels of poverty between countries as national poverty lines vary considerably.

Source: World Development Indicators (http://databank.worldbank.org/data/views/variableSelection/selectvariables.aspx?source=world-development-indicators# [Accessed 11 May 11 2017])

the success of the Green Revolution in Asia, *have* played a role in raising rural incomes and, therefore, reducing rural poverty. But in many areas even more important for improving rural livelihoods has been the role of non-farm and non-rural activities, as noted. As Dercon says at a global level, "it is unlikely to be true that investing only or even largely in agriculture and smallholders is going to be sufficient for sustained poverty reduction anywhere in the world" (2013, 186). The implication, and in this regard Dercon echoes the World Bank's (2007) *Agriculture for development* report, is that sustained poverty reduction is dependent on deepening engagement with non-farm work.

The figures in Table 4.4 are money-metric measures of poverty based on income/consumption. They do not tell us very much about well-being, and one of the concerns is that while incomes are increasing and poverty (measured in terms of income) declining, the quality of growth has been such that well-being has not improved to the same degree and, in some respects, may have been compromised (McGregor 2008). Migration has separated parents from their children, the elderly are 'left' in villages with declining levels of family and community support, environmental conditions are eroding, and the moral economy of the peasant has been worn down. These negative well-being trends are, furthermore, intimately associated with those processes that have brought material prosperity and declining poverty. Although this is contentious, and it should certainly not be read as indicative of the 'failure' of the rural development project, the pursuit of affluence may, in important ways, compromise human flourishing.

Rural livelihoods and climate change

Poverty and climate-related hazards intertwine in the lives of smallholders in rural Southeast Asia. The livelihoods and poverty chapter of the IPCC Assessment Report (AR5) states that "climate-related hazards exacerbate other stressors, often with negative outcomes for livelihoods, especially for people living in poverty" (Olsson et al. 2014, 796). These stressors are social, political and economic in nature, and they complicate or amplify the impacts of climate variability or extremes so that vulnerabilities become socially constituted and underscored by the presence (or absence) of appropriate and responsive institutions (Ribot 2010).

Efforts aimed at reducing rural poverty and promoting rural development often exacerbate climate-related problems. The transformation of peatlands in insular Southeast Asia into farms or plantations, for example, has led to massive methane release to the atmosphere (Hergoualc'h and Verchot 2013). Deforestation is the leading cause of peatlands loss driven by logging activities, resulting in increased fire hazards, and large-scale oil-palm and pulpwood plantation developments (Miettinen et al. 2012). The goal of development may be to address poverty, but the fear is that the impacts of climate change will whittle away the gains that have been achieved in alleviating poverty over the last two decades.

As noted, while the contribution of the agricultural sector to GDP may be declining across the region, the sector still accounts for a large share of employment, and is particularly important for the poor for whom agriculture is a major source of income (see Table 4.1). Thus, any decline in output due to climate change is likely to have significant welfare implications for the poor. This will be particularly acute, it is projected, for Indonesia, the Philippines, Thailand and Vietnam where welfare losses are expected to be considerable (Zhai and Zhuang 2009). Modelling undertaken for a USAID-funded project in the Mekong Basin predicts increased flooding and significant increases in agricultural drought in the southern and eastern portions of the basin (ICEM 2013). Furthermore, such climate change-related hazards intersect with other, unrelated processes such as land and water grabs, which further accentuate the risks for poor smallholders (Global Witness 2012; Oldenburg and Neef 2014; Schönweger et al. 2012).

While great store – and great faith – is placed in the latent adaptability of the Southeast Asian smallholder in the face of climate change, the fear is that many of the changes evident across the region – marketization, land grabbing and concentration, chemicalization of production and export crop dependence – are making poor smallholders more rather than less exposed to climate-related risks (Rigg and Oven 2015). This is a point that Cannon and Müller-Mahn (2010) note in their paper, when they rhetorically pose the question: "Can 'development' help to reduce risks, including the ones created by climate change, or is it itself – at least to some extent – responsible for the manufacturing of risks?"

Rural futures: predictions and uncertainties

What does the future hold for rural livelihoods in Southeast Asia? There are elements that would seem to be quite firmly established: intensified engagement with the market, further diversification and increased levels of mobility. However, there are also some elements of rural livelihoods where it is unclear what will occur:

- Will smallholders and smallholdings continue to dominate the rural landscape? Or will we see the farm size transition begin to take hold?
- Will migrant children establish themselves permanently away from natal villages or will they, in later life, return 'home' and take up farming?
- Will dispossession continue to be a feature of some countries and areas or will popular protest and public pressure stem the trend?
- How will smallholders adapt and adjust to climate change?

To some extent the answers to these questions will depend on policy decisions taken by individual states. East Asia has famously protected small farms; some countries in Southeast Asia, although not to the same degree or with the same success, have also put in place schemes to protect small farmers and support their livelihoods. This includes Indonesia (McCulloch and Timmer 2008), Malaysia (Henley 2012, S32–S34) and, more recently, Thailand with its controversial rice pledging scheme (Pongkwan Sawasdipakdi 2014). The Philippines, through its 1988 Comprehensive Agrarian Reform Program (CARP), notionally set a five hectare ceiling on land ownership (Adamopoulos and Restuccia 2013, 34) while in Vietnam the state ultimately owns all land, acting as a brake on land transfer.

The political economy of farming in Asia does, in these ways, go some way to explaining the persistence of the smallholder against the backdrop of rapid economic growth and structural change. Should national governments feel impelled to further protect smallholders, this will have a significant bearing on the direction of change and on its speed.

We can, however, be fairly clear on two issues – at least at a general level. First, should smallholdings continue to dominate the agrarian context then the only way that rural livelihoods can be sustained at a reasonable level (bearing in mind that the land frontier in most countries cannot be further extended) will be through continuing and likely intensified engagement with non-farm work. Farms for many are sub-livelihood in size and while there is some scope for increases in yields, productivity and returns to farming, these are unlikely to be sufficient to keep pace with rising needs. If, on the other hand, and second, the economics of agriculture enables and the politics of farming permits the amalgamation of holdings then there will be a re-orientation of large numbers of households toward livelihoods that no longer have a farm element, and may not even be rural in their location.

Of course, all these permutations will, in practice, exist (as they do at the moment). Rather, the question concerns the balance of the agrarian transition, within and between countries. At a broader level still, there is the important question not of the amount or quantity of rural development, but its quality. Growing numbers of commentators are highlighting the negative, often unintended, repercussions of the transitions underway. Rural livelihood transformations are occurring, in other words, at considerable social and environmental cost.

References

Adamopoulos, T. and Restuccia, D. (2013). *The size distribution of farms and international productivity differences.* Working Paper 480, Toronto, Department of Economics, University of Toronto.

Anh, N. T., Rigg, J., Huong, L.T.T. and Dieu, D.T. (2012). Becoming and being urban in Hanoi: Rural-urban migration and relations in Viet Nam, *Journal of Peasant Studies*, 39(5), pp. 1103–1131.

Baird, I. G. (2011). Turning land into capital, turning people into labor: Primitive accumulation and the arrival of large-scale economic land concessions in the Lao People's Democratic Republic. *New Proposals: Journal of Marxism and Interdisciplinary Inquiry*, 5(1), pp. 10–26.

Barney, K. (2009). Laos and the making of a 'relational' resource frontier. *The Geographical Journal*, 175, pp. 146–159.

Barney, K. (2012). Land, livelihoods and remittances: A political ecology of youth out-migration across the Lao – Thai Mekong border. *Critical Asian Studies*, 44(1), pp. 57–83.

Caces, F., Arnold, F., Fawcett, J. T. and Gardner, R. W. (1985). Shadow households and competing auspices: Migration behavior in the Philippines. *Journal of Development Economics*, 17(1–2), pp. 5–25.

Cannon, T. and Muller-Mahn, D. (2010). Vulnerability, resilience and development discourses in context of climate change. *Natural Hazards*, 55, pp. 621–635.

Cortes, P. (2013). The feminization of international migration and its effects on the children left behind: Evidence from the Philippines. *World Development*, 65, pp. 62–78.

Delang, C. O., Toro, M. and Charlet-Phommachanh, M. (2013). Coffee, mines and dams: Conflicts over land in the Bolaven Plateau, southern Lao PDR. *The Geographical Journal*, 179(2), pp. 150–164.

Dercon, S. (2013). Agriculture and development: Revisiting the policy narratives. *Agricultural Economics*, 44(s1), pp. 183–187.

Douglass, M. (2014). Afterword: Global householding and social reproduction in Asia. *Geoforum*, 51(C), pp. 313–316. doi:10.1016/j.geoforum.2013.11.003

Glassman, J. (2006). Primitive accumulation, accumulation by dispossession, accumulation by 'extra-economic' means. *Progress in Human Geography*, 30(5), pp. 608–625.

Global Witness. (2012). *Dealing with disclosure: Improving transparency in decision-making over large-scale land acquisitions, allocations and investments.* London: Global Witness, International Land Coalition, and The Oakland Institute.

Hart, G. (2006). Denaturalizing dispossession: Critical ethnography in the age of resurgent imperialism. *Antipode*, 38(5), pp. 977–1004.

Hazell, P. B. R. (2005). Is there a future for small farms? *Agricultural Economics*, 32, pp. 93–101.

Henley, D. (2012). The agrarian roots of industrial growth: Rural development in South-East Asia and sub-Saharan Africa. *Development Policy Review*, 30, pp. s25–s47.

Hergoualc'h, K. and Verchot, L. V. (2013). Greenhouse gas emission factors for land use and land-use change in Southeast Asian peatlands. *Mitigation and Adaptation Strategies for Global Change*, 19, pp. 789–807.

Huijsmans, R. (2014). Becoming a young migrant or stayer seen through the lens of 'householding': Households-in-flux and the intersection of relations of gender and seniority. *Geoforum*, 51, pp. 294–304. doi:10.1016/j.geoforum.2012.11.007.

ICEM. (2013). *USAID Mekong ARCC climate change impact and adaptation study.* Main Report. Prepared for the United States Agency for International Development by ICEM – International Centre for Environmental Management, Bangkok: USAID Mekong ARCC Project.

Kelly, P. F. (2011). Migration, agrarian transition, and rural change in Southeast Asia. *Critical Asian Studies*, 43(4), pp. 479–506.

Kenney-Lazar, M. (2012). Plantation rubber, land grabbing and social-property transformation in southern Laos. *The Journal of Peasant Studies*, 39(3–4), pp. 1017–1037.

Klatt, W. (1972). Agrarian issues in Asia: I. land as a source of conflict. *International Affairs (Royal Institute of International Affairs 1944–)*, 48(2), pp. 226–241.

Koning, J. (2005). The impossible return? The post-migration narratives of young rural women in Java. *Asian Journal of Social Sciences*, 33(2), pp. 165–185.

Lahiri-Dutt, K., Alexander, K. and Insouvanh, C. (2014). Informal mining in livelihood diversification mineral dependence and rural communities in Lao PDR. *South East Asia Research*, 22(1), pp. 103–122.

Li, T. M. (2002). Local histories, global markets: Cocoa and class in upland Sulawesi. *Development and Change*, 33(3), pp. 415–437.

Li, T. M. (2007). *The will to improve: Governmentality, development, and the practice of politics*. Durham, NC and London: Duke University Press.

Li, T. M. (2014). *Land's end: Capitalist relations on an indigenous frontier*. Durham and London: Duke University Press.

Lipton, M. (2010). From policy aims and small-farm characteristics to farm science needs. *World Development*, 38(10), pp. 1399–1412.

Locke, C., Thi Thanh Tam, N. and Thi Ngan Hoa, N. (2014). Mobile householding and marital dissolution in Vietnam: An inevitable consequence? *Geoforum*, 51, pp. 273–283. doi:10.1016/j.geoforum.2013.03.002

McCulloch, N. and Peter Timmer, C. (2008). Rice policy in Indonesia: A special issue. *Bulletin of Indonesian Economic Studies*, 44(1), pp. 33–44.

McGregor, J. A. (2008). Wellbeing, development and social change in Thailand. *Thammasat Economic Journal*, 26(2), pp. 1–27.

McKay, D. (2005). Reading remittance landscapes: Female migration and agricultural transition in the Philippines. *Geografisk Tidsskrift-Danish Journal of Geography*, 105(1), pp. 89–99.

Miettinen, J., Shi, C. and Liew, S. C. (2012). Two decades of destruction in Southeast Asia's peat swamp forests. *Frontiers in Ecology and the Environment*, 10, pp. 124–128.

Mills, M. B. (1997). Contesting the margins of modernity: Women, migration, and consumption in Thailand. *American Ethnologist*, 24(1), pp. 37–61.

Mills, M. B. (1999). *Thai women in the global labor force: Consumed desires, contested selves*. New Brunswick: Rutgers University Press.

Mortimer, R. (1973). *Showcase state: The illusion of Indonesia's accelerated modernisation*. Sydney: Angus and Robertson.

Nguyen, T. A., Rigg, J., Huong, L. T. T. and Dieu, D. T. (2012). Becoming and being urban in Hanoi: Rural-urban migration and relations in Viet Nam. *Journal of Peasant Studies*, 39(5), pp. 1103–1131.

Oldenburg, C. and Neef, A. (2014). Reversing land grabs or aggravating tenure insecurity? Competing perspectives on economic land concessions and land titling in Cambodia. *Law and Development Review*, 7, pp. 49–77.

Olsson, L., Opondo, M., Tschakert, P., Agrawal, A., Eriksen, S. H., Ma, S. et al. (2014). Livelihoods and poverty. In: C. B. Field, V. R. Barros, D. J. Dokken, K. J. Mach, M. D. Mastrandrea and T. E. Bilir et al., eds., *Climate change 2014: Impacts, adaptation, and vulnerability. Part A: Global and sectoral aspects: Contribution of working group II to the fifth assessment report of the intergovernmental panel of climate change*. Cambridge and New York: Cambridge University Press, 793–832.

Otsuka, K. (2012). *Food insecurity, income inequality, and the changing comparative advantage in world agriculture*. Presidential Address at the 27th International Conference of Agricultural Economists, Foz do Iguaçu, Brazil.

Otsuka, K., Liu, Y. and Yamauchi, F. (2014). *The future of small farms in Asia*. Paper [Draft] for Future Agricultures Consortium. Available at: www3.grips.ac.jp/~esp/wp-content/uploads/2014/11/112014.pdf. Accessed 2 August 2015.

Parreñas, R. S. (2005a). *Children of global migration: Transnational families and gendered woes*. Stanford, CA: Stanford University Press.

Parreñas, R. S. (2005b). Long distance intimacy: Class, gender and intergenerational relations between mothers and children in Filipino transnational families. *Global Networks*, 5, pp. 317–336.

Piper, N. (2008). Feminisation of migration and the social dimensions of development: The Asian case. *Third World Quarterly*, 29(7), pp. 1287–1303.

Pongkwan Sawasdipakdi, P. (2014). The politics of numbers: Controversy surrounding the Thai rice pledging scheme. *The SAIS Review of International Affairs*, 34(1), pp. 45–58.

Ribot, J. (2010). Vulnerability does not fall from the sky: Toward multiscale, pro-poor climate policy. In: R. Mearns and A. Norton, eds., *Social dimensions of climate change: Equity and vulnerability in a warming world*. Washington, DC: International Bank for Reconstruction and Development/World Bank, 47–74.

Rigg, J. and Oven, K. (2015). Building liberal resilience? A critical review from developing rural Asia. *Global Environmental Change*, 32(5), pp. 175–186.

Rigg, J. and Salamanca, A. (2011). Connecting lives, living, and location: Mobility and spatial signatures in Northeast Thailand, 1982–2009. *Critical Asian Studies*, 43(4), pp. 551–575.

Rigg, J., Salamanca, A. and Parnwell, M. J. G. (2012). Joining the dots of agrarian change in Asia: a 25 year view from Thailand. *World Development*, 40(7), pp. 1469–1481.

Roque, A. (2015). College grads energize PH rice farming, *The Inquirer*. Makati, Philippines: The Daily Inquirer Inc., 26 April. Available at: http://newsinfo.inquirer.net/687731/college-grads-energize-ph-rice-farming. Accessed 19 July 2017.

Schönweger, O., Heinimann, A., Epprecht, M., Lu, J. and Thalongsengchanh, P. (2012). *Concessions and leases in the Lao PDR: Taking stock of land investments*. Bern and Vientiane: Centre for Development and Environment (CDE), University of Bern and Geographica Bernensia.

Scott, J. C. (1985). *Weapons of the weak: Everyday forms of peasant resistance*. New Haven, CT: Yale University Press.

Shafiai, M. H. M. and Moi, M. R. (2015). Financial problems among farmers in Malaysia: Islamic agricultural finance as a possible solution. *Asian Social Science*, 11(4): 1–16.

Sulthani, L. (2011). Ageing farmers threaten Indonesian food security. Reuters, 10 June. Available at: http://www.reuters.com/article/us-indonesia-farmers-idUSTRE7591FD20110610. Accessed: 19 July 2017.

Thapa, G. (2009). *Smallholder farming in transforming economies of Asia and the Pacific: challenges and opportunities*. Discussion Paper prepared for the side event organized during the thirty-third session of IFAD's Governing Council, 18 February 2009. Rome, International Fund for Agricultural Development (IFAD).

Thapa, G. and Gaiha, R. (2011). *Smallholder farming in Asia and the Pacific: challenges and opportunities*. IFAD Conference on New Directions for Smallholder Agriculture. Rome, International Fund for Agricultural Development.

UNDP. (2009). Thailand Human Development Report 2009: Human security, today and tomorrow. Bangkok: United Nations Development Programme.

Walker, A. (1999a). *The legend of the golden boat: Regulation, trade and traders in the borderlands of Laos, Thailand, China and Burma*. Richmond. Surrey, UK: Curzon Press.

Walker, A. (1999b). Women, space, and history: Long-distance trading in northwestern Laos. In: G. Evans, ed., *Laos: Culture and society*. Chiang Mai and Thailand: Silkworm Books, 79–99.

White, B. (2012). Agriculture and the generation problem: Rural youth, employment and the future of farming. *IDS Bulletin*, 43(6), pp. 9–19.

White, B. (2015). Generational dynamics in agriculture: Reflections on rural youth and farming futures. *Cahiers Agricultures*, 24, pp. 330–334.

World Bank. (2007). *World development report 2008: Agriculture for development*. Washington, DC: World Bank.

Zhai, F. and Zhuang, J. (2009). *Agricultural impact of climate change: A general equilibrium analysis with special reference to Southeast Asia*. Manila: Asian Development Bank.

Acknowledgments

This chapter was written while the first author was Professor-at-large in the Institute of Advanced Studies at the University of Western Australia. He is grateful to the support of the IAS and UWA and particularly to Susan Takao and Matthew Tonts.

5
'NATURE' EMBODIED, TRANSFORMED AND ERADICATED IN SOUTHEAST ASIAN DEVELOPMENT

Victor R. Savage

Introduction

> The failure of social scientists to recognize both the radical historicity of human society and the radical historicity of nature thus leads to a failure to address the ecological crisis of our time with the realism, dialectical understanding, urgency, and commitment to revolutionary transformations in human society that it requires.
> (Foster et al., 2010, 37)

The above quote is a timely reminder that the environmental crisis cannot be addressed in separate academic arenas of science and social sciences, instead realistic solutions demand interdisciplinary research. In social science, formal and informal regions like Southeast Asia provide the best spatial containers of interdisciplinary dialogue on environmental challenges and climate change. This paper situates the human dimension of the region's human-nature relationships and underscores the importance of social science interventions. While Jared Diamond and James Robinson's (2010) *Natural Experiments in History* undergirds this interdisciplinary human-nature perspective, Joachim Radkau's (2014) *The Age of Ecology* amplifies the global multi-disciplinary consciousness of environmentalism. In my view, three broad themes embed green consciousness and action in developing regions like Southeast Asia.

First, as Nobel Prize winner Paul Crutzen has defined, Gaia is in a new geological phase, the *Anthropocene* or the "New Human" that focuses on Homo sapiens as a major driver of environmental and climatic change. Armed with science, technology, possibilist confidence and modernity mind-sets, Southeast Asian governments and political leaders believe in the power of human beings in shaping environments, cultural landscapes and development. Consumed with the immediate situation of growth, many do not see the profound ramifications in the changing human-nature equation. Ironically, if uncontrolled human activities play themselves out in the *Anthropocene*, this geological age will not be about the 'New Human' era but the end of the human species. We are repeatedly reminded that we are at the 'tipping point' of global environmental disaster in this century and that Gaia's "newly attained consciousness – which is made possible only by our global civilisation – will vanish, perhaps to be lost forever" (Flannery 2009, 12–13). Over 50 years ago, the Japanese social anthropologist Kinji Imanishi (2002) in his thought-provoking book, *The World of Living Things*, using a Marxian analysis, discussed how at various point of Earth's ecological history, a dominant 'ruling' species ruled the world and later died out from extinction. The irony this time is that Homo sapiens, the current dominant species, despite their intelligence, social and cultural systems, and technological prowess will subject themselves to mass suicide of their species. Elizabeth Kolbert's (2014) *The Sixth Extinction* argues that Gaia is experiencing the sixth mass species extinction in the last half billion years and this time human beings might be part of the species casualties. Anthony Giddens (2009, 10) wonders if "other civilizations have come and gone; why should ours be sacrosanct?"

Second, the 193 states and political entities in the world and the 11 states in Southeast Asia are locked into the developmental paradigm based on a linear idea of progress, material betterment, increasing standards of living and higher quality of life. Such unbridled state developmental trajectories undermine human-nature relationships. Development must instead be seen in terms of what Edward Wilson (1984, 1) defines as 'biophilia' or the "innate tendency to focus on life and lifelike processes." Biophilia underscores our broader concept of development in terms of sustaining nature and all its organisms. Without a healthy understanding of human biophilic preservation, long-term development in the region will be undermined. This view in some ways underscores Manuel Castells' (2004) argument of what underlies the global green movement and human identity: it is about socio-biological identity or what he calls "green culture" – to recognize the interconnectedness of all living creatures and to respect the value of each thread in the vast web of life (Castells 2004, 185). All cultures are thus woven into a human hypertext of "historical diversity and biological commonality" (Castells 2004, 185). To a large extent religions and spiritual movements are custodians of this ecological interconnectedness in nature – as found now in Deep Ecology, Sacred Nature, the Gaia thesis, Land Ethics and Deep Green Theory (Curry 2011). In Southeast Asia, where people are still deeply spiritual both in animistic beliefs and world religions, these spiritual human-nature interventions still resonate in specific terrestrial places and cosmic planetary space in varied ways.

Third, more than any other human catalyst and driver of development, capitalism and the drive for modernisation (modernity) are not only perceived panaceas for development but more so the predominant wedge in separating human beings from nature (Foster 2002; Sarkar 1999). Capitalism under modernization has been a powerful force in the conquest of nature. Social scientists have modified the rise of capitalism through the conceptual prism of ecological modernization, which translates to bending nature and green issues to fit human interests, desires and behavior. We are the predominant species in a hurry for self-gratification and self-actualization regardless of its ecological ramifications. And Southeast Asians, as dominant nouveau riche communities, are torchbearers of materialism, consumerism and status enhancements that threaten to undermine their national ecosystem and short-change their sustainable goals.

The three themes encased in the current global environmental challenge play themselves out in different ways with varied impacts on societies and communities. Unfortunately environmental and climate change do not have common outcomes in states, cities, villages and communities, hence their political and social manifestations differ. All three themes are pertinent issues in the 11 states in the Southeast Asian region. Using selective case studies drawn from states, cities and communities, human-nature relationships are unraveled in this chapter's discussion of state social changes, economic programs and cultural influences. This chapter is meant to reflect on the current environmental state in Southeast Asia as products of historical continuities that unfold in the current nexus between state development and nature. These interventions are meant to foreground the plight of the region's 'nature' capital, its environmental degradation and its ecological sustainability. At the end of the day the question that puzzles social scientists is why some communities, societies and kingdoms against all odds seem to sustain themselves over centuries while others collapse and end up in the rubbish dump of history.

Trade, colonialism and capitalism: nature valued and appraised

While Southeast Asia has not been a pivotal region in global history on its own terms according to historian Wang Gungwu (Ooi 2015), it was a region that gained de facto global recognition because of three factors. First, Anthony Reid (1988, 5) depicts Southeast Asia as having a "common physical environment" that underscored common features in food, drinks, diet and betel chewing. For centuries in Western eyes the term 'plenitude' was used to define the region's rich biodiversity (Savage 1984). An Indonesian proverb emphasizes biodiversity as: "different fields, different grasshoppers, different pools, different fish." Andre Gunner Frank (1998) sums up the region's biodiversity in economic terms: "Southeast Asia was one of the world's richest and commercially most important regions."

Second, Southeast Asia for centuries prior to the western marine connections in Asia was an active participant in Indian Ocean trade and cultural diffusion. The thalassic kingdoms of (Oc-Oe of Funan, Linyi [Champa], SriVijaya, Temasek-Singapura and Malacca) demonstrated Southeast Asia's Indian Ocean economic linkages. The region was attractive to other Asian regions for varied natural resources, food delicacies, precious stones, marine resources and agricultural produce. Trade networks in the Indian Ocean ecumene diffused not only natural resources and artefacts between ports, thalassic kingdoms and *natios* (enclaves of separate jurisdiction apart from kingdoms) but also culture and religions (Barendse 2002). Revisionist historians (Wolters, Manguin and Gunn) are now giving more credit to Southeast Asian seafarers for "opening the entire sea route from India to China" (Hall 2011, 45). Felipe Fernandez-Armesto (2002, 402), in his engaging book, *Civilizations*, argues the Indian Ocean was a "major avenue for the transmission and transaction of culture" and the "world's most influential ocean" despite Atlantic-talk and Pacific-talk.

Third, the region is saddled between two enduring 'Great Traditions,' India and China, and hence was greatly influenced by and intertwined with both civilizations in varied ways. Such populous countries provided markets for Southeast Asian produce for centuries. While the Chinese Imperial court viewed the peoples of Southeast Asia as 'barbarians of the south,' China's huge appetite for natural resources and gastronomic delicacies led to the cultural appraisal of the region's diverse natural resources – from bird's nest to sea cucumber, from rhinoceros horn to hard woods, from pearls to jade. India's influence in the region was more cultural, religious and political, though gold and the spices of the region were early imports. India's Sanskritization according to Sheldon Pollock (2006) made no distinction between India and Southeast Asia – the cultural influence covered both areas as one seamless region of different kingdoms and cosmic states. India's cultural, religious and political footprint in the region was the most lengthy and enduring and continues today in mainland Southeast Asia and in pockets in Java, Bali, Lombok, Kalimantan and Sumatra.

The downside of being sandwiched between two enduring civilizations is that scholars and the informed public tend to downplay any importance of the Southeast Asian region in favor of India and China. Yet there are growing positive appraisals of prehistoric peoples in the region making major technological developments surpassing or equaling those of its giant neighbors: the lignic phase (vegetative civilization); outrigger boats; blow pipes, agricultural origins and plant/animal domestication; socketed tools; Dongson drums; terrace agriculture; a variety of social systems; and folk medicines and houses on stilts to name some (Oppenheimer 1999; Bellwood 1985; Wolters 1999).

Southeast Asia remained the source for the vast and varied demands of natural resources from many ecosystems and indigenous communities – hunters and gatherers in the tropical forests to the *orang laut* or 'sea nomads' and from shifting cultivators to sedentary sawah 'peasants.' It was this continual foreign demand for resources that led to the variety of indigenous ports, cities and kingdoms that succeeded one another at various points in time. These changes are not the clichéd rise and fall of kingdoms, but reflect what Janet Abu-Lughod (1991) posits as the "restructuring" and "substitution" of varied world systems.

The capitalistic system: is it sustainable?

Any realistic appraisal of human-nature relationships must deal with capitalism, trade and land tenure systems. All three have played a major influence over the last 500 years in the way nature has been turned into natural resources, ecosystems monetized, land valued as property, and trade expanding economic transactions and ecological footprints. Dealing with capitalism is a tricky subject. Here I accept Rodney Stark's (2005, 56) relatively neutral and somewhat comprehensive definition of capitalism as "an economic system wherein privately owned, relatively well organised, and stable firms pursue complex commercial activities within a relatively free (unregulated) market, taking a systematic long-term approach to investing and reinvesting wealth (directly or indirectly) in productive activities involving a hired workforce, and guided by anticipated and actual returns."

While one might accept that in the thirteenth century, "a variety of protocapitalist systems coexisted in various parts of the world, none with sufficient power to outstrip the others" (Abu-Lughod 1991, 371), by the late fifteenth century the Western Age of Exploration changed the global trading landscape. The period of cultural and economic diffusion through trade for two centuries in the region needs no rehearsing here as it is best captured in Reid's (1988) two-volume work, *Southeast Asia in the Age of Commerce 1450–1680*. It would be misleading to believe that European trade opened the Indian Ocean region to a global economy, when in fact trade via a maritime route (the Maritime Silk route) was already well developed before the Europeans arrived in the late fifteenth century (Gunn 2011; Hall 2011; van Luer 1955; Miksic 2013).

The region's 'spices' (cloves, mace) in the Moluccas initiated Western colonization, emboldened capitalism, created intense Western economic competition in the region and led to one of the most concerted colonial state trading traditions in the world. The formation of colonial trading powerhouses, The Dutch East India Trading Company (Vereenigde Oostindische Compagnie, VOC), The English East India Trading Company, Austrian East India Company, the Swedish East India Company and the French East India Trading Company, was devoted to extracting natural resources and in some cases colonizing lands, and extending colonial hegemony through pin-prick colonialism from the 16th to 19th centuries. This European expansion into Asia is what Geoffrey Gunn (2011, 183) calls the "Eurasian Exchange" between East and West or what Oliver Wolters (1967) refers to as the "single ocean" trade. But clearly the Western intervention in Indian Ocean trade created intense competition for the region's natural resources and thereby initiated the mass undermining of the region's natural environment through exploitation, capitalism and ecological destruction because of the sheer demand by European and expanding Asian markets. In other words, the region became one Western theater of what Grove calls 'ecological imperialism' (Fernandez-Armesto 2002, 384).

The seeds of capitalism grew out of robust East-West trade but soon took root in Southeast Asia with a variety of impacts: 1) It created changing demands for natural resources, which left local communities reeling from boom and bust situations; 2) It provided varying values of natural resources and raw materials, which led to disparities of wealth between the colonialists, Chinese trading middlemen and the native producing communities; 3) The demand for the region's natural resources and raw materials were greatly enhanced as a result of the industrial revolution – the development of mining (tin, bauxite, iron, copper and coal) and plantation agriculture (rubber, oil palm, cinchona, coffee, tea and pineapples) were direct outcomes of the industrial revolution; and 4) it enabled rapid expansion of land colonization and major transportation inroads (roads, railway) into major natural resource and plantation areas. However, if one accepts capitalism was an Eastern innovation, Asian critics cannot blame 'Westerners' for depleting their natural resources and degrading the natural environment.

These capitalistic outcomes had impacts on changing human-nature relationships. The idea of the 'village' common lands changed with the capitalistic idea of property and the colonial government. Whether you accept James Scott's (1979) moral economy (safety-first ethics) thesis or Samuel Popkin's (1979) 'political economy' (rational peasant) argument, human-nature relationships were severely altered and eroded under colonialism. In many colonies in Malaya, Indonesia, Burma and Vietnam, the colonial administration took over all forest and vacant land as 'Crown lands' belonging to the colonial administration. The most impactful aspect of colonialism and capitalism was its creation of landless peasants (Elson 1997). More startling is Elson's (1997) thesis that by 1910, the region witnessed an end to peasantry with all its ramifications of changing cultural and social systems and their land tenure implications. For historian Craig Lockard (2009, 200–201) capitalism with its emphasis on competition underscored a profound social impact – it "undermined the traditional values of cooperation and community."

If one fast forwards the capitalistic system today the prospect of neo-colonialism seems to haunt the developing countries. While capitalism is the dominant economic regional system the prognosis of its tendency to widen disparities of wealth is disturbing. Thomas Picketty's (2014) *Capital in the Twenty-First Century* argues that capitalism will not reduce economic inequality, lessen income gaps and reduce social inequity. In his analysis, "wealth accumulated in the past grows more rapidly than output and wages. This inequality expresses a fundamental logical contradiction." Picketty (2014) continues to argue that no cultural and political intervention can change this economic situation. In short, "the past devours the future" (Picketty 2014, 571). The current flood of refugees from poorer African and Asian (Bangladesh and Rohingas from Myanmar) countries is testimony to this growing situation of inequality.

Colonialism: land colonization, land tenure and environmental impacts

Pre-colonial Southeast Asian countries had rather loosely defined land tenure systems based on the Indian Laws of Manu or indigenous customary laws such as *adat* based on village usufruct rights, which was rarely endorsed by the prevailing political authority. Since the implementation of land tenure laws under the colonial Roman legal system, the colonial law system that defined the territorial boundaries of the state was prejudicial against the indigenous customary land tenure systems creating many conflicts over village 'common lands' and forested areas.

Rita Lindayati (2003) provides an interesting case study of how national politics has shaped land tenure laws in Indonesia and how it affected the use of forested areas. Forests in indigenous communities before colonialism was the village 'commons' dictated by *adat* law or *hukum adat* and this changed under Dutch colonial rule and maintained under the New Order (1967–1998) of President Soeharto. However, in the post-New Order Period under reformasi, forested lands were once again accepted as 'common lands' under village *adat* laws. As Lindayati (2003, 257) argues, the policy shift to community-based forest management (under *adat*) is "a good start for both pragmatic (e.g., local people deal with natural resources on a day-to-day basis) and social justice (e.g., forest-dependent communities are usually poor) reasons." In colonial Malaya, however, the British authorities were more enlightened and preserved the customary land tenure areas of the Malay peasants by converting them under Malay reservations that were not subject to colonial land tenure laws (Wong 1977). In short, land under colonialism was capitalized and peasants were displaced, common lands (forests) were nationalized under Crown lands, fixed land taxes were imposed creating peasant indebtedness, plantations became widespread and the rest of 'nature' was valuated and lost its irreplaceable value in ecological terms. In small places like Singapore, the value of cash crops like gambier for tanning of leather led to the massive deforestation of the island. Estimates show that 90 percent of the forest cover in Singapore was gone in 50 years in the nineteenth century (O'Dempsey 2014, 35). Unfortunately, regional deforestation is a recurrent theme for different reasons. The massive forest fires in 1997 in the region and its accompanying haze underscores once again the anthropocentric driven and capitalist incentives for deforestation as evident in the 2001 ASEAN Report Response Strategy (Asian Development Bank 2001, xiv–xv).

The colonial period was a long period of applying science and technology in changing landscapes, modifying nature and engineering environments in the name of development and what the French in Indo-China called the *mission civilisatrice*. Armed with 'possibilist' ideological orientations (science, technology, planning, progress and rational administration), colonial governments turned the region into one experimental environmental management laboratory based on plantation agriculture, dam building, infrastructure development (road, rails and bridges), botanical gardens, processing industries, hydro-electric power, land reclamation, and urban development and planning (see for example Biggs 2009 on the French and Vietnamese government interventions in the Mekong Delta).

Nature in indigenous knowledge: is it relevant?

> Each society has its own way of organising the world, and our own perspectives are so ingrained in us that we normally take them completely for granted. Often this classification of reality has its roots in the ancient past but major changes can also take place.
> (Douglas Davies 1994, 3)

Jared Diamond (2012) in his book *The World Until Yesterday* gives us many case studies of why it is beneficial to study traditional societies; for him he found a "huge range of traditional human experience" to learn from and adapt in modern living. For example in Papua New Guinea, multilingualism is an important adaptive mechanism for indigenous peoples as a survival mechanism for trading, developing alliances, providing access to resources and even getting a spouse (Diamond 2012, 383). One of the key areas of benefit from studying traditional societies often cited by environmentalists is the human-nature relationship experience. As Carolyn Merchant (2005, 1) argues, radical ecology seeks a new "ethic of the nurture of nature and the nurture of people." These qualities in human society are often exemplified in indigenous communities where co-existence with nature, human-nature harmonious relationships and respect for nature are embedded culturally (Peterson 2001).

In studying indigenous groups in prehistory and over time in Southeast Asia, there are three conclusions one can make with regard to human-nature relationships:

a) The region has a unique prehistory that reflects in many ways the ecological soundness of human-nature relationships in a region with biodiverse abundance and formerly sparse populations.

b) Wilhelm Solheim's (1970) classification of the 'lignic' period or the vegetative phase in Southeast Asian prehistory is unique in global human prehistory. Southeast Asians more than any other peoples in prehistory have the largest experience and reservoir of knowledge of their vegetative environment and their ecological importance. The peoples of the region through innovation and indigenous genius have converted the forest resources and nature into an amazing variety of goods and domesticated crops (Sauer 1969; Savage 2012). While for most of Southeast Asian prehistory, indigenous communities were illiterate they were, however, 'earth literate.'

c) Most indigenous peoples in the region tend to adopt 'aterritorial' behavior (Savage 2009). Whether on land or at sea, mobility seems to be a characteristic feature of communities in the region. Given the sparse populations, communities have always been accustomed to free movements and the process of culturally adapting to varied ecosystems on terrestrial areas, the coastal ecotones and the seas. Even amongst sedentary ethnic groups the value of migration is an in-build cultural attribute. The sedentary Minangkabaus have a cultural endorsed propensity for migration called "merantau" in which its menfolk are encouraged to merantau to 'see the world' just as the Dayaks have "meratus" mobility, a form of semi-nomadism which "offer the pleasures of autonomy as well as the stigma of disorder" (Tsing 1993, 61).

The Southeast Asian region is home to a couple of thousand tribal and indigenous groups who have lived in relative 'harmony' with their ecosystem for centuries. These ethnographic studies across different ecosystems provide insights into how varied communities and indigenous groups adapted to environments, exploited natural resources, managed nature and sustained livelihoods. Craig Reynolds (1995) contends the Southeast Asian 'agency' is their inventiveness, their genius to adapt, their flexibility and adaptability to adjust to varying conditions environmentally and socially. Southeast Asians in general have a "tolerance" for foreign culture and an "outward looking attitude and openness." One can argue that in general Southeast Asians have a culture that is malleable to different conditions. While many societies and communities might embed their culture with history and past recollections, most Southeast Asian communities live in the present and are concerned with the "now" (Reynolds 1995) and adopt rather pragmatic means to living.

While critics denigrate swidden cultivators as wasteful, environmentally damaging and unproductive, the sustainability of tribal groups in the region over thousands of years tells us another story. Both Clifford Geertz (1963) and James Scott (2009) provide illuminating views on swidden adaptation to mountainous ecosystems and sedentary state administrations. In his thought-provoking book *The Art of Not Being Governed*, Scott (2009) provides lessons of how hundreds of tribal swidden groups in mainland Southeast Asia far from being primitive are "better seen as on a long view as adaptations designed to evade both state capture and state formation." They are "political adaptations to nonstate peoples to a world of states, that are, at once, attractive and threatening" (Scott 2009, 9).

It is not surprising the practice of ecological responsible behavior still exists amongst the many pockets of indigenous peoples in Southeast Asia. The best sample of this is Harold Conklin et al.'s (1980) masterly atlas and analysis of the Ifugao in the Philippines; their sustainable subsistence over several thousand years was based on woodlots, swidden agriculture, hunting and gathering, and sedentary terraced rice farming in difficult mountainous terrain. The downside of this is that many indigenous communities are being eliminated, transferred to settled agricultural areas and given 'modernized' settlements and national education.

The stark examples of the elimination of indigenous communities lost to government-driven national development programs in Southeast Asia are numerous but the most poignant ones are best captured by Georges Condominas (1977, xiv) on the 'ethnocide' of the Mnong Gar villagers (Montagnard) in Central Vietnam where the government wanted the Mnong Gar culture to "totally disappear, to uproot every trace of their ethnicity." In Malaysia under former Prime Minister Mahathir, the government systematically removed the Penans from their natural habitat in Sarawak in the name of development. Aiken and Leigh (2015, 87) in surveying dam building in Malaysia note that "indigenous peoples' lands and resources have been increasingly appropriated, enclosed and commodified for mainly private and state accumulation." For the Malaysian government the forest was too economically valuable to be kept for the Penans.

Two studies of different sets of indigenous peoples give the cultural logic of human-nature relationships and the sustainability of peoples in the region. Edmund Leach's (1964) ethnographic work on the Kachin in mainland Southeast Asia demonstrates that cultural identity is not a permanent ethnic identity but one that is negotiable depending on environmental, social and political circumstances. In order to survive, the Kachin oscillated between three different modes of human-nature relationships: they could be care-free roaming groups, swidden cultivators or sedentary cultivators tied to rice growing Shans – they were at different times, gumsa, gumlao and Shan.

Leach's student Geoffrey Benjamin (2002) examined these tri-partite relationships on a broader regional scale. Using Malaysia's indigenous groups as a basis, Benjamin criticized Heine-Geldern's (1942) eight wave migration thesis of peoples from China descending in the region and the folk-scholarly "*kuih lapis*" (layer cake) ethnology. In the Heine-Geldern categorization, different production systems (hunters and gatherers, swidden, and sedentary sawah cultivators) are viewed as products of different stages and periods of migrations from the Yunnan, China. Benjamin's thesis refutes this temporal argument in Chinese migration into the region; he proposed that each production group (hunters and gatherers – Semang; swidden communities – Senoi; and sedentary farmers – Malays) evolved in situ from within their own distinctive ecosystems (dense tropical forests; hilly-mountainous terrain; low-lying valley-delta areas) and developed a human-nature relationship best suited to exploit natural resources and culturally adapt to their ecosystems.

Reflections

> The more we remain preoccupied with current events, the more that individuals and their choices matter; but the more we look out over the span of centuries, the more their geography plays a role.
>
> (Robert D. Kaplan 2012, xx)

Taking a broad sweep of history and analyzing the geopolitical issues over the centuries, Robert Kaplan's (2012, 34) quote above endorses what he advocates as "geography matters" in understanding global events, localisms and specific landscapes. Unfortunately, Kaplan's geography matters thesis is locked in a Cold War mind-set of competing powers, rising states, balance of power analysis and national strategies. The geography that will matter in the twenty-first century will need to address more global issues arising from climate and environmental change, income inequalities between and within states, an urbanized Gaia and planetary sustainability. The future shift in geographical analysis must come not just from spatial analysis (*a la* Lefebvre's production of space) but more so from understanding and managing human-nature relationships. The massive impact of climate and environmental changes on human activity and cultural landscapes has foregrounded human-nature relationships in global and local interdisciplinary interventions.

Time is an important variable in addressing the region's human-nature balance in development programs and policies. Southeast Asian communities are caught between the indigenous circulatory and cyclical notions of time and the current Western linear views of temporal progress. Presenjit Duara (2015) argues that one needs to view 'circulatory histories' (based on transnational and trans-local flows) in which natural resources, ecosystems and environmental goods are valued, appraised and transformed at various times.

Despite the negative views on environment and development in the region, there are three positive signs that policy makers can count on. First, the Southeast Asian region is one of the few regions in the developing world with a strong regional organization, ASEAN. For a region that espouses some 'cultural coherence' and common geographical matrix (Wolters 2008, 1999), these commonalities can be leveraged by ASEAN to cement coordination and cooperation amongst states for the common good of communities within the region. One positive development is that ASEAN countries have grown politically more mature and now accept international-bound legal interventions to solve many bilateral border and trans-boundary issues. ASEAN countries must recognize that 'global value chains' that spur development are indeed regionally based. Thriving regions are those where 'foreign value-added' exports are based on high intra-regional trade: Europe with 72 percent intra-regional trade and East Asia with 56 percent compared to unsustainable South America where intra-regional trade is only 30 percent (*The Economist* 2015, 35).

Second, Singapore is a shining example from within the region having moved from Third World to First World in 50 years (Lee 2000). Singapore's exemplification of stable, progressive development provides a substantive model and practical icon for national development within the region, albeit from a unique city-state perspective. Singapore's 'city in a garden' is another vote of confidence about the relationship of nature and development. The most important ecological message for not only ASEAN countries but many others globally is how Singapore has reduced its water footprint (Tan 2009; Tortajada et. al. 2013).

Third, given the extensive impacts of climate change on current communities and societies in Southeast Asia, there is growing public and government recognition of environmental issues. We applaud the NGOs in trying to save nature, biodiversity, specific animals (*orang utans*, elephants, tigers and rhinoceroses), marine organisms, river, forest and lake ecosystems in the region. The model for environment has to be a cooperative endeavor between government, private organizations and civic groups and NGOs. In Southeast Asia, green spaces should not be ecological exceptions but the norm of biophilic relationships amongst all living organisms including human beings.

References

Abu-Lughod, J. L. (1991). *Before European hegemony*. Oxford: Oxford University Press.
Aiken, S. R. and Leigh, C. H. (2015). Dams and indigenous peoples in Malaysia: Development, Displacement and Resettlement. *Geografiska Annaler*, 97(1), pp. 69–93.
Asian Development Bank. (2001). *Fire, smoke, and haze: The ASEAN response strategy*. Manila: Asian Development Bank.
Barendse, R. J. (2002). *The Arabian seas: The Indian Ocean world of the seventeenth century*. New Delhi: Vision Books.
Bellwood, P. (1985). *Prehistory of the Indo-Malayan archipelago*. North Ryde and Sydney: Academic Press Australia.
Benjamin, G. (2002). On being tribal in the Malay World. In: G. Benjamin and C. Chou, eds., *Tribal communities in the Malay world*. Singapore: ISEAS, 7–76.
Biggs, D. (2009). *Contested waterscapes in the Mekong region: Hydropower, livelihoods and governance*. London and Sterling, VA: Earthscan.

Castells, M. (2004). *The power of identity: Volume II*. Oxford: Blackwell Publishing.
Condominas, G. (1977). *We have eaten the forest: The story of the Montagnard Village in the highlands of Vietnam*. New York: Harmondsworth.
Conklin, H. C., Lupāih, P. and Pinter, M. (1980). *Ethnography Atlas of Ifugao: A study of environment, culture, and society in Northern Luzon*. New Haven, CT: Yale University Press.
Curry, P. (2011). *Ecological ethics: An introduction*. 2nd ed. Cambridge: Polity Press.
Davies, D. (1994). *Attitudes to nature*. London: Pinter Publishers.
Diamond, J. (2012). *The world until yesterday*. London: Allen Lane.
Diamond, J. and Robinson, J. (eds.) (2010). *Natural experiments of history*. Cambridge, MA: Harvard University Press.
Duara, P. (2015). *The crisis of global modernity*. Cambridge: Cambridge University Press.
The Economist. (2015). Learning the lessons of stagnation. *The Economist*, 415(8944), pp. 34–36.
Elson, R. E. (1997). *The end of the peasantry in Southeast Asia*. London: Palgrave Macmillan.
Fernandez-Armesto, F. (2002). *Civilizations*. New York: A Touchstone Book.
Flannery, T. (2009). *Now or never*. New York: Atlantic Monthly Press.
Foster, J. B. (2002). *Ecology against capitalism*. New York: Monthly Review Press.
Foster, J. B., Clark, B. and York, R. (2010). *The ecological rift*. New York: Monthly Review Press.
Frank, A. G. (1998). *ReOrient: Global economy in the Asian age*. Berkeley: University of California Press.
Geertz, C. (1963). *Agricultural involution: The processes of ecological change in Indonesia*. Berkeley: University of California Press.
Giddens, A. (2009). *The politics of climate change*. Cambridge: Polity Press.
Gunn, G. C. (2011). *History without borders: The making of an Asian world region, 1000–1800*. Hong Kong: Hong Kong University Press.
Hall, K. R. (2011). *A history of early Southeast Asia*. New York: Rowman & Littlefield Publishers, Inc.
Heine-Geldern, R. (1942). Conceptions of state and kingship in Southeast Asia. *Far Eastern Quarterly*, 2, pp. 15–30.
Imanishi, K. (2002). *A Japanese view of nature: The world of living things*. London: Routledge Curzon Taylor and Francis Group.
Kaplan, R. D. (2012). *The revenge of geography: What the map tells us about coming conflicts and the battle against fate*. London: Random House.
Kolbert, E. (2014). *The sixth extinction*. London: Bloomsbury.
Leach, E. R. (1964). *Political systems of highland Burma: A study of kachin social structure*. London: G. Bell and Sons Ltd.
Lee Kuan, Y. (2000). *From third world to the first: The Singapore story: 1965–2000, memoirs of Lee Kuan Yew*. Singapore: Times Edition, Singapore Press Holdings.
Lindayati, R. (2003). Shaping local forest tenure in national politics. In: N. Dolsak and E. Ostrom, eds., *The commons in the new millennium*. Cambridge, MA: MIT Press, 221–264.
Lockard, C. A. (2009). *Southeast Asia in world history*. Oxford: Oxford University Press.
Merchant, C. (2005). *Radical ecology*. New York: Routledge.
Miksic, J. N. (2013). *Singapore & the silk road of the sea 1300–1800*. Singapore: NUS Press Singapore & National Museum of Singapore.
O'Dempsey, T. (2014). Singapore's changing landscapes since c. 1800. In: T. P. Barnard, ed., *Nature contained*. Singapore: NUS Press Singapore, 17–48.
Ooi, K.B. (2015) *The Eurasian Core and its Edges: Dialogues with Wang Gungwu on the History of the World*, Singapore: Institute of Southeast Asian Studies
Oppenheimer, S. (1999). *Eden in the East: The drowned continent of Southeast Asia*. London: Phoenix.
Peterson, A. L. (2001). *Being human*. Berkeley: University of California Press.
Picketty, T. (2014). *Capital in the twenty-first century*. Cambridge, MA: The Belknap Press.
Pollock, S. (2006). *The language of the gods in the world of men: Sanskrit, culture, and power in premodern India*. Berkeley: University of California Press.
Popkin, S. L. (1979). *The rational peasant*. Berkeley: University of California Press.
Radkau, J. (2014). *The age of ecology*. Cambridge: Polity Press.
Reid, A. (1988). *Southeast Asia in the age of commerce 1450–1680*. Chiang Mai: Silkworm Books.
Reynolds, C. J. (1995). A new look at old Southeast Asia. *The Journal of Asian Studies*, 54(2), pp. 419–446.
Sarkar, S. (1999). *Eco-socialism or eco-capitalism? A critical analysis of humanity's fundamental choices*. London: Zed Books.
Sauer, C. (1969). *Agricultural origins and dispersals*. Cambridge: MA: MIT Press.

Savage, V. R. (1984). *Western impressions of nature and landscape in Southeast Asia*. Singapore: University of Singapore Press.

Savage, V. R. (2009). *A question of space: From aterritorial communities to colonised states in Southeast Asia*, unpublished Conference Paper.

Savage, V. R. (2012). Southeast Asia's indigenous knowledge: The conquest of the mental terra incognitae. In: A. Bala, ed., *Asia, Europe and the emergence of modern science*. New York: Palgrave Macmillan, 253–270.

Scott, J. C. (1979). *The moral economy of the peasant*. New Haven, CT: Yale University Press.

Scott, J. C. (2009). *The art of not being governed*. New Haven, CT: Yale University Press.

Solheim, W. G. II. (1970). Northern Thailand, Southeast Asia and world prehistory. *Asia Perspectives*, 13, pp. 145–162.

Stark, R. (2005). *The victory of reason*. New York: Random House Trade Paperbacks.

Tan, Y. S. (2009). *Clean, green and blue*. Singapore: ISEAS.

Tortajada, C., Joshi, Y. and Biswas, A. K. (2013). *The Singapore water story: Sustainable development in an urban city-state*. London: Routledge.

Tsing, A. L. (1993). *In the realm of the diamond queen*. Princeton, NJ: Princeton University Press.

van Luer, J. C. (1955). *Indonesian trade and society: Essays in Asian social and economic history*. The Hague and Bandung: W. van Hoeve.

Wilson, E. O. (1984). *Biophilia*. Cambridge, MA: Harvard University Press.

Wolters, O. W. (1967). *Early Indonesian commerce: A study of the origins of Sri Vijaya*. Ithaca, NY: Cornell University Press.

Wolters, O. W. (1999). *History, culture, and region in Southeast Asian perspectives*. Ithaca, NY: Southeast Asia Program Publications and Singapore: ISEAS.

Wolters, O. W. (2008). *Early Southeast Asia selected essays*, ed. C. J. Reynolds. Cornell: Cornell University Press.

Wong, D. S. Y. (1977). *Tenure and land dialogue in the Malay states*. Singapore: Singapore University Press.

PART 2

Development institutions and economies in Southeast Asia
Introduction

Andrew McGregor, Lisa Law and Fiona Miller

A wide range of institutions are engaged with development in Southeast Asia. Each pursues a particular vision of development and is involved in the circulation of associated economies. For multilateral institutions like the World Bank and the International Monetary Fund (IMF), the pursuit of neoliberal forms of development provides a primary rationale for existence. For the Association for Southeast Asian Nations (ASEAN), development provides an important legitimizing role for political and economic decision-making, such as the recent establishment of the ASEAN Community. At the village scale the pursuit of development may come in the form of community enterprises, collectives and businesses that reflect local engagements with broader development flows – often in the form of changing access to technology, policy or infrastructure. In this section, some of the key institutions involved in development in Southeast Asia are reviewed. We have purposely cast our net wide to include not only core development institutions like the World Bank, IMF and ASEAN, but also lesser known but increasingly important development actors like the International Labour Organization (ILO), Chinese aid and community-scale organizations. While many of these are involved in what could be considered the formal architecture of the development industry – the business of grants and loans in pursuit of development – we also survey shifts in the tourism industry, urban manufacturing and community economies – that provide alternative means of pursuing and understanding development.

From this eclectic mix at least three themes are prominent in the papers that follow: the influence of neoliberalism, the pursuit of social justice and the growing influence of China as a development actor. Toby Carroll's careful analysis of financial flows from three Multilateral Development Organisations (MDOs) – the World Bank, the IMF and the Asian Development Bank (ADB) – provides insights into the shifting approaches to neoliberalism adopted by the development community. He tracks the uneven application of three market-oriented development modalities – the Washington consensus, the post-Washington consensus and, since the mid-2000s, what he refers to as 'deep marketization' – the increasing focus on enabling investment climates and direct private sector financial support amidst increasing donor competition, including that emerging from China. Regional experiences have been uneven, shaped by the relative leverage of MDOs and Southeast Asian states over time; however Carroll emphasises that it is hard to underestimate the huge influence of these MDOs on how development is articulated, "For almost four decades, neoliberalism has completely reinterpreted what it means to 'do development,'"

Ford, Gillan and Thein then trace how the ILO is repositioning itself in the region in relation to these neoliberalizing processes. Drawing on case studies from Myanmar and Timor-Leste their research suggests the ILO has shifted away from the increasingly crowded space of labor standards to a more market-oriented, or at least influenced, development actor. The ILO in Myanmar, for example, has encouraged investment by transnational corporations by guaranteeing particular labor standards, which minimizes risks to corporate reputation and kick-starts patterns of export-oriented development. The authors raise the troubling prospect that in adopting a new role as development facilitator some of its traditional responsibilities for workers may wain: "Questions persist as to whether employment growth, skills development and industrial diversification so neatly dovetail with social protection and labor standards without trade-offs between these objectives." The ILO is treading a delicate path as it experiments with these new roles in neoliberalizing economies.

Rachel Hughes' careful analysis of international justice tribunals in post-conflict Cambodia and Timor-Leste also finds interesting links between the functions of these institutions and market-oriented development. Hughes traces the role and priorities of tribunals in each country and the important roles they play in pursuing justice and enabling national healing in countries scarred by physical violence. Hughes argues, however, that the design of the tribunals, such as the silence surrounding issues of ongoing structural violence and the framing of issues as national problems but requiring international expertise, neatly avoids addressing international culpability and associated reparations, and in doing so brings on "healthy conditions for investment." She asks the tantalizing question as to the extent to which "these tribunals serve to have 'cultural violence' writ large, so as to further increase the penetration of global capital as a 'civilizing' force." This is not the intention of the tribunals or those working in them, but a potential outcome of their structure and implementation.

Kelly Gerard traces neoliberal reforms through ASEAN, focusing on the newly formed ASEAN Economic Community and the strategies of elite groups to protect their interests. She shows how existing political economies shape the interpretation and articulation of neoliberal policy with greater benefits concentrated amongst higher income earners. She also details how the good governance reforms of the post-Washington consensus have created new spaces for civil society to engage in dialogue with ASEAN leaders, while simultaneously shrinking those spaces through restrictions on who participates (few development or advocacy organizations but many professional organizations like the ASEAN Cosmetics Association!), the form of participation and the issues discussed, in what seems a clear example of post-politics. Despite these failings and the barriers they create for more equitable forms of development Gerard remains hopeful that the regionalization of advocacy will strengthen civil society voices in contesting shared injustices.

Delik Hudalah and Adiwan Aritenang provide a more traditional economic geographic analysis of the shifting spaces of manufacturing in the region, focusing on how changing patterns of investment are transforming the location of industry in Jakarta. They trace how manufacturing industries are moving out of core zones to the cheaper land and labor of the suburbs. The impact of these shifting patterns of capital investment are reshaping the social fabric of cities, contributing to gated economies which are more tied to global production chains than local economies. They conclude that governments should seek to attract capital for small and medium industries over larger ones as they are more resilient in financial crises and bring more benefits to local economies.

Katherine Gibson, Ann Hill and Lisa Law also place their hope in smaller scale initiatives, providing an important counterpoint to the capitalocentric analyses that dominate this section.

Rather than critique the very real neoliberalizing processes spreading through the region they instead highlight the many other diverse institutions and economies that also exist at more local scales, and the key contributions they make to the quality and processes of everyday life. Southeast Asia is presented as a region of immense diversity and possibility, where hope and justice lie in learning from and building with these diverse economies. Perhaps best epitomizing their desire to unsettle conventional thinking on Southeast Asia is their recasting of patron–client relationships in potentially much more positive ways than what is current orthodoxy.

Alternative approaches to development also underpin May Tan-Mullins' examination of Chinese aid in Southeast Asia. Eschewing terms like donor in favor of south-south collaborations and equal partnerships, China prides itself on doing aid differently to Western donors. China has a long history of supporting communist states initially avoided by the OECD and concentrates on infrastructure development usually involving Chinese workers. In differing from the neoliberal modalities described by Carroll it has attracted criticism particularly around issues of transparency, conditionalities and good governance. Tan-Mullins suggests that some of these issues have arisen from its relatively recent transformation into a large aid actor and that improvements are in process. Certainly the entry of a new development actor in the region is creating new opportunities, as well as risks, for the pursuit of justice in Southeast Asia.

The rise of China is also evident in ongoing transformations of tourism industries – one of the main sectors of economic growth in many Southeast Asian countries – and the focus of John Connell's final chapter in this section. Chinese tourists, with their own sets of expectations and demands, are now the largest source of non-ASEAN visitors to the region. As a form of development perhaps the most significant shift has been the diversification and maturing of tourist destinations from entrepreneurial community-run guesthouses and experiences to much more organized tours with heightened expectations and accompanying infrastructure. Market logics dictate that many local benefits dry up when this occurs as handicrafts become mass produced and imported from elsewhere and visitors stay in foreign-owned hotels eating imported food isolated from local economies. Connell writes, "The prominence of the informal sector almost everywhere, and undercurrents of tension in sex tourism and voluntourism, emphasizes that tourism has not removed poverty but is often intimately associated with it."

As a section the chapters capture some of the dynamism pervading development institutions and their associated economies in Southeast Asia. They demonstrate just how influential neoliberalism has become in shaping development policy, locations and initiatives, at the expense of other possible models. Alongside the rise of neoliberalism has been the rise of China as an important development actor in the region, with new flows of aid, foreign direct investment and large amounts of tourists, diverting the region from established North-South development flows. Amidst these changes remains a desire for justice, where current patterns of uneven development are embedding injustices, inspiring opposition and the search for alternative economies. The challenge for institutions at multiple scales is to balance these three disjunctive development flows – and the possibilities and risks they entail – in pursuit of place-specific and -appropriate modes of development.

Andrew McGregor, Lisa Law and Fiona Miller

6
NEOLIBERALISM AND MULTILATERAL DEVELOPMENT ORGANIZATIONS IN SOUTHEAST ASIA

Toby Carroll

Introduction

Over the last three and a half decades, multilateral development practice – the ideas and modalities operationalized by multilateral organizations in the name of development – has undergone important shifts propelled by the fraught context of neoliberalism and late capitalism.[1] Multilateral organizations such as the World Bank Group, the Asian Development Bank (ADB) and the International Monetary Fund (IMF) have been central in both developing and contesting development practice and the role of the state in relation to both development and economic growth. Southeast Asia – a region of over 600 million people, the world's third largest developing country (Indonesia), and large numbers of poor and marginal populations – has been at the forefront of innovations in the activities of multilateral development organizations (MDOs), while also being an important site of popular social protest and resistance to these innovations. It is also a region of increasing economic importance in global and regional value chains (production, assembly and services), attracting private investment capital as well as relatively large flows of official development assistance (ODA) (see Tables 6.1–6.4). This has made the region an important locale in terms of contestation and cooperation between governments, civil society and MDOs over the very nature of development practice.

This chapter focuses on the work of the most important MDOs – the World Bank Group (including the International Finance Corporation), the ADB and the IMF – in extending neoliberal policy agendas, from the 1980s on, into Southeast Asia in the name of development. I present a *political economy of development* approach that – drawing upon a tradition of thought that prioritizes competing interests (and in particular class interests) and ideologies in understanding particular institutional agendas and their outcomes (see for example Rodan et al. 2001) – seeks to explain the form of different phases of MDO activities in the region.[2] I argue that MDO activities in Southeast Asia over the last three and a half decades reflect a combination of the ideological shifts in development policy that have emanated from MDO headquarters in Washington and Manila – often in response to emergent contradictions and conflict over and around policy – and the relative leverage, interests and ideational dispositions of governments in the region within a consolidating and contradictory world market (Cammack 2016). This historical context has seen the increasing roll-out by MDOs of market-oriented policy as a default 'fix-it' while also demanding ongoing modifications to the precise policy suites endorsed by MDOs as legitimacy and other challenges have emerged.

Three important phases of neoliberal developmental policy are worthy of attention (see Table 6.1), each having seminal, though geographically uneven, impacts on reshaping MDO practice in the region: the Washington consensus (from the early 1980s to the mid- to late 1990s); the post-Washington consensus (PWC) (from the mid- to late 1990s to the mid-2000s); deep marketization (from the mid-2000s to present). The analysis presented here is significantly based upon a comprehensive survey of approved MDO projects and other activities (predominantly using publicly accessible MDO databases and annual reports) in Southeast Asia during the three aforementioned phases. Projects were initially categorized by their sectoral and thematic MDO demarcations, with project documents then subjected to further analysis.[3]

The chapter begins by characterizing the region's experience with the first two phases of neoliberal development policy advanced by MDOs in Southeast Asia since the 1980s – the Washington consensus and the PWC. I illustrate that on the back of impressive growth and reasonable fiscal health, much (though not all) of the region was able to significantly postpone the impacts of the Washington consensus until the crisis of 1997–98. However, the crisis constituted a critical juncture for the region. It served to simultaneously expose more countries in the region to 'structural adjustment' lending *and* then-new institution building/'good governance' prescriptions of the PWC, pushes which further drew countries into the world market (less on their terms than before) and exposed them to the more all-encompassing market discipline that this entailed. The final section presents an overview of the latest MDO agendas being promoted in the region. Here, I point to the operation of a *new politics of development*, with states often enjoying more (though still highly varied and regularly highly constrained) leverage with MDOs than previously, on the back of broader and deeper capital markets, changes in foreign direct investment flows, more globalized forms of production and the entrance of new development actors (such as NGOs, nation-states, such as China, and moves to establish new MDOs such as the Asia Infrastructure Investment Bank). This said, rather than signaling the death knell of neoliberalism and neoliberal MDO activities, neoliberal development policy's latest form caters explicitly to this context, and subsequently has enjoyed significant expansion. Moreover, the modalities of deep marketization play important roles in subject countries to further world market discipline, constraining policy choices and amplifying exposure to contradiction.

Leverage and the first two phases of neoliberal development policy in Southeast Asia: the Washington consensus and the post-Washington consensus

In the 1980s and early 1990s, both Northeast and Southeast Asia were at the center of development discourse and often intense battles over development policy. Following on the heels of Japan's rise to industrial prominence as the world's second largest economy, Asia's 'newly industrializing countries' (NICs) – comprising the four 'Asian Tigers' of Hong Kong, Singapore, South Korea and Taiwan – and 'newly industrializing economies' (NIEs), such as Indonesia, Malaysia and Thailand, gained the attention of development scholars and practitioners. Notably, following Chalmers Johnson (Johnson 1982), a whole host of 'statist' scholars famously drew attention to the neo-mercantilist role of the state in successful late industrializing countries in Asia, which often exhibited not only sustained growth but also growth that was deemed (including by the World Bank) more equitable than elsewhere (World Bank 1993a, v).

However, following the travails of 'the Chicago Boys' in Pinochet's Chile and the rise of Thatcher and Reagan in the United Kingdom (UK) and the United States (US), the new orthodoxy gaining ground – and rapidly concretizing within both the World Bank and the IMF – centered on liberal economic assumptions and an aggressively liberal market-oriented reform agenda. Constituting what some described as a veritable 'counter revolution' in development policy (Toye 1987, 22), the new focus in development theory was not on the importance of aid, physical capital, market protection and subsidies for industry, for example, but rather economic growth and its apparent determinants (Robison and Hewison 2006, xi). In this view, the state was increasingly maligned as both a repository of 'rent seeking' by 'predatory elites' and for engaging in 'distorting prices,' interfering with the much-vaunted price-setting mechanism in the allocation of resources and inhibiting the flourishing of entrepreneurial activity.

John Williamson dubbed the general agreement over policy endorsed by the leading institutions within 'technocratic Washington' – the World Bank, the IMF and US Treasury – throughout the 1980s as 'the Washington consensus'(Williamson 1990, 7). Mirroring the broad pro-market sentiments emanating out of the US and UK, development policy's first neoliberal phase pushed a policy set that sought to generally remove the state from interfering with and impeding market forces. When combined with 'conditional lending' (lending attached to policy and institutional change, often within a period of crisis) from the IMF and World Bank, these policies were tantamount to what came to be known in relatively popular development parlance as 'structural adjustment.' The policy recommendations included fiscal discipline, privatization (of public utilities and other entities), the provision of secure property rights, redirecting public spending toward 'pro-growth' areas and areas to improve income distribution, tax reform (encouraging lower marginal rates and broader tax bases through efforts such as value added taxes, for example), liberalization of interest rates, trade and foreign direct investment (FDI) flow, and competitive exchange rates (Williamson 2000, 252–253; see Table 6.1).

Yet, as statists pointed out, despite the alacrity with which the Washington consensus was adopted in MDO circles, the approach sat in marked tension with many (though not all) of the policies adopted by ostensible development success stories in Asia, including those in Southeast Asia. Here, statists and critical political economists identified development and growth as the products of more complicated narratives that included the influence of Japanese colonialism, post-colonial/Cold War/post-Plaza Accord relations of production, often-extensive state intervention and planning (associated with the embrace and promotion of import-substitution industrialization and export-oriented industrialization), culture, lashings of leftist/labor repression, and forms of comparative advantage centered upon cheap labor and or natural resources. Indeed, consistent and often impressive growth in much of Southeast Asia and the importance of maintaining Cold War allies likely did much to contain (though far from entirely) the influence of the Washington consensus in MDO activities in the region throughout the consensus' heyday of the 1980s (especially relative to its impact in Sub-Saharan Africa and Latin America), while also providing a congenial environment to often repressive regimes and emergent capitalist classes (see for example Hewison 2001, 81–86; Hutchison 2001, 46–49). Nonetheless, as the 1980s proceeded, many of the elements above were progressively accompanied by neoliberal-friendly patterns of financial deregulation and economic opening as strategic support from Western governments and MDOs within the context of the Cold War dried up, forcing countries to compete within the global market for financing and market share to further propel growth and industrialization (Rodan Hewison and Robison 2001, 15–16).

This said, the combination of economic vulnerability and the ideological orientation of leaders such as Corazon Aquino and Fidel Ramos saw the World Bank, the IMF and Washington consensus structural adjustment having a significant impact in the Philippines prior to the 1997–1998 crisis. The debt crisis in the 1980s, which had its roots in the late 1970s and the blame for which the World Bank notably laid squarely at the feet of 'pervasive state intervention and overprotection of industry'(Hutchison 2001, 49; World Bank Independent Evaluation Group (IEG) 1996), provided the context for a series of significant structural adjustment interventions by the World Bank, culminating in the 1990 Debt Management Loan Program (US$200 million), which complemented numerous stand-by arrangements by the IMF and bilateral efforts. Indeed, of the 49 approved World Bank projects in the Philippines during the 1980s (totaling US$3.66 billion),[4] many of the biggest commitments to the country were structural adjustment loans. Projects during this period and up to the mid-1990s specifically supported neoliberal agendas designed to reduce trade protection and public ownership and increase central bank independence (World Bank 1993b). Significant World Bank lending to the Philippines throughout this period also included a mix of infrastructure (energy, power, water, ports and roads), 'social lending' (health and education) and disaster relief (for the 16 July 1990 earthquake, for example), albeit with the infrastructure and social lending often exhibiting the increasing influence of burgeoning preferences for institution building, monitoring and evaluation, non-governmental organization (NGO) participation and decentralization, which would become important components within the PWC (see the next section and Table 6.1). For example, of the three main components that comprised the 1989 Health Development Project, one component focused upon strengthening institutional capability and monitoring, while a second component was designed to foster partnerships between government, NGOs and other community-based organizations (World Bank 1989; see also World Bank 1993c).

The ADB, yet to be as neoliberal and policy-oriented as the World Bank in terms of its lending, continued to provide considerable loans to the Philippines throughout the same period up to the crisis for infrastructure-related projects, such as coal-based power plants (the Masinloc Thermal Power project) and roads, and disaster relief. However, indicative of the country's fiscal situation, in 1985 a "shortage of counterpart funds and consequent slowdown in the implementation and loan disbursements for nearly all the [ADB]'s ongoing projects in the country," saw the ADB not approving a single loan for the Philippines in 1985 – with only a small technical assistance program of US$1.3 million (ADB 1986, 53–54). Moreover, and in line with the neoliberal positions of the administrations of the time, it should be noted that both the World Bank and the ADB were considered important among government circles as suppliers of market-oriented technical assistance for the concretizing privatization push in the country (Carroll 2010, 119–122).

In Indonesia (the largest cumulative regional borrower for both the World Bank and ADB, see Tables 6.2–6.4), the relationship with MDOs during the 1980s and up until the crisis of 1997–98 was of a different nature, reflecting the relative leverage enjoyed by that country within the context of the Cold War and the impressive economic growth achieved (albeit with persistent and often pervasive development challenges). Notably, many of the World Bank's projects continued to evoke a more 'classically' developmental 'basic needs' quality – indicative of the Bank under Robert McNamara – than that of the structural adjustment programs in the Philippines. Of the nearly 190 approved World Bank projects valued at over US$21.4 billion between 1980 and August of 1997 many of the biggest sums in terms of commitment were for infrastructure, with considerable sums going toward large-scale power and energy projects (24 projects worth over US$4.7 billion), education and training (24 projects; US$1.8 billion), highway projects (eight projects worth over US$1.6 billion), telecommunication projects (four projects; US$1.1 billion), industrial sector support (including foreign exchange financing), and

Table 6.1 Three phases of neoliberal development policy

	The Washington Consensus Early 1980s to mid-1990s	The post-Washington Consensus Mid- to late 1990s to mid-2000s	Deep marketization Mid-2000s to present
Policy and reform foci	Fiscal discipline Privatization Secure property rights Redirecting public spending toward 'pro-growth' areas and areas to improve income distribution Tax reform (encouraging lower marginal rates and broader tax bases through efforts such as value added taxes) Liberalization of interest rates, trade and foreign direct investment flows Competitive exchange rates	Market-oriented institution building and 'good governance' Public–private partnerships The promotion of forms of 'participation', 'partnership' and 'empowerment' (involving sections of the state and civil society) in the market-building process Market-oriented modes of social development (linked to institution building and 'social capital') Budget support Microfinance Ongoing usage of Washington consensus policies, often with modifications	The promotion of 'enabling environments' via benchmarking exercises Public–private partnerships 'Access to finance' and other forms of micro, small and medium enterprise (MSME) support Selective ongoing usage of Washington consensus and PWC policies
Key policy instruments and advocacy documents	World Bank and IMF conditional ('structural adjustment') lending Project and program lending World Bank Country Assistance Strategies World Development Reports	Project and program lending Technical assistance and capacity building Poverty Reduction Strategy Papers World Bank Country Partnership Strategies The Comprehensive Development Framework World Development Reports; Asian Development Outlook	Multilateral investment in and lending to the private sector Multilateral risk mitigation support for the private sector Project and program lending Technical assistance and capacity building Doing Business Report Series; World Development Reports; Asian Development Outlook
Key events and dynamics	The rise of neoliberalism in the US and UK and the demise of non-neoliberal (Keynesian, social democratic, neo-mercantilist, socialist and communist) development strategies	The fallout of the transition experience in Eastern and Central Europe The aftermath of the 1997–98 crisis The rise of local and global social movements against multilateral organizations and the policies of the Washington consensus and structural adjustment	Expanding and deepening capital markets and other processes of financialization The celebration of 'emerging and frontier markets' and increasing economic integration

(Continued)

Table 6.1 (Continued)

	The Washington Consensus Early 1980s to mid-1990s	The post-Washington Consensus Mid- to late 1990s to mid-2000s	Deep marketization Mid-2000s to present
	Crises in Latin America The fall of the Berlin Wall and the experience of 'transition countries' The 1997–98 'Asian' Crisis	Diminishing leverage of multilateral organizations	Further diminished leverage for the traditionally significant multilateral organizations Increased importance of private sector-oriented multilateral organizations, such as the International Finance Corporation and the European Bank for Reconstruction and Development
Impact in Southeast Asia	Moderate, with important exceptions, until the 1997–98 crisis (which entailed a dramatic scaling up of Washington consensus interventions)	Significant, with large-scale institution building programs and other new modalities and strategies piloted and 'scaled up'	Significant and ongoing

Source: Toby Carroll

Table 6.2 ADB 'sovereign' approvals (including loans, grants and official co-financing and technical assistance) for Southeast Asian developing member countries

	1980	1985	1990	1995	2000	2005	2010	2014
Cambodia	0	0	0	55	114	69.6	203.15	295.7
Indonesia	284.6	502.1	950.4	1067	813.4	1486.5	901.63	850.11
Lao PDR	7	0	28	99	68.1	140.52	211.93	90.35
Malaysia	83.75	134.5	17.1	74	0	0	0	0.375
Myanmar	50.5	0	0	0	0	0	0	96.4
Philippines	178.3	1.3	699.6	578.1	476.4	204.72	708.01	1,534.44
Thailand	170	168.9	116.1	335	2.6	3.35	509.11	2.725
Vietnam	0	0	0	244.6	232.6	616.49	1,246.48	1,758.05
Regional Total	774.15*	806.8	1,811.2	2,452.7	1,707.1	2,521.18	3,780.31	4,628.15

* Singapore had a loan for US$19 million approved to support vocational and industrial training in 1980.

Source: Compiled by the author from ADB Annual Reports (and related appendices), 1980–2014. Amounts are in millions of dollars.

Table 6.3 World Bank (International Development Association and International Bank for Reconstruction and Development) commitments for Southeast Asian developing member countries

	1980	1985	1990	1995	2000	2005	2010	2014
Cambodia	0	0	0	37	41.7	38	5	0
Indonesia	754	972.7	1,632.8	1,417	133.4	917.2	2,986.4	1,077.6
Lao PDR	13.4	0	44.7	19.2	0	76	124.3	102.8
Malaysia	50	89.8	154.2	0	0	0	0	0
Myanmar	160	32.3	0	0	0	0	0	281.5
Philippines	412	254	941.8	18	277.5	99	685	1279
Thailand	542	112.5	144	513.1	400	0	79.3	0
Vietnam	0	0	0	415	285.7	698.8	2,129.3	1541
Regional Total	1,931.4	1,461.3	2,917.5	2,419.3	1,138	1,829	6,009.3	4,281.9

Source: Compiled by the author from World Bank Annual Reports (and related appendices), 1980–2014. Amounts are in millions of dollars.

Table 6.4 Total historical cumulative MDO allocations to Southeast Asian countries (as of 2014) in millions of dollars^

	World Bank	Asian Development Bank
Cambodia	899	1,728
Indonesia	49,469	24,430
Lao PDR	1,291	1,577

(*Continued*)

Table 6.4 (Continued)

	World Bank	Asian Development Bank
Malaysia	4,151	1,414
Myanmar	1,639	980
Philippines	16,676	12,815
Singapore	181	144
Thailand	9,268	5,214
Vietnam	18,530	11,882
Regional Total	*102,104	**60,184

* IDA and IBRD total cumulative
** Includes sovereign and non-sovereign funds (the latter being 'loans, guarantees, B loans and risk transfer' to or for majority state-owned entities); amounts rounded to nearest million. Compiled by the author from 2014 ADB and World Bank annual reports and associated appendices.
^ The cumulative totals presented here are historical totals derived from the different initial borrowing dates of the respective countries.

ongoing (and controversial) transmigration projects (three projects worth US$324.5 million), the latter continuing processes started during Dutch colonial times and which were also funded by the ADB. This said, the creeping (though still limited) influence of neoliberal reform could be seen (especially from the 1990s on) in some of these infrastructure projects, which often built in 'technical assistance' components that sought to establish regulatory regimes for promoting competition in utility provision. Moreover, during this time both explicit and more subtle efforts at adjustment lending did take place, for example, in the form of a sectoral adjustment loan (SECAL) for water supply reform and a loan dedicated toward 'modernizing' the telecommunication sector (World Bank 1995).

The ADB's work in Indonesia throughout the 1980s and early 1990s had a strong emphasis on infrastructure and education, an emphasis which – along with broader efforts by the ADB throughout the region – would soon be reoriented toward and/or overlaid with more overt neoliberal policy-oriented elements as the late 1990s and 2000s rolled on. 'Priority projects' were supported in agriculture, energy (including energy diversification and efficiency), education and infrastructure. Notably, ADB annual reports throughout the 1980s document significant agriculture and energy support, followed by transport and communications, and education as attracting the biggest cumulative lending sums, with cumulative agriculture lending (US$2.35 billion) more than double that of energy (US$978 million) (ADB 1986, 44, 1991, 69).

Malaysia, in line with its relatively positive development trajectory and leverage, exhibited only modest draws on both World Bank and ADB resources throughout the 1980s, with these largely tapering off in the 1990s (see Tables 6.2 and 6.3). Projects that were funded by the World Bank also largely mirrored the profile above of those funded for Indonesia, with no overt structural adjustment projects out of the 40 projects totaling US$2.37 billion approved between 1980 and 1997, with the biggest projects funding infrastructure (including US$562 million for power/energy projects), education (US$482 million) and agriculture. Once again though, the ideational drive to instill neoliberal policies could be seen increasing toward the mid-1990s in a fairly limited sense in some project documents, with concerns over creating 'demand driven' and decentralized education systems, for example.

Thailand's portfolio of approved World Bank projects throughout the 1980s and early 1990s included significant commitments to power, education, agriculture and industrial sector support, but it also received two structural adjustment loans prior to the crisis (in 1982 and 1983, for US$150 million and US$175.5 million respectively). Tellingly though, the tone of the loan documents – which point to the alignment with existing government plans and a congenial working relationship between the Bank and the government – typically eschew the more aggressive elements of the Washington consensus (which was still very much in gestation) (World Bank 1982, 1983). For ADB commitments to Thailand during the period, considerable attention was given to physical infrastructure (large amounts on energy, transport, communications, and water supply and sanitation), albeit often with increasing references to 'promoting private sector investment' and increasing state enterprise efficiency, portending a push that would only gain more momentum over time (ADB 1991, 82).

Not surprisingly, given the mix of regimes and politics operating, both the ADB and the World Bank had (mostly small) commitments to Laos and Myanmar in the 1980s, with Vietnam and Cambodia, still yet to be brought back within the MDO fold. However, by 1995 – nearly a decade after the initiation of *doi moi* market reforms and the later normalizing of diplomatic relations that included the lifting of US restrictions on the flow of multilateral aid to the country – Vietnam had begun attracting increasing commitments from both the World Bank and ADB (see Tables 6.2 and 6.3).

The 1997–98 crisis and the start of a new era of MDO activity in Southeast Asia

While fast growth in the region meant that many countries were able to stave off grand MDO structural adjustment in the 1980s, the crisis of 1997–98 changed that. Crucially, the crisis also dovetailed with arrival of then-new 'PWC' agendas, which broadly took root during James Wolfensohn's presidency (1995–2005) at the World Bank and under the influence of people such as Joseph Stiglitz, who was briefly Chief Economist of the Bank during Wolfensohn's tenure. Here, efforts toward facilitating 'participation,' 'partnership,' 'community-driven development,' 'empowerment' and 'country ownership' were paired with 'good governance' institution-building programs, the latter being marshaled toward addressing common charges of 'crony capitalism' during the crisis and more substantive attempts to remake state and society in a neoliberal image (see Table 6.1). What these promising-sounding foci of participation, partnership, empowerment and country ownership actually meant in the hands of the MDOs was open to significant critique, with questions raised over the degree to which partnerships were equal and the degree of autonomy that countries enjoyed in 'owning' what ended up in Poverty Reduction Strategy Papers (PRSPs), for instance (Hatcher 2014, 33–35). However, within some MDO circles, such as the social development unit of the World Bank, relatively progressive factions were trying to use this new push that had emerged out of the serious legitimacy challenges raised by activists, academics and politicians to challenge – within the confines of neoliberalism – the World Bank's traditional ideological 'power bloc' of orthodox economists while also trying to change poor governance in country contexts such as Indonesia (Guggenheim 2006). However, rather than challenging the underlying economic assumptions of the Washington consensus, in practice the PWC reflected recognition within MDO circles of the difficulties of structural adjustment *implementation* – especially in terms of the politics of reform – and, not unrelated, a renewed role for the state then being drafted by neoliberals influenced by new institutional economics toward addressing 'transaction costs,' 'imperfect information' and 'market failure' (Carroll 2010). Importantly for our purposes here, though, the onset of the crisis heralded a new phase of neoliberal MDO activity in the region, that combined both Washington consensus and these then-newly emerging PWC elements, with ADB operations (despite that MDO's reputation for being 'infrastructure oriented') increasingly mirroring (and indeed actively harmonizing with) much of what was emanating from the World Bank.

In mid- to late 1997, beginning in Thailand and quickly moving to Indonesia, countries in the region that had selectively liberalized in a bid to continue growth and industrialization experienced serious economic and financial turmoil. In several prominent cases this was followed by conditional structural adjustment lending from both the IMF and the World Bank (often supported by the ADB) and, in some cases, dramatic political and social turmoil and upheaval (Robison 2001; Rodan Hewison and Robison 2001, 16–17). As Thailand's descent into crisis unfolded, the IMF extended a US$17 billion conditional package that demanded large-scale Washington consensus-style reform and restructuring that mirrored much of what had earlier unfolded in Latin America, Africa and the economies of the former Soviet Union (Hewison 2001, 92). Between September 1997 and October 1999 the World Bank made the last of what would be its biggest commitments to Thailand (indicative of the country's increasing leverage in relation to MDOs) for over a decade. Many of these loans had a strong structural adjustment quality to them, such as the Finance Companies Restructuring Loan and the Economic and Financial Adjustment Loans 1 and 2 (all three of which were prepared in close coordination with the IMF). These loans, and another Public Sector Reform adjustment loan, totaled US$1.75 billion. Also notable at the time – and broadly emblematic of the PWC fusing of community development, 'empowerment,' decentralization and 'demand driven'/market-oriented approaches to poverty reduction – was a relatively large 'Social Investment Project' (US$300 million) and a large energy project to support 'restructuring, corporatization, and privatization' efforts associated with the Electricity Generating Authority of Thailand (EGAT). Closely complementing these efforts, the ADB's efforts focused on 'restoring stability in the economy; strengthening competitiveness and sustainable growth; and improving Thailand's quality of life, with a focus on social and environmental goals' (ADB 1998, 116). Concrete ADB loan commitments at the time included a financial market sector reform loan of US$300 million (part of the combined crisis-related rescue package put together by multilateral and bilateral agencies) with very significant amounts of attendant commercial co-financing (US$1 billion), and the Rural Enterprise Credit Project (US$200 million), which continued a shift toward support for microfinance and market-oriented institutional strengthening (ibid.; ADB 2001, 2). The financial sector reform loan, which was part of immediate stabilizing efforts after the onset of the crisis, was also designed toward, among other purposes, broadening, deepening and improving the efficiency of financial markets in Thailand, in part through very PWC concerns relating to market regulation and supervision, and improving the investment climate and SME support (the latter two themes soon to become central components within neoliberal development policy) (ADB 2003, 3–8, 13).

In Indonesia the reforms requested by the IMF would deal significant blows to the family-centered networks of power and, indeed, the country's greater social fabric.[5] The rapidly declining rupiah (which continued to fall after its floatation), and a quickly approaching banking, corporate and, indeed, fiscal crisis, led the government to call in the IMF, which demanded a raft of reforms targeting large industrial projects (including national car and aircraft projects), state trade monopolies in a range of staple foods and 'the liquidation of insolvent banks'(Robison 2001, 118). Conditional structural adjustment lending began in November of 1997 with a US$10 billion stand-by facility (augmented with another US$1.8 billion in mid-1998) and large pledges (US$26 billion) by multilateral donors (including US$3.5 billion by the ADB) and bilateral organizations (Asian Development Bank 1998; International Monetary Fund 2000).[6] Given what was at stake for the regime and its most prominent beneficiaries (big corporations and 'politico-business families'), which had benefited from state guaranteed 'frameworks of protection and privilege,' intense conflict and negotiation – including halting the disbursement of funds and new concessions – between the IMF and the government was to be expected (ibid.). IMF demands to control spending led to an austerity budget that would impact the beneficiaries above and, via dramatic cuts to fuel and food subsidies (including those for rice) at a time of drought, have dreadful consequences for Indonesia's poorest, culminating in riots, large-scale (often student-led) demonstrations and Suharto's resignation in May 1998 (Gill 1999; Lane 2008, 169–171).

For the World Bank, which previously enjoyed a close Cold War-nurtured relationship with the Suharto regime and which had its portfolio heavily impacted by the crisis with massive allegations of corruption circulated just before the crisis (Winters 2002, 102; 26), the time was an extremely active one. In keeping with the nature of the crisis and its fallout, and the evolving confluence of Washington consensus and PWC elements, significant structural adjustment and other neoliberal policy-oriented lending in areas such as financial and banking reform, corporate restructuring, 'governance' and (limited) social safety nets dominate portfolio commitments to Indonesia after the onset of the crisis. A series of Policy Reform Support Loans (PRSLs) and projects targeting banking reform and corporate restructuring (totaling well over US$2 billion in commitments), which were to be complemented by more significant commitments in the form of 'Development Policy Loans' (no less than ten worth US$5 billion in commitments), and other adjustment loans throughout the noughties are worthy of attention here, signaling a trend toward policy-based lending that continues up to the present. Notably these were coupled with technical assistance (TA) projects promoting more PWC-esque orientations such as marketizing utility provision (including the promotion of public-private partnerships, PPPs), decentralization and – importantly – then-novel social development and community-driven development programs such as the soon-to-be massive Kecamatan Development Program (KDP), which drawing on burgeoning interest in social institutions and social capital, sought to reshape patterns of governance, from the 'bottom up.' The palatability of such programs was perhaps not surprising given the ongoing cries over 'crony capitalism' (and its apparent association for many neoliberals with the crisis), not to mention the broader PWC push around 'good governance.' Notably, various iterations of the highly influential KDP and its offspring totaled well over US$4.5 billion in commitments, with the project reportedly impacting every village in the country and being replicated in countries within the region (the Philippines) and beyond.

Importantly, throughout the 1990s and early 2000s, the ADB's efforts in Indonesia, the Philippines, and elsewhere in the region began to mirror those of the World Bank, starting with crisis coordination around structural adjustment with its multilateral peers, and intensified efforts around 'governance' (the latter increasingly understood as synonymous with market-related institutional integrity), 'reaching out to civil society' (which, as with the World Bank, mostly meant incorporating amenable NGOs into operations), supporting decentralization and the promotion of public-private partnerships (PPPs), and increasing amounts of direct support for the private sector. Yet as growth returned, the leverage tables were once again turning for MDOs, presenting a further challenge to the attractiveness of the often expensive and controversial institutional reform programs of the PWC. However, the increasing presence of countries such as Vietnam and Laos within multilateral development portfolios (see Tables 6.2 and 6.3) in the context of the celebration of 'emerging' and 'frontier markets' heralded a new era of neoliberal development policy in the region that would manifest in increased commitments toward facilitating 'policy dialogue' (often via generous forms of budget support), risk mitigation strategies centered around co-financing, MDO loan and equity investments directly to the private sector and the scaling up of PPPs and 'access to finance' projects. Moreover, while neoliberal institution building would continue, it would do so in a different form, with the good governance agenda of the PWC morphing into agendas to foster 'enabling environments' for capital. By the mid noughties this deep marketization agenda would be consistently reflected in both World Bank and ADB portfolios throughout the region.

The current state of pay in MDO activity: a new politics of development and the promotion of deep marketization in Southeast Asia

If the 1980s was an era in which growth largely permitted postponing conditional structural adjustment, and the late 1990s to mid-2000s were to see a fusion of structural adjustment (especially around the crisis) and PWC neoliberal institution building agendas, the present era constitutes something different again. Indeed, a *new politics of development* has been further reshaping MDO activities toward the promotion of *deep marketization* (see Carroll and Jarvis 2015, 295–298), reflecting both the ongoing interring of countries into the global political economy *and* (not altogether unrelated) changes, once again, in leverage between MDOs and member countries. After several decades of the neoliberal policy agendas outlined above, elements of which have been employed willingly by states, some less willingly, capital is now perhaps better positioned than ever before to enjoy spatial flexibility in its pursuit of new repositories of cheap labor, new investment opportunities and new markets. Increased FDI flows, deeper and broader capital markets, and the increasing capacity for emerging market firms and countries to issue bonds, reflect both the interest of capital in comparatively fast growth rates of 'emerging economies' and – in some significant cases – a lessening of country dependence on conditional lending from MDOs. Added to this are the entrance of new actors such as China and the related efforts to create new development banks, such as the Asian Infrastructure Investment Bank (AIIB) and the New Development Bank (NDB; formerly the BRICS Development Bank), all of which present some potential competition for traditional MDO operations based around conditional lending or the more timid 'capacity/institution building' projects of the PWC.

This said, institutional reform toward establishing 'enabling environments' for capital and 'improving the investment climate' is now promoted through MDO-led benchmarking efforts, such as the IFC/World Bank *Doing Business* report series, which assess and rank countries in relevant areas and send signals to capital about the investment worthiness of a given locale. Further, considerable ongoing policy-based lending that promotes further neoliberal reform also carries on the institutional reform agendas started under the Washington consensus and PWC, suggesting that some within national governments are convinced of the merits and/or the ostensible inevitability of the need for continued reform efforts within the current international political economy. Importantly, many elements of the current MDO agenda are being vigorously pushed in countries that are the newest entrants to the neoliberal MDO club, often through direct-to-sector private sector support. Here, the IFC and the ADB (Cammack 2016; Carroll 2015) have been at the forefront of recrafting neoliberal development agendas in the region around financialized risk-mitigating/risk-allocating modalities designed to promote private sector activity generally. This promotion has come in the form of supporting PPPs, providing loans to and taking equity in private entities, 'access to finance' programs designed to foster micro, small and medium enterprises (MSMEs) and promote 'financial inclusion' and other projects (around land titling, for example) that seek to extend the breadth and depth of private property ownership.[7]

Looking at recent commitments by the World Bank and ADB, key deep marketization elements and the continuing influence of the two earlier phases of reform are all clearly evident. In the case of Indonesia, which in 2014 was the eighth largest borrower from the World Bank's 'hard lending' (i.e., non-grant/non-credit-based) window, the International Bank for Reconstruction and Development (IBRD), the large and sustained tranches for PWC 'social/community driven development' projects, such as the KDP and its progeny, remain. Accompanying these are significant sums for 'development policy lending' (of approximately 20 projects totaling US$10.65 billion since 2005) for projects that emphasize many neoliberal demands including broad institutional reform related to finance, infrastructure, taxation, climate change, 'economic resilience' and social assistance. This continues the trajectory largely begun during the 1997–98 crisis, albeit in quite a different economic environment, although the recent economic slowdown and depreciating currencies, on the back of China's slowdown, may once again challenge this. Significant commitments supporting decentralization and local government, the promotion of transparency and accountability, and power and education are also present, often exhibiting pronounced marketizing elements. Moreover, the IFC has been making important advances in Indonesia, partnering with local and international banks to promote access to finance and improve competitiveness via infrastructure investments, while also focusing on encouraging private sector investment, in part through work to 'improve the investment climate'. The ADB's foci in Indonesia also mirror much of the above, including significant sums for development policy lending with strong PWC elements (see for example ADB 2006, 99), and infrastructure support (the latter also now including significant neoliberal policy elements). However, the increased focus by the ADB on enhancing competitiveness, direct private sector support and 'access to finance' style projects stands out as being of particular note, emblematic of contemporary MDO preferences more broadly (see for example ADB 2015, 34–36).

In the Philippines (the sixth largest IBRD borrower in 2014), recent development policy lending commitments – for example, toward "increasing physical and human capital investment; tackling regulatory barriers in land, labor, and capital markets; all in the context of ensuring fiscal sustainability and boosting fiscal governance and transparency (World Bank 2014b)" – are also overt in the portfolio, as are significant commitments for crisis response and very PWC-esque 'community driven/social development' projects, and 'social protection' and education projects. Notably, the ADB's 2011–16 strategy for the Philippines prioritizes the core deep marketization foci of improving 'the investment climate' and direct private sector support, in part via policy-based lending and technical assistance (ADB 2011, 5–6).

Yet perhaps one of the most important shifts in MDO activity in Southeast Asia in the contemporary era has been regional commitments to 'emerging' and 'frontier' markets, often using the modalities outlined above. Notable here, in 2014 Vietnam was ranked as the sixth biggest recipient of World Bank International Development Association ('soft lending') funds (US$1.34 billion) (World Bank 2014a, 11). On the back of its impressive development scorecard and interest in 'policy dialogue,' the country has also been (perhaps for some) an unlikely recipient of large multilateral (and bilateral) amounts of grant-bolstered budget support, which served to make the country one of the world's leading recipients of ODA by the mid-1990s (Carroll 2010, 157). The IFC, which established its office in Vietnam in 1997, currently focuses on promoting access to finance in the country, along with supporting structural reform – including in the form of taking equity in a state-owned bank to pave "the way for other foreign strategic investors to invest in state-owned commercial banks and [support] their partial privatization"– and "improving the business climate" (via its support for the Vietnam Business Forum, for example). The ADB closely complements this work, for example through its US$320 million program to improve transparency and efficiency in state-owned enterprises (SOE) and another US$230 million project to improve competitiveness and the business environment through the promotion of a range of reforms to banking and fiscal policy, public sector administration and public investment, and SOEs (ADB 2015, 34).

While the World Bank's relationship with Cambodia has been plagued by concerns over corruption, and the amounts committed small-to-non-existent (see Table 6.3 and Carroll 2010, chapter 7), the ADB's activity in that country (the MDO is the country's largest multilateral partner) has been more significant (see Table 6.2). Here, the ADB supports many of the deep marketization modalities detailed above – with direct private sector support to the energy and finance sectors receiving significant attention – in addition to infrastructure, education, agriculture and natural resources. In Myanmar, the transition and opening have seen new levels of interest from MDOs, with the IFC early into the country to promote the virtues of 'business-enabling infrastructure' and 'access to finance,' significantly challenging the rest of the World Bank and ADB with a country portfolio expected to be around US$400 million for FY 2015, rising to US$1 billion over the coming three years (Aung 2014). This said, the World Bank's 2015 US$400 million National Community Driven Development Project (significantly echoing the KDP and other earlier CDD projects) and relatively large commitments for a mix of reform projects, including a US$440 million project with strong neoliberal requirements oriented toward clearing the country's arrears with the World Bank and hence 'normalizing' World Bank-Myanmar relations, are worthy of note, as are generally increasing commitment figures. In Laos, MDO commitments (including, once again, the IFC, which has provided more than US$60 million in loans and equity to private companies operating in the country) have been uneven though still important (as Tables 6.2–6.4 show). Much of the deep marketization agenda that dominates contemporary development practice is again apparent, with the IFC promoting access to finance to SMEs, improving the investment climate and promoting PPPs and private sector participation in power generation and distribution (IFC 2015). The ADB has also strongly supported private sector participation in Laos, with loans and guarantees (for example two private sector energy projects with commitments of US$321 million), while also making significant commitments to infrastructure, governance and public sector capacity building.

Conclusion

This chapter has made the case that three key phases of neoliberalism – the Washington consensus, the PWC and deep marketization – are important in understanding neoliberal development policy in Southeast Asia. The degree of rollout of projects and programs associated with these phases has been significantly linked to the relative leverage and interests of countries and MDOs at different points in time and, relatedly, important evolutions and crises within the global political economy under late capitalism. Moreover, the genesis of each of these three phases also needs to be understood as the product of key ideological and material interests (including those of the MDOs and fractions of capital) that have had to respond to perennial issues of legitimacy and competition within the context of intensifying patterns of globalization and crisis.

Direct correlation of MDO activities with precise quantifiable outcomes in terms of shifts in GDP growth, impacts upon GINI coefficients and degrees of liberalization and ideational change, for example, is difficult. However, neoliberal development policy as operationalized by MDOs has had, and continues to have, an undeniable impact in Southeast Asia. As the analysis above attests, waves of policies attached to significant funding have promoted marketization and economic integration for three and a half decades, dovetailing and, indeed in many instances, directly underpinning *actual marketization and levels of economic integration* (via trade and investment liberalization and particular privatizations, for example). And while the *new politics of development*, characterized by the relative empowerment of 'emerging and frontier markets' on the backs of increased capital flows and new development actors, such as China, has been heralded as a challenge to the power of MDOs, in reality this has been an important factor in empowering MDOs advocating pro-private sector modalities and encouraging a further reinvention of neoliberal modalities. Here, organizations such as the IFC and ADB have played catalytic roles in mitigating risk to capital (including foreign capital), fostering congenial institutional environments for capital accumulation and diffusing market social relations generally. For almost four decades, neoliberalism has completely reinterpreted what it means to 'do development.' In a highly financialized and economically integrated world, the ongoing (albeit politically contested) influence of market-oriented MDOs and the policies that they promote is all but assured, with the current turmoil filtering through China and the underdeveloped world perhaps only making that more likely.

Notes

1 By 'neoliberalism,' I refer to the broad and changing policy sets that – based upon the assumptions of liberal economics – have been deployed over the last four decades or so to subject state and society to market and market-like discipline ('marketization'). By 'late capitalism,' I am merely referring to capitalism in its most recent form and in no way mean to imply that capitalism is near terminal.
2 Given the limitations of space and the broad scope of the chapter, attention is given largely to the *form* that neoliberal development policy has taken in Southeast Asia, rather than the specific impacts of policy in terms of material, political and ideological shifts.
3 This approach to surveying MDO activities was supported by 15 years' worth of research by the author on both MDO operations in Southeast Asia and neoliberal development policy more broadly.
4 All MDO figures are based upon publically available data from MDOs (from a combination of annual reports and project databases) and reflect published commitments approved by MDO boards.
5 Indonesia's crisis encounter with the IMF was in part immortalized by the January 1998 image of the Fund's Managing Director, Michel Camdessus, standing cross-armed behind a soon-to-be ousted President Suharto, as the late dictator signed a 'Letter of Intent' (one of a number at the time).
6 A further US$5 billion three-year extended arrangement was negotiated in early 2000.
7 For a more detailed exploration of the deep marketization agenda, see Carroll 2012 and 2015.

References

Asian Development Bank. (1986). *Annual report 1985*. Manila: ADB.
Asian Development Bank. (1991). *Annual report 1990*. Manila: ADB.
Asian Development Bank. (1998). *Annual report 1997*. Manila: ADB.
Asian Development Bank. (2001). *Rural enterprise credit project completion report*. Manila: ADB.
Asian Development Bank. (2003). *Program completion report on the financial markets reform program (Loan 1600-THA)*. Manila: ADB.
Asian Development Bank. (2006). *Annual report 2005*. Manila: ADB.
Asian Development Bank. (2011). *Country partnership strategy*. Manila: ADB.
Asian Development Bank. (2015). *Annual report 2014*. Manila: ADB.
Aung, H.T. (2014). IFC Myanmar investment to rise to $1 billion in three years. *Reuters*. Available at: www.reuters.com/article/2014/12/12/myanmar-ifc-idUSL3N0TW3RK20141212. Accessed 19 August 2015.
Cammack, P. (2016). World market regionalism at the Asian Development Bank. *Journal of Contemporary Asia*, 46(2), pp. 173–197. doi:10.1080/00472336.2015.1086407
Carroll, T. (2010). *Delusions of development: The world bank and the post-Washington consensus in Southeast Asia*. London: Palgrave MacMillan.
Carroll, T. (2012). Working on, through and around the state: The deep marketisation of development in the Asia-Pacific. *Journal of Contemporary Asia*, 42(3), pp. 378–404. doi:10.1080/00472336.2012.687628
Carroll, T. (2015). "Access to Finance" and the death of development in the Asia-Pacific. *Journal of Contemporary Asia*, 45(1), pp. 139–166. doi:10.1080/00472336.2014.907927
Carroll, T. and Jarvis, D. S. L. (2015). The new politics of development: Citizens, civil society, and the evolution of neoliberal development policy. *Globalizations*, 12(3), pp. 281–304. doi:10.1080/14747731.2015.1016301
Gill, S. (1999). The geopolitics of the Asian crisis. *Monthly Review*, 50(10). Available at: https://monthlyreview.org/1999/03/01/the-geopolitics-of-the-asian-crisis/. Accessed 18 July 2017.

Guggenheim, S. (2006). Crises and contradictions: Understanding the origins of a community development project in Indonesia. In: A. Bebbington, M. Woolcock, S. Guggenheim and E. Olson, eds., *The search for empowerment: Social capital as idea and practice at the World Bank*. Bloomfield: Kumarian, 111–144.

Hatcher, P. (2014). *Regimes of risk*. Basingstoke: Palgrave MacMillan.

Hewison, K. (2001). Thailand's capitalism: Development through boom and bust. In: G. Rodan, K. Hewison and R. Robison, eds., *The political economy of South-East Asia, second edition*. Melbourne: Oxford University Press, 71–103.

Hutchison, J. (2001). Crisis and change in the Philippines. In: G. Rodan, K. Hewison and R. Robison, eds., *The political economy of South-east Asia: Conflict, crises and change*. Melbourne: Oxford University Press, 42–70.

International Finance Corporation. (2015). *IFC in Lao PDR*, IFC. Available at: www.ifc.org/wps/wcm/connect/f2f13800478bfe5b9507f7752622ff02/IFC+in+Laos.pdf?MOD=AJPERES. Accessed 19 August 2015.

International Monetary Fund. (2000). *Recovery from the Asian crisis and the role of the IMF*. Washington, DC: IMF.

Johnson, C. (1982). *MITI and the Japanese miracle – The growth of industrial policy, 1925–1975*. Stanford, CA: Stanford University Press.

Lane, M. (2008). *Unfinished nation*. London: Verso.

Robison, R. (2001). Indonesia: Crisis, oligarchy and reform. In: G. Rodan, K. Hewison and R. Robison, eds., *The political economy of South-East Asia, second edition*. Melbourne: Oxford University Press, 104–137.

Robison, R. and Hewison, K. (2006). East Asia and the trials of neoliberalism. In: K. Hewison and R. Robison, eds., *East Asia and the trials of neoliberalism*. Abingdon: Routledge vii–xx.

Rodan, G., Hewison, K. and Robison, R. (2001). Theorising South-East Asia's boom, bust and recovery. In: G. Rodan, K. Hewison and R. Robison, eds., *The political economy of South-East Asia, second edition*. Melbourne: Oxford University Press, 1–41.

Toye, J. (1987). *Dilemmas of development*. Oxford: Basil Blackwell Ltd.

Williamson, J. (1990). What Washington means by policy reform. In: J. Williamson, eds., *Latin American adjustment: How much has happened?* Washington, DC: Institute for International Economics, 5–20.

Williamson, J. (2000). What should the world bank think about the Washington consensus? *The World Bank Research Observer*, 15(2), pp. 251–264.

Winters, J. (2002). Criminal debt. In: J. Pincus and J. Winters, eds., *Reinventing the World Bank*. Ithaca, NY: Cornell University Press, 101–30.

World Bank. (1982). *Report and recommendation of the president of the IBRD to the executive directors on a structural adjustment loan to the Kingdom of Thailand*. Washington, DC: World Bank.

World Bank. (1983). *Report and recommendation of the president of the IBRD to the executive directors on a structural adjustment loan in an amount equivalent to US$175.5 Million to the Kingdom of Thailand*. Washington, DC: World Bank.

World Bank. (1989). *Loan agreement: Health development project*. Washington DC: World Bank.

World Bank. (1993a). *The East Asian miracle: Economic growth and public policy*. Oxford: Oxford University Press.

World Bank. (1993b). *Philippines – Reform program for government corporations*. Available at: http://documents.worldbank.org/curated/en/1993/04/735295/philippines-reform-program-government-corporations. Accessed 28 July 2015.

World Bank. (1993c). *Urban health and nutrition project*. Washington, DC: World Bank.

World Bank. (1995). *Indonesia – Telecommunications sector modernization*. Available at: www.worldbank.org/projects/P004001/indonesia-telecommunications-sector-modernization?lang=en. Accessed 29 July 2015.

World Bank. (2014a). *Annual report 2014*. Washington, DC: World Bank.

World Bank. (2014b). *Third Philippines development policy loan*, World Bank. Available at: www.worldbank.org/projects/P147803?lang=en. Accessed 18 August 2015.

World Bank Independent Evaluation Group (IEG). (1996). *Debt management program in the Philippines*. Available at: http://lnweb90.worldbank.org/oed/oeddoclib.nsf/InterLandingPagesByUNID/D4773C1117006DB285256BD4006635A2 Available at 28 July 2015.

7
THE INTERNATIONAL LABOUR ORGANIZATION AS A DEVELOPMENT ACTOR IN SOUTHEAST ASIA

Michele Ford, Michael Gillan and Htwe Htwe Thein

Introduction

Typically, the International Labour Organization (ILO) is discussed in narrow terms with specific reference to its role in setting labor standards and the success or otherwise of its attempts to convince governments and employers to respect them. Yet over several decades it has also sought to engage in other aspects of the world of work including knowledge production and employment generation through projects more readily associated with international development organizations or even grassroots non-governmental organizations. Although there is by no means a consensus among either scholars or practitioners about the efficacy of these interventions, it is clear that the ILO has managed to embed its concept of 'decent work' not only into contemporary discourse concerning the rights of workers and the duties of employers and states to respect them, but also that around economic development.

In this chapter, we discuss the extent to which the ILO can be understood as a development actor, how its emphasis on development has evolved over time, and how its development agenda has been pursued by means of various strategic initiatives and programs. Our analysis is grounded in Southeast Asia, which as a region is important in terms of size, population and economic weight but also because of the profound changes in political and economic status experienced by its 11 states. Not surprisingly, the ILO has played an important and sometimes controversial role in reshaping the legal and institutional framework of labor regulation in the post-authoritarian states of Cambodia, Indonesia, Timor-Leste and now Myanmar. Less recognized is the fact that this work is part of an integrated development agenda that has included measures to create opportunities for employment through large-scale development projects, entrepreneurship initiatives and deeper integration into global production networks, increasingly carried out in conjunction with influential member states, most notably the United States of America, and International Financial Institutions (IFIs) including the International Monetary Fund (IMF) and the World Bank.[1] We illustrate these different aspects of the ILO's role as a development actor with reference to these different post-authoritarian states. Particular attention is paid to Myanmar, where the ILO's enactment of its development agenda has benefited both from its special status in that country but also from its experiments elsewhere in the region.

Challenges to the ILO

The ILO is unique among the agencies of the United Nations (UN) because of its tripartite structure, which requires representation from the governments of member states and from recognized employer and worker organizations, both of the latter having access to separate bureaus within the technical apparatus of the ILO that support their general activities and institutional participation. This characteristic has driven the internal politics that determine its agenda and despite internal and external critics of the narrowness of tripartism – and a failed attempt at the turn of the millennium to allow greater formal participation and representation by civil society organizations – its governance structure has remained largely impervious to change (Baccaro and Mele 2012, 207–210).

The core purpose of the ILO's governance structure has been to establish and monitor the observance of a system of international conventions and recommendations approved by two-thirds of delegates at its annual International Labour Conference and designed to protect a wide range of labor rights. While ILO recommendations are intended to merely "guide national and international policy," conventions are "detailed legal provisions that become part of the corpus of national law if a country ratifies them" (Baccaro and Mele 2012, 197). In theory, the fact that the ILO's instruments have been forged through international tripartite dialogue means that member states will observe its recommendations and ratify and implement its conventions. In practice, however, this reliance on the willingness of member states to comply has led to an uneven pattern of ratification and an even more disappointing record of implementation.

As Maupain (2005, 123) observes, the intensification of economic globalization and international competition from the 1970s onwards has made the "dilemma of voluntarism and efficacy in promoting the ILO's objectives more acute." As a global organization with a mandate for labor regulation and policy, the ILO has been challenged by market-driven logics of free trade and deregulation that run counter to a broad-based commitment to tripartite dialogue and consensual regulation. The position of the ILO as the arbiter of labor standards has also been tested by a plethora of new organizations and multilateral initiatives – from social accountability standards to pacts, accords and voluntary corporate social responsibility initiatives that lay claim to relevance and effect in regulating different forms of work and labor standards in various industries, production networks and supply chains – that have challenged what Maupain (2015) refers to as its 'magisterial' function.

Insofar as they often make reference to core labor standards, these initiatives acknowledge the status of the ILO and thus can be interpreted as successfully embedding its principles in the international order. Alternately, however, they can be construed as competing instruments of labor regulation that have flourished because of the 'governance gaps' inherent in national and international labor regulation regimes. What is clear, however, is that the ILO is no longer a sole actor but rather one of many in a crowded field (Mundlak 2015, 81) and that this new reality has provided impetus for it to attempt to redefine its role.

The ILO as a development actor

The ILO's first-line response to these challenges to its status as the primary arbiter of labor standards was its 1998 Declaration on Fundamental Principles and Rights at Work, which identified and delimited a set of 'core' labor standards and established a requirement for states, by virtue of their membership, to respect the principles and rights elaborated within them and attempted to embed these, and the relevance of the ILO itself, in the discourse and practice of economic development. Because demonstrating respect for core labor standards does not necessitate the ratification of associated recommendations and conventions, the 1998 Declaration has been interpreted as signifying a shift toward the logics of 'soft' regulation. The extent to which this strategic redirection was necessary and its consequent effects have been the subjects of ongoing debate (Maupain 2005, 2009; Baccaro and Mele 2012). However, far less attention has been paid to the extent to which the 1998 Declaration, and the associated focus on the concept of 'decent work,' have signaled an acceleration in the ILO's gradual effort to position itself as a development actor.

The ILO's increasing focus on development through its Decent Work Agenda is evident in the *2014 World of Work Report*. Tellingly subtitled "Developing with Jobs," the report states that because "jobs, rights, social protection and dialogue are integral components of development," decent work should be a "central goal" (ILO 2014a, xxiii). According to Standing (2008, 356), the ILO has been "from the outset . . . to some extent, a development agency, in the sense that it anticipated that, as they developed, the 'colonies' and 'primitive' economies would adopt the standards, policies and institutions forged in the 'advanced' countries." It was not, however, until the late 1960s that the ILO began focusing explicitly on the relationship between economic development, poverty reduction and the nature of employment. This stance was first formalized at its World Economic Conference of 1976, but only fully realized two decades later with the 1998 Declaration and the subsequent launch of the Decent Work Agenda, which confirmed its desire "to connect with debates both within and outside the United Nations on poverty, globalization and the Millennium Development Goals" (Standing 2008, 370).

Analysts have argued that the ILO's decision to engage with, rather than resist, the Washington Consensus was largely driven by its fraught relationship with the United States of America (Alston 2004; Standing 2008). According to Alston (2004, 493), the genesis of the 1998 Declaration lay in the United States' 1984 Generalized System of Preferences (GSP) Renewal Act, which he argues introduced the concept of a core set of "internationally recognized workers' rights," a concept subsequently "refined and extended" in order "to link respect for labour rights to eligibility for investment, trade and development assistance." The United States used the allocation of most favored nation status under the GSP as a means of pressuring the governments of developing countries to allow freedom of association and to guarantee labor rights. In the case of Indonesia, this tactic was pivotal in the survival of the Indonesian Prosperous Labor Union, an alternative union established to challenge the Suharto regime's one-union policy (Glasius 1999).[2] The United States also experimented with the inclusion of labor standards in regional and bilateral free trade agreements, beginning with the North American Free Trade Agreement of 1994. It was, however, with the signing of the US-Cambodia Bilateral Textile Trade Agreement in 1999 that such an initiative for the first time combined trade incentives with compliance with international labor standards and factory-level inspections, and the first time that the ILO had been involved directly in the latter (Arnold and Toh 2010; Hughes 2007; Kolben 2004).[3]

Core to the Clinton Administration's decision to involve the ILO directly in the US-Cambodia Bilateral Textile Trade Agreement was its desire "to prove that trade and export-orientated policies could reduce poverty and play a central role in the development process" (Arnold and Toh 2010, 407). Despite a lack of enforcement mechanisms and incentive structures at the factory level (Kolben 2004), what became known as the Better Factories Cambodia Project gained traction because of interest, and ultimately financial buy-in, from international brands intent on improving their corporate image, such as Gap, H&M, Nike, Reebok and Disney (Hughes 2007). For the ILO's part, the Cambodian experiment provided an "example of a successful strategy, the underlying principles of which could provide inspiration for the elaboration of a global strategy to promote fair globalization in the post-MFA environment" (cited in Arnold and Toh

2010, 408). While it is beyond the scope of the present discussion to assess its ultimate utility, it is clear that Better Factories Cambodia – the model for Better Work programs in Bangladesh, Haiti, Indonesia, Jordan, Lesotho, Nicaragua and Vietnam (Better Work 2015) – was an important turning point in the ILO's push toward a greater emphasis on economic development. The fact that this expansion was made in partnership with the International Finance Corporation, a key agency of the World Bank group, is also salient as it shows the increasing alignment between these global institutions.

A second key aspect of the ILO's foray into development involved a conscious decision to broaden its focus from formal sector workers to unregulated waged workers, home workers and the self-employed. In doing so, it embraced the prospect of engaging not only in the promotion of employment growth but also in supporting the development of micro-enterprises and other forms of non-waged income generation. An example of the latter is a program called Start and Improve Your Own Business (SIYB), which has provided training and institutional support for micro and small to medium sized businesses in more than 100 developing nations in Africa, Latin America and Asia (ILO, n.d.f).[4] The inclusion of such initiatives within the Decent Work Agenda was by no means uncontroversial. As Baccaro and Mele (2012, 211–212) have noted, there was fundamental disagreement between employers focused on "employment creation" and unions, with their focus on "fundamental rights, most particularly the right of workers to freely associate." Yet, despite these criticisms, subsequent major ILO initiatives have continued this emphasis on local participatory economic development and job creation.

A key example of the trend toward job creation and other forms of economic activity was the 2008 Declaration on Social Justice for a Fair Globalization. Indeed, one comment on the Declaration suggested that it went "further than any of the existing constitutional texts by recognizing the essential role played by enterprises and entrepreneurship (private and public) in job creation and the necessity of creating an environment conducive to their sustainable development" (Maupain 2009, 834). The ILO also participated in the United Nations-led Open Working Group to develop Sustainable Development Goals, which culminated in the approval of the 2030 Agenda for Sustainable Development at a UN Summit in September 2015. Alongside 16 other development goals, the 2030 Agenda includes the objective of achieving "inclusive and sustainable economic growth, employment and decent work for all" which, as stated on an ILO resource page, is "reinforced by specific targets on the provision of social protection, eradication of forced and child labor, increasing productivity, addressing youth employment, SMEs and skills development" (ILO n.d.b). Notably, trade unions have continued to question this approach: a briefing document for the governing body of the ILO on the process recorded that trade unions "regretted the lack of dedicated goals on decent work and social protection and of references to international labour standards and the issues of wages and green jobs" (ILO 2014d, 3).

The ILO's attempt to reposition itself as an organization able to contribute to a global development agenda is nowhere more evident than in post-authoritarian Southeast Asia. In Cambodia, the Decent Work Agenda has been vigorously pursued through livelihoods initiatives involving indigenous people and people living with disabilities, women's entrepreneurship, community-based enterprise development and microfinance (ILO 2015a). Similar approaches have been adopted in Indonesia and Timor-Leste, with a particular focus on vocational education and business development projects through the SIYB initiative. The ILO has also pursued less intuitive approaches to livelihoods and employment creation in these countries through large projects focused on the construction and maintenance of roads. Between 2008 and 2016, the ILO's Mission to Timor-Leste managed three road building initiatives valued collectively at US$51.5 million. The first of these, called TIM-Works, ran from 2008 and 2012 and was supported by multiple donors (ILO n.d.d), while the second, called Enhancing Rural Access (ERA), ran between 2011 and 2015 with funding from the European Commission (ILO n.d.c). In 2012, the ILO was recruited by Australian AID, which had been part of the consortium that had supported the TIM-Works initiative, as the designated implementing agency for the third, called Roads for Development (R4D). Valued at US$30 million, this was the largest of the three projects, aiming to rehabilitate 450 kilometers, construct an additional 40 kilometers and conduct routine maintenance of 2000 kilometers of rural roads, and make available some 4.7 million person days of work (ILO n.d.e). The ILO justified its engagement in these initiatives on the basis of its alignment with the emphasis on increasing rural employment through infrastructure projects that encourage economic development (Interview with ILO Head of Mission to Timor-Leste, July 2013). Road-building projects in Indonesia to the value of US$11.8 million in post-tsunami Aceh and Nias have been similarly described as seeking to "facilitate aid delivery and economic recovery" through the "rehabilitation and improvement of the rural road network," using "employment-intensive approaches" (ILO n.d.a).[5]

Even the ILO's more traditional work in the area of labor regulation has been recast as supporting economic development in the region. In Indonesia and Timor-Leste, the bulk of ILO work on labor law reform and building the capacity of government officials, trade unions and employers' associations was funded by the United States under an umbrella cooperative agreement designed to help countries realize the objectives of the 1998 Declaration on Fundamental Principles and Rights at Work (Interview with former ILO Indonesia Country Director, August 2015). The stated objectives of the ILO/USA Declaration Project in Indonesia was to create a "sound, harmonious and fully functioning industrial relations system aimed at promoting economic growth while guaranteeing workers' rights" (ILO 2006, 3). This emphasis is also evident in the final evaluation of Strengthening and Improving Labour Relations in Timor-Leste, which concludes that the project "partially achieved its development objective of contributing to Timor-Leste's social and economic progress through the establishment and operation of an effective labour relations system" (Scanteam 2006, 2). As discussed below, the same approach has been used in Myanmar, where there has been significant emphasis on livelihoods and 'responsible business' strategies and on labor regulation as means of promoting economic development.

The ILO in Myanmar

Then known as Burma, Myanmar became a member state of the ILO in 1948. However, it was not until the 1990s that the country came into focus as a consequence of frequent reports of widespread use of forced labor and other labor rights abuses. The ILO's engagement since that time has been significant, if not fundamental, to the development of labor regulation and institutions in that nation; it has also been important for the ILO itself as a counter to claims regarding its declining effectiveness and weakened role in international labor standard setting and monitoring (Maupain 2005; Tapiola and Swepston 2010). Its unusually interventionist approach to Myanmar first emerged in 1998 when a special Commission of Inquiry was convened to investigate the claims surrounding forced labor. The findings of the inquiry paved the way for the invocation of Article 33 of the ILO Constitution at the 2000 International Labour Conference, censuring the regime for its failure to prevent or discontinue forced labor rights violations and requesting that all member states review their relations with the country – an unprecedented step in the history of a body often criticized for lacking the enforcement powers necessary to uphold labor standards (Maupain 2005; Tapiola and Swepston 2010).

Having refused to cooperate in the inquiry or accept the ILO's censure, the regime later agreed to allow the ILO to send an investigatory team to the country in 2001 and then to establish a Liaison Office in Yangon in 2002 (Tapiola and Swepston 2010). In accordance with the 2000 decision to limit Myanmar's participation in the ILO, the mandate of the Liaison Office was restricted to dealing with forced labor. After considering a report on the lack of progress in addressing forced labor rights violations, the ILO began to discuss the reactivation of Article 33 sanctions in 2005, leading to a serious escalation of tension and threats of withdrawal by the regime. In Yangon, several government-backed demonstrations were staged to denounce the ILO and its activities and the Chief Liaison Officer received a series of anonymous death threats (ILO 2005). In an attempt to de-escalate the conflict, an official dialogue was established, resulting in a breakthrough agreement in 2007 under which the regime granted the ILO unique, in many respects extraordinary, authority to receive and investigate complaints related to forced labor (Horsey 2011).

While the initial actions of the ILO could be interpreted as an international political intervention, its subsequent engagement with the Government of Myanmar has been characterized by a pragmatic approach that emphasized its provision of 'technical' assistance. This approach was held to be more strategic than reproaches directed to the military government, as it allowed for the establishment of a level of trust that ultimately reduced resistance to its engagement with the issue of forced labor (Tapiola and Swepston 2010). Through its work on forced labor, the ILO had positioned itself as an international intermediary – a role thought to be especially important because of the absence of other international rights organizations in the country – capable of building a local institutional presence and a level of credibility that could translate into ongoing effect (Tapiola and Swepston 2010, 524). This indeed proved to be the case when the Government of Myanmar, led by President Thein Sein, initiated a series of political and institutional reforms in an effort to re-engage with the international community from 2011 (Cheesman et al. 2014).

As the only independent formal actor in the labor domain, the ILO was uniquely well placed to serve as a source of knowledge, training and technical assistance following the announcement of the reforms. As a consequence, the scope of its operations expanded considerably, most notably with the establishment of a Freedom of Association project as part of a US$15 million program on Decent Work, to run from 2012 to 2014 (ILO 2012). Close to half of the projected program budget was reserved for initiatives to combat forced labor. However, an allocation of US$6.1 million was made for an 'inception phase' of a series of projects that aimed to build technical and institutional capacity in the rights and standards domain, including labor legislation, labor market governance and labor migration. This program also incorporated initiatives related to socially responsible enterprise development, enhancing employment opportunities, and the employment dimension of trade and investment, which had an obvious economic development and market-supporting orientation (ILO 2012).

By 2014, the ILO's activities had expanded to encompass three primary objectives: promoting fundamental principles and rights at work, strengthening employers' and workers' organizations and labor market institutions, and enhancing employment opportunities and social protection (ILO 2014b). Reflecting its development orientation, the latter includes peace-building and livelihood initiatives in conflict-prone areas; the development of small and medium enterprises (SMEs), especially in the tourism industry; 'responsible business' strategies including value chain analysis of the garment and fisheries industries; and planning strategies for productivity improvements, in addition to support for the development of social protection policies (ILO 2014b). However, it has been through its work on labor regulation and foreign investment that the ILO has made its most influential interventions in Myanmar.

Labor regulation as a development strategy

One key plank in the ILO's labor regulation work in Myanmar has been quite traditional, namely seeking to ensure that the country's regulatory and institutional arrangements meet international expectations with regard to freedom of association and trade union involvement in tripartite mechanisms of policy-making and dispute resolution. As in Cambodia, Indonesia and Timor-Leste, a strong focus of these efforts has been on the establishment of a 'package' of laws that embed the principles of tripartism and social dialogue in the legal bedrock of industrial relations (Ford 2016; Ford and Sirait 2016; Ward and Mouyly 2016). With the ILO's assistance, the Government of Myanmar drafted and passed a series of new labor laws enabling the formation of trade unions and employer organizations and establishing institutions and regulatory provisions to allow for dispute resolution, strikes and collective bargaining. A series of related laws and regulatory provisions sought to reform social security, occupational health and safety, and minimum wages (Gillan and Thein 2016).

The ILO has emphasized the need to develop functioning employer and worker representation structures to give substance to these new laws. Through its Freedom of Association project (ILO 2012), it ran 145 activities involving 5,449 participants in the two years from September 2012 to September 2014 in Yangon and Mandalay, the cities with the largest concentration of formal employment and industries such as garment manufacturing, as well a number of mid-sized cities (ILO 2014b). Project officers and other ILO staff also met with activists from labor NGOs and the rapidly growing trade unions on a regular basis in order to assist in the development of trade union leaders and further the goal of "strengthening trade union coordination at sectoral, regional and national level" (ILO 2012). In addition, the ILO has worked closely with government agencies in various other technical areas, including the implementation of a National Labor Force Survey in 2015; improving labor inspection regimes; and building the capacity of the newly formed industrial relations dispute resolution bodies in order to streamline their processes and improve the quality and consistency of decision-making. It has also collaborated closely with the World Bank and the World Food Programme to gather information and work with government, business and civil society to identify and discuss gaps in social protection institutions and regulation (ILO 2015b).[6]

Many imperfections remain in the labor regulatory framework (Gillan and Thein 2016). Myanmar is, moreover, experiencing many of the same challenges of implementation faced by Cambodia, Indonesia and Timor-Leste in their attempts to establish a functioning system of labor relations (Ford 2016; Ford and Sirait 2016; Ward and Mouyly 2016). However, in a context where workers, employers and government agencies had no understanding or experience of freedom of association or collective bargaining, the ILO has played a crucial role in creating a formal architecture for industrial relations that recognizes the right of workers to collective representation.

Facilitating foreign investment

Importantly, the ILO sought to link these labor reforms with economic growth and global economic reintegration. The relationship between labor relations and foreign investment is explicitly recognized in the context of its Labour Law Reform and Institutional Capacity program, funded by Denmark, Japan and the United States. In a joint statement by the donors, the ILO and the Government of Myanmar, Myanmar's "labor regime" is presented as an "important component of its investment environment," which, in conjunction with other measures, can assist in repositioning Myanmar as an "attractive sourcing and investment destination" and advance "overall sustainable growth and development" (ILO 2014d). It is within this context also that the ILO has sought to facilitate foreign investment in Myanmar in ways that support employment generation and development goals while also seeking to meet international labor standards. The ambit of the 'responsible business' program has been particularly broad in recognition of the low level of capacity in the country's formal economy to respond to an upsurge of interest from potential investors, including many multinational corporations in the consumer goods sector, since the beginning of the transition toward civilian rule.

As noted earlier, the ILO has focused on micro, small and medium enterprises in many developing country contexts in its role as a development actor. In Myanmar – as in Cambodia – it has also sought to work with large-scale enterprises, industry associations and government to support economic development and, through it, employment generation. In an attempt to enhance the capacity of industries with a current or potential export profile such as manufacturing and fisheries to articulate with global trade and supplier networks, it has initiated value chain assessments and industry development plans in conjunction with government and business. It has also begun groundwork for the development of factory-level training and measurement programs, modeled in part on the Better Factories Cambodia program, as well as various activities designed to promote 'social dialogue' and 'awareness raising' around CSR and labor standard issues in value chains (ILO 2014b). One concrete example of such an initiative is a project overseen by the ILO and funded by the Swedish International Development Agency and the Swedish fashion multinational corporation H&M to improve industrial relations in the garment industry by offering factory level training to managers and employees, as well as working with the Myanmar Garment Manufacturing Association and trade unions (ILO 2014b).

Through these and other initiatives, the ILO has taken on a direct intermediary role in Myanmar's reintegration with global production networks. Well-known international brands have consulted with the Liaison Office prior to investing in order to mitigate or lessen the potential reputational risk of expanding production or distribution within the country (Interview with the ILO Chief Liaison Officer, January 2014). Firms originating in the United States are also required by the United States Government to provide reports on the environmental and social impacts of investments in Myanmar. As labor rights and standards are a particular focus both for these reporting requirements and in Myanmar's attempts to regain trade preferences from the US Government, the ILO's role as a regulatory and standard-setting body has been especially relevant for US-based investors. The apparel firm GAP, for example, has participated in ILO 'stakeholder forums' on labor law and standards as well as publicly stating its commitment to the labor law reform process and its implementation according to ILO standards and principles (GAP 2015). As such statements suggest, the ILO's provision of technical assistance and advice on the emergent labor laws and institutions have not only been a very significant aspect of their work in Myanmar, it has also been deeply integrated with attempts to kick-start an export-oriented economy.

Conclusion

The ILO's emphasis on the link between labor reform and global economic reintegration as an engine of economic growth in Myanmar aligns with its general model of development and its interest in working with international institutions with a market-building and -supporting function such as the World Bank. It is also pragmatic in the sense that the framing of labor standards as a necessary precondition of global economic reintegration and growth appealed to, rather than threatened, the Thein Sein government and indeed is compatible with the call of Aung

San Suu Kyi in her role as leader of the National League for Democracy, which won the 2015 national election, for foreign investment that is ethical, responsible and underpinned by respect for labor and other social standards.

Whether these normative statements translate to actual practice and development outcomes remains unresolved. Questions persist as to whether employment growth, skills development and industrial diversification can so neatly dovetail with social protection and labor standards without trade-offs between these objectives, particularly in the context where other developing nations have adopted a 'low road' pathway to economic growth (Seekings 2015). With regard to labor rights and bargaining power, for example, Caraway (2006, 210) has argued that while positive to the extent that it ended the monopoly of the state-sanctioned union in Indonesia, the ILO's liberal interpretation of freedom of association, inspired by the policy orientations of the World Bank and the IMF, led to "the formation of 'free' as opposed to powerful trade unions" after the transition from authoritarianism. Similarly, it is by no means certain that Myanmar's nascent institutions and political and civil society organizations, including trade unions – which the ILO has played a significant role in shaping – will be effective, representative and able to contribute to social equity or whether these institutions and organizations will simply reproduce existing social and economic inequalities.

The case of the ILO in Myanmar is also illustrative of the way in which the role of international institutions must be understood with reference to the dynamism of their interaction with the shifting landscape of domestic politics and government. Over time, the ILO has moved from a sanctioning and monitoring role focused on forced labor to a more diverse and expansive series of interventions across the labor standards and development spheres not unlike those found in other post-authoritarian states in the region. Its ability to take on this expanded role is in part a consequence of the deep in-country experience, knowledge, relationships and credibility developed through its dealings with the government over forced labor, which has positioned it as an important development actor in the context of democratization and associated reform programs. Clearly, however, its approach has also been driven by its quest to reposition itself as a development intermediary for other international institutions, national and international businesses, as well as state agencies and policy-makers. It may be difficult to assess the impact of the Decent Work Agenda in Myanmar or elsewhere (cf. Baccaro and Mele 2012), but the ILO's approach in post-authoritarian Southeast Asia since the late 1990s suggests that its development agenda is well and truly here to stay.

Notes

1 Authors are listed alphabetically and have contributed equally to the chapter, the writing of which was supported by Australian Research Council grants DP130101650 and FT120100778.
2 See Hadiz (1997) and Ford (2009) for detailed discussions of the Indonesian trade union movement in this period.
3 See Kolben (2004) for a detailed analysis of the ILO's initial proposal, which adopted a more traditional emphasis on capacity building in industrial relations, and in particular in the Cambodian labor inspectorate, the US counter-proposal, designed by the Department of Labor with significant input from the United States Trade Representative, the Department of State, AFL-CIO and UNITE, and the final agreement.
4 See Rodríguez-Pose (2002) on the role of the ILO in supporting and engaging with local economic development and an assessment of some of its challenges in coordinating effective and integrated development initiatives.
5 The international labor movement was also heavily involved in development projects in Aceh, including the construction of community centers (Ford and Dibley 2012). For further discussion of the Aceh case, and of trade union aid as a form of mediated diffusion, see Ford and Dibley (2011). For a discussion of

the ILO's conscious attempt to boost its significance as a development agency in order to supplement its limited budgets and personnel capacity, see Standing (2008).
6 This initiative was represented as a model example of cooperation and coordination between international development agencies, in alignment with the announcement in mid-2015 of a global commitment for ongoing collaboration between the ILO and the World Bank for a "mission and plan of action towards universal social protection" (ILO 2015b).

References

Alston, P. (2004). 'Core Labour Standards' and the transformation of the international labour rights regime. *European Journal of International Law*, 15(3), pp. 457–521.

Arnold, D. and Toh, H. S. (2010). A fair model of globalisation? Labour and global production in Cambodia. *Journal of Contemporary Asia*, 40(3), pp. 401–424.

Baccaro, L. and Mele, V. (2012). Pathology of path dependency? The ILO and the challenge of new governance. *Industrial & Labor Relations Review*, 65(2), pp. 195–224.

Better Work. (2015). *Countries*. Available at: http://betterwork.org/global/?page_id=314. Accessed 18 August 2015.

Caraway, T. (2006). Freedom of association: Battering Ram or Trojan Horse? *Review of International Political Economy*, 13(2), pp. 210–232.

Cheesman, N., Farrelly, N. and Wilson, T. (eds.) (2014). *Debating democratization in Myanmar*. Singapore: Institute of Southeast Asian Studies.

Ford, M. (2009). *Workers and intellectuals: NGOs, trade unions and the Indonesian labour movement*. Singapore: NUS/Hawaii/KITLV.

Ford, M. (2016). The making of industrial relations in Timor-Leste. *Journal of Industrial Relations*, 58(2), pp. 243–257.

Ford, M. and Dibley, T. (2011). Developing a movement? Aid-based mediated diffusion as a strategy to promote labour activism in post-tsunami Aceh. *Asian Journal of Social Science*, 39(4), pp. 469–488.

Ford, M. and Dibley, T. (2012). Experiments in cross-scalar labour organizing: Reflections on trade union-building work in Aceh after the 2004 Tsunami. *Antipode: A Radical Journal of Geography*, 44(2), pp. 303–320.

Ford, M. and Sirait, M. (2016). The state, democratic transition and employment relations in Indonesia. *Journal of Industrial Relations*, 58(2), pp. 229–242.

GAP. (2015). *Statement by H&M and gap Inc. at the conclusion of the ILO labor capacity building forum in Yangon*. Available at: www.gapinc.com/content/gapinc/html/media/pressrelease/2015/med_pr_GPS_HM_jointstatement.html. Accessed 22 August 2015.

Gillan, M. and Thein, H. H. (2016). Employment relations, the state and transitions in governance in Myanmar. *Journal of Industrial Relations*, 52(2), pp. 273–288.

Glasius, M. (1999). *Foreign policy on human rights: Its influence on Indonesia under Soeharto*. Antwerpen: Intersentia.

Hadiz, V. (1997). *Workers and the state in new order Indonesia*. London and New York: Routledge.

Horsey, R. (2011). *Ending forced labour in Myanmar: Engaging a Pariah regime*. Abingdon and New York: Routledge.

Hughes, C. (2007). Transnational networks, international organizations and political participation in Cambodia: Human rights, labour rights and common rights. *Democratization*, 14(5), pp. 834–852.

ILO. (2005). Developments concerning the question of the observance by the government of Myanmar of the Forced Labour Convention, 1930 (No. 29). GB.294/6/2294th Session Governing Body Geneva, November.

ILO. (2006). ILO/USA Declaration Project Indonesia: Snapshot 2001–2006. Jakarta: ILO.

ILO. (2012). *Decent work in Myanmar, ILO programme framework: November 2012– April 2014*. Available at: www.ilo.org/wcmsp5/groups/public/-dgreports/-exrel/documents/publication/wcms_193195.pdf. Accessed 18 August 2015.

ILO. (2014a). *World of work report 2014: Developing with jobs*. Geneva: ILO.

ILO. (2014b). *National tripartite dialogue: Presentation on ILO's programme of work in Myanmar, ILO Liaison office in Myanmar*, 4 December 2014. Available at: www.ilo.org/yangon/whatwedo/events/WCMS_329889/lang–en/index.htm. Accessed 18 August 2015.

ILO. (2014c). The post-2015 sustainable development agenda: Update. GB.322/INS/6, Governing Body 322nd Session, Geneva, November.

ILO. (2014d). *New initiative to improve labour rights in Myanmar: Joint statement of the Republic of the Union of Myanmar, the United States of America, Japan, Denmark and the International Labour Organization (ILO), Yangon, Myanmar*, 14 November. Available at: www.ilo.org/yangon/info/public/speeches/WCMS_319811/lang-en/index.htm. Accessed 18 August 2015.

ILO. (2015a). *Cambodia resources*. Available at: www.ilo.org/asia/countries/cambodia/facet/lang–en/nextRow-0/index.htm?facetcriteria=TYP=Project. Accessed 18 August 2015.

ILO. (2015b). *Successful example of collaboration on social protection in Myanmar* [Press Release], 28 July. Available at: www.ilo.org/wcmsp5/groups/public/-asia/-ro-bangkok/-ilo-yangon/documents/pressrelease/wcms_386783.pdf. Accessed 18 August 2015.

ILO. (n.d.a). *Creating jobs: Capacity building for local resource-based road works in selected districts in NAD and Nias*. Available at: www.ilo.org/jakarta/whatwedo/projects/WCMS_145289/lang-en/index.htm. Accessed 18 August 2015.

ILO. (n.d.b). *Decent work and the 2030 Agenda for sustainable development*. Available at: www.ilo.org/global/topics/sdg-2030/lang--en/index.htm. Accessed 18 August 2015.

ILO. (n.d.c). *Enhancing Rural Areas (ERA) project brief*. Available at: www.ilo.org/wcmsp5/groups/public/-asia/-ro-bangkok/-ilo-jakarta/documents/project/wcms_221367.pdf. Accessed 18 August 2015.

ILO. (n.d.d). *Investment budget execution support for rural infrastructure development and employment generation (TIM Works) project – AUS (Timor-Leste)*. Available at: www.ilo.org/jakarta/whatwedo/projects/WCMS_163724/lang-en/index.htm. Accessed 18 August 2015.

ILO. (n.d.e). *Roads for development (R4D) (Timor-Leste)*. Available at: www.ilo.org/jakarta/whatwedo/projects/WCMS_184617/lang-en/index.htm. Accessed 18 August 2015.

ILO. (n.d.f). *Start and improve your business programme*. Available at: www.ilo.org/empent/areas/start-and-improve-your-business/lang-en/index.htm. Accessed 18 August 2015.

Kolben, K. (2004). Trade union monitoring, and the ILO: Working to improve conditions in Cambodia's garment factories. *Yale Human Rights and Development Law Journal*, 7, pp. 79–107.

Maupain, F. (2005). Is the ILO effective in upholding workers' rights? Reflections on the Myanmar experience. In: P. Alston, ed., *Labour rights as human rights*. Oxford and New York: Oxford University Press, 85–142.

Maupain, F. (2009). New foundation or new façade? The ILO and the 2008 declaration on social justice for a fair globalization. *The European Journal of International Law*, 20(3), pp. 823–852.

Maupain, F. (2015). Revisiting the future. *International Labour Review*, 154(1), pp. 103–114.

Mundlak, G. (2015). In search of coherence? *International Labour Review*, 154(1), pp. 79–84.

Rodríguez-Pose, A. (2002). *The role of the ILO in implementing local economic development strategies in a globalised world*. Geneva: ILO.

Scanteam. (2006). *Final evaluation of strengthening and improving labour relations in East Timor (SIMPLAR)*. Oslo: Scanteam.

Seekings, J. (2015). Does one size fit all? A comment on the ILO's 2014 world of work report. *Global Social Policy*, 15(2), pp. 192–194.

Standing, G. (2008). The ILO: An agency for globalization? *Development and Change*, 39(3), pp. 355–384.

Tapiola, K. and Swepston, L. (2010). ILO and the impact of labor standards: Working on the ground after an ILO commission of inquiry. *Stanford Law and Policy Review*, 21(3), pp. 513–526.

Ward, K. and Mouyly, V. (2016). Employment relations and political transition in Cambodia. *Journal of Industrial Relations*, 58(2), pp. 258–272.

8
JUSTICE PROCESSES AND DISCOURSES OF POST-CONFLICT RECONCILIATION IN SOUTHEAST ASIA

The experiences of Cambodia and Timor-Leste

Rachel Hughes

Introduction

The year 2015 saw the 40th anniversary of the 1975 Indonesian invasion of Timor-Leste. The anniversary was marked with calls for an end to impunity for those Indonesian military leaders responsible for orchestrating and carrying out criminal acts during the invasion, the subsequent 24-year occupation of the country and during the 1999 independence vote.[1] In the same year, Cambodia also observed a 40th anniversary, of the fall of Phnom Penh to the Khmer Rouge and the beginning of the mass political violence of 1975–1979. An internationalized criminal tribunal examining Khmer Rouge crimes has been underway in the country for nearly ten years. Also in 2015, a People's Tribunal was convened in The Hague to examine the crimes committed within Indonesia from 1965 against members or suspected members of the communist party (KPI).[2]

What do these calls for justice, and formal and semi-formal justice processes, tell us about Southeast Asian development? First, they recall the brutal events that have, within living memory, affected these places and communities, their politics, economics, spiritual and social lives, and livelihoods. They must also serve to remind us of similar violent events experienced elsewhere in the region but in places that have not chosen, or been in a position to choose, the formal reconciliation or legal mechanisms seen in relation to Timor-Leste, Cambodia or Indonesia.[3] Second, they make clear that a significant number of survivors and supporters thirst for adequate recognition of still-silenced crimes in these (and other) post-conflict contexts. Third, they give a sense of the urgency of this justice 'deficit,' as each anniversary sees the passing on of survivors, witnesses and perpetrators. Fourth, they suggest ways in which injustices in one national context might be related to the conduct of actors or states elsewhere, such as Indonesian militarism at home and as a colonial power, or how the fight against communism in United States foreign policy produced political and developmental support (or the withdrawal of such support) in different parts of southeast Asia, at different times, to significant effect.[4] But these justice processes also make obvious the hopefulness that is engendered in survivors and their supporters that perpetrators will be brought to account for criminal acts.

This chapter deals in detail with two of the three contexts just introduced, Cambodia and Timor-Leste, where new justice processes have emerged within wider discourses of post-conflict national reconciliation. I argue, however, that these justice processes need to be understood as negotiated outcomes of other spatially extensive processes, actors and events, and as such cannot be considered exclusively in terms of the national context with which they are generally identified. In the remainder of this introduction, brief background discussions aim to orient the reader to the specificities of these cases. The chapter then proceeds through four further discussions: discourses of post-conflict reconciliation, tribunal justice, victim participation, and reparations and non-state practices.

Cambodia's mass political violence under Khmer Rouge rule began in 1975 and ended in 1979. While efforts were made to find justice for victims during the reconstruction of the country in the 1980s and 1990s, the hybrid or internationalized criminal tribunal (the Extraordinary Chambers in the Courts of Cambodia or ECCC) was only instituted in the last decade (fully operational since 2007). Violence in Timor-Leste continued throughout Indonesia's occupation from 1975 to the 1999 referendum, with the vote in favor of independence seeing an escalation and intensification of violence on the part of the Indonesian military and pro-Indonesian Timorese militia (CAVR 2005). Two formal justice processes have been attempted since, the first being the trials of the internationalized Special Panels for Serious Crimes (SPSC) in Dili, which examined the crimes of 1999, and the second being a truth and reconciliation commission, the Comissão de Acolhimento, Verdade e Reconciliação (CAVR or Commission for Reception, Truth and Reconciliation), which examined the occupation period as well as the violence in 1999, and was conducted in various parts of the country but headquartered in Dili.

These justice processes necessarily work in and through the politics and economics of the respective post-conflict periods, as well as in relation to earlier experiences of conflict and rule, including French and Portuguese colonialisms, pre-1975 civil conflict and Cold War regional interventions (for an overview, see McGregor 2008). While there is little space to explore all this here, it is necessary to note the artificial temporality and spatiality that the designations 'post-conflict Cambodia' and 'post-conflict Timor-Leste' entail.

'Post-conflict Cambodia' has multiple start times, just as the designation of the beginning of Cambodia's 'conflict' is difficult to pinpoint. For some, Cambodia's major period of violence came to an end when the Khmer Rouge regime was toppled from power in January 1979. For others, peace came with the relative stability of the United Nations-sponsored elections of 1993, but then civil conflict continued between the government in Phnom Penh and remnant Khmer Rouge in the west and north of the country into the late 1990s. For those who remember this later fighting, there was no peace until controversial government amnesties brought the last few Khmer Rouge leaders into the national fold. In Timor-Leste, while the independence referendum in 1999 ended the often covert and structural violence of Indonesian occupation of the country, 1999 itself saw widespread violence at unprecedented intensities committed in large part by the ousted Indonesian military and pro-Indonesia militia. This post-conflict timeline is complicated by unrest in 2006 that threatened further widespread violence.

There are also multiple spatialities of the states under review, not least Portuguese and Indonesian territorial annexation of Timor-Leste, the covert and overt military incursions into Cambodia by the United States and Vietnam, and the almost complete isolation of Cambodia under the Khmer Rouge. Conflict in and near to Timor-Leste and Cambodia made their national borders porous to both incoming states' interests and actors, and their own outgoing populations moving under duress. Population flight gave rise to Cambodian and Timorese diasporas, and these diasporas came to play a role in the life of the post-conflict nation that further disrupted national borders, unsettled national identities and generated new forms of political leadership.

Post-conflict reconciliation?

Discourses of 'national reconciliation' developed in Cambodia in the late 1980s, in the context of ongoing civil war between the post-1979 government and remnant Khmer Rouge. At this time, Vietnamese troops and advisors who had supported the new government since 1979 were being withdrawn back to Vietnam. In 1987, a 'Declaration on the Policy of National Reconciliation' discussed the repatriation of Cambodian refugees from the border areas and the government's willingness to "meet with other groups of Khmers and their leaders, except the criminal Pol Pot and his close associates" (quoted in Gottesman 2003, 278). In the lead-up to the Paris Accords of 1991, the Cambodian government moved toward economic liberalization, which, as will be discussed, influenced reconciliation efforts (Gottesman 2003, 345; Ledgerwood 2008). As the civil war continued, Cambodian 'national reconciliation' became typified by the amnesties granted to those Khmer Rouge leaders who defected to the government. This amnesty program provided some defectors with positions in the Royal Cambodian Armed Forces, others with land and financial assistance; physical safety, the right to work and the security of property were also assured (Linton 2010, 43–44). These defections eventually led to the disintegration of the Khmer Rouge as an effective movement.

Within this self-declared period of 'reconciliation,' the United Nations (UN) peacekeeping mission UNTAC (UN Transitional Authority in Cambodia) arrived in Cambodia to facilitate free and fair elections in the country (elections that would nonetheless aim to include the Khmer Rouge). The Khmer Rouge eventually boycotted the electoral process and resumed their violent campaign. By the time the amnesties offered by the Cambodian government were bringing an end to the civil war in the late 1990s, the international community was decrying this deal-making with the Khmer Rouge, despite the fact that where the 1993 elections might have legitimized and reinvigorated the group, the amnesties sought to neutralize them. With the key defections of Ieng Sary (in 1996) and Khieu Samphan and Noun Chea (in 1998), and the death of Pol Pot (also in 1998), the ruling Cambodian People's Party (CPP) considered the reconciliation process complete. In a 1999 ceremony at the Choeung Ek 'killing field' genocide memorial, for example, Phnom Penh Deputy Governor and CPP member Chea Sophea directly addressed victims' remains interred inside the memorial *stupa*:

> I am here today to inform all of you who died that owing to the win-win policy of the Prime Minister Hun Sen, all Khmer people have reconciled and united. The Khmer people are now at peace, and the Kingdom of Cambodia has become a full member of ASEAN.
>
> (Chea quoted in Mydans 1999)

Choeung Ek's dead were apparently to be comforted by the knowledge that their compatriots were so reconciled that they could become a member of the preeminent regional economic grouping. Reconciliation is here seen as an important step in bringing about the conditions for regional acceptance and the stability required for development investment.

Caroline Hughes notes of post-conflict contexts like Timor-Leste and Cambodia that policies of accountability for past atrocity are an important means by which "idealized conceptions of the *bounded* political community" are promoted, whilst the *unbounded* "expan[sion of] the possibilities for far-reaching [economic] intervention in the post-conflict state and society" is simultaneously enacted (C. Hughes 2009, 72, my emphasis). In this analysis, accountability (or reconciliation in policy terms) becomes part of "the tendency to shrink internationalized conflict [...] into a national box" (C. Hughes 2009, 71). The moral responsibility for atrocity is thus localized, but the necessary expertise needed to address past (and feared future) atrocity necessitates further international economic and developmental intervention and dependence.

Joseph Nevins has similarly argued that the representations of mass violence in Timor-Leste between 1975 and 1999 perpetuated by the international community serve particular political and geopolitical interests, and result in impunity for the Indonesian military and "sectors of various national governments long supportive of Jakarta" (Nevins 2002, 525; see also Nevins 2005). This impunity inhibits national reconciliation between supporters of independence and those who worked with the Indonesian military, guards against proper reparations thus perpetuating impoverishment, and allows members of the international community complicit in the destruction of Timor-Leste to appear charitable when they might otherwise appear responsible (Nevins 2002, 534–535). Lia Kent has also and more recently linked structural violence and calls for justice and reconciliation, observing that justice activists (among others) need to "consider how justice advocacy could become part of a political call for the broader transformation of the poor economic conditions, health and education of the disadvantaged and marginalized" (Kent 2012b, 1044).

In this expanded view, state policies of reconciliation and accountability in post-conflict contexts are often problematic. Arguably, they lay responsibility for decades of violence at the feet of too few; they insist on the symbolic coming together of groups with little material transformation of suffering and inequality; they respond superficially where comprehensive investigation, recognition and reparation is urgently needed; and they can provide a smokescreen behind which new economic violence is wrought. Nonetheless, policies of reconciliation and accountability can also raise and debate difficult questions around survivor, victim and perpetrator identities, past events and the ways in which they are memorialized, and meaningful ways of living with the losses of the past.

Tribunal justice

Internationalized tribunals are often supported and directed by post-conflict national governments seeking wider legitimacy, and funded by powerful external states with their own strategic interests and versions of the past. Given these deep-seated interests, they may have little real effect on the impunity enjoyed by the majority of culpable actors within and beyond the post-conflict nation. But such internationalized justice processes also potentially work on and through the social, political and economic connections between the post-conflict nation and other places and processes, making these connections more visible, palpable and responsive. In this section, I attempt to think about tribunals in ways that exceed the 'national box' (C. Hughes 2009, 71) to which they too are often ascribed. This kind of consideration of tribunal justice has also been termed a 'critical geopolitics of justice' approach – one that enquires into the relationship between present and prior claims to justice and memory, and that is attentive to the extra-national processes (of negotiation, funding, functioning, etc.) that have facilitated but perhaps also limited these justice forms (see R. Hughes 2015; Jeffrey 2011).

Justice for Cambodian and Timorese victims has been sought through international criminal law and international humanitarian law, with the establishment of 'hybrid' (internationalized) tribunals in both places. Hybrid tribunals bring a mix of international law and national law to bear upon the alleged crimes of the defendants (see Cohen 2007; Mendez 2009; Williams

2012), and are held in situ within affected countries and their legal systems. These tribunals are often considered to be highly novel developments in international criminal law, ones that have learned the lessons of the ad hoc courts – the International Criminal Tribunal for the former Yugoslavia, and the International Criminal Tribunal for Rwanda – which, although important in many ways, have been lengthy, expensive endeavors situated at physical and popular remove from affected communities.

In Timor-Leste, the SPSC was created in 2000 and was invested with exclusive jurisdiction over what were defined as 'serious crimes': genocide, war crimes and crimes against humanity committed during any time period, and additionally, murder and sexual offences committed between 1 January and 25 October 1999. As a special panel of the Dili District Court, the SPSC was located within the nascent legal system of the newly-independent nation, and was presided over by one national and two international judges. Between 2000 and 2005, the SPSC completed 55 trials involving 87 defendants, convicting 84 and acquitting three, with one of the acquitted subsequently convicted by the Court of Appeal (Cohen 2007, 9). As David Cohen attests, however, the SPSC suffered from a lack of 'ownership' on the part of either the Timorese government or the various UN missions in Timor-Leste over its period of operation. This lack of ownership led to the evasion of responsibility on key provisions such as electricity for the court, witness protection, warrant enforcement and qualified legal personnel (Cohen 2007, 9–10). Crucially, the UN was unwilling (and unable) to bring pressure to bear on Indonesia to cooperate; this meant that the SPSC only prosecuted low-level militia members while the vast majority of indictees, including senior members of the Indonesian military, remained beyond the court's jurisdiction (Cohen 2007; Nevins 2002).

Cambodia's hybrid tribunal, the ECCC, is trying 'senior leaders and those most responsible' for serious crimes committed in Cambodia between April 1975 and January 1979. It is a "national court with international characteristics" (Sok An quoted in Heindel 2009, 87), established by domestic Cambodian law pursuant to a 2003 Agreement between the UN and the Cambodian government. In many ways this means the ECCC is a legal descendant of the SPSC, and certainly legal commentators have been concerned that the Cambodian courts learn from the SPSC experience. Only a small number of crimes are being prosecuted by the ECCC. Those taken from international criminal law include crimes against humanity, genocide and grave breaches of the Geneva Conventions (war crimes), destruction of cultural property and crimes against diplomatically protected persons; those taken from Cambodian law include homicide, torture and religious persecution (see ECCC Law 2004).

In the first case to come before the ECCC, a single defendant, Kaing Guek Eav (alias 'Duch') was found guilty of crimes against humanity and grave breaches of the 1949 Geneva Convention. The substantive hearing of evidence in a second case began in late 2011. The two remaining defendants in this case are Nuon Chea and Khieu Samphan, former leaders of the regime, both now very elderly. The judgment in the first mini-trial of the second case was handed down in August 2014; both defendants were found guilty of crimes against humanity (see ECCC Trial Chamber Judgment in Case 002/01).[5] The life convictions of these three individuals across two cases along with, as Anne Heindel (2009) argues, the entire jurisprudence of the tribunal, constitute *judicial legacies* of the ECCC that are both nationally and internationally significant. Cambodians have welcomed this retributive justice, and support for the tribunal has increased over its operation. In 2010, national-level research by the Human Rights Centre of the University of California Berkeley found that three in four Cambodians believed that the ECCC was fair, neutral, promoted national reconciliation and was bringing about positive impacts on society (see Pham et al. 2011). These latter two *non-judicial legacies* will be discussed further below.

Victim participation

In Cambodia, it is widely accepted that ECCC has interdependent and mutually reinforcing retributive, restorative and procedural justice roles (Ciorciari 2009). That is, in addition to the retributive justice processes and successes of international criminal law just described, additional roles for these large and long-term (but not permanent) legal entities are currently demanded. Significant to the ECCC's restorative justice role is the tribunal's novel provision for victims to join the proceedings as 'civil parties.' Victims of the Khmer Rouge may choose to participate at the ECCC in two ways: first, they may be recognized as a complainant by submitting useful information to the tribunal; second, if they can show that they have been directly affected by the crimes under consideration, they can become a 'civil party.' This provision is the first of its kind internationally and a feature of the tribunal that was adopted from Cambodia's civil law tradition, itself derived from the French civil law tradition (not the English common law tradition). Until now, victims of the crimes being heard in the ad hoc and hybrid tribunals (including Timor-Leste's SPSC) have participated only as witnesses (where they have been called to appear) and as distant observers of these lengthy and expensive legal undertakings.

ECCC civil parties are represented in court as parties to the proceedings (just as the Prosecution and Defense are represented in court). More than 3,800 Cambodians have participated in ECCC Cases 001 and 002 as civil parties (see Pham et al. 2011; Poluda et al. 2012; Stover et al. 2011), and more have applied to be joined to potential future cases (003 and 004). Victim participation at the ECCC has been pioneering but also dogged by uncertainty, and so by anxiety and suspicion (Jarvis 2014; see also Sperfeldt 2013; Stover et al. 2011). At the ECCC, civil parties may assist the Prosecution and make reparations claims for 'moral and collective' reparation, but not monetary or individual reparation, of which more will be said below.

The ECCC, and victim participation in the tribunal, have seen increased public interest in the tribunal and in survivors' experiences of the Khmer Rouge regime. Talk of the rights of victims to participate, however, and of trauma in largely individual-psychological terms in media and civil society discourse, has emphasized the individual victim's access to, and role within the ECCC. One of the effects of this has been to propel some individuals, civil parties, to greater national and international prominence. Another effect of ECCC participation has been the development of victim associations, such as *Ksaem Ksan* and the Association of Khmer Rouge Victims in Cambodia; these visible and vocal socio-political groupings have not hitherto existed in Cambodia. While these successes are significant, a number of vocal civil parties have emerged from their participation disillusioned with the tribunal. Victim participation in Cambodia finds both parallels and differences in Timorese experiences.

In Timor-Leste, it was the CAVR, not a tribunal process that attempted to extensively and sensitively engage with the experiences and justice claims of victims. The terms of reference and structure of CAVR were developed with far greater input by Timorese political leadership and non-governmental organizations than for the SPSC (Kent 2012a, 47). The CAVR was mandated to inquire into and establish the truth regarding human rights violations during the political conflicts in Timor-Leste between 25 April 1974 and 25 October 1999, prepare a comprehensive report, formulate recommendations concerning reforms and initiatives designed to prevent the recurrence of human rights violations and to respond to the needs of victims, to recommend prosecutions where appropriate, promote reconciliation, assist in restoring the dignity of victims, and to promote human rights (CAVR 2005). Although it was developed quite separately from the Serious Crimes Process, the CAVR was nonetheless expected "to play a complementary role and, in particular, the CRP was expected to relieve the pressure on an overburdened and underdeveloped East Timorese legal system" (Kent 2012a, 15).

The community reconciliation process developed by the CAVR "allowed for an alternative form of accountability, one which drew from *lisan* [traditional local customs and laws] and formal justice, [which] was worked out with the participation of the local community, and [was] ultimately formalized in an agreement which was lodged with the local district court" (Linton 2010, 86). Ultimately, this nationwide process "adjudicated almost 1400 cases of minor crimes, and its attempts to merge aspects of local dispute resolution methods into its reconciliation framework are perceived by many commentators as having contributed to its local legitimacy" (Kent 2012a, 16). As the final section discusses, however, there is still significant unevenness across this experience of victim participation, in terms of the nature of that participation and its effects, and many victims are increasingly seeking out informal, non-state practices in their 'dealing with the past' in Timor-Leste.

Reparations and non-state practices

Calls have been made for the CAVR report to be developed into "an interim secondary school curriculum" in light of the existing gap in Timor-Leste's education syllabus concerning the period 1974–1999 (see Leach 2010, 129). Similarly, in Cambodia, the historical record established by the ECCC has been seized upon as of significant potential use in educational domains. In Cambodia, due to the political priorities and development realities of the decades following the end of Khmer Rouge rule, details of the events and crimes of the 1970s have generally gone unreported or misreported. Lack of access to secondary schooling, and low levels of public media access are still experienced in Cambodia, especially in rural areas. A textbook for high school students has been produced and distributed by a private research organization, the Documentation Centre of Cambodia (DC-Cam) (see Dy 2007).

In partnership with NGOs (see Sperfeldt 2013) and through its own legal outreach program, the ECCC is working to facilitate knowledge and understanding beyond familial and community-based accounts of the period and its crimes. Various forms of documentation from and about the tribunal are important potential remedies to this situation. Experts have argued that documentation from the proceedings and from popular commentary should be preserved and made accessible in perpetuity, as should materials that capture the experiences of how individuals and society were affected by the tribunal in its time (Cohen in ECCC and CHRAC 2012). The tribunal's Public Affairs Section has to date produced thousands of images, recordings and documents pertaining to a great range of public and procedural activities undertaken since 2006. Recently, the Cambodian government, with support from the government of Japan, has established the Legal Documentation Centre relating to the ECCC (LDC-ECCC). This key archive currently holds hard and soft copies of public documents from ECCC Case 001, and it is envisaged that the full public records of all the tribunal's cases will eventually be housed and made publically accessible at the LDC. Similar developments have occurred in Timor-Leste, where the headquarters of the CAVR, a former prison known as Comarca Baliade, has been renovated as a permanent site for the CAVR archives, a library, and exhibition and meeting spaces.

Public memorialization of the Khmer Rouge period has been given new and significant impetus by the tribunal. Choeung Ek, a crime site of ECCC Case 001, and the Tuol Sleng Museum, a crime site and long-time museum, now function as stopping-off points for the ECCC's successful legal outreach program known as the Khmer Rouge Study Tour (see Elander 2014). This program provides free transport to hundreds of Cambodians to visit the tribunal and hear about its work. In this way the tribunal has delivered a significant increase in visitor numbers and greater national public profile for the Choeung Ek and Tuol Sleng sites. As well, a new memorial *stupa* (chedi) for Tuol Sleng was inaugurated in 2015. The *stupa* was developed by the ECCC in conversation with victims (part of the tribunal's non-judicial measures), and was funded by the German government (Jarvis 2015). Both Tuol Sleng and Choeung Ek are also major international tourism destinations and have seen significant recent improvements in terms of physical access to the sites and new, high-quality audio-tours (see R. Hughes 2008; Jarvis 2015).

Reparations measures that have already been delivered by the ECCC include the naming of civil parties in the published judgment for the first case and a compilation (published on the ECCC website) of the statements of apology made by the defendant Duch during that trial; however many victims considered these reparations less than satisfactory. In the course of ECCC Case 002, the tribunal required that its own Victim Support and Civil Party Lead Co-Lawyers sections – in dialogue with civil parties, donors, and Cambodian and international NGOs – design and secure funding for reparations as part of final submissions to the Trial Chamber (prior to the judgement). The 13 reparations measures ultimately set out in the 002/01 Trial Chamber Judgment included: projects to build education centers in multiple locations within Cambodia, funding to publish the court's full judgment in Khmer, and funding for cross-cultural testimonial therapy and the development of self-help groups amongst victims. The Royal Cambodian Government recently declared an annual Day of Remembrance in direct response to Case 002 reparations calls (see Hughes and Elander 2016).

In Timor-Leste, calls for reparations have repeatedly been made in a context in which "endemic poverty is experienced as a shattering of expectations and popular beliefs that good things would flow from independence" (Kent 2011, 444). Compounding this disappointment and frustration is the lack of political reception given to the CAVR report. The parliament has repeatedly postponed discussion of draft laws that propose the establishment of a reparations program and an 'Institute of Memory' to implement other CAVR recommendations. There has also been little formal recognition of its substantive findings, findings that drew on widespread testimonial labor. One of the explicit recommendations of the CAVR was that Indonesia, along with nation-states that provided financial, military and diplomatic support to Indonesia during its occupation of Timor-Leste, should now provide reparations to the country (CAVR 2005, 42).

Within Timor-Leste, state-supported memorialization has largely sought to 'valorize' those considered 'veterans' of the struggle for independence, in particular Falintil[6] forces, without threatening good relations between Timor-Leste and Indonesia (see Leach 2010). These measures and activities have "excluded those who are not deemed to have participated in the resistance struggle, including women, self-described victims, and the rural poor" and their present-day needs (Kent 2011, 440–441; see also Loney 2012). Arguably, while stability and development aspirations are being served, some needs for wider justice forms are not. Although not uncommon to post-conflict nation-states, this formal course of action risks producing slow burn dissent on the part of those who perceive themselves to be excluded by such policies.

As such, there is an emergent geography of symbolic reparations, remembrance and customary practice being undertaken by communities that "unsettles official attempts to mark a clear break between the past and the future" (Kent 2011, 436). The inauguration of 'minor' memorials not only honors and cares for the dead, but also provides opportunities for communities to restate specific or variant versions of 'national historical' or 'international historical' events. Judith Bovinsiepen contributes an account from remote Funar in Timor-Leste, where people are "re-engaging with the recent past by reinvigorating ancestral practices" in ways that are not immune from "disputes and contestations" (Bovinsiepen 2016, 16–17). Where these sites, practices and commemorations enlist the support of elite political representatives, they may also form strategic grounds for requesting greater respect, services or amenities from the central government.[7]

In Cambodia too, communities spend precious financial and material resources on fulfilling their customary obligations to the dead. In Cambodia these practices center on the annual festival of the ancestors known as *pchum ben*, when offerings are brought to the temple in order to feed the spirits of those who have passed on. Local level memorials to victims of 'the Polpotists' built in the immediate post-1979 period, usually in the form of interred human remains preserved inside a *stupa* and located inside a temple, often provide a focus for these rituals (Jarvis 2015).

Conclusion

The slogan 'no reconciliation without justice' continues to rally those in Timor-Leste demanding prosecutions of Indonesian military figures and thoroughgoing reparations, campaigns that are indicative of "widespread community disillusionment" with the justice process so far attempted (Kent 2012a, 16). In Cambodia, where prosecutions and convictions of high-ranking former Khmer Rouge have occurred, and where there is widespread support for the ECCC, emphasis appears to be shifting to a consideration of the legacies of this tribunal.

The ECCC undoubtedly plays its own role as a 'development agent,' given the economic boost that the tribunal has provided to the Phnom Penh economy (albeit within existing inequalities between the expatriate and local economy). There is an emphasis on building local 'capacity' in legal professionals and legal support staff. There is also a perception, especially from international perspectives, that the tribunal, as an internationally legitimized process for 'dealing with the past,' will increase Cambodia's attractiveness to international investors. This particular development discourse was given voice by the then US Ambassador to Cambodia, Joseph A. Mussomeli, at the time of the ECCC's inception, when he stated that:

> A Khmer Rouge Tribunal is a necessary first step to healing the three-decades-old wound that continues to fester. There will remain severe limitations on how far Cambodia can progress and reform until some degree of justice is rendered.
> (Mussomeli 2006, 51)[8]

Apparently mired in the past, Cambodia in this Orientalist view is like a patient beleaguered by a festering sore, where only the balm of justice (externally rendered) will bring on healthy conditions for reform and development.

More than a decade later, a large part of the ECCC legacy discussion centers on the reparations that have been recognized by the Trial Chamber and supported by various sections of the ECCC. Reparations are taking the form of donor-dependent projects, because the ECCC has no dedicated funds for reparations, and no wider Trust Fund or government funding mechanism exists. Only reparations that have already found external funding have been recognized in the most recent judgment. But a donor-funded, project-based approach is arguably alien to established legal and sociolegal understandings of reparation, and risks the hard-won popular understanding of the tribunal as a Cambodian legal process, not a development process. The public comments of one victim association, *Ksaem Ksan*, are instructive in this regard. The association's statement reads: "The vision of our association is neutral, not on the government side or the NGO side" (Chum 2012, 49). The decision to have reparations negotiated between the tribunal, NGOs (and their donors) and victims, rather than developed, funded and enforced from 'within' the tribunal, risks at least the perception that such measures primarily benefit 'the NGO side.' On this issue, however, it is far too soon to say.

One of the major differences between the two country contexts discussed here is that Timor-Leste's processes were initiated at the height of UN mainstreaming of so-called 'transitional justice' approaches into state building and peace building interventions. Timor-Leste was considered to be 'transitioning' to democracy, and transitional justice literatures, mechanisms and experts were enthusiastically deployed. Lia Kent has chronicled the particular effects and affects of this approach in Timor, noting that current memory practices are, in many parts of the country, "far more dynamic, contested and open-ended than narratives of retribution, restoration, transition and 'closure' imply" (Kent 2011, 455). This revival and adaptation of customary practices is also seen in other spheres of society and politics in Timor-Leste, whether as claim to environmental governance (see Palmer and de Carvalho 2008), in re-establishing lived connections with ancestral land (Bovinsiepen 2016), or in what Lisa Palmer has more recently termed 'spiritual ecology' (Palmer 2015).

Cambodia's tribunal was negotiated on very different terrain, of neither state building nor peace building. As well, the negotiating parties were a more tribunal-experienced UN and a national government that saw the reconstruction of Cambodia not only as complete but as a self-made achievement, often reminding the UN negotiators of the organization's failure to act sooner on the question of Khmer Rouge crimes post-1979 (see Fawthrop and Jarvis 2004). Narratives of 'transitional justice' have not taken root in the same way amongst the international non-governmental organizations supporting the work of the tribunal, many of which have preferred instead to 'monitor' its progress and remain largely agnostic about its worth and effectiveness (but see Hinton 2013). Thus it may be that the kind of disappointment widely expressed in Timor-Leste will not be felt in Cambodia because its national community was, in the main, stable and nearly three-decades 'reconstructed' at the outset of its most major legal process, the ECCC.

Finally, both Cambodia and Timor-Leste raise questions of the fit between growing collective memory practices and a national development context in which increasing atomization and individualization is economically rewarded (C. Hughes 2009). And what too of the fit between ideas of culpability beyond borders in a regional context in which ASEAN promotes "conflict avoidance but not conflict resolution" (Narine in McGregor 2008, 89). Future research would do well to marry a critical geopolitics of justice approach with critiques of neoliberalizing development, given that at least some of the donor support for the justice processes herein described appears to be founded on a perception of future economic benefit. If Orientalism lives on in perceptions of 'cultures of violence' in places like Cambodia (Springer 2009), then to what extent do these tribunals serve to have 'cultural violence' writ large, so as to further license the penetration of global capital as a 'civilizing' force. This is to return to Caroline Hughes' thesis of the accountability and economic intervention nexus to ask whether this condition might extend beyond peacekeeping periods to the keeping of economic peace that so much of international governance now seems bent toward.

Notes

1 These calls have included those of the families of Australian journalists who were caught up and killed in the 1975 invasion. See www.abc.net.au/news/2015-10-16/balibo-five-relatives-call-for-investigation-to-be-reopened/6859058. Accessed 12 November 2015.
2 The violent repressions and killings that began in 1965 in Indonesia, and continued for many years after, have yet to be fully chronicled (but see Kammen and Mcgregor 2012). The recent International People's Tribunal 1965, held in The Hague, noted that "500,000 to one million people accused of being members or supporters of the Communist Party of Indonesia (PKI) were murdered, and many hundreds of thousands of people were detained without trial, perished or exiled." The Tribunal's mandate was to "examine the evidence for these crimes against humanity, develop an accurate historical and scientific

record and apply principles of International Law to the collected evidence." See http://1965tribunal.org/about/press-release/. Accessed 24 November 2015.
3 Other instances include the violence of colonial withdrawals across the region, the conflict in southern parts of the Philippines, and state violence in West Papua and Myanmar. The actions of the US and its allies in Vietnam during the early part of the Vietnam War were, however, examined by the Russell Tribunal on Vietnam, a people's tribunal convened in Paris in 1966 and 1967 (see Boyle and Kobayashi 2015).
4 In Cambodia, in the secret bombing campaign of the country prior to Khmer Rouge rule (see Owen and Kiernan 2006), and in Indonesia through financial and material support before, during and after its occupation of Timor-Leste (see Nevins 2002).
5 See *Nuon Chea and Khieu Samphan sentenced to life imprisonment for crimes against humanity* by the ECCC Trial Chamber. Available at: www.eccc.gov.kh/en/articles/nuon-chea-and-khieu-samphan-sentenced-life-imprisonment-crimes-against-humanity. Accessed 18 January 2015.
6 Forças Armadas da Libertação Nacional de Timor-Leste, originally the military wing of the Frente Revolucionária de Timor-Leste Independente, the 'Fretilin' political party.
7 Lia Kent discusses these practices, as well as the establishment of victim groups, in Covalima, Liquica and Lautem, as well as Dili (see Kent 2011).
8 Speech of Joseph A. Mussomeli, quoted in *Searching for the Truth,* special English edition, first quarter 2006, DC-Cam, Phnom Penh. In an "Ask the Ambassador" web forum just over two years later, however, Mussomeli noted that "Cambodia recently has had one of the highest GDP growth rates in the world and has one of Asia's most liberal regimes for foreign investors." See http://2001-2009.state.gov/r/pa/ei/ask/99222.htm. Accessed 1 May 2016.

References

Bovinsiepen, J. (2016). *The land of gold: Post-conflict recovery and cultural revival in Independent Timor Leste.* New York: Cornell University Press.
Boyle, M. and Kobayashi, A. (2015). In the face of epistemic injustices? On the meaning of people-led war crimes tribunals. *Environment and Planning D: Society and Space,* 33(4), pp. 697–713.
Chum, M. (2012). *Survivor: The triumph of an ordinary man in the Khmer Rouge genocide,* Translated by S. Sim and S. Kimsroy. Phnom Penh: Documentation Center of Cambodia.
Ciorciari, J. D. (2009). Introduction. In J. D. Ciorciari and A. Heindel, eds., *On trial: The Khmer rouge accountability process.* Phnom Penh: Documentation Center of Cambodia, 13–29.
Cohen, D. (2007). Hybrid justice in East Timor, Sierra Leone, and Cambodia: Lessons learned and prospects for the future. *Stanford Journal of International Law,* 43, pp. 1–38.
Comissão de Acolhimento, Verdade e Reconciliação (CAVR). (2005). [Commission for Reception, Truth and Reconciliation in Timor-Leste] 'Part 2: The Mandate of the Commission' in *Chega!* [Stop, enough, no more!] *The Report of the CAVR.* Available at: www.cavr-timorleste.org/chegaFiles/finalReportEng/02-The-Mandate-of-the-Commission.pdf. Accessed 30 November 2015.
Dy, K. (2007). *A history of democratic Kampuchea.* Phnom Penh: Documentation Center of Cambodia.
Elander, M. (2014). Education and photography at the Tuol Sleng Museum. In: P. D. Rush and O. Simic, eds., *The arts of transitional justice.* Springer: New York, 43–62.
Extraordinary Chambers in the Courts of Cambodia (ECCC). (2004). *Law on the establishment of the extraordinary chambers in the courts of Cambodia, with inclusion of amendments as promulgated on 27 October 2004 (NS/RKM/1004/006).* Phnom Penh, ECCC.
Extraordinary Chambers in the Courts of Cambodia (ECCC) and CHRAC. (2012). *Hybrid perspectives on legacies of the ECCC.* Phnom Penh: Conference Report and Recommendations, ECCC.
Fawthrop, T. and Jarvis, H. (2004). *Getting away with genocide: Elusive justice and the Khmer Rouge Tribunal.* London: Pluto Press.
Gottesman, E. (2003). *Cambodia after the Khmer Rouge: Inside the politics of nation building.* New Haven, CT and London: Yale University Press.
Heindel, A. (2009). Overview of the extraordinary chambers. In: J. D. Ciorciari and A. Heindel, eds., *On trial: The Khmer rouge accountability process.* Phnom Penh: Documentation Center of Cambodia, 85–124.
Hinton, A. (2013). Transitional justice time: Uncle San, Aunty Yan and outreach at the Khmer Rouge Tribunal. In: D. Mayersen and A. Pohlman, eds., *Genocide and mass atrocities in Asia: Legacies and prevention.* London and New York: Routledge, 185–208.

Hughes, C. (2009). *Dependent communities: Aid and politics in Cambodia and East Timor*. Ithaca, NY: Cornell University Press.

Hughes, R. (2008). Dutiful tourism: Encountering Cambodia's genocide. *Asia Pacific Viewpoint*, 49(3), pp. 318–330.

Hughes, R. (2015). Ordinary theatre and extraordinary law at the Khmer Rouge tribunal. *Environment and Planning D: Society and Space*, 33(4), pp. 714–731.

Hughes, R. and Elander, M. (2016). Justice and the past: The Khmer Rouge Tribunal. In: K. Brickell and S. Springer, eds., *The handbook of contemporary Cambodia*. London and New York: Routledge, 42–52.

Jarvis, H. (2014), "Justice for the deceased": Victims' participation at the extraordinary chambers in the courts of Cambodia. *Genocide Studies and Prevention: An International Journal*, 8(2), pp. 19–27.

Jarvis, H. (2015). Powerful remains: The continuing presence of victims of the Khmer Rouge regime in today's Cambodia. *Human Remains and Violence*, 1(2), pp. 36–55.

Jeffrey, A. (2011). The political geographies of transitional justice. *Transactions of the Institute of British Geographers*, 36, pp. 344–359.

Kammen, D. and Mcgregor, K. (eds.) (2012). *The contours of mass violence in Indonesia, 1965–68*. Singapore: NUS Press.

Kent, L. (2011). Local memory practices in East Timor: Disrupting transitional justice narratives. *The International Journal of Transitional Justice*, 5, pp. 434–455.

Kent, L. (2012a). *The dynamics of transitional justice: International models and local realities in East Timor*. London and New York: Routledge.

Kent, L. (2012b). Interrogating the 'gap' between law and justice: East Timor's serious crimes process. *Human Rights Quarterly*, 34(4), pp. 1021–1044.

Leach, M. (2010). Writing history in post-conflict East Timor. In: M. Leach, N. Canas Mendes, A. B.da Silva, A. da Costa Ximenes and B. Boughton, eds., *Hatene kona ba Compreender Understanding Mengerti: Timor Leste/Proceedings of the 'Understanding Timor-Leste' Conference*. Melbourne: Timor-Leste Studies Association/Swinburne Press, 124–130.

Ledgerwood, J. (2008). Ritual in 1990 Cambodian political theatre: New songs at the edge of the forest. In: A. Hansen and J. Ledgerwood, eds., *At the edge of the forest: Essays on Cambodia, history and narrative, in honor of David Chandler*. Ithaca, NY: Cornell Southeast Asia Program Publications, 195–220.

Linton, S. (2010). Post conflict justice in Asia'. In: M. Cherif Bassiouni, ed., *The pursuit of international criminal justice: A world study on conflicts, victimisation and post-conflict justice*. Brussels: Intersentia Nv, 515–753.

Loney, H. (2012). Women's activism in Timor-Leste: A case study on fighting women. In: M. Leach, N. Canas Mendes, A. B.da Silva, B. Boughton and A. da Costa Ximenes, eds., *Peskiza foun kona ba Timor-Leste: New Research on Timor Leste/Proceedings of the Timor-Leste Studies Association Conference, Centro Foramacao Joao Paulo II, Comoro, Dili, Timor-Leste, 30 June–1 July 2011*. Melbourne: Timor-Leste Studies Association/Swinburne Press, 265–269.

McGregor, A. (2008). *Southeast Asian development*. London and New York: Routledge.

Mendez, P. K. (2009). The new wave of hybrid tribunals: A sophisticated approach to enforcing international humanitarian law or an idealistic solution with empty promises? *Criminal Law Forum*, 20, pp. 53–95.

Mussomeli, J. A. (2006). *The Khmer rouge genocide and the need for justice' searching for the truth first quarter, English edition*. Phnom Penh: DC-Cam. Available at: www.d.dccam.org/Projects/Magazines/Image_Eng/pdf/1st_Quarter_2006.pdf. Accessed 3 May 2016.

Mydans, S. (1999). Choeung Ek journal: A word to the dead: we've put the past to rest. *The New York Times*, 21 May. Available at: www.nytimes.com/1999/05/21/world/choeung-ek-journal-a-word-to-the-dead-we-ve-put-the-past-to-rest.html. Accessed 24 November 2015.

Nevins, J. (2002). (Mis)representing East Timor's past: Structural-symbolic violence, international law, and the institutionalization of injustice. *Journal of Human Rights*, 1(4), pp. 523–540.

Nevins, J. (2005). *A not so distant horror: Mass violence in East Timor*. Ithaca, NY: Cornell University Press.

Owen, T. and Kiernan, B. (2006). Bombs over Cambodia. *The Walrus*, October, 62–69. Available at: www.yale.edu/cgp/Walrus_CambodiaBombing_OCT06.pdf. Accessed 6 July 2015.

Palmer, L. (2015). *Water politics and spiritual ecology: Custom, environmental governance and development*. London and New York: Routledge.

Palmer, L. and de Carvalho, D. (2008). Nation building and resource management: The politics of 'nature' in Timor Leste. *Geoforum*, 39, pp. 1321–1332.

Pham, P. N., Vinck, P., Balthazard, M., Hean, S. and Stover, E. (2011). *After the first trial: A population-based survey on knowledge and perception of justice and the Extraordinary Chambers in the Courts of Cambodia*. Berkeley: University of California Human Rights Centre.

Pham, P. N., Vinck, P., Balthazard, M., Strasser, J. and Om, C. (2011). Victim participation at the trial of Duch at the extraordinary chambers in the courts of Cambodia. *Journal of Human Rights Practice*, 3(3), pp. 264–287.

Poluda, J., Strasser, J. and Chim, S. (2012). Justice, healing and reconciliation in Cambodia. In: B. Charbonneau and G. Parent, eds., *Peacebuilding: Memory and reconciliation*. London and New York: Routledge, 91–109.

Sperfeldt, C. (2013). The role of Cambodian civil society in the victim participation scheme of the extraordinary chambers in the courts of Cambodia. In: T. Bonacker and C. Safferling, eds., *Victims of international crimes: An interdisciplinary discourse*. The Hague: T. M. C. Asser Press, 345–372.

Stover, E., Balthazard, M. and Koenig, K. A. (2011). Confronting *Duch*: Civil party participation in Case 001 at the Extraordinary Chambers in the Courts of Cambodia. *International Review of the Red Cross*, 93, pp. 503–546.

Williams, S. (2012). *Hybrid and internationalized criminal tribunals: Selected jurisdictional issues*. Oxford: Hart Publishing.

9
CIVIL SOCIETY PARTICIPATION IN THE REFORMED ASEAN

Reconfiguring development

Kelly Gerard

Amid much fanfare, the ASEAN Economic Community was signed into existence in December 2015 by the grouping's ten heads of state. This event was the final stage of a 15-year transformation, prompted by the regional economic crisis of the late 1990s. The ASEAN Economic Community entails the free movement of goods and services and the freer movement of capital and labor, and has been promoted as transforming member states into "a single market and production base, a highly competitive economic region, a region of equitable economic development, and a region fully integrated into the global economy" (ASEAN 2008).

Alongside the creation of the single market, ASEAN elites sought to dispel the grouping's image as a 'club of dictators,' attained through decades of regional governance being characterized by closed-door meetings and tacit agreements, alongside the various tactics employed by the grouping's authoritarian regimes to silence dissent. With this aim, ASEAN elites promoted the grouping's reform as being 'people-oriented,' and committed to strengthening good governance, the rule of law and human rights in pivotal agreements such as the ASEAN Charter.

To ascertain the development implications of this transformation, this chapter examines changing modes of political participation in regional governance. The first section charts the context for this transformation, describing the regional economic crisis and subsequent reforms to ASEAN's governance structure, notably the development of a regulatory framework. The chapter then looks to the strategies employed to re-legitimize this political project, examining the participatory channels established to engage civil society organizations. This section demonstrates how these channels have been structured to include groups that are amenable to the interests of ASEAN elites, and marginalize dissenting voices.

The chapter highlights how ASEAN's reform is characteristic of neoliberal development policy, where liberalization and the deepening engagement of international capital and global markets has been pursued alongside measures to socially embed these processes and manage the conflicts generated (see Carroll and Jarvis 2015). The chapter argues that despite rhetoric of inclusion, this political project remains calibrated to defending powerful interests, thereby furthering, rather than challenging, regional inequalities. However, despite its limited potential to advance more equitable and empowering forms of development, ASEAN's reform has provided a rallying point for civil society organizations across the region, with its conflicts and contradictions driving resistance.

Reconfiguring regional governance

The regional economic crisis of the late 1990s had far-reaching and long-lasting impacts. For those ASEAN countries that were hardest hit – Thailand, Indonesia and Malaysia – it resulted in widespread unemployment, a decline in real wages and a spike in poverty. The sudden impacts of the crisis on people's livelihoods and welfare generated widespread criticism of governments, and people mobilized in support of greater transparency and accountability. The most significant of these domestic upheavals was in Indonesia, where the Suharto regime collapsed after 30 years of rule. This was fueled by the violent riots that took place in many parts of Indonesia in May 1998, as people organized in opposition to the spike in prices and widespread unemployment in a context where Suharto's opponents within the government and military were vying for power. In Malaysia the dismissed Deputy Prime Minister Anwar Ibrahim led the *Reformasi* movement, advocating for the resignation of Prime Minister Mahathir Mohamad and the end of corruption and cronyism, with this conflict continuing to reverberate through Malaysian politics over the subsequent decade. Similarly, protests in Thailand forced Chavalit Yongchaiyudh to resign and eventually made way for Thaksin Shinawatra's rise to prime minister after the Thai Rak Thai Party was established in 1998. Elites in these and other affected countries pursued some social and political reforms to draw investors back to the region and restore domestic stability. ASEAN's newer members,[1] however, faced little imperative for reform because of the weakness of opposition forces in these countries (see Jones 2015).

These political mobilizations against governments emerged alongside ASEAN's legitimacy crisis. The organization's conspicuous absence from the recovery, together with the involvement of the IMF, raised questions and criticisms. These concerns were publicly acknowledged by ASEAN elites, including the secretary-general at the time, Rodolfo Severino, who noted: "The same commentators who used to assume a future of continuous growth for ASEAN now seem to believe that ASEAN can do nothing right – or can just do nothing" (Ahmad and Ghoshal 1999, 759).

To encourage the return of capital to the region and promote economic recovery, ASEAN elites embarked on a market-building program. In the post-crisis years elites agreed to a series of reforms intended to intensify regional integration and create an integrated and liberal market – the ASEAN Economic Community. To facilitate these market-building reforms, regional governance was reconfigured around a regulatory framework where a diversity of state and non-state actors collaborate through networks to harmonize domestic standards. This regional regulatory framework is embodied in the networks that have been established across a widening set of issues, including transnational crime, forest law enforcement and governance, consumer protection, and competition policy. Elliott notes that governments reconfigured regional governance around regulatory networks so as to increase their influence and the efficacy of activities: "Member states have *instigated* these arrangements to enhance their authority and the quality of formal rules. ASEAN policy-makers have made explicit strategic and political claims for the advantages of transgovernmental network arrangements" (Elliott 2012, 49, emphasis in original).

These various regulatory networks were situated in one of three 'communities.' ASEAN's restructure was agreed to through the ASEAN Charter, this being the grouping's constituent instrument that elites committed to developing after the crisis. While the Charter was a nod to a more rule-based form of governance, its 'opt out' clauses in dispute settlement mechanisms suggest these continue to function as political rather than legal instruments, indicating the persistence of 'rule by law' rather than 'rule of law' (Gerard, forthcoming). The Charter outlined ASEAN's restructure around three 'communities,' replicating the language employed in the European integration process, with the ASEAN Economic Community sitting alongside the

ASEAN Political-Security Community and the ASEAN Socio-Cultural Community. Regulatory networks are housed in one of these three communities and report to the relevant sectoral body. For example, the ASEAN Intergovernmental Commission on Human Rights that comprises academics, bureaucrats and civil society representatives is situated in the ASEAN-Political Security Community and reports to the ASEAN Foreign Ministers.

This shift to a regional regulatory framework mirrored transformations in statehood across members and their embedded 'politics of competitiveness,' where "the orientation of these regimes reflects national trajectories that have seen the outflanking or defeat of more radical class and developmental projects by enthusiastic proponents of capitalist development" (Cammack 2009, 269; see also Carroll and Jarvis 2015). Since the late 1980s, under the guidance of leading development institutions including the World Bank, 'good governance' directives increasingly shaped state management, with technocrats ever more involved in decision-making. New technocratic forms of government that were intended to enhance those institutions that aid markets increasingly characterized states' operations, "replacing politics with 'good governance'" (Robison 2012, 9–10; see also Carroll this volume). States' new political projects were increasingly characterized by 'authoritarian liberalism,' where a strong state apparatus is combined with a liberalizing economy that is organized around regulatory modes of governance, and the state serves as "guardian of the market order" to ensure market credibility (Jayasuriya 2003, 205).

ASEAN's restructure around a regulatory framework reflected some of these transformations in state management. Jayasuriya (2015) notes that regulatory regionalism leads not to supranational governance but to the creation of regional governance spaces both within and between states. These regional governance spaces cut across national territorial boundaries, resulting in "the molding and shaping of new forms of institutional spaces within the state," and thus transforming domestic political processes (Jayasuriya 2015, 518). Crucially, by encompassing a diversity of actors, including representatives from state agencies, the private sector, civil society organizations, think tanks, academia and scientific communities, these regulatory networks enable a greater diversity of social groups to shape regulatory regimes, raising the question of who participates in these spaces, and how.

The regulatory reforms to establish the single market entail the rescaling of issues to the regional scale, with these governed through regulatory networks comprising a mix of state and non-state actors. The rescaling of governance to regional regulatory networks is significant because each scale (whether regional, national, subnational or local) presents a different configuration of actors, resources and political opportunities. Rescaling the governance of an issue can privilege particular interests at the expense of others, and social actors consequently attempt to rescale issue governance to the scale that is most compatible with their interests (see Hameiri and Jones 2015; Keil and Mahon 2009).

For the ASEAN Economic Community, the conflicts arising from this attempt to reorganize the regional economy are reflected in the very uneven liberalization that has occurred, with a few sectors highly integrated and others remaining nationally bound. The various assessments published by the ASEAN Secretariat, governments and private actors note that the process of economic integration has generally been lackluster, with governments failing to meet various targets. As described by Jones (2015), this variegated integration reflects the region's political economy and the conflicts that have emerged around this attempt to rescale governance. Establishing an integrated regional market entails extensive regulatory reform that impacts the distribution of power and resources among the region's politico-business alliances. Historical processes of state-led development in Southeast Asia saw the fusing of state and business interests, such that the latter has developed a vast influence over public policy (see Rodan et al. 2006).

These predatory arrangements made particular forms of economic growth possible in the 1980s and 1990s; however this degree of influence has also meant that this recent attempt to reorganize the regional economy by liberalizing cross-border flows of goods, services, labor and capital has been strongly contested by relevant coalitions, with their support or opposition determined by the potential to gain from this process.

The governance of labor migration through the ASEAN Economic Community illustrates how conflicts over regulatory reforms have resulted in a highly uneven integration. This unevenness can be seen in, first, sectoral differences across agreements to facilitate high-wage migration; and second, in the differences across attempts to regulate high- and low-wage migration. On sectoral differences in high-wage migration, the ASEAN Economic Community entails 'freer' movement of labor, referring specifically to 'skilled' labor.[2] The movement of high-wage labor has been facilitated through Mutual Recognition Arrangements where governments have agreed the conditions under which people employed in specific professions can move between countries. Eight have been negotiated, including one for architects and one for medical professionals. While the Mutual Recognition Arrangement for architects facilitates labor movement, the one for medical professionals is a protectionist arrangement, safeguarding medical jobs in each country. These differences in the regional regime governing high-wage labor movement reflects the political economy of each sector, where each government's bargaining position has been determined by whether the sectoral national body for each country has a direct interest in supporting labor migration (Jones 2015). In the case of architects, regional labor movement builds on existing arrangements that are supported by national bodies, and they have subsequently designed their Mutual Recognition Arrangement to facilitate labor movement. However, national bodies for medical professionals, while rhetorically supporting labor movement, have developed their Mutual Recognition Arrangement to protect medical jobs in each country (Sumano 2013). National coalitions that would be negatively impacted by integration have opposed the required regulatory reforms while those that stand to gain have supported reforms, resulting in a highly uneven integration.

The case of labor migration also highlights the project's uneven integration through the differences in attempts to regulate high- and low-wage migration. Importantly, these differences are also indicative of how the ASEAN Economic Community is organized around the substantial inequalities that exist across communities in Southeast Asia, and that this project furthers these inequalities, rather than challenges them. The movement of high-wage labor builds on existing national regulatory regimes where countries seek to attract high-wage migrants by structuring entitlements according to income. For example, Singapore's 'Personalized Employment Pass' is for individuals earning in excess of SG$18,000 per month prior to commencing work in Singapore, and they are granted the flexibility to change jobs, bring family members and apply for permanent residency. For people in the categories of 'Work Permits for Foreign Workers,' employed in the construction, manufacturing, marine, process and service sectors, and 'Work Permits for Foreign Domestic Workers,' they are employed on a transient basis, are tied to a specific employer and are not permitted to marry a Singaporean citizen or deliver a child in Singapore (Ministry of Manpower 2015). This revolving door for low-wage workers enables governments to increase numbers during economic expansions and decrease the intake during economic declines (Kaur 2010, 10).

While high-wage migration is being facilitated in some sectors through Mutual Recognition Arrangements, attempts to regionally regulate low-wage migration – and, crucially, to improve working conditions – have stalled. ASEAN's governance of migrant workers' rights is of key relevance for intra-ASEAN migration, given that a majority of the substantial increase in intra-ASEAN migration in recent decades has comprised low-wage migration, and much of it irregular.

Negotiations for an instrument governing the rights of migrant workers commenced in 2007 with the establishment of a network to design and implement regulatory reforms, the ASEAN Committee on the Implementation of the ASEAN Declaration on the Promotion and Protection of Migrant Workers. However, negotiations deadlocked over key issues, notably whether the instrument would be legally binding and if it would cover irregular migrant workers (Bacalla 2012; Forum Asia 2013). These are not minor points of disagreement and would determine whether the instrument would have scope to improve conditions for low-wage migrants. Hence, despite elites' rhetoric on inclusive development through the ASEAN Economic Community, the regulatory reforms for labor migration are beneficial for high-wage workers and do not yet improve the conditions faced by low-wage migrants. Labor migration governance through the ASEAN Economic Community enables regional elites to assert they are working to address the issues associated with the significant expansion in intra-ASEAN low-wage migration of recent years, while continuing to facilitate flows and in doing so, enjoy the benefits this trend has conferred.

The differing approaches to governing high- and low-wage migration thus highlight how economic integration remains calibrated to defending powerful interests and furthers, rather than challenges, regional inequalities. As a consequence, civil society groups have increasingly sought to influence regional governance, responding to the conflicts and contradictions of this project. The subsequent section examines the participatory channels that have been established for civil society organizations, and how these channels have been structured to facilitate representation, but not contestation.

Widening, yet limiting, regional governance[3]

Through ASEAN's restructure, a greater diversity of actors now contributes to regional governance, and civil society organizations[4] are one such actor. This section first charts the inclusion of civil society organizations in regional governance, and then demonstrates how this widening of policymaking has not enabled dissenting voices to shape policy or advance alternatives.

Prior to the regional economic crisis, civil society organizations paid little attention to ASEAN given its limited impact on their activities and its highly elitist and exclusive mode of governance. This is evident in ASEAN elites' limited interactions with non-state actors prior to the crisis. ASEAN elites engaged with, first, the ASEAN Chamber of Commerce and Industry that was established in 1972 and comprised the apex business chamber of each state. Second, they consulted with some professional bodies, such as the Federation of ASEAN Public Relations Organizations, through the affiliation system that was established in 1979. Third, ASEAN elites engaged with economic and security technocrats from the late 1980s through the ASEAN Institutes of Strategic and International Studies (ASEAN-ISIS), with these engagements driven by the need for ASEAN elites to address issues arising from the declining presence of the United States in the region along with the dislocation and disruption being experienced because of rapid industrialization.

This elitist and opaque mode of governance throughout ASEAN's first few decades occurred in the context of a highly atomized and weak civil society across the region. As charted by Hewison and Rodan (2012), the Left was highly significant in the nationalist struggles of the 1930s and after the Second World War. However, the space for political expression was attacked and dissenting views repressed, first by colonial authorities and then by authoritarian states, such that radical attempts for socio-economic and political transformation abated. The Cold War then presented a decidedly inhospitable environment for the political Left.

The rise of 'authoritarian liberalism' across the region from the late 1980s brought a rapid and predatory mode of capitalist development, along with a different set of conflicts over political representation. Those whose fortunes improved with the region's economic development began demanding political participation, which saw the renewal of civil societies. These forces were vocal critics of governments after the regional economic crisis, leading calls for measures to address corruption and greater transparency and accountability. Ahmad and Ghoshal note: "The shock of the crash prompted widespread challenges to the political and social status quo, with a bolder and better-educated middle class challenging the paternalistic order of the past" (1999, 767). However, these new demands for political representation were largely compatible with market reforms. Activism centered on the protection of rights, liberty and representative forms of government, and was detached from the region's historical movements that had organized around reforming the structural sources of social inequality (Hewison and Rodan 2012, 25; see also Clammer 2003). Activism was thus reinvigorated over this period, but collective action was focused on supporting and improving market reforms, rather than challenging them. Civil society organizations were thus revitalized with the region's economic development in the late 1980s, albeit along far less radical agendas than during the 1930s and 1940s, and these groups were a critical organizing force after the crisis.

ASEAN elites sought to re-legitimize their regional project, and its reform, with these discontented social forces. Vision 2020 was released in 1997, and it committed member states to creating a "community of caring societies", where "civil society is empowered and gives special attention to the disadvantaged, disabled and marginalized and where social justice and the rule of law reign," and countries are "governed with the consent and greater participation of the people with its focus on the welfare and dignity of the human person and the good of the community" (ASEAN 1997). The participatory component of ASEAN's reform agenda was then affirmed in the Vientiane Action Program, signed in 2004, where governments endorsed the "effective participation of family, civil society, and the private sector in tackling poverty and social welfare issues" (ASEAN 2004). Finally, with the Charter of 2008, ASEAN elites agreed, "ASEAN may engage with entities which support the ASEAN Charter, in particular its purposes and principles" (ASEAN 2007).

From the early 2000s, ASEAN elites began complementing this inclusive rhetoric with new participatory channels for civil society organizations. The affiliation system, noted above, commenced from 1979, and organizations – predominantly professional associations – could apply to become affiliated to ASEAN and in doing so, access some participatory functions. However, in the early 2000s ASEAN elites also endorsed two new modes of participation: informal consultations on specific issues, with the most high-profile of these being the consultations conducted for the Charter; and five annual sectoral forums intended to facilitate dialogue between elites and civil society organizations. These are known as the GO-NGO forums (meaning government organization-non-governmental organization) and are held on migrant labor, rural development and poverty eradication, agriculture and forestry, social welfare and development, equal opportunity for people with disabilities, and children's issues.

ASEAN elites' claims to inclusive and participatory governance along with their shift to focus on issues relevant to the region's civil society organizations – including environmental management, human rights and migrant workers' rights – attracted much interest. ASEAN's reform was an important catalyst in the regionalization of advocacy, and some groups subsequently sought to shape regional governance in alignment with their objectives, with elites' 'people-oriented' rhetoric suggesting their contributions would be welcomed. Groups began collaborating across countries, with their common experiences organizing around issues arising from states' narrow pursuit of growth providing fertile ground for collaboration. Issues arising as a consequence of this approach to state management, such as land evictions, deforestation, election monitoring, child trafficking and sex tourism, provided rallying points for collaboration.

Regional networks were established, and ASEAN-focused activities within existing networks developed. A key network that drove the regionalization of advocacy was the Solidarity for Asian People's Advocacy (SAPA) network. SAPA was formally established in February 2006 at its first regional consultation held in Bangkok, and it comprises approximately 100 organizations, both national and regional. Its activities cover the entire Asian region, and it is organized around three sub-regions – Northeast Asia, South Asia and Southeast Asia – with working groups targeting the multilateral processes in each sub-region. In the case of Southeast Asia, the SAPA Working Group on ASEAN forms an organizational umbrella under which various task forces operate. These task forces align with a relevant regulatory network or issue, such as the SAPA Task Force on ASEAN and Human Rights that was formed in August 2007 to unite rights advocates in lobbying the ASEAN Intergovernmental Commission on Human Rights. Similarly, the SAPA Task Force on ASEAN and Migrant Workers was formed in April 2006 to bring together relevant groups to lobby the ASEAN Committee on the Implementation of the ASEAN Declaration on the Promotion and Protection of the Rights of Migrant Workers. Others include the SAPA Task Force on ASEAN and Burma and the SAPA Task Force on ASEAN and Freedom of Information. The SAPA Working Group on ASEAN was the key civil society actor in the consultations that ASEAN conducted during the drafting of the Charter (see Gerard 2015), and it played a central role in the early years of the ASEAN Civil Society Conference,[5] which has become the central parallel summit (see Gerard 2013). In bringing together its very geographically dispersed and issue-specific membership, the SAPA network has contributed in unifying organizations in their attempts to influence ASEAN policy.

While signaling a substantial shift in regional governance, the channels established by ASEAN elites to include civil society organizations in fact continue their prior practice of a highly exclusive and elitist mode of governance. This is because these participatory channels offer greater opportunities for representation, but not opportunities for groups to contest policy or advance alternatives. This narrow space for political contestation is a consequence of three characteristics of these participatory channels. First, there are strict controls over who can participate. In the case of the GO-NGO forums, an informal system of affiliation operates where each country compiles a list of relevant organizations that they wish to invite to participate. Every member of the ASEAN Committee of Permanent Representatives considers each list, and if any of the nominated organizations are deemed contentious by a member state, they are removed and the remaining groups are invited to participate (Gerard 2015). Hence, all participants must receive the endorsement of all member states, making participation in these forums highly exclusive. Many of the participants in these forums are subsequently GONGOs (government-organized non-governmental organizations), these being organizations that are established and/or maintained by states. While the distinction between a GONGO and an independent organization varies, as does the extent of governmental control over an organization's activities, groups with a more contentious agenda are typically not represented in the GO-NGO forums. This mode of participation has been dominated by those groups that are nationally accredited, with this system of selecting participants biased toward groups with formalized and legalized systems of operation and those that do not contest national policies and maintain the favor of governments.

Second, there are strict controls over the nature of participation in these spaces. In the case of the affiliation system, the only form of participation that is guaranteed is the submission of written statements to the ASEAN Committee of Permanent Representatives. All other activities – such as presenting information to an agency, attending meetings or obtaining access to documents – must be requested in writing, and there are no mechanisms for recourse if an application is rejected. The guidelines governing the affiliation system were first agreed upon in 1979, and then revised in 2006 and again in 2011, with this most recent version adopted by the Committee of Permanent Representatives on 5 November 2012. This latest revision did not lead to more substantive ways for civil society organizations to participate in policy-making. Guidelines were revised largely to align them with ASEAN's restructure, such as by referring to the ASEAN Committee of Permanent Representatives, rather than the ASEAN Standing Committee. The revised guidelines also introduced two additional obligations for affiliated organizations. First, their activities must comply with the laws of member countries (ASEAN 2012, article 8). Second, affiliated organizations are to be reviewed every three years by the Committee of Permanent Representatives, and any organization that does not meet its obligations or undertakes activities contrary to ASEAN and member states' aims and principles will have its affiliation revoked (ASEAN 2012, articles 10–11). Unsurprisingly, the list of 58 organizations continues to be dominated by professional associations, such as the ASEAN Cosmetics Association and the ASEAN Association of Radiologists (ASEAN 2016).

Third, spaces for civil society participation in ASEAN are determined by the issue under discussion, limiting the range of political activities by activists. Issues aligned with ASEAN's market-building agenda, such as rural development and poverty eradication, have been the focus of the GO-NGO forums and the informal ad hoc consultations between policymakers and civil society organizations. However, groups working on issues deemed contentious by elites because they directly challenge powerful interests – such as land evictions in Cambodia or the enforced disappearance of environmental activists – have been excluded from all forums for civil society engagement. These three characteristics of participatory channels established by ASEAN – determining who participates, the nature of participation and the issues discussed – narrow the possibilities for those seeking to shape, and contest, regional governance.

Alongside these three participatory channels, ASEAN elites also began to interact with civil society groups through external forums, such as the ASEAN People's Assembly and the ASEAN Civil Society Conference.[6] However, ASEAN elites have not institutionalized these forums, meaning that officials attend only on their terms, utilizing these activities in accordance with their interests. In the case of the ASEAN Civil Society Conference, a regional steering group collaborates with national organizations and, as necessary, the ASEAN chair for that year.[7] This event has two components, one being a forum comprising plenaries and workshops where participants discuss advocacy strategies and collaborate in drafting a 'People's Statement' addressed to ASEAN leaders. At this forum participants also appoint a representative for each country that participates in the second component of the conference, this being the interface meeting where heads of state and civil society representatives engage in some form of exchange, ranging from an informal dialogue to simply the presentation of the 'People's Statement.'

ASEAN elites have over the years expanded their repertoire of tactics to direct the conference and recast it according to their preferences. One strategy governments have employed is replacing the civil society-appointed representatives for the interface meeting with an individual of their choosing. For example, at the interface meeting in Cha-Am in October 2009 the government of Burma replaced Aung Myo Min, the director of the Human Rights Education Institute of Burma, with Police Colonel Sitt Aye, recently appointed as the head of President Thein Sein's legal advisory team (ACSC 2011). Another strategy governments have employed is setting the agenda for the interface meeting, rather than allowing civil society representatives to develop it, drawing on the conference proceedings. For example, in Jakarta in May 2011, the agenda of the interface meeting was confined to the rather non-controversial issue of 'Health toward achieving the Millennium Development Goals and/or Poverty Reduction.' A further strategy has been for governments to apply pressure to the venues where the conference has been held. For example, in Phnom Penh in March 2012 under pressure from the Cambodian government the management of the conference venue opposed some of the scheduled workshops, threatening to cut power and padlock the venue if particular workshops were not canceled, specifically those on Myanmar's human rights abuses, land evictions, the expansion of monoculture plantations and the protection of ethnic minorities' rights to land (Tupas 2012). Through these various strategies, ASEAN elites have recast the boundaries of civil society participation in the conference, seeking to direct this event according to their preferences. These growing intrusions into the event limit civil society groups' potential to advance their objectives and indicate the hollowness of ASEAN's 'people-oriented' commitments.

ASEAN elites have thus complemented their market-building program with channels for civil society organizations to participate in policymaking. These participatory channels, however, have been organized around excluding groups that are not supportive of market-building reforms. By creating spaces for some civil society representatives to contribute their views, while restricting their ability to contest policy, ASEAN elites have created opportunities for representation, but not contestation.

Conclusion: contesting regional governance

This chapter has considered the boundaries of some of the new modes of political participation that have emerged through ASEAN's reconfiguration over the past 15 years so as to ascertain the development implications of this project's reform. The participatory channels established by ASEAN elites function in socially embedding the single market by providing spaces through which conflicts can be managed, and mitigated. Permitting selected civil society organizations to participate and limiting their ability to contest policy, these participatory mechanisms are structured to legitimize ASEAN elites' market-building program, including amenable groups and marginalizing dissenting voices. The limited opportunities for contestation through these channels highlight the importance of examining the implementation and ongoing practice of elites' transformative claims, and also draws attention to the role of ASEAN elites in structuring these new political channels (Gerard 2015). These participatory spaces did not emerge from a vacuum, but were established in response to specific conflicts and structured by ASEAN elites, making it unsurprising that they are organized toward advancing elites' interests (Jayasuriya and Rodan 2007).

With limited opportunities for contestation through these channels, many groups avoid engaging ASEAN altogether and advocate outside of such channels, where they also contest the limited formal opportunities to influence regional governance through ASEAN structures. ASEAN's transformation over the past 15 years has, however, provided a rallying point for groups across the region and a catalyst for the regionalization of advocacy. Various online platforms, including Google and Facebook groups, function in connecting groups across countries. These platforms allow activists to share information on official developments and advocacy strategies, enabling them to continue to exert pressure, and to do so across governance scales.

The ASEAN Civil Society Conference, despite governments' growing incursions into its organization, continues to provide an annual opportunity for groups to 'stocktake' their progress and jointly plan activities. The conference raises awareness of activists' activities through the reporting of the event in both the official media and social media, aided by the creative means through which activists advocate their agendas, such as the various art displays and theater performances held at the 2015 conference in Kuala Lumpur as well as the conference's first protest march. The conference is also a space for sharing ideas about alternative means of shaping ASEAN policy. For example, at the 2015 event, in multiple workshops participants canvassed the notion of a regional strike or day of action, where people would withdraw their labor as a means of communicating their opposition to the ASEAN Economic Community and its inequalities.

Similarly, in response to the Laos government's increasing restrictions over civil society activities along with suspicions regarding its involvement in the disappearance of environmental activist Sombath Somphone in December 2012, the regional steering committee controversially decided not to hold the ASEAN Civil Society Conference in Laos in 2016, as scheduled. The group agreed to instead hold it in Dili in solidarity with civil society organizations in Timor-Leste, given that the country has been seeking to join ASEAN, and its civil society organizations have been involved in the ASEAN Civil Society Conference since 2004 (THC 2016). The ASEAN Civil Society Conference thus highlights activists' persistent presence, and challenge, to the trajectory of Southeast Asian regionalism.

While ASEAN's transformation has not led to more equitable and empowering forms of development for communities in Southeast Asia, and it will likely exacerbate existing inequalities, it has provided impetus for the regionalization of advocacy. Attempts to influence regional governance through direct engagement with ASEAN elites have thus far yielded little. However, the conflicts and contradictions that this project has generated have driven regional collaboration among activists and the development of innovative strategies to collectively contest it.

Notes

1 Termed the 'CLMV countries,' Cambodia, Laos, Myanmar and Vietnam joined ASEAN from 1995 to 1999.
2 ASEAN documentation employs the category of 'skilled' labour. The author acknowledges that the usage of this category and its counterparts – 'unskilled'/'semi-skilled'/'low-skilled' – can often be incorrect, given that it is not unusual for migrant workers to be employed in jobs that do not reflect their qualifications. The use of the terms 'high-wage' and 'low-wage' in this article draws from the Singaporean and Malaysian governments' categorization of migrants' entitlements according to their monthly salary (see Ministry of Manpower 2015; Nah 2012).
3 This section substantively draws on Gerard 2014.
4 ASEAN documentation refers to 'civil society organizations,' and this chapter adopts the same term. For a critical discussion of the use of this term in ASEAN documentation and by regional actors see Gerard 2014: 14–16.
5 This forum has been held under various titles over its existence, including the 'ASEAN People's Forum' and the 'ASEAN Civil Society Conference/ASEAN People's Forum.' This chapter refers to these events using their original title, the 'ASEAN Civil Society Conference.'
6 The ASEAN People's Assembly was first held in 2000 and established by the ASEAN-ISIS network as a forum for dialogue between civil society organizations and government officials, while the ASEAN Civil Society Conference was first held in 2005, led by regional civil society organizations in collaboration with the Malaysian government, given that Malaysia was the ASEAN Chair in that year. For a detailed discussion of political participation at these two events, see Gerard 2013.
7 The ASEAN chairmanship rotates annually between member states, generally moving in alphabetical order. The ASEAN chair serves as the host government and manager of key meetings, including the Leaders' Summits.

References

ACSC [ASEAN Civil Society Conference]. (2011). Statement of representatives of 2011 ASEAN Civil Society Conference/ASEAN People's forum to the informal meeting between ASEAN leaders and civil society. *Burma Partnership*, 7 May. Available at: www.burmapartnership.org/2011/05/statement-of-representatives-of-2011-asean-civil-society-conferenceasean-peoples-forum-to-the-informal-meeting-between-asean-leaders-and-civil-society/. Accessed 1 June 2011.

Ahmad, Z. H. and Ghoshal, B. (1999). The political future of ASEAN after the Asian crisis. *International Affairs*, 75(4), pp. 759–778.

ASEAN. (1997). *ASEAN vision 2020*. Jakarta: ASEAN Secretariat.

ASEAN. (2004). *Vientiane action program*. Jakarta: ASEAN Secretariat.
ASEAN. (2007). *Charter of the association of Southeast Asian nations*. Jakarta: ASEAN Secretariat.
ASEAN. (2008). *ASEAN economic community Blueprint*. Jakarta: ASEAN Secretariat.
ASEAN. (2012). *Guidelines on Accreditation of Civil Society Organisations (CSOs)*. Jakarta: ASEAN Secretariat.
ASEAN. (2016). *Register of accredited Civil Society Organisations (CSOs)*. Available at: www.asean.org/wp-content/uploads/2012/05/Accredited-Civil-Society-Organisations-as-of-25-May-2016.pdf. Accessed 3 June 2016.
Bacalla, T. (2012). ASEAN urged to set up mechanism for migrant rights. *Vera Files*, 10 September. Available at: http://verafiles.org/asean-urged-to-set-up-mechanism-for-migrant-rights/. Accessed 19 October 2014.
Cammack, P. (2009). The shape of capitalism to come. *Antipode*, 41, pp. 262–280.
Carroll, T. and Jarvis, D. S. L. (2015). Markets and development: Civil society, citizens, and the politics of neoliberalism. *Globalizations*, 12(3), pp. 277–280.
Clammer, J. (2003). Globalisation, class, consumption and civil society in South-East Asian cities. *Urban Studies*, 40(2), pp. 403–419.
Elliott, L. (2012). ASEAN and environmental governance: strategies of regionalism in Southeast Asia. *Global Environmental Politics*, 12(3), pp. 38–57.
Forum Asia. (2013). *ASEAN committee on migrant workers*. Available at: http://humanrightsinasean.info/asean-committee-migrant-workers/about.html. Accessed 12 October 2014.
Gerard, K. (2013). From the ASEAN people's assembly to the ASEAN civil society conference: The boundaries of civil society advocacy. *Contemporary Politics*, 19(4), pp. 411–426.
Gerard, K. (2014). *ASEAN's engagement of civil society: Regulating dissent*. Basingstoke, Hampshire and New York, NY: Palgrave Macmillan.
Gerard, K. (2015). Explaining ASEAN's engagement of civil society in policy-making: smoke and mirrors. *Globalizations*, 12(3), pp. 365–382.
Gerard, K. (forthcoming). ASEAN as a "rule-based" community: Business as usual. *Asian Studies Review*.
Hameiri, S. and Jones L. (2015). *Governing borderless threats: Non-traditional security and the politics of state transformation*. Cambridge: Cambridge University Press, 2015.
Hewison, K. and Rodan, G. (2012). Southeast Asia: The left and the rise of bourgeois opposition. In: R. Robison, ed., *Routledge handbook of Southeast Asian politics*. Abingdon and Oxon: Routledge, 25–39.
Jayasuriya, K. (2003). Introduction: Governing the Asia Pacific: beyond the "new regionalism". *Third World Quarterly*, 24(2), pp. 199–215.
Jayasuriya, K. (2015). Regulatory regionalism, political projects, and state transformation in the Asia-Pacific. *Asian Politics & Policy*, 7(4), pp. 517–529.
Jayasuriya, K. and Rodan, G. (2007). Beyond hybrid regimes: More participation, less contestation in Southeast Asia. *Democratization*, 14(5), pp. 773–794. doi:10.1080/13510340701635647
Jones, L. (2015). Explaining the failure of the ASEAN Economic Community: The primacy of domestic political economy. *The Pacific Review*, 29(5), pp. 1–24, 647–670.
Kaur, A. (2010). Labour Migration in Southeast Asia: Migration Policies, Labour Exploitation and Regulation. *Journal of the Asia Pacific Economy*, 15(1), pp. 6–19.
Keil, R. and Mahon, R. (2009). *Leviathan undone? Towards a political economy of scale*. Vancouver: University of British Columbia Press.
Ministry of Manpower. (2015). *Work passes and permits*. Available at: http://beta.mom.gov.sg/passes-and-permits. Accessed 3 February 2015.
Nah, A. M. (2012). Globalisation, sovereignty and immigration control: The hierarchy of rights for migrant workers in Malaysia. *Asian Journal of Social Science*, 40(4), pp. 486–508.
Robison, R. (2012). Interpreting the politics of Southeast Asia: debates in parallel universes. In Richard Robison, ed., *Routledge Handbook of Southeast Asian Politics*. Abingdon, Oxon: Routledge, 5–22.
Rodan, G., Hewison, K. and Robison, R. (2006). Theorising markets in Southeast Asia: Power and contestation. In: G. Rodan, K. Hewison and R. Robison, eds., *The political economy of Southeast Asia: Markets, power, and contestation*. Oxford: Oxford University Press, 1–38.
Sumano, B. (2013). *Explaining the liberalisation of professional migration in ASEAN*, PhD thesis, School of Politics and International Relations, Queen Mary, University of London.
THC [The Habibie Centre]. (2016). *ASEAN civil society conference/ASEAN people's forum (ACSC/APF) 2016 in Timor-Leste: Potentials and constraints for the future of CSOs' engagement*, Jakarta, 21 June.
Tupas, J. M. (2012). Phnom Penh hotel bars presentation on Myanmar situation, gov't hand eyed. *InterAksyon*, 30 March. Available at: www.interaksyon.com/article/28230/pnom-penh-hotel-bars-presentation-on-myanmar-situation-govt-hand-eyed. Accessed 31 March 2012.

10
INDUSTRIAL ECONOMIES ON THE EDGE OF SOUTHEAST ASIAN METROPOLES

From gated to resilient economies

Delik Hudalah and Adiwan Aritenang

Introduction

Urban deconcentration, suburbanization and the emergence of urban elements in the outskirts of established cities are common spatial development patterns worldwide. Moreover, in Southeast Asia, more of the urban population will reside in suburbs rather than city cores of its megacities such as Jakarta, Bangkok and Manila (Murakami et al. 2005). In the past few decades, new towns, industrial parks and other urban-scale megaprojects have transformed the suburban structure of these megacities. Dick and Rimmer (1998) have suggested that the rebundling of urban elements in Southeast Asian suburbs were triggered by the rise of the middle class as well as trends in the global economy such as the transnational relocation of Foreign Direct Investment (FDI) in manufacturing. According to some estimates, the manufacturing sector's contribution to Southeast Asia's economic growth is 3–8 times higher than that of non-manufacturing (Felipe 1998). Over the past four decades, the manufacturing sector has played a vital role in the formation of Southeast Asia's major cities such as Singapore, Kuala Lumpur, Jakarta, Manila and Bangkok.

This chapter focuses upon shifting trends in the manufacturing sector in Indonesia. The sector comprises 24.7 percent of Indonesia's Gross Domestic Product (GDP) (Nehru 2013), and can be divided into large and medium industry (LMI) and small and micro industry (SMI). LMIs consist of industries with at least 20 employees, and are concentrated in a few metropolitan areas with significant market shares such as the Jakarta Metropolitan Area (JMA). Between 2001 and 2005, LMIs contributed 29 percent to Indonesia's GDP. The latest data by *Badan Pusat Statistik* (BPS) show that LMI output grew by 11 percent, while the number of industries declined 1.3 percent annually between 2008 and 2013. On the other hand, SMIs are firms with one to 19 employees (Badan Pusat Statistik 1996). According to the Ministry of Industry (2015) the contribution of SMIs to Indonesia's GDP gradually increased from 33.91 percent in 2011 to 34.56 percent in 2014. Another analysis indicates that a large share of SMI employment is found in JMA (16.6 percent) followed by Temanggung (14.5 percent) and Greater Surabaya (8.1 percent) (Kuncoro 2007).

This chapter investigates how SMIs and LMIs have shifted in intensity from Southeast Asia's metropolitan cores to their respective suburbs, thus reshaping entire metropolitan spatial and economic structures. Past studies of Southeast Asian urban forms have highlighted the growth of the so-called 'gated community' and 'urban enclave,' giving attention to the social implications of these processes at the community and local levels (Hudalah 2017; Leisch 2002; Pow 2011). This chapter argues that at the metropolitan level these 'gated' and 'enclave' qualities have implications for suburban spatial reorganization. To this end, they need to be extended in economic terms by linking physical reorganization with regional economic restructuring.

The central argument is organized as follows. The next section revisits the urban and industrial deconcentration literature emphasizing the decline of employment in the city core and corresponding growth in the suburbs. The subsequent section overviews the growth of foreign direct investment (FDI) and LMIs – focusing on the JMA as a case study. The JMA, also referred to as Jabodetabek, encompasses the city of Jakarta and its seven surrounding districts and municipalities known as Bodetabek (Bogor-Depok-Tangerang-Bekasi). It is the largest and most suburbanized metropolitan area in Southeast Asia (Murakami et al. 2005), covering an area of 5,898 km^2 with a total population of more than 22 million (Hudalah et al. 2013). The dynamics of SMIs in JMA are subsequently investigated, highlighting the spatial and economic implications of industrial growth and expansion. The chapter concludes that the major suburban industrial concentrations have captured most of the manufacturing employment that has dispersed from central Jakarta. While foreign-induced and large-scale industries have increasingly created gated suburban structures, locally based and small-scale industries are seen to be a more interlinked and resilient sector in the face of turbulent economies.

Industrial relocation, agglomeration and urban deconcentration

This section explores theories of industrial location at the metropolitan scale. Location Theory, derived from regional science, emphasizes industrial concentration or agglomeration results from industrial growth seeking lower production costs through resources sharing and labor-pooling (Myrdal 1957; Perroux 1950). The economic gain from agglomeration is generated from two channels: increasing returns and externalities (Fujita and Thisse 2002). The increasing returns concept argues that agglomeration allows the sharing of resources that minimize production costs and provide labor pooling and specialization. Agglomeration also has positive externalities in the form of economic scale, specialized input and the accumulation of human capital, and the provision of infrastructure. In this sense, industrial agglomeration refers to a grouping of production activities at a particular site to improve efficiency.

In terms of industry type, there are two types of agglomeration: economic localization and economic urbanization. Economic localization is an agglomeration of activity that occurs due to backward and forward linkages in industrial production. Economic urbanization, in contrast, is an agglomeration that emerges following a variation of economic activities and in unrelated industries. Economic urbanization suggests that different types of production activities lead to innovative and new means of production that, in the long run, determine economic growth. This economic growth benefits from a dense and diverse economic structure that expands local knowledge stocks and accelerates radical innovations across sectors (Rodríguez-Pose and Hardy 2016). Examples of these economies include fashion, publishing and FIRE (finance, insurance and real estate) that tend to be found in large metropolitan areas (Day and Ellis 2013).

Urban deconcentration is the relocation and agglomeration of urban economic activities to the suburbs due to urban expansion or a declining Central Business District (CBD). Industrial growth in the suburbs may lead to the emergence of a polycentric metropolitan structure with more than one urban center and the formation of regional specialization (Berry and Kim 1993; Bogart and Ferry 1999; Bontje 2001; Bourdeau-Lepage and Huriot 2005; Gottdiener and

Kephart 1991; Guillain et al. 2006). There are two types of urban deconcentration: population deconcentration and employment or job deconcentration (Carlino and Chatterjee 2002; Hudalah et al. 2013). While population deconcentration is characterized by increasing population flows to the suburbs, this chapter focuses on the second type, employment or job deconcentration – referring to the movement of employment from the metropolitan core to the suburbs. Employment or job deconcentration indicates the spatial restructuring of the metropolitan economies. The process of job deconcentration can be measured using two calculations. First, the industrial employment growth rate in the core city can be used to explore levels of deindustrialization (Zhou and Ma 2000; Gilli 2009). Second, industrial employment in the suburbs is measured to represent industrial relocation and growth in the metropolitan suburbs (Feng et al. 2008; Gilli 2009; Gottdiener and Kephart 1991; Rustiadi and Panuju 1999; Zhou and Ma 2000).

FDI, industrial expansion and regional economic performance

Clark (2006) argues that there are three factors regarding suburban industrial expansion that are unique to Southeast Asia: infrastructure networks, the role of transnational corporations and foreign direct investment (FDI). First, infrastructure networks such as the dense local road and canal systems have existed since the pre-industrial period. Subsequent regional transportation development simply improved and extended these lines of transport. Second, transnational corporations (TNC) have organized new international divisions of labor to strengthen global production processes. TNCs penetrate local and international markets by using low-cost labor in labor-intensive manufacturing industries. The third unique factor distinguishing Southeast Asian suburban development is the FDI flowing through TNCs and industrial clusters. All these factors, especially the third, have accelerated the deconcentration of industrial growth and development into suburban regions.

The role of FDI in manufacturing on economic growth has been widely studied (Canfei 2006; Lipsey and Sjöholm 2011; Rahmaddi and Ichihashi 2013). FDI in Southeast Asia, especially in ASEAN, has been particularly prominent since the early 1990s with the establishment of several policy initiatives such as the ASEAN Investment Area (AIA), the ASEAN Free Trade Agreement (AFTA) (Freeman 2001), and the more recent formation of the ASEAN Economic Community (see Gerard this volume). The region's ability to attract FDI has been one of the main drivers of Southeast Asia's economic development (McGregor 2008). In his book, McGregor (2008) argues that FDI has primarily boosted low skill and labor-intensive industries in urban and peri-urban areas in Southeast Asia. It also fosters economic division within urban and metropolitan areas. For instance, middle-class occupations in the financial, retail and service sectors are located in core zones, whilst older light industries such as garment, textile and footwear industries are found in areas surrounding these zones. The outer rings of industrial activity include chemical, automotive and heavy-machinery industries that are labor-intensive and attract migrants and low-skill workers.

Such trends are apparent in Indonesia, which experienced a significant increase of FDI in the mid-1980s: from US$847.6 million in 1986 to US$1.5 billion in 1987 (Thee 1994). During this period, the government issued the Presidential Regulation (*Keputusan President*) No. 53/1989 on Industrial Parks – a regulation that allows domestic and foreign companies to invest in industrial park development. In 2000, there were 203 industrial park companies comprising 66.8 Ha in 20 provinces. Four were in Jakarta and 103 in West Java (52.7 percent national total) (Yustika 2014). Indonesia's large market and relatively low production cost has continued to attract FDI. As an illustration, the share of *domestic* investment in Java has declined from 87 percent in 2002 to 64.6 percent in 2008. In contrast, there has been an increasing share of FDI in Java from 88.6 percent to 91.2 percent (Soekarni et al. 2010). FDI is concentrated in downstream industries, i.e., consumer goods, such as clothing, food and beverages, automotive and electronic goods (Tambunan 2010).

Hudalah and Firman (2012) indicate that in the case of JMA, FDI relocation has taken place at the (metropolitan) regional scale, making JMA one of the fastest growing global metropolitan areas in Southeast Asia. This relocation of global capital inflows has contributed to the deconcentration of LMIs from the metropolitan core (Jakarta) to its neighboring suburban districts and municipalities (Hudalah et al. 2013).

Hudalah (2017) shows that since the early 1990s, the flows of FDI to JMA have fluctuated. In the early 1990s most FDI increasingly located in the suburbs. FDI flowing to the suburbs peaked in 1996 (more than US$12 million), before it sharply declined during the 1997 Asian Financial Crisis. In the early 2000s FDI in the metropolitan core started to grow again. Meanwhile, the introduction of decentralization policies in 2001 was unable to quickly facilitate the recovery of foreign investments in the suburbs. It took a decade for foreign investment to grow again in the suburbs; the late 2000s marked another rapid growth of FDI in the suburbs, reasserting its attraction over the metropolitan core. As an illustration, in 2013 the investment gap between the suburbs and the core reached the highest margin of more than US$7 million (Hudalah 2017). These trends imply that as the socio-political system began to stabilize, FDI returned to the suburbs for its lower costs of production and labor.

These trends also show that the suburbs have recently become the prime destination for industrial investment. The investment and industrial relocation follows shifts in industrial policy reforms coupled with increasingly more expensive land and property, traffic congestion, stifling pollution and high rates of criminality in Jakarta. Hudalah et al. (2013) emphasize that suburban industrial park development is also partly a consequence of the scarcity of new industrial spaces in the metropolitan core – industrial land prices in Jakarta may be as high as threefold that of the suburbs.

The increasing price of industrial land suggests there is competition between the city core as the 'center' or node and the suburbs in the economic activity of JMA. In the past, the suburbs played a supporting role to the main city of Jakarta, especially in the provision of cheap and comfortable homes. Due to transport infrastructure improvements, however, many suburbs grew as independent entities and industrial activities now link to other parts of the country and the world without having to do so intensively through the core city (Hudalah 2017).

The dynamics and resilience of SMIs

Table 10.1 shows the number of SMIs in the core and suburbs of JMA for the years 1986, 1996 and 2006. Initially there were more SMIs in the city core compared to the suburbs. A decade later, more than a half of SMIs in the JMA were in the suburbs. This trend continued after the 1997 Asian Financial Crisis and decentralization, with almost 65 percent of SMIs – 62,000 firms – located in the suburbs by 2006. The only exception to this deconcentration trend of SMIs in Jakarta (Ansar 2013) was West Jakarta which still hosts 30 percent of SMIs in the region.

The 1997 Asian Financial Crisis affected SMIs less so than LMIs. This is evident in the continuing growth of total SMIs in JMA although the pace slowed during 1996–2006. Two possible reasons for economic resilience of SMIs in the JMA are that most goods produced by SMIs are consumer goods, with insignificant backward and forward linkages with LMIs, and less financial support from banks (Ansar 2013). These limited linkages with global industrial and financial systems prevented SMIs from collapsing during global economic turmoil. The crisis only seems to have affected SMI performance

Table 10.1 Total number of SMIs in Jakarta Metropolitan Area

No	Municipality/district	1986	1996	2006
1	South Jakarta	1,505	6,958	6,257
2	East Jakarta	1,752	8,101	6,518
3	Central Jakarta	1,070	4,947	4,219
4	West Jakarta	2,507	11,591	12,706
5	North Jakarta	1,019	4,710	4,134
Core		**7,852**	**36,307**	**33,834**
6	Bogor District	818	18,876	27,072
7	Bekasi District	445	10,272	6,156
8	Tangerang District	500	8,967	15,710
9	Bogor District	34	782	3,030
10	Tangerang Municipality		2,555	3,463
11	Bekasi Municipality			3,700
12	Depok Municipality			2,758
Suburban		**1,797**	**41,452**	**61,889**
JMA		**9,649**	**77,759**	**95,723**

Source: analyzed from Badan Pusat Statistik (1986, 1996, 2006)

Table 10.2 Total employment in SMIs in Jakarta Metropolitan Area

No	Municipality/district	1986	1996	2006
1	South Jakarta	11,262	28,877	25,968
2	East Jakarta	11,972	30,696	24,698
3	Central Jakarta	8,664	22,214	18,945
4	West Jakarta	27,162	69,646	76,346
5	North Jakarta	9,341	23,952	21,023
Core		68,401	175,386	166,980
6	Bogor District	8,556	56,729	79,904
7	Bekasi District	6,762	44,836	29,663
8	Tangerang District	6,951	35,868	62,840
9	Bogor District	472	3,131	12,130
10	Tangerang Municipality	*)	10,220	13,852
11	Bekasi Municipality	**)	**)	13,357
12	Depok Municipality	***)	***)	9,745
JMA		22,741	150,783	221,491
TOTAL		91,142	326,169	388,471

Note:
*) Tangerang Municipality was part of Tangerang District
**) Bekasi Municipality was part of Bekasi District
***) Depok Municipality was part of Bogor District

Source: Analyzed from Badan Pusat Statistik (1986, 1996, 2006)

in the metropolitan core, where SMI numbers declined from 36,307 in 1986 to 33,834 in 2006. The major effects of the economic crisis on the fluctuating performance of SMIs in the core were the higher price of imported raw materials coupled with reduced purchasing power in domestic markets.

Another possible explanation for the declining number of SMIs in the core was the issuance of local government policies requiring the synchronization of business owners' residential locations with their business sites (Ansar 2013). Because most SMIs in Indonesia are home-based industries, these policies forced them to move out from the city core. In addition, urban land use policy increasingly facilitated the growth of commercial and service activities in the core rather than manufacturing industries.

Table 10.2 depicts a rapid increase of SMI employment across JMA in the last three decades. In 1986 SMI employment in the core parts of JMA was twice that of the suburbs. This continues in 1996 but since the early 1990s, there was a very significant increase in suburban SMI employment. In fact, the number surpassed that of the core. This trend continued with the decentralization policy with employment in the suburbs surpassing 220,000 people by 2006.

SMI employment benefited from the centralized development policies of Suharto's administration that effectively attracted permanent and seasonal labor migration from other parts of Indonesia into the JMA to work in the urban sectors and manufacturing industries. As most LMIs were unable to absorb migrant labor due to a lack of appropriate skills and qualifications, migrant workers took up employment in low skill SMIs and thus contributed to the growing number of SMI workers.

The spatial pattern of industrial development

Industrial deconcentration

Hudalah et al. (2013) argue that since the early 1990s most LMI employment has deconcentrated from the metropolitan core toward the suburbs. They emphasize that deconcentration has not only weakened the role of the core but also extended the boundaries of the metropolitan region toward the eastern outer suburbs (Cilegon and Serang) and western outer suburbs (Karawang and Cikampek) (see Figure 10.1). Expanding this account, we analyze the spatial redistribution of SMIs in the metropolitan area. Following Hudalah et al. (2013), Coffey and Shearmur (2002), Bogart and Ferry (1999), and Guillain et al. (2006), we use the employment-population ratio (E/P) to indicate employment 'density' as a measurement of the spatial concentration and/or deconcentration of SMIs. A high E/P ratio indicates that a sub-district or cluster of sub-districts is a location for SMIs' concentration. The SMI employment data are prepared by Ansar (2013) based on the economic censuses conducted by the National Bureau of Statistics or BPS (Badan Pusat Statistik 1986; Badan Pusat Statistik 2006; Badan Pusat Statistik 1996).

Table 10.3 shows the city core as the center of SMI employment during the 1980s. However, in the mid-1990s, the concentration of SMIs started to decline in most parts of the core. As an exception, the employment-population ratio for Central Jakarta Municipality was still considerably high because the denominator (the population) has declined.

The same table also shows that the suburbanization of SMI employment has occurred since the late 1980s, continuing to increase during the 1990s–2000s. Bekasi was the only suburban district with a significant declining employment density. This exceptional case can be explained by the growing number of LMIs, particularly in the industrial parks in Cikarang, which largely absorbed skilled labors from its surrounding suburbs. When compared with Hudalah et al. (2013), it becomes clear that the increased number of LMIs in Cikarang is in inverse proportion to the growth of SMIs in Bekasi (the district in which Cikarang is located). This comparison suggests there has been a lack of spatial linkages between LMIs and SMIs. Furthermore, in the analysis we use *kecamatan* (sub-district) as the unit of analysis for the employment-population ratio. Based on this more detailed analytical scale, it is evident that SMIs did not disperse in all

Table 10.3 The concentration of SMIs in Jakarta Metropolitan Area

(Group of) municipality or district	Employment density (Employment/population)		
	1986	1996	2006
Metropolitan core			
South Jakarta	7.04	15.12	12.64
East Jakarta	8.02	15.50	10.23
Central Jakarta	7.43	19.95	21.24
West Jakarta	21.65	46.71	35.83
North Jakarta	9.87	21.49	14.48
Suburbs			
Bogor	2.86	16.02	15.77
Tangerang	3.74	12.39	15.31
Bekasi	4.86	15.35	10.42
Bogor	1.97	4.66	13.40

Source: Analyzed from Badan Pusat Statistik (1986, 1996, 2006)

kecamatans. Instead, they were 'clustered' around several principal kecamatans including Ciomas (Bogor District), Cikarang (Bekasi District), Serpong (South Tangerang Municipality) and Cikupa (Tangerang District).

Figure 10.1 presents the overlays between the locations of the identified SMIs clusters with what Hudalah et al. (2013) term 'industrial centers' or the concentration of LMIs. The map confirms that the industrial employment has deconcentrated from Jakarta toward its neighboring districts and municipalities. Furthermore, the map suggests that the locations of most SMI clusters do not correspond with that of LMIs. They only intersect with each other in three sub-districts located in Bogor and Bekasi Districts.

Industrial specialization

Analyzing the specialization of SMIs among the identified clusters, we employ Location Quotient (LQ) analysis. The method has been widely used to identify job specialization in suburban centers (Bogart and Ferry 1999; Bourdeau-Lepage and Huriot 2005; Guillain et al. 2006). The formula used is as follows:

$$LQ^{R_i} = \frac{E_i^R / E^R}{E_i^N / E^N}$$

Notes:

LQ : *Location Quotient* or Hoovver-Balassa coefficient
E_i^R : number of SMI jobs in subsector i in the cluster of kecamatans R
E^R : total number of SMI jobs in the cluster of kecamatans R
E_i^N : number of SMI jobs in the subsector I in all identified clusters in JMA
E^N : total number of SMI jobs in all identified clusters in JMA

LQs with a value greater than 1 indicate a subsector concentration within a cluster of kecamatans compared to all identified clusters in the JMA. Meanwhile, an LQ value of less than 1 indicates that the area lacks of subsector specialization.

From gated to resilient economies

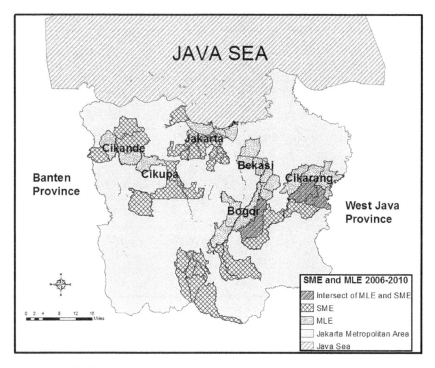

Figure 10.1 Map of industrial concentration in Jakarta Metropolitan Area
Source: Authors' analysis

The results of the LQ analysis are summarized in Table 10.4. First, the Ciomas cluster has an LQ value of 1.85 for leather industries. Ciomas is a well-known center for leather products such as shoes and sandals. Its proximity to the markets in Jakarta, Bogor and other West Java cities has strengthened the production linkages among the SMIs. Meanwhile the Cikarang cluster has become a concentration of wood and furniture industries. Interestingly, and despite the large number of consumer goods, a very small number of raw materials are produced by SMIs in Cikarang. This may indicate a lack of production linkages between the industrial sectors in Cikarang and an inter-regional supply chain of production network. At the same time, Cikarang is home to Indonesia's largest concentration of private industrial parks and FDI in manufacturing (Hudalah and Firman 2012). Despite the large number of LMIs, there is a lack of production linkages between the LMIs and the SMIs in Cikarang. The analysis suggests that the large-scale industrial economies in Cikarang tend to operate as 'gated' or 'enclave' entities, linking directly with the global production network rather than the local economies.

In a similar fashion, the sub-districts next to Cikupa are where the first generation of private industrial parks were built in the suburbs (Hudalah et al. 2013). As in Cikarang, the concentration of LMIs around Cikupa largely absorbs the supply of workers from the surrounding sub-districts. The LQ analysis reveals that Cikupa hosts a concentration of SMIs in raw materials industries such as chemicals, textiles, metals, machineries and electronics.

Serpong is the prime location for JMA's new towns such as *Bumi Serpong Damai* (BSD) and Lippo Karawaci. The Serpong cluster shows high LQ indexes in food, clothing and printing industries. These high LQ indexes for consumer goods suggest that the cluster serves the

Table 10.4 Location Quotient of SMI employment clusters in Jakarta Metropolitan Area

Subsector		Cluster				
		Ciomas	Cikarang	Serpong	Cikupa	Core
Food	Foods and beverage	0.28	0.08	1.17	0.80	0.62
Clothing	Clothes	1.27	0.09	1.91	0.49	1.19
	Leather	1.85	0.25	0.00	0.00	0.46
Shelter	Wood and furniture	0.12	1.29	0.94	1.06	0.79
Printing	Paper	0.00	0.00	0.36	1.38	0.99
	Printing	0.66	0.00	1.39	0.77	0.82
Equipment	Rubbers and Plastics	0.14	0.17	0.13	0.92	1.14
Raw	Chemicals	0.58	0.00	0.26	1.50	0.71
Material	Textiles	0.35	0.00	0.71	1.38	0.17
	Mining	0.00	0.00	0.00	0.00	0.68
	Metals	0.28	0.27	0.87	1.54	0.85
	Machineries	0.00	0.51	0.20	1.57	1.11
	Electronics	0.00	0.00	0.00	3.67	0.00
	Others	0.55	0.00	0.16	1.43	1.40

Source: Analysed from Badan Pusat Statistik (2006)

demands of the Jakarta and Tangerang urban populations. Overall, the analysis suggests that in the past two decades the manufacturing industries have deconcentrated to the peripheries of JMA. The SMIs are clustered in a number of suburban sub-districts with different sector specializations.

Conclusion

This chapter illustrates trends of industrial deconcentration from Southeast Asian metropolitan cores to their respective suburbs by analyzing the case of the JMA. The high E/P ratio and LQ analysis in the spatial and economic analysis confirms work by Harmadi and Brodjonegoro (2003) who show that agglomeration in the JMA is shaped by localization economies as a company's decisions to locate is based on production proximity with similar industries and economics of scale, rather than market proximity. Furthermore, this finding also extends the application of Marshall's argument that the presence of labor market pooling and production input sharing allows the reduction of production costs (Marshall 1890).

The chapter also emphasizes the declining role of metropolitan cores, in combination with an increasing role of the respective suburbs, which has reshaped metropolitan spatial and economic structures. The firm and employment density analyses show that there has been a significant growth of both SMIs and LMIs in Jakarta's suburbs. Interestingly, if we overlay these spatial patterns, there are districts that experience an increasing growth of LMIs whilst at the same time losing SMIs. This suggests a lack of spatial linkages between FDI-financed LMIs and locally based SMIs. It illustrates that foreign-induced manufacturing suburbanization in Southeast Asia tends to encourage the emergence of a 'gated economy,' which becomes an integrated part of global production chains without significant links with the local economies.

The analysis also suggests the important role that national economic policies play in shaping metropolitan industrial landscapes, hence these metropolitan areas reflect the socio-political and economic dynamics at national levels. In Indonesia, for example, the end of the oil and gas sector boom in the late 1980s shifted national economic policies toward industrialization. As a result, a growing number of firms and industrial parks were established across Indonesia's major metropolitan areas. In the following decade, the financial crisis in the late 1990s contracted the growth of both LMIs and SMIs. Nevertheless, our analysis shows that the SMIs are more resilient in passing through such global economic turbulence. A similar scenario was also found at the national level during the peak of the 1997 Asia Financial Crisis, when the manufacturing value added in LMI declined 27.2 percent whilst the SMI grew 34.9 percent (Berry et al. 2001).

Governments in Southeast Asian emerging countries should therefore focus not just on attracting large-scale and foreign investments, but also on facilitating the growth and spatial distribution of SMIs. A more balanced industrial policy would increase the resilience of the national economies in the face of increasingly more volatile global economies and markets. Furthermore, a better spatial linkage between LMIs and SMIs should also be promoted to increase the impact of FDI on local and regional economic development.

References

Ansar, Z. (2013). *Dekonsentrasi Industri Kecil Di metropolitan Jabodetabek*, MSc Thesis, Bandung, Institut Teknologi Bandung.
Badan Pusat Statistik. (1986). *Sensus Ekonomi*. Jakarta: Badan Pusat Statistik.
Badan Pusat Statistik. (1996). *Sensus Ekonomi*. Jakarta: Badan Pusat Statistik.
Badan Pusat Statistik. (2006). *Sensus Ekonomi*. Jakarta: Badan Pusat Statistik.
Berry, A., Rodriguez, E. and Sandee, H. (2001). Small and medium enterprise dynamics in Indonesia. *Bulletin of Indonesian Economic Studies*, 37(3), pp. 363–84. doi:10.1080/00074910152669181
Berry, B. J. L. and Kim, H-M. (1993). Challenges to the monocentric model. *Geographical Analysis*, 25(1), pp. 1–4. doi:10.1111/j.1538–4632.1993.tb00275.x
Bogart, W. T. and Ferry, W. C. 1999. Employment centres in greater Cleveland: Evidence of evolution in a formerly monocentric city. *Urban Studies*, 36(12), pp. 2099–2110. doi:10.1080/0042098992566
Bontje, M. (2001). Dealing with deconcentration: Population deconcentration and planning response in polynucleated urban regions in North-West Europe. *Urban Studies*, 38(4), pp. 769–785. doi:10.1080/00420980120035330
Bourdeau-Lepage, L. and Huriot, J-M. (2005). On poles and centers: Cities in the French style. *Urban Public Economics Review*, 3, pp. 13–36.
Canfei, H. E. (2006). Regional decentralisation and location of Foreign direct investment in China. *Post-Communist Economies*, 18(1), pp. 33–50. doi:10.1080/14631370500505131
Carlino, G. and Chatterjee, S. (2002). Employment deconcentration: A new perspective on America's postwar urban evolution. *Journal of Regional Science*, 42(3), pp. 445–475. doi:10.1111/1467–9787.00267
Clark, J. (2006). *Peri-urban agglomerations in Southeast Asia*. Exurban Change Project White Paper. Columbus, OH: The Ohio State University. Available at: http://sri.osu.edu/sites/sri/files/d6/files/periurban_se_aggl.pdf
Coffey, W. J. and Shearmur, R. G. (2002). Agglomeration and dispersion of high-order service employment in the Montreal metropolitan region, 1981–96. *Urban Studies*, 39(3), pp. 359–378. doi:10.1080/00420980220112739
Day, J. and Ellis, P. (2013). Growth in Indonesia's manufacturing sectors: Urban and localization contributions. *Regional Science Policy & Practice*, 5(3), pp. 343–68. doi:10.1111/rsp3.12015
Dick, H. W. and Rimmer, P. J. (1998). Beyond the third world city: The new urban geography of South-East Asia. *Urban Studies*, 35(12), pp. 2303–2321. doi:10.1080/0042098983890
Felipe, J. (1998). The role of the manufacturing sector in Southeast Asian development: A test of Kaldor's first law. *Journal of Post Keynesian Economics*, 20(3), pp. 463–485. doi:10.1080/01603477.1998.11490164
Feng, J., Zhou, Y. and Wu, F. (2008). New trends of suburbanization in Beijing since 1990: From government-led to market-oriented. *Regional Studies*, 42(1), pp. 83–99. doi:10.1080/00343400701654160
Freeman, N. J. (2001). ASEAN investment area: Progress and challenges. In: M. Than, ed., *ASEAN beyond the regional crisis: Challenges and initiatives*. Singapore: Institute of Southeast Asian Studies, 80–125.
Fujita, M. and Thisse, J-F. (2002). *Economics of agglomeration: Cities, industrial location, and globalization*. Cambridge: Cambridge University Press.

Gilli, F. (2009). Sprawl or reagglomeration? The dynamics of employment deconcentration and industrial transformation in greater Paris. *Urban Studies*, 46(7), pp. 1385–1420. doi:10.1177/0042098009104571

Gottdiener, M. and Kephart, G. (1991). The multinucleated metropolitan region: A comparative analysis. In: R. Kling, S. C. Olin and M. Poster, eds., *Postsuburban California: The transformation of orange county since world war II*. Berkeley: University of California Press, 31–54.

Guillain, R., Le Gallo, J. and Boiteux-Orain, C. (2006). Changes in spatial and sectoral patterns of employment in Ile-de-France, 1978–97. *Urban Studies*, 43(11), pp. 2075–2098. doi:10.1080/00420980600945203

Harmadi, S. H. B. and Brodjonegoro, B. P. S. (2003). Analisis aglomerasi industri manufaktur besar Dan Sedang Di DKI Jakarta Tahun 1975–1998. *Jurnal Ekonomi Dan Pembangunan Indonesia*, 3(1), pp. 1–13. doi:10.21002/jepi.v3i1.406

Hudalah, D. (2017). Governing industrial estates on Jakarta's peri-urban fringe: From shadow government to network governance. *Singapore Journal of Tropical Geography*, 38(1), pp. 58–74.

Hudalah, D. and Firman, T. (2012). Beyond property: Industrial estates and post-suburban transformation in Jakarta metropolitan region. *Cities*, 29(1), pp. 40–48.

Hudalah, D., Viantari, D., Firman, T. and Woltjer, J. (2013). Industrial land development and manufacturing deconcentration in greater Jakarta. *Urban Geography*, 34(7), pp. 950–971. doi:10.1080/02723638.2013.783281

Kuncoro, M. (2007). *Ekonomika Industri Indonesia: Menuju Negara Industri Baru 2030*. Yogyakarta: Penerbit Andi.

Leisch, H. (2002). Gated communities in Indonesia. *Cities*, 19(5), pp. 341–350.

Lipsey, R. E. and Sjöholm, F. (2011). Foreign direct investment and growth in East Asia: Lessons for Indonesia. *Bulletin of Indonesian Economic Studies*, 47(1), pp. 35–63. doi:10.1080/00074918.2011.556055

Marshall, A. (1890). *Principles of economics: An introductory volume*. London: Palgrave Macmillan.

McGregor, A. (2008). *Southeast Asian development*. New York: Routledge.

Ministry of Industry. (2015). *2014 Annual report of the performance of the ministry of industry*. Jakarta: The Planning Bureau of the Ministry of Industry.

Murakami, A., Medrial Zain, A., Takeuchi, K., Tsunekawa, A. and Yokota, S. (2005). Trends in urbanization and patterns of land use in the Asian mega cities Jakarta, Bangkok, and metro Manila. *Landscape and Urban Planning*, 70(3), pp. 251–259.

Myrdal, G. (1957). *Economic theory and under-developed regions*. London: University Paperbacks. Available at: www.worldcat.org/title/economic-theory-and-under-developed-regions-by-gunnar-myrdal/oclc/459757773

Nehru, V. (2013). Manufacturing in India and Indonesia: Performance and policies. *Bulletin of Indonesian Economic Studies*, 49(1), pp. 35–60. doi:10.1080/00074918.2013.772938

Perroux, F. (1950). Economic space: Theory and applications. *The Quarterly Journal of Economics*, 64(1), pp. 89–104. doi:10.2307/1881960

Pow, C-P. (2011). Living it up: Super-rich enclave and transnational elite urbanism in Singapore. *Geoforum, Themed Issue: Subaltern Geopolitics*, 42(3), pp. 382–393. doi:10.1016/j.geoforum.2011.01.009

Rahmaddi, R. and Ichihashi, M. (2013). The role of Foreign direct investment in Indonesia's manufacturing exports. *Bulletin of Indonesian Economic Studies*, 49(3), pp. 329–354. doi:10.1080/00074918.2013.850632

Rodríguez-Pose, A., and Hardy, D. (2016). Firm competitiveness and regional disparities in Georgia. *Geographical Review*, May, pp. n/a–n/a. doi:10.1111/j.1931-0846.2016.12180.x

Rustiadi, E. and Panuju, D. R. (1999). *Suburbanisasi Kota Jakarta*, Bogor. Available at: http://repository.ipb.ac.id/handle/123456789/25064

Soekarni, M., Hidayat, A. S. and Suryanto, J. (2010). Peta Penanaman Modal Asing (PMA) Dan penanaman modal dalam Negeri. *Jurnal Ekonomi Pembangunan*, 18(1), pp. 1–20.

Tambunan, T. (2010). Perkembangan Industri Nasional Dan Peran PMA. *Jurnal Ekonomi Pembangunan*, 18(1), pp. 21–36.

Thee, K. W. (1994). *Industrialisasi di Indonesia: beberapa kajian*. Jakarta: LP3ES.

Yustika, A. E. (2014). *Tapak pengembangan industri nasional*. Bogor: IPB Press.

Zhou, Y. and Ma, Laurence, J. C. (2000). Economic restructuring and suburbanization in China. *Urban Geography*, 21(3), pp. 205–236. doi:10.2747/0272-3638.21.3.20

11
COMMUNITY ECONOMIES IN SOUTHEAST ASIA
A hidden economic geography

Katherine Gibson, Ann Hill and Lisa Law

Introduction

Researchers have long recognized practices of mutual aid, reciprocity and sharing as prevalent features of everyday community life in Southeast Asia. Such practices are often represented as persistent vestiges of pre-capitalist societies and variously categorized as aspects of 'informal economies,' 'patron-client' relations or 'social capital.' In debates about capitalist development these 'relict' practices are seen as standing in the way of modern economic growth, as something to be overcome or enrolled into the mechanics of transition to market capitalism – that is, they are harnessed into a narrative of either decline or transcendence. However, such a framing obscures the valuable role mutual aid, reciprocity and sharing may have played in shaping responses to social, economic, political and environmental threats over the long durée. It is clear that these practices contribute to local social safety nets and act to support households in the event of misfortune or calamity, even today (Ong and Curato 2015). While they may be ill fitted to capitalist development trajectories, they are well suited as survival strategies and may potentially contribute to development trajectories more suited to life in the Anthropocene, the age we have entered in which human systems have become a geological force capable of destabilizing earth systems (Steffen et al. 2015). This chapter outlines an intellectual framing that situates mutual aid, reciprocity, sharing and other 'community economic practices' within a diverse economy in which the trajectory of change is not dominated by the capitalist development narrative but is up for negotiation.

In the first section of the chapter we introduce the diverse economy framing and describe a sample of local economic practices that occur in Southeast Asia drawing on contemporary and historical studies. This section acts as an incomplete inventory that sketches out the landscapes of radical heterogeneity that are unleashed when we employ fine-grained modes of differentiating economic practices and read for diversity. The next section of the chapter reviews the ways that economic diversity has been accommodated within literatures that take capitalist economic relations to be the norm or the goal of economic development. Here we discuss studies of local economic practices that have been theorized in terms of 1) informality, 2) patron-client relations and moral economies and 3) social capital. We argue that these studies employ a capitalocentric discourse of economy, the effect of which is to undermine or render 'non-credible' (Santos 2004) the vitality and potentiality of much of what supports life in non-Western contexts. In the last section of the chapter we identify ways in which scholars are extending thinking on community economic practices and suggest some avenues for research on what has remained until now a hidden economic geography of Southeast Asia.

Diverse economic practices and community economies in Southeast Asia

Economic geographers highlight the phenomenal growth of Southeast Asia over the last half century through the development of commodity markets, including waged labor markets; capitalist industrialization, especially in export processing zones; and the financialization of life. Asia has played a key role in economic globalization. Many scholars have traced the course of agricultural mechanization, monetization, commoditization, proletarianization and labor out-migration – what are usually represented as the key drivers of capitalist transformation (Gibson et al. 2010). But accompanying these processes we have seen the persistence of many economic practices that are not so easily grasped within the capitalist development narrative. When we shift the focus from new and emerging trends to the economic geographies of practice in Southeast Asia we see the continual evolution of a diverse array of non-capitalist economic activities alongside and interconnected with those associated with capitalism. To bring this landscape of heterogeneity to visibility we must cultivate an eye for economic difference and read against the grain of powerful discourses that organize events/realities into seemingly predictable trajectories (Gibson-Graham 2014).

The diverse economy framing developed by Gibson-Graham (1996, 2006) and Gibson-Graham et al. (2013) can guide this reading (see Figure 11.1). Economic diversity is theorized not in terms of different product sectors or primary, secondary, tertiary and quaternary industries, but in terms of the *economic relations* by which goods and services are produced and distributed, and by which well-being and wealth, as well as exploitation and inequality, are generated. These economic relations are enacted through practices of labor, enterprise, transactions, property and finance. The diverse economy includes not only those activities usually associated with the 'capitalist' economy, that is, waged and salaried labor, capitalist enterprise, formal commodity markets, private property and mainstream bank and credit institution finance (in the top cells of each column in Figure 11.1), but also many other categories of activity that operate according to non-capitalist or alternative capitalist logics and dynamics.

Researchers have used this framing to identify the mix of economic practices that combine to create economies in place. For example, at the community scale Michelle Carnegie (2008) has documented the array of diverse market and non-market transactions engaged in by Christian farmers and Muslim traders and seafarers in Oelua on the eastern Indonesian island of Roti. Sarah Wright (2010) has employed the diverse economies framing to show how three residents of Puno on Panay Island in the Philippines combine a complex mix of labor practices, tangible and intangible exchanges, work for money and work for direct consumption to forge a living.[1] In each locality there are specific names for many of the alternative and non-market, unpaid and non-capitalist activities. In the Philippines, for example, reciprocal farming labor exchange is referred to as the *dagyaw* system on Puno (Wright 2010, 304) and as *hungus* in Jagna (Gibson et al. 2010, 8). As far as we know there has been no systematic recording or mapping of this language of economic diversity across Southeast Asia, although there are detailed place-based accounts of practices of mutual exchange, sharing and reciprocity in countless village level anthropological studies. There is, thus, a hidden economic geography that awaits analysis.

What kind of analysis might be conducted on this hidden economic geography remains an open question. A number of scholars have mistaken the practice of inventorying diverse

ENTERPRISE	LABOR	PROPERTY	TRANSACTIONS	FINANCE
CAPITALIST Family firm Private unincorporated firm Public company Multinational	**WAGE** Salaried Unionized Non-union Part-time Contingent	**PRIVATE** Individually owned Collectively owned	**MARKET** Free Naturally protected Artificially protected Monopolized Regulated Niche	**MAINSTREAM MARKET** Private banks Insurance firms Financial services Derivatives
ALTERNATIVE CAPITALIST State owned Environmentally responsible Socially responsible Non-profit	**ALTERNATIVE PAID** Self-employed Cooperative Indentured Reciprocal labor In-kind Work for welfare	**ALTERNATIVE PRIVATE** State-owned Customary (clan) land Community land trusts Indigenous knowledge	**ALTERNATIVE MARKET** Fair and direct trade Alternative currencies Underground market Barter Co-operative exchange Community supported agriculture, fishing etc.	**ALTERNATIVE MARKET** State banks Cooperative banks Credit unions Govt. sponsored lending Community-based financial institutions Micro-finance Loan sharks
NON-CAPITALIST Worker cooperatives Sole proprietorships Community enterprise Feudal enterprise Slave enterprise	**UNPAID** Housework Family care Volunteer Neighborhood work Self-provisioning Slave labor	**OPEN ACCESS** Atmosphere Water Open ocean Ecosystem services Outer Space	**NON-MARKET** Household sharing Gift giving State allocations/appropriations Hunting, fishing Gleaning, gathering Sacrifice Theft, piracy, poaching	**NON-MARKET** Sweat equity Rotating credit funds Family lending Donations Interest-free loans Community supported business

Figure 11.1 A diverse economy framing

economic practices with blind advocacy for 'alternative' non-capitalist practices and are keen to point out that many of the activities in a diverse economy are undesirable, exploitative and inflected with oppressive power relations (Aguilar 2005; Kelly 2005; Wright 2010; Turner and Schoenberger 2012; Roy 2011). These are points that diverse economy proponents Gibson-Graham (2005), Cahill (2008) and Gibson et al. (2010) have also made, for example, by drawing attention to the multiple ways in which power is enacted in Jagna and by highlighting transactions involving theft, work involving child labor, enterprise that relies on feudal surplus extraction and local lending at extortionate rates of interest in Jagna's Diverse Economies Inventory. The point of the diverse economy inventory is not to value diversity *as such*, but to develop a nuanced appreciation of the potential for *certain* diverse economy practices to enact interdependence in ways that might help build more vibrant and nourishing *beyond capitalist* economies, to use Wright's term.[2] Exactly which activities contribute to survival and well-being and which activities undermine 'surviving well together' (Gibson-Graham et al. 2015) cannot be answered by a simple inventory. This kind of analysis calls for a more detailed examination of what kind of ethical concerns motivate practices in the first place.

We identify a *community economy* as a space in which interdependence with each other and with earth others is honored and negotiated. Within any diverse economy *community economic practices* involve ethical negotiations around:

- *laboring* or working to survive well,
- *business* or enterprise of any kind that generates and shares surplus,
- *markets* and transactions of any kind that encounter others with respect,
- commoning *property* by sharing access and responsibility for its care, and
- *financing* and investing in any form in collective futures.[3]

A number of studies have attempted to identify community economic practices that involve ongoing negotiations and ethical decision-making in particular contexts. In the peri-urban context of Opol in Northern Mindanao, for example, Hill (2011) has documented the work of citizens in communal gardens who produce vegetables that are gifted to the municipality's schools to be fed to undernourished children as part of a school feeding program (see also Hill 2013). Gardeners agree to receive in-kind payments in vegetables, herbs, seeds and seedlings for their labor. An estimated 40 percent of the vegetables they produce become a social surplus that spreads the ability to survive well to children across the municipality. The communal gardens are underpinned by negotiations between the municipal government and local landowners to temporarily 'common' their privately owned land. Access to land is gifted free of charge with the understanding it will be cared for – idle ground will become productive, the soil will be improved, and the sites will be free of illegal squatting and rubbish dumping. The whole exercise is further supported by volunteer labor such as that of women from Rural Improvement Clubs who visit communal gardening sessions and offer free advice and farmers who offer to transport excess produce from communal gardens to the market on market day, free of charge. In Opol the creation of a community economy around *surviving well* has been a consciously organized project that draws upon a shared valuing of mutual support practices.

In other localities the negotiation of ethical economic practices are a longstanding but less orchestrated part of village life. In the rural Bohol municipality of Jagna, for example, Gibson et al. (2010) identified a range of existing community economic practices. Work that supported *surviving well* included: sharing the labor of harvest with landless poor in return for in-kind payments of a negotiated fraction of the fish, corn or rice harvest; and allowing the landless to glean from harvested lands. Household *surplus* (generated in small amounts from many different activities and enterprises) was shared with the community during the yearly fiesta. There were many *transactions* that were guided by an ethic of respect and care for others: barter between upland and lowland underpinned the interdependence between different ecosystems; ritual offerings to the spirits of land and trees were made to respect the natural environment and its guardians and keep alive a recognition of human-earth other connection; gifting and pooling of resources to help others and spread the burden of major of costs such as weddings and funerals also ensured community survival and well-being. *Commonly shared* infrastructure such as irrigation channels and roads, houses and neighborhood meeting halls was maintained by volunteering and community work that is unpaid. Finally credit was advanced in many ways to support livelihoods and to meet unexpected demands.

Many of the community economic practices across Southeast Asia are so much a part of everyday life that they are considered unremarkable. Others have been studied, but within a framing that demotes their value as contributors to active and dynamic economies. It is to these studies that we now turn to read against the grain and begin to marshal a greater understanding of community economic practices and their potential.

Finding community economic practices within capitalocentric framings of economic diversity in Southeast Asia

Modernist knowledge systems render many of the activities of the diverse economy non-credible and thus largely invisible (Santos 2004). How does this happen? We contend that the available framings for understanding economic diversity are structured by a capitalocentric logic that undermines the identification of community economies. A capitalocentric discourse positions relations of economic difference as either the same as, the opposite of, an alternative to, or contained within capitalism (Gibson-Graham 1996, 35). As such, economic practices that do not take the form of 'free' waged labor, commoditized exchange, for-profit enterprise, privatized property and market interest rate bearing are seen as:

- the 'same as,' or tending toward a capitalist form – for example, micro-private enterprise is an incipient for-profit capitalist business;
- the 'opposite of,' and thus less efficient, modern or even viable – for example, reciprocal labor exchange is embedded in pre-modern obligatory relations that prevent the free movement of waged labor that underpins its efficiency/replace-ability/cheapness;
- an 'alternative to,' that given the dominance of capitalism, is destined to be a minor player in the economy – for example, cooperative enterprises that value people over profits and are therefore not foot-loose or able to attract the entrepreneurial innovators who are motivated by personal gain; or
- 'contained within,' and thus floating in a vast ocean of capitalism, susceptible to co-option and unable to act independent of capitalist dynamics – for example, feudal tenancy on export oriented agricultural plantations.

In each positioning the dominant signifier is the capitalist economy with its component relationships and dynamics. As the following sections illustrate, community economic practices in Southeast Asia can be found subsumed within discourses of the informal economy, patron-client relations and social capital.

The informal economy

The informal sector is a grab bag category that refers to underground activities, vulnerable workers, the self-employed, home-based workers and micro-enterprises with or without hired workers (Sethuraman 1992). In Hart's original designation, the informal economy was made up of activities conducted by an urban 'sub-proletariat,' that is, workers who were not part of the 'organized labor force' (1973, 61). Writing in an earlier era, Geertz referred to these activities as the 'bazaar economy' (1963). In Asia today the term informal is widely used to refer to the unregistered entrepreneurialism that a vast majority of people in urban and rural areas deploy to survive. Within this huge category of activities there are many types of diverse economic practices each operating with different logics and according to non-state forms of regulation.[4]

Informal micro-enterprise, small family businesses, hawking etc. (what Roy 2011 terms 'subaltern' informal activities) have been described as "necessity-driven survivalist business" (Dahles and Prabawa 2013, 242). The profit margin is minimal or non-existent, but the activity makes just enough for a family or trader to live on. While represented as highly individualistic, operators in the informal sector often rely on strong personal ties to suppliers and customers that involve negotiated agreements to deal with uncertainty and loyalty. Dahles and Prabawa (2013) note, for example, how pedi-cab drivers in Yogyakarta established relationships with certain shops and restaurants, bringing their tourist customers only to these establishments in return for an agreed upon percentage of any sales made. At the same time these drivers maintained close relations with their village communities to which they returned each night, participating in "gift giving to extended family and neighbours, attending and paying for rituals and ceremonies, joint saving clubs and donations to the needy" (2013, 255).

Early on, informal sector activities were seen as something to be eradicated as they stood in the way of modern urban development – but more recently they have been targeted by development agencies for promotion of micro-enterprise-led market growth. Yet necessity driven, survival oriented entrepreneurs are a far cry from capitalist entrepreneurs – some might make the transition to formal capitalist business but most do not (see for example the study of food catering in Iloilo City by Barth and Kuo 1984). If we released these informal activities from their discursive positioning as 'other' to formal markets, or 'becoming' capitalist enterprise, we might see many opportunities for strengthening community economies. State support via, for example, a basic income payment for informal sector workers, as proposed by Ferguson (2015), would allow them to continue to produce much needed goods and services in complex urban environments under less impoverished and desperate conditions.

Patron-client relations

Patron-client relationships involve dependency between a powerful patron who uses influence and resources to provide security for a less powerful agent in return for personal services, support and loyalty that in turn legitimates the power of the patron (Scott 1972, 92). These relations involve both coercion and voluntarism. Political analyses of Southeast Asia foreground patron-client relations, but they also play a role in economic analysis as it is through forms of patronage that resources are redistributed (Polanyi 1957). Traditionally patron power in Southeast Asia was lodged in ownership of land and the ability to grant or deny access by the landless to become tenant farmers.

In this rural context patron-client relations were embedded in a delicately balanced 'moral economy' of subsistence security. Access and entitlement by the poor to essential necessities was linked to agreeing with the patron on what constituted 'just' rents and taxes, and 'just' "access to land, gleaning or fishing rights, rights of way across landowners properties, and redistributive mechanisms and forms of reciprocity that linked peasants with elites and with each other" (Edelman 2005, 332).[5]

More recently patron-client structures have been "closely linked to the national level with jobs, cash, and petty favors flowing down the network, and votes or support flowing upward" (Scott 1972, 105). In both their traditional and contemporary guises patron-client relations are definitely positioned as 'other' to the modern, transparent, economic relations between independent, individualized agents in a capitalist economy. They are seen to flourish in the absence of modern institutions of governance and protection. Indeed, from the capitalocentric perspective of mainstream economics, patron-client relations signal corruption and are a barrier to modern economic development.

The actual economic practices of patronage and client loyalty take many forms. In large part they have been represented as mechanisms of "exploitation of the poor by the mighty or the rich" (Pelras 2000, 341), but this belies the delicate negotiation of benefit and protection that patron-client economic relations entail. In his study of South Suluwesi, Pelras, for example, notes how exchanges of labor and tribute within patron-client bonds contributed to "the smoothing of economic differences" and "helped reinforce social solidarity and cohesion among strata of unequal rank" (2000, 342). A capitalocentric approach to patron-client relations in Southeast Asia blinds us to the diverse nuance of economic relations in which the powerful and the less powerful are bound together. Pelras shows that the detailed vocabulary of patronage exchanges in agriculture, aquaculture, fishing, sea trade and artisanry reveals a complex mesh of negotiated interdependencies that are not grasped by the language of exploitation.

As is increasingly clear, and contrary to earlier views, patron-client political relations do not stand in the way of rampant capitalist development, but have been deployed in many contexts by elites to hasten processes of capitalist transition and capture its benefits (see Springer this volume). So if patron-client relations can be harnessed to capitalist growth, might they also be enrolled in non-capitalist development? In a pioneering research project in Cambodia, Lyne (2016) argues that local patrons might be seen as taking the lead as social entrepreneurs, fostering the development of values-led economic development that supports widespread well-being. While not condoning the corruption that is often associated with patron-client relations, it is worth revisiting the diverse economic practices through which patronage and loyal tribute flow and liberating those with promise for building community economies to play a role in a non-capitalocentric developmental narrative.

Social capital

Social capital is defined as "the information, trust, and norms of reciprocity inhering in one's social networks" (Woolcock 1998, 153) "that enable people to act collectively" (Woolcock and Narayan 2000, 226). As the term 'capital' denotes, social capital is framed as something that can be 'invested' in to enhance economic performance – its 'stock' can be increased and the investor gain greater benefits in return. The concept embodies a slippage between the *individual's* propensity to trust, cooperate and discipline those who renege, what Carpenter et al. term 'behavioral social capital,' and the *community* level networks that enforce rules of behavior or 'associational social capital' (2004, 534).

The origins and development of the term by Bourdieu, Coleman and Putnam among others are multiple and varied, and heavily debated in terms of inclusionary or exclusionary potential and exploitative ends. Geertz (1963), for example, notes how successful businessmen in Bali were constantly assaulted by loan- and job-seeking kinsmen such that social capital networks and local norms about mutual assistance worked against economic progress (Portes and Landolt 2000). Despite contestations, the World Bank and the aid and development sector at large have mobilized the concept with the conviction that "indicators of social capital (such as membership in civic associations) correlate positively with indicators of political democracy and economic growth (such as voting rates and per capita income)" (Rankin 2002, 4). Too little social capital is seen as an obstacle to effective capitalist economic growth and development, and too much is seen as obstructing the efficient operations of markets and the state.

The diverse economic practices by which associational social capital is built up are conducted according to the kinds of ethical negotiations we associate with the enactment of a community economy. Woolcock (1998) categorizes these practices in terms of their contribution to 'getting by' or 'getting ahead':

> 'Bonding social capital' refers to the intra-community ties that enable poor people in a village setting to 'get by' (e.g., monitoring property rights, labour exchange, emergency assistance, rotating savings groups, provision of communal facilities). 'Linkage'

and 'bridging' social capital refer to the extra-community networks that enable individuals and groups to tap outside sources of information, support, and resources, not just enabling them to 'get by' but to 'get ahead' (e.g., links to traders and financier, extension agents, non-governmental organisations).

(Cramb 2005, 214–215)

Development interventions are aimed at mobilizing social capital toward 'getting ahead,' not just 'getting by.' There is evidence, however, that interventions designed to stimulate linking and bridging social capital can have unintended detrimental effects on bonding social capital. In a controlled study of community-driven infrastructure development that took place over a three-year period involving 2,100 households across 11 municipalities in the Philippines, Labonne and Chase (2008, 5) observed that while participation in assemblies and trust in strangers increased, participation in informal (*bayanihan*) collective actions decreased. Similarly in Cramb's study of community landcare groups in Mindanao it was found that soil conservation groups developed easily in communities with strong customary systems of *alayon* or labor exchange (i.e., high levels of bonding social capital), but that the bridging and linking that was facilitated by involvement in the Landcare Program encouraged individuals to 'go it alone' and use their knowledge "gained from experience in nursery management through the communal landcare nurseries ... to develop private nurseries and pursue commercial outlets for their planting materials" (2005, 223).

The analysis of social capital within a capitalocentric world view deflects attention away from the non-capitalist logics and customary or community dynamics of interdependence that are generated and maintained (and sometimes undermined) through ethical negotiation. This point is illustrated with reference to the landcare study cited above. Cramb (2005) notes that the initial motivation for landcare groups to work on establishing contour hedgerows was not to maximize farm income (and thus to 'get ahead'), but to ensure soil erosion was reduced, soil fertility was maintained and terraces were formed (217). The formation of landcare groups between 1998 and 2002 built upon an established ethic of earth care in the community and tapped into the existing traditional local value of *pakikisama* or community solidarity (2005, 222). Cramb describes how landcare participants expressed feelings of "enhanced pride and purpose in being part of landcare" which only served to confirm a traditional sense of earth 'stewardship,' especially in indigenous communities (222). The more nuanced and fine grained attention to associations of interdependence that a community economies perspective offers, allows for the possibility that social capital might be mobilized for 'other than capitalist' development in which 'getting ahead' does not undermine longstanding and resilient practices of 'getting by.'

Conclusion: new research directions for community economies in Southeast Asia

Contemporary scholars of Southeast Asia are increasingly drawn toward imagining and enacting regional development futures that are diverse and different (Rigg 2015). In such imagining community economic practices are at the fore. In this chapter we have shown how the diverse economic geography of the region has been hidden to academic scholarship. Economic practices in which everyday ethical negotiations take place have been downplayed, ignored or devalued by being enrolled into capitalocentric discourses (even by those critical of capitalist development). As the capitalist development trajectory is increasingly recognized as not only unsustainable but as actively destroying the conditions for life as we have known it on earth, it is important that new pathways for equitable development are found.

In 2005 Gibson-Graham proposed that a post-development agenda centered on community economies would involve:

- sustaining and strengthening the diverse practices that support subsistence and produce well-being directly;
- reclaiming, safeguarding and enlarging the commons that provides a base for survival, subsidizing subsistence and creating community; and
- generating surplus and marshalling and distributing it to foster expansion of the productive base and increase standards of living (2005, 17).

To this list we might now add (some 11 years later and in the light of new scholarship in the region that has been inspired by a community economies perspective) the following:

- fostering translocal community encounters – between regions, between nations;
- increasing the role of the state in supporting community economies;
- working with different forms of power; and
- rekindling and strengthening ethical relationships that enhance earth stewardship.

Research has a valuable role to play, not only in foregrounding economic diversity but also in bringing to light what can be done to strengthen and grow community economies in the Anthropocene. Moving forward, a key challenge is to better understand how to support and foster community economic development in the face of increasing climate and resource uncertainty. The interdependencies conceived in a community economy perspective open up possibilities for understanding interdependencies with ecological systems in need of being repaired and rethought. The revelation of hidden economic geographies in Southeast Asia is just the first step in this bigger 'project' of cultivating new pathways for more equitable and sustainable development.

Notes

1. See also applications of the diverse economies framing in Flores, Indonesia (Curnow 2008); Jagna, (Gibson-Graham 2005) and Northern Mindanao (Hill 2011) in the Philippines; Hanoi, Vietnam (Turner and Schoenberger 2012); and Kampong Cham, Cambodia (Lyne 2016).
2. This image of 'another economy' is variously named 'post-capitalist,' 'more than capitalist' or 'other than capitalist,' which is, ironically, a reflection of the power of capitalist discourse to dominate future imaginaries. Indeed, it should be noted that capitalocentric discourse includes those who are pro- and anti-capitalist, with the latter perhaps inadvertently contributing to the prominence of this particular understanding of the economy.
3. This list is a modified version of that developed by Gibson-Graham, Cameron and Healy (2013).
4. According to Roy, by deploying the term 'informal' (with its connotations of illegality) an institutionalized form of violence is condoned whereby "subaltern informalities" are criminalized while "elite informalities" involving, for example, illegal land grabs by corporations and mafia-like organizations to build mega shopping complexes or high rise housing, are valorized (2011: 233).
5. Peasant communities, according to Scott (1976), would do everything in their power to resist threats to this order and rhythm of subsistence living coupled with reciprocity practices on demand. One of the threats he studied was the peasant rebellions (in Vietnam and Burma) in the face of state control and colonial rule.

References

Aguilar, F. V. (2005). Excess possibilities? Ethics, populism and community economy. *Singapore Journal of Tropical Geography*, 26(1), pp. 27–31.

Barth, G. A. and Kuo, M. J. (1984). *Crossing the gap between microeconomic activities and small scale food-catering enterprises*. Washington, DC: Equity Policy Center.

Cahill, A. (2008). Power over, power to, power with: Shifting perceptions of power for local economic development in the Philippines. *Asia Pacific Viewpoint*, 49(2), pp. 294–304.

Carnegie, M. (2008). Development prospects in Eastern Indonesia: Learning from Oelua's diverse economy. *Asia Pacific Viewpoint*, 49(3), pp. 354–369.

Carpenter, J. P., Daniere, A. M. and Takahashi, L. M. (2004). Cooperation, trust, and social capital in Southeast Asian urban slums. *Journal of Economic Behaviour and Organization*, 55, pp. 533–551.

Cramb, R. A. (2005). Social capital and soil conservation: Evidence from the Philippines. *Australian Journal of Agricultural and Resource Economics*, 49(2), pp. 211–226.

Curnow, J. (2008). Making a living on Flores, Indonesia: Why understanding surplus distribution is crucial to economic development. *Asia Pacific Viewpoint*, 49(3), pp. 370–380.

Dahles, H. and Prabawa, T. S. (2013). Entrepreneurship in the informal sector: The case of the pedicab drivers of Yogyakarta, Indonesia. *Journal of Small Business & Entrepreneurship*, 26(3), pp. 241–259.

Edelman, M. (2005). Bringing the moral economy back in . . . to the study of 21st century peasant movements. *American Anthropologist*, 107(3), pp. 331–345.

Ferguson, J. (2015). *Give a man a fish: Reflections on the new politics of distribution*. Durham, NC: Duke University Press.

Geertz, C. (1963). *Peddlers and princes: Social development and economic change in two Indonesian towns*. Chicago: University of Chicago Press.

Gibson, K., Cahill, A. and McKay, D. (2010). Rethinking the dynamics of rural transformation: Performing different development pathways in a Philippines municipality. *Transactions of the Institute of British Geographers*, 35(2), pp. 237–255.

Gibson-Graham, J. K. (1996). *The end of capitalism (As We Knew It): A feminist critique of political economy*. Oxford: Blackwell.

Gibson-Graham, J. K. (2005). Surplus possibilities: Post-development and community economies. *Singapore Journal of Tropical Geography*, 26(1), pp. 4–26.

Gibson-Graham, J. K. (2006). *A postcapitalist politics*. Minneapolis: University of Minnesota Press.

Gibson-Graham, J. K. (2014). Rethinking the economy with thick description and weak theory. *Current Anthropology*, 55(S9), pp. S147–S153.

Gibson-Graham, J. K., Cameron, J. and Healy, S. (2013). *Take back the economy: An ethical guide for transforming our communities*. Minneapolis: University of Minnesota Press.

Gibson-Graham, J. K., Cameron, J. and Healy, S. (2015). Pursuing happiness: The politics of surviving well together. In: D. Pike, C. Nelson and G. Ledvinka, eds., *On happiness: New ideas for the twenty-first century opens in a new window*. Perth: University of Western Australia Press, 116–131.

Hart, K. (1973). Informal income opportunities and urban employment in Ghana. *The Journal of Modern African Studies*, 11(1), pp. 61–89.

Hill, A. (2011). A helping hand and many green thumbs: Local government, citizens and the growth of a community based food economy. *Local Environment*, 16(6), pp. 539–553.

Hill, A. (2013). *Growing community food economies in the Philippines*, Unpublished PhD, College of Asia and the Pacific, Australian National University, Canberra.

Kelly, P. (2005). Scale, power and the limits to possibilities. *Singapore Journal of Tropical Geography*, 26(1), pp. 39–43.

Labonne, J. and Chase, R. S. (2008). *Do community-driven development projects enhance social capital? Evidence from the Philippines*. World Bank Policy Research Working Paper Series, July.

Lyne, I. (2016). *Social enterprise in Cambodia: Practice and theory*, Unpublished PhD, Institute for Culture and Society, Western Sydney University, Sydney, NSW.

Ong, J. C. and Curato, N. (2015). Disasters can lift veils: Five issues for sociological disaster studies. *Philippine Sociological Review*, 63(1), pp. 1–25.

Pelras, C. (2000). Patron-client ties among the Bugis and Makassarese of South Sulawesi. *Bijdragen tot de Taal-, Land-en Volkenkunde*, 156(3), pp. 393–432.

Polanyi, K. (1957). The economy as instituted process. In: K. Polanyi, C. M. Arensberg and H. W. Pearson, eds., *Trade and market in the early empires*. Glencoe: Free Press, 243–270.

Portes, A. and Landolt, P. (2000). Social capital: Promise and pitfalls of its role in development. *Journal of Latin American Studies*, 32(2), pp. 529–547.

Rankin, K. N. (2002). Social capital, microfinance, and the politics of development. *Feminist Economics*, 8(1), pp. 1–24.

Rigg, J. (2015). *Challenging Southeast Asian development: The shadows of success*. London: Routledge.

Roy, A. (2011). Slumdog cities: Rethinking subaltern urbanism. *International Journal of Urban and Regional Research*, 35(2), pp. 223–238.

Santos, B. de S. (2004). The world social forum: Towards a counter-hegemonic globalisation. In: J. Sen, A. Anand, A. Escobar and P. Waterman, eds., *World social forum: Challenging empires*. New Dehli: The Viveka Foundation, 146–156.

Scott, J. C. (1972). Patron-client politics and political change in Southeast Asia. *American Political Science Review*, 66(1), pp. 91–113.

Scott, J. C. (1976). *The moral economy of the peasant: Rebellion and subsistence in Southeast Asia*. New Haven, CT: Yale University Press.

Sethuraman, S.V. (1992). *Urban informal sector in Asia: An annotated bibliography*, International Labour Bibliography No. 13, International Labour Office, Geneva.

Steffen, W., Broadgate, W., Deutsch, L., Gaffney, O. and Ludwig, C. (2015). The trajectory of the Anthropocene: The great acceleration, *The Anthropocene Review*, 2(1), pp. 81–98.

Turner, S. and Schoenberger, L. (2012). Street vendor livelihoods and everyday politics in Hanoi, Vietnam: The seeds of a diverse economy? *Urban Studies*, 49(5), pp. 1027–1044.

Woolcock, M. (1998). Social capital and economic development: Toward a theoretical synthesis and policy framework. *Theory and Society*, 27(2), pp. 151–208.

Woolcock, M. and Narayan, D. (2000). Social capital: Implications for development theory, research, and policy. *The World Bank Research Observer*, 15(2), pp. 225–249.

Wright, S. (2010). Cultivating beyond-capitalist economies. *Economic Geography*, 86(3), pp. 297–318.

12
IMPLICATIONS OF NON-OECD AID IN SOUTHEAST ASIA
The Chinese example[1]

May Tan-Mullins

Introduction

In 2009 China surpassed the United States and Japan to become Southeast Asia's most important trade partner, comprising 14 percent of total external trade for ASEAN in 2013 (Miller 2015, 1). China will become an even more important trading partner, particularly with the completion of the ASEAN-China Free Trade Agreement in November 2015 (Li and Xu 2016). As an immediate neighbor, Southeast Asia represents an enormous market for Chinese trade, goods and influence. The establishment of the Asian Infrastructure Investment Bank (AIIB), spearheaded by the Chinese government and focusing on the Asia Pacific region, also signals the increasing Chinese role in financing aid-related projects (Kamal and Gallager 2016).

The Development Assistance Committee (DAC) of the Organization of Economic Cooperation and Development (OECD) comprises 24 industrialized countries and the European Union. These countries make up more than two-thirds of external finance for the least developed countries (OECD 2015a) and have a strong interest in promoting global norms of good governance through the disbursement of official development assistance (ODA). China is not a member of the DAC, and although current Chinese aid levels are low compared to traditional OECD donors, the estimated combined figures for Chinese preferential export buyers' credit, grants, interest-free or concessional loans reached US$14.1 billion in 2013 (Kitano 2014, 301). As such, there has been increasing scholarly research on the levels and impacts of Chinese overseas aid, particularly in Africa (Bräutigam 2009; Davies et al. 2008; Moyo 2009), and at global and regional scales (Brautigam 2009, 2011; Kitano 2014). However, there is currently little data on Chinese foreign aid to Southeast Asia and its developmental impacts within the region. The absence of information is compelled by the lack of understanding of the different types of Chinese aid as well as a lack of transparency and information.

This chapter responds to this gap in knowledge by examining the historical trajectory of Chinese aid in the region and its wider implications for Southeast Asia. The data used for the paper were collated from interviews with academics, researchers from think tanks, government officials and diplomats, in addition to government policy papers and other secondary sources. Section two draws on this information to trace the historical development of Chinese aid and its modalities, both in a general sense and at the Southeast Asia regional scale. Section three will examine two topical issues surrounding Chinese foreign aid with reference to Southeast Asia:

a) its difference to OECD aid and (b) concerns regarding good governance. Finally, the chapter will conclude with an assessment of the wider implications of the growing influence of Chinese aid and Chinese roles and responsibilities as a global aid donor.

A brief history of Chinese aid globally and in Southeast Asia

China, as an aid giver, is considered part of a wider group of 'emerging' donors (Woods 2008). However, as Kragelund (2008) noted, China and many others deemed 'emerging' have been active donors for most of the Cold War period and beyond. Indeed, as early as 1950, the People's Republic of China (PRC) commenced its aid program by providing material assistance to Asian countries such as the Democratic People's Republic of Korea (DPRK) and Vietnam, to help them to achieve national independence and develop their economies (State Council 2011, 7). In addition, in 1956 after the Bandung conference, China gave Cambodia an outright grant of 800 million riyals (US$22.4 million) to build four cement and steel factories that created 5,000 new jobs (Marsot 1969, 195–196).

Communist states, such as Vietnam in Southeast Asia, received most of the initial aid from China (Marsot 1969, 189). The early years of Chinese aid were ideologically influenced, with catchphrases such as 'mutual common ground' and 'equality.' Foreign aid was officially guided by the Eight Principles for Economic Aid and Technical Assistance proposed by Premier Zhou Enlai in late 1963. Among the more prominent axioms was the call for equality, mutually beneficial relationships and a no strings attached policy. Aid projects were to be conducted:

1. in an equal and mutual beneficial relationship,
2. with no conditions or privileges,
3. to lighten the burden of the recipient countries,
4. to help recipient countries to achieve self-reliance and independent development,
5. to yield quick results with less investments,
6. to provide best quality equipment and materials,
7. to ensure transfer of technology and skills, and
8. Chinese experts are not allowed special demands or enjoy special amenities.

Over time the focus gradually shifted toward economic pragmatism in the 1980s, in line with the open door policy launched in 1979 and later the 'Go Out' policy in 2002 (see below). Aid to Southeast Asia has been influenced by the desire to win political allies (Ambassador, interview in Beijing, April 2011), with a particular focus on Myanmar, Cambodia and Vietnam. Contrary to the Chinese rhetoric that no conditions are attached to aid, a key principle in exchange for Chinese aid is that no diplomatic relations are to be formed by the receiving country with Taiwan (State Council 2011).

In 1993, the Chinese government established 'the foreign aid fund for joint ventures and cooperative projects.' Subsequently, from 1995, China with the China ExIm Bank began to provide medium- to long-term low-interest loans to developing countries. From 2010 to 2012, China gave US$14.41 billion comprised of grants, interest-free loans and concessional loans (State Council 2014, 1). Chinese aid is often project-based such as Engineering, Procurement and Construction (EPC) rather than taking the form of program aid (Davies et al. 2008), and in concrete terms there is a blurring of the boundaries between aid and investment. The Chinese usually pay for part of their oil and other resources in infrastructure, which means there is less free-floating cash for unscrupulous diversion. Aid and investment are channeled through a select group of Chinese corporations in line with the Chinese government's 'Go Out' Policy (Reilly and Na 2007). This policy was officially introduced in China's Tenth Five Year Plan (2001–2005), which aimed to raise the rate of outflow foreign direct investment (FDI). The 'going out' strategy was then comprehensively implemented in the Eleventh Five Year Plan drafted in 2006 (CCPIT 2010). Some 180 companies have been designated by the Chinese state to benefit from preferential finance, tax concessions and political backing in order to 'go global' and become true multinationals (Alden and Davies 2006).

In terms of Chinese foreign aid policy, China has been criticized for the lack of information and directions. This resulted in the release of two White Papers on Chinese Foreign Aid, the first in April 2011 and second in July 2014. The first paper looks at the 60 years of history of Chinese foreign aid, focusing very much on the south-south cooperation and self-development rhetoric. There was also mention of utilizing a regional framework for disbursing of large aid amounts, such as the ASEAN framework. The strategic partnership between China and ASEAN was announced in 2003, and between 2010–2012, China has stepped up its efforts to provide funding through multiple channels, with particular emphasis on infrastructure construction. From this partnership, China has trained over 5,000 officials and technicians in fields such as agriculture and health (State Council 2014). The second White Paper is focused on the 2010–2012 period and two key themes of 'helping improve peoples' livelihood' and 'promoting economic and social development' (Brant 2014).

The White Papers indicate that Asia is the second largest regional recipient of Chinese aid, attracting 30.5 percent of total aid. As illustrated in Table 12.1, out of a total of 121 recipient countries, 30 are in Asia, 51 in Africa, 19 in Latin America and the Caribbean, nine in Oceania and 12 in Eastern Europe (State Council 2014). China has also canceled 41 debts from 10 Asian countries amounting to 59.9 billion Yuan by the end of 2009 (State Council 2011). China is actively engaging with the regional frameworks and multilateral organizations such as the Greater Mekong Sub-regional Economic Cooperation Program and the Asian Development Bank to build the Kunming Bangkok Highway. One of the most impressive Chinese aid packages to Southeast Asia as a region was in April 2009, when Beijing offered an aid package of US$25 billion to help Southeast Asia countries to cope with the financial crisis, in addition to US$39.7 million of special aid to Cambodia, Laos and Myanmar (Reilly 2011, 89).

It is difficult to gauge exactly how much aid individual Southeast Asian countries are receiving due to the lack of accurate statistics. Some of the figures are also conflated with investment capital (as illustrated below). However, from numerous reports and research papers, the main Chinese aid recipients are Low Income Countries (LIC) such as Myanmar, Laos, Cambodia and Vietnam. According to the 2008 Congressional report (Lum et al. 2008), Myanmar is the largest recipient of Chinese aid. According to Genser (2006), China has pledged nearly US$5 billion in

Table 12.1 Geographical distribution of China's foreign assistance funds, 2010–2012

Regions	Percentage
Africa	51.8%
Asia	30.5%
Europe	1.7%
Latin America and the Caribbean	8.4%
Oceania	4.2%
Others	3.4%

Source: State Council 2014

Table 12.2 Timeline of Chinese aid with particular reference to Southeast Asia

Year	Programs/Milestones	Comments
1950	Start of material assistance/foreign aid program for Democratic Republic of Korea and Vietnam	Ideology-driven
1955	Asian-African Conference in Bandung	South-south cooperation, ideology driven
1964	Eight principles adopted	As proposed by Premier Zhou Enlai
1993	Foreign Aid Fund for Joint Ventures and Cooperative Projects were set up	
1995	China ExIm Bank began to disburse low-interest loans	
1997	Premier Zhu Rongji increased assistance to Southeast Asia after the conflict with Philippines over reefs in South China Sea	
From 2000s	Rapid increase in Chinese foreign aid volume	Coincided with its increasing financial reserves relating to accession to World Trade Organization and global trade
2003	China and ASEAN signed Joint Declaration on Strategic Partnership for Peace	Partly due to the South China Sea conflict
2006	China pledges nearly US$5 billion in loans, equipment and investment in minerals and power sectors for the Burmese	
2007	Combined value of Chinese aid to Southeast Asia is US$14.8 billion	

Source: May Tan-Mullins

loans, equipment and investment in minerals and power sectors for the Burmese. From 1991 to 2006, approximately US$883 million in grants, interest-free and subsidized loans and debt relief were disbursed to Myanmar (Reilly 2011). Projects are usually in the sector of infrastructure, power and agriculture such as hydropower dams, roads and factories.

China also continues to provide considerable aid to Laos and Cambodia. From 2000 to 2009, China provided Cambodia with US$204 million in grants and US$500 million in concessional loans. Much of it has gone to support infrastructural projects, which enhance market access and provide valuable experience and economies of scale for Chinese state-owned companies, thus strengthening their positions in the Cambodia market (Reilly 2011, 79). In 2006, China pledged US$600 million in aid and loans to Cambodia and forgave Cambodia's entire debt to China (Kurlantzick 2006). Chinese aid to Laos has also increased gradually since 1999. By 2009, China's total ODA was estimated at US$46.5 million with numerous prestigious projects, such as the Tonchan palace, five star hotels in Vientiane and Lao National Stadium in its list of achievements (Reilly 2011, 82).

Between 2002 and 2007, combined value of aid and related investment projects to Southeast Asia amounted to US$14.8 billion, of which 43 percent was on infrastructure and public works, 32 percent in natural resource extraction and 3 percent to military, humanitarian and technical assistances (Turner and Wu 2011). Other forms of assistance include the China-ASEAN action plan on comprehensive food productivity enhancement plan to set up 20 experimental stations of improved crop varieties, and technology transfer and capacity building (State Council 2014).

Through constructing infrastructural projects, such as roads linking the various countries, China has managed to improve access to raw materials from these countries, and also consolidate export markets for Chinese products (Goh 2011). The timeline of major milestones of Chinese aid policy is summarized in Table 12.2.

Chinese aid and OECD aid

One of the debates surrounding Chinese aid is how it is different from OECD aid in terms of its definitions, types and delivery mode. As Ambassador Liu Zhemin (Permanent Mission of the People's Republic of China to the UN, 2007) indicated, the Chinese aid program is indeed different from the Western aid agencies as it is a developing country and not a developed rich nation. A lack of understanding is compounded by the lack of information. This section will explore the similarities and differences between Chinese and OECD aid through assessing the delivery mode, definitions, modalities and discourses.

According to the Chinese State Council (2011), Chinese aid is mainly delivered bilaterally, through forms such as grants, interest-free loans and concessional loans, which are divided into eight categories, ranging from financial and technical assistance for key investments through to medical aid. Grants are mainly used for welfare and public projects such as schools, hospitals and water wells. Other projects utilizing grants include technical cooperation and emergency humanitarian aid. Interest-free loans are used for infrastructural projects such as roads and public facilities. These loans usually have the five-five-ten rule (which means five years of use, five years of grace and ten years of repayment).

The most contested form of Chinese aid is concessional loans. This form of aid is used to help recipient countries to build medium to large size infrastructural projects, generating both economic and social benefits (State Council 2011). It is also disbursed mainly through engineering, procurement and construction (EPC) projects. This usually means aid is disbursed to the recipient through infrastructural projects, such as roads, stadiums and power stations, mainly built by the Chinese contractors (Tan-Mullins et al. 2010). Between 2010 and 2012, there were 550 completed projects in sectors ranging from public services such as hospitals and schools, economic infrastructure such as roads and power plants to agriculture and industry sectors, such as demonstration farms and factories in 80 countries (State Council 2014).

In terms of definitions, the OECD defines official development assistance (ODA) as: "Flows of official financing administered with the promotion of the economic development and welfare of developing countries as the main objective and which are concessional in character with a grant element of at least 25 percent (using a fixed 10 percent rate of discount)." However, this definition is highly irrelevant in today's context due to changing banking realities. Low interest rates in today's global economy and lending context make the above percentage exorbitant and unrealistic. As Brautigam (2011, 3) writes "When market interest rates hovered in the range of 1–4 percent, the DAC definition provides for a perverse effect: official loans could have a grant element of over 25 percent even if the interest rates being charged is twice." However, the OECD (2015b) has recently modernized this definition where under the new system, loans to LDCs and LICs must reach a grant element of at least 45 percent (instead of 25 percent previously) to be reportable as ODA, while Lower Middle Income Countries will require a 15 percent minimum grant element and Upper Middle Income Countries a 10 percent minimum grant element. Finally, the maximum ODA interest rates permitted have been lowered for all country categories and nearly halved for LDCs and LICs.

The Chinese also do not use the same categories of aid or ODA as the OECD, using some of their assistance to support joint ventures between Chinese firms and those in developing countries, including military aid (Tan-Mullins et al. 2010, 862). In some aspects Chinese foreign assistance resembles ODA, but in others it shares characteristics of foreign investment (Lengauer 2011, 38). Large lines of credit offered by Chinese policy banks, represented as Other Official Financing (OOF) (Brautigam 2011), are also included in the Chinese aid calculation. These include buyers' credits, suppliers' credits and also investment capital provided by the China Africa Development Fund, which theoretically do not count as ODA. Similarly, China does not include scholarships as part of ODA, whereas DAC counts them as part of development assistance (Brautigam 2011). As a result, it is difficult to provide a comparison due to the different items included in the numbers, which Brautigam (2009) termed as 'comparing apples to oranges.' A historic lack of domestic transparency compounds the uncertainties about what can be considered to be aid (Lancaster 2007).

Chinese aid generally has been perceived by critics as involving massive amounts, and possibly undermining global reforms in this area due to the different ways it is disbursed to developing countries. One of the main reasons accounting for the lack of transparency could be the administrating structure of Chinese aid in overseas contexts. Generally, the wide array of stakeholders and agencies involved with aid hinders the easy flow of information and transparency of management. Today, officially, the Ministry of Commerce (under the guidance of the State Council), where the Department of Foreign Aid is located, is the principal agent of foreign aid programs (Kitano 2014). However, there are other agents working on various projects concurrently, such as Ministry of Foreign Affairs (through its foreign-based embassies), Ministry of Finance (through its debt relief programs), China ExIm Bank (through concessional loans), provincial governments (through sister cities programs) and state-owned enterprises (Corporate Social Responsibilities programs). Kitano (2014) listed 13 different agencies involved in Chinese aid programs. Yet, the division of responsibilities between these actors is not clear, as there are many agencies involved and a lack of coordination is resulting in an overlap of duties between the stakeholders. As such, there is an absence of accurate and comprehensive data on foreign aid. Indeed, according to one respondent, China does not have a strong monitoring and evaluating system in place, which is a big challenge (interview with a Chinese think tank, 19 April 2011). Undoubtedly, the aid accounting system is currently evolving; perhaps it is not the intent of the Chinese government to avoid publishing aid figures, instead it is a direct result of no centralized institution coordinating aid activities.

Chinese aid also differs in its discourse surrounding modalities and practices. The objections from critics tend to be that Chinese aid delivery is fragmented, comes through a confusing array of modalities, places too much pressure on recipient states and increases transaction costs (see Birdsall 2008; Collier 2006; De Renzio 2006). Negotiation of projects is moreover very bilateral and high level without any public participation and usually focuses on infrastructure. However, the clearest distinction is over tied aid – requiring that a certain percentage of aid funds to be used to purchase services or items from the donor country. China explicitly requires only Chinese companies complete the infrastructural projects funded by its concessional loans (Reilly 2011, 76–77). As such, the concessional foreign aid loan program operated by China ExIm Bank is often viewed as a mix diplomacy, development and business (Brautigam 2011, 755), and rendered by some as an exploitative strategy to obtain market shares in overseas contexts, especially in LICs of Southeast Asia.

Another difference is the rhetoric of Chinese aid. The Chinese government does not consider itself as a 'donor' but as an 'aid-deliverer,' as it promotes an equal relationship instead of 'indebted' relations. 'Sharing best practices' and 'equal partners' are phrases often evoked in discussions of foreign aid relations. According to a respondent from a government-linked think tank based in Beijing: "We are reluctant to use these terms [donors], it is unfair. We say we are

Table 12.3 Similarities and differences between OECD and Chinese aid

Issues	OECD	Chinese aid	Comments
Bilateral aid	Yes	Yes	Chinese prefer to give aid bilaterally instead of through development agencies, regional organizations or NGOs.
Financing through development agency	Yes	No	
Concessional or favorable lending terms	Yes	Yes	Loans are often given out by the China ExIm Bank.
Grant element of at least 25%	Yes	No	
Payment in kind (including resources)	No	Yes	Payment by resources is often referred as the 'Angola mode' of lending, as Angolans sometimes pay in oil barrels.
Conditionalities	Yes	Yes	Human rights, governance and transparency for OECD, One-China policy for China. In some projects, the Chinese government also stipulates only Chinese companies are qualified as builders and contractors.

Source: May Tan-Mullins

partners. We are not just giving money; China always considers aid as part of the south-south cooperation" (interview with a government-linked think tank, Beijing 14 April 2011). Indeed, the relationship is also built on terms such as south-south brotherhood, win-win, mutual benefits and mutual respect. Similarly, the avoidance of the term 'donor' allows China to justify its practices through the 'Chinese way,' which currently deviates from the DAC norms of accounting aid volume and assessing its effectiveness. According to the Chinese respondents, their development experience is different from Western countries as they have developed their own rules and ways to deal with international affairs and relations. Some of the basic differences between OECD and Chinese aid are summarized in Table 12.3 below.

Good governance

Debates as to whether China is a rogue donor supporting corrupt regimes, exploiting natural resources and perpetuating the 'resource curse,' and implementing projects with poor human rights, labor and environmental records, is one of the most contentious issues surrounding Chinese foreign aid. For some, Chinese aid has been regarded as a new form of colonialism, exploitative in nature through employment of its own firms and national workers instead of hiring local labor. Naim (2007) describes China as an irresponsible rising power, non-democratic and non-transparent in nature. Brautigam (2011, 753) also indicated critics believe that China's aid program is focused primarily on propping up pariah regimes or smoothing the way for Chinese companies to gain access to resources.

China offers assistance without the conditions that other donors frequently place on aid (i.e., democratic reforms, market opening and environmental protection) (Lum et al. 2008, 5). Western donors feel particularly threatened by China's foreign aid policy, as an increasing number of developing countries engage in projects with China rather than with traditional donors (Lengauer 2011). They worry that Chinese practices challenge hard-won reforms in the area of aid and official finance (Brautigam 2011, 753). Furthermore, China is not a member of the OECD and not bound by any rules or expectations to follow their guidelines on foreign aid (Lengauer 2011, 38; interview, 18 April 2011). Chinese justification through the south-south and developing country rhetoric of disbursing aid differently from the OECD may also compound fears.

In Southeast Asia, as elsewhere, Chinese foreign assistance and investment diverge from internationally accepted norms emphasizing good governance, transparency and conditionality. Many observers fear that China's unconditional and non-transparent aid efforts and growing economic integration in Southeast Asia negate efforts by Western nations to promote political and economic reform, reduce corruption and protect the environment in mainland Southeast Asia (Lum et al. 2008, 5). Criticisms of how Chinese aid-funded projects undermine good governance are most prominent in Myanmar and Cambodia, where numerous Chinese hydropower projects, such as the Myitsone Dam on the Irrawaddy River in Burma, were suspended due to local and national opposition, lack of transparency and information, in addition to cultural and ecological reasons. In Cambodia, the completed Kamchay dam in the Kampot Province also demonstrated poor planning regarding environmental and social changes. Our recent study on Chinese hydropower dams in Africa and Asia demonstrated there is very little compliance to the Environment Impact Assessment process, and weak mitigation strategies toward social and environmental implications arising from these dams, especially in the case study of Cambodia's Kamchay dam (Urban et al. 2015).

Whereas mainstream global development practice tends to impose conditionality for assistance or loans, China's rhetoric of "non-interference" dominates its way of aid giving in the region and resonates strongly within the countries of ASEAN – an organization that holds dear principles of mutual non-interference and cooperation (Kalathil 2012, 2–3). Indeed, non-interference is the cornerstone of Chinese overseas aid policy, which leads to the bigger issues of transparency and good governance, especially in terms of volume and allocation in receiving countries.

The inadequacy of information regarding foreign aid is attributed to the absence of transparency and good governance in Chinese overseas aid activities. Issues of transparency and good governance were first picked up by the international media and have slowly infiltrated the domestic media space.

International concerns arise over the type of regimes China deals with accompanied by Corporate Social Responsibility (CSR) issues (concerning labor rights and environmental protection. However, as demonstrated in some works (Tan-Mullins and Mohan 2013; Tan-Mullins et al. 2010), the outcomes and effectiveness of Chinese aid in LICs vary and are highly dependable on country, sector of engagement, local government and civil society leadership. For example, in Reilly's (2011) paper, he demonstrated that Chinese aid is more effective in Cambodia and Laos, due to an extensive international aid organizations and civil society presence, compared to Myanmar, which has little or no civil society participation. Similarly, Tan-Mullins and Mohan (2013) and Tan-Mullins et al. (2010) show that Chinese projects are more transparent and have better governance mechanisms in Ghana compared to Angola, due to the more democratic nature of Ghana and the strong presence of NGOs in the country.

Conclusions: China as a rising power as a developmental partner

Chinese soft power diplomacy has been popular with Southeast Asia and other developing countries as it is an alternative to international finance institutions and OECD-DAC donors bankrolling loans with conditionalities. Southeast Asia, as a region, has experienced first-hand the harsh conditionality tied to aid and loans disbursed by the Western institutions during the 1998 and 2008 financial crisis (Wing 2000; Sussangkarn 2011). Past experiences taught that harsh conditionalities for loans might not be the best way to enable recovery and development.

It is not surprising that Southeast Asia countries are increasingly more willing to turn to China, due to its economic clout and also as an alternative to Western influences. Moreover, with the set-up of the Asian Infrastructural Investment Bank (AIIB), China's increasing role as a leading aid partner allows ASEAN states to 'triangulate' and play donors off against each other, and negotiate the best deal for the country in terms of conditionalities and benefits. To summarize, China and Chinese foreign aid is here to stay, and ASEAN states and OECD countries need to engage constructively with China in this changed aid landscape.

Providing large amounts of aid is a relatively new role for China and it is in a transitional phase in many aspects. On the one hand, China stresses the distinctiveness of its approach. On the other hand, it also expresses the desire to contribute to, or be part of, global aid efforts. There is an increased openness to advice and willingness to learn from other countries. While preaching non-interference in domestic politics, China's interventions have undoubtedly exacerbated existing political problems in some countries, either by design or by default. China's increasing willingness to engage with multilateral organizations such as OECD and Asian Development Bank are good signs of intention to exchange ideas and knowledge.

The most recent example of this willingness was the release of Chinese due diligence guidelines for responsible mineral supply, co-designed and implemented by Global Witness (an international NGO) and the Chinese Chamber of Commerce of Metals, Minerals and Chemical Importers and Exporters (see global *witness 2015* for more information). This demonstrates how China has intensified efforts to promote international cooperation in development assistance to learn effectively from international experiences, improve efficiency of aid and enrich assistance forms. Another example is China and World Bank holding joint workshops on international cooperation featuring capacity development and shared experience in the field (State Council 2014). The above sentiments are captured by a senior official from a DAC-donor country: "China has not been a donor country for very long, so it is still learning for itself how it wants to do things, and China will be more open as the years go by'" (interview with an ambassador, Beijing, 18 April 2011).

Capitalizing on these positive moves instead of excessive criticism and scrutiny on Chinese principles of lending is likely to be a constructive approach to engaging the Chinese state. As demonstrated above, the international community such as NGOs, academics and researchers have been successful in liaising with China and can step up engagement through helping China to improve evaluating and monitoring systems. Although China has the financial power to become a giant in the aid architecture, it is still very weak in terms of technology, know-how and good practices. There is also a need to strengthen cultural understanding between China and the international community, especially in overcoming cultural differences and misperceptions.

It is important to view China as a collaborator in the aid regime, and explore means and ways to cooperate and coordinate in terms of aid giving. OECD countries could perhaps explore how China and DAC donors could work together and coordinate complementary aid activities, such as China working on the infrastructure (hardware) and DAC donors focusing on human development (software) for recipient countries. It is only through focusing on the similarities, instead of emphasizing only differences, that China and OCED donors will be able to work together to provide effective assistance to the less fortunate in Southeast Asia and elsewhere.

Note

1 Acknowledgements: I would like to thank Mr. Stephan Stewart for proofreading the chapter.

References

Alden, C. and Davies, M. (2006). *Chinese multinational corporations in Africa*. Chinese National Offshore Oil Corporation. Available at: www.cctr.ust.hk/materials/conference/china-africa/papers/Chris_Alden_Chinese_Multinational_Corporations.pdf. Accessed 23 June 2016.

Birdsall, N. (2008). Seven deadly sins: Reflection on donor failings. In: W. Easterly, ed., *Reinventing foreign aid*. Cambridge, MA: MIT Press, 515–551.

Brant, P. (2014). *China's foreign white paper quick overview*. Lowy Institute for International Policy, 10 July 2014. Available at: www.lowyinterpreter.org/post/2014/07/10/China-Foreign-Aid-White-Paper-overview.aspx. Accessed 26 October 2015.

Brautigam, D. (2009). *The dragon's gift: The real story of China in Africa*. Oxford: Oxford University Press.

Brautigam, D. (2011). Aid 'with Chinese Characteristics: Chinese foreign aid and development finance meet the OECD-DAC aid regime. *Journal of International Development*, 23(5), pp. 752–764.

CCPIT (China Council for the Promotion of International Trade). (2010). *Zhong Guo Zou Chu Qu Zhan Lu:e De Xing Cheng: The making of our country's 'Go Out' strategy'*. Available at: www.ccpit.org/Contents/Channel_1276/2007/0327/30814/content_30814.htm. Accessed 23 June 2016.

Collier, P. (2006). Is aid oil? An analysis of whether Africa can absorb more aid. *World Development*, 34(9), pp. 1482–1497.

Davies, M., Edinger, H., Tay, N. and Naidu, S. (2008). *How China delivers development assistance to Africa*. South Africa: Centre for Chinese Studies, University of Stellenbosch.

De Renzio, P. (2006). Aid, budgets and accountability: A survey article. *Development Policy Review*, 24(6), pp. 627–645.

Genser, J. (2006). China's role in the world: The China-Burma relationship [Testimony as a Visiting Fellow 2006–2007, National Endowment for Democracy], U.S. Economic and Security Review Commission, 3 August 2006. Available at: www.uscc.gov/sites/default/files/06_08_3_4_genser_jared_statement.pdf. Accessed 26 October 2015.

Goh, E. (2011). *Limits of Chinese power in Southeast Asia*, Yale Global, 26 April 2011. Available at: http://yaleglobal.yale.edu/content/limits-chinese-power-southeast-asia. Accessed 26 October 2015.

Kalathil, S. (2012). Influence for sale? China's trade, investment and assistance policies in Southeast Asia. *East and South China Seas Bulletin*, 4. Centre for a New American Security. Available at: www.cnas.org/files/documents/publications/CNAS_ESCA_bulletin4.pdf. Accessed 26 October 2015.

Kamal, R. and Gallager, K. (2016). *China goes global with development banks*. Brettonwoods project, April 2016. Available at: www.brettonwoodsproject.org/2016/04/20508/. Accessed 8 April 2016.

Kitano, N. (2014). China's foreign aid at a transitional stage. *Asian Economic Policy Review*, 9(2), pp. 301–317.

Kragelund, P. (2008). The return of Non-DAC donors to Africa: New prospects for African development? *Development Policy Review*, 26(5), pp. 555–584.

Kurlantzick, J. (2006). China's 'charm offensive in Southeast Asia', *currently history*, September 2006, 270–276. Available at: http://carnegieendowment.org/files/Kurlantzick_SoutheastAsia_China.pdf. Accessed 26 October 2015.

Lancaster, C. (2007). *The Chinese aid system*, Centre for Global Development. Available at: www.cgdev.org/content/publications/detail/13953/. Accessed 27 October 2015.

Lengauer, S. (2011). China's foreign aid policy: Motive and method. *Bulletin of the Centre for East-West Cultural & Economic Studies*, 9(2), pp. 35–81.

Li, X. and Xu, Y. (2016). A preview of China-Southeast Asia relations in 2016. *The Diplomat*. Available at: http://thediplomat.com/2016/01/a-preview-of-china-southeast-asia-relations-in-2016/. Accessed 13 April 2016.

Lum, T., Morrison, W. and Vaughn, B. (2008). *China's "Soft Power" in Southeast Asia*. [Report for Congress], Order Code RL34310. Congressional Research Service. 4 January 2008. Available at: http://fas.org/sgp/crs/row/RL34310.pdf. Accessed 26 October 2015.

Marsot, A. (1969). China's aid to Cambodia. *Pacific Affairs*, 42(2), pp. 189–198.

Miller, M. (2015). China's relations with Southeast Asia: Testimony for the U.S. China economic and security review commission. National Bureau of Asian Research. Available at: www.uscc.gov/sites/default/files/Miller_Written%20Testimony_5.13.2015%20Hearing.pdf. Accessed 13 April 2016.

Moyo, D. (2009). *Dead aid: Why aid is not working and how there is a better way for Africa*. London: Allen Lane.

Naim, M. (2007). Rogue aid. *Foreign Policy*, March/April 2007, 159, 95–96. Available at: http://foreignpolicy.com/2009/10/15/rogue-aid/). Accessed 26 October 2015.

OECD. (2015a). *Development aid stable in 2014 but flows to poorest countries still falling.* Available at: www.oecd.org/dac/stats/development-aid-stable-in-2014-but-flows-to-poorest-countries-still-falling.htm. Accessed 1 April 2016.

OECD. (2015b). *Modernising Official Development Assistance (ODA) concessional loans before and after the high level meeting.* Available at: www.oecd.org/dac/stats/documentupload/ODA%20Before%20and%20After.pdf. Accessed 23 June 2016.

Permanent Mission of the People's Republic of China to the UN. (2007). *Statement by H.E. Ambassador Liu Zhenmin, deputy permanent representative of China to the UN, at the specific meeting focused on development of the 62nd UNGA*, New York, 6 December 2007. Available at: www.fmprc.gov.cn/ce/ceun/eng/gywm/czdbt/liuzhenminhuodong/t388771.htm. Accessed 26 October 2015.

Reilly, J. (2011). A norm taker or a norm-maker? Chinese aid in Southeast Asia. *Journal of Contemporary China*, 21(73), pp. 71–91.

Reilly, J. and Na, W. (2007). China's corporate engagement in Africa. In: M. Kitissou, ed., *Africa in China's global strategy*. London: Adonis & Abbey Publishers Ltd, 132–155.

State Council. (2011). Full text: China's foreign aid. *Xinhua News Agency*, 21 April 2011. Available at: http://news.xinhuanet.com/english2010/china/2011-04/21/c_13839683.htm. Accessed 26 October 2015.

State Council. (2014). Full text: China's foreign aid. *Xinhua News Agency*, 10 July 2014. Available at: http://news.xinhuanet.com/english/china/2014-07/10/c_133474011.htm. Accessed 26 October 2015.

Sussangkarn, C. (2011). Chiang Mai initiative multilateralization: Origin, development and outlook. *Asian Economic Policy Review*, 6(2), pp. 203–220.

Tan-Mullins, M. and Mohan, G. (2013). The potential of corporate environmental responsibility of Chinese state-owned enterprises in Africa. [Special issue] *Environment, Development and Sustainability*, 15(20), pp. 265–284.

Tan-Mullins, M., Mohan, G. and Power, M. (2010). Redefining 'Aid' in the China–Africa context. *Development and Change*, 41(5), pp. 857–881.

Turner, J. and Wu, A. (2011). *Step lightly: China's ecological impact on Southeast Asia.* Wilson Centre, 7 July 2011. Available at: www.wilsoncenter.org/publication/step-lightly-chinas-ecological-impact-southeast-asia. Accessed 26 October 2015.

Urban, F., Siciliano, G., Sour, K., Lonn, P. D., Tan-Mullins, M. and Mang, G. (2015). South-South technology transfer of low carbon innovation: Chinese large hydropower Dams in Cambodia. *Sustainable Development*, 23(7–8), pp. 232–244.

Wing, T. W. (2000). The Asian financial crisis: Hindsight, insight, foresight. *ASEAN Economic Bulletin*, 17(2), pp. 113–119.

Woods, N. (2008). Whose aid? Whose influence? China, emerging donors and the silent revolution in development assistance. *International Affairs*, 84(6), pp. 1205–1221.

13

'TIMELESS CHARM'? TOURISM AND DEVELOPMENT IN SOUTHEAST ASIA

John Connell

There isn't anywhere else on Earth where people confluence – like the way they do in Singapore. From visitors and transients to citizens, this progressive island and gateway to South East Asia is a dynamic mix of cultures, ideas and histories neither ethnic nor exotic, instead, the essence of modern Asia – sparklingly savvy, with a touch of old school. The first-time visitor should expect to be surprised, confused, and, charmed.

Anita Kapoor, TV and Travel Host[1]

Introduction

This chapter examines recent changes in the tourism industry of Southeast Asia and especially the emergence of particular niches in the market, and the extent to which these trends have contributed to socio-economic development or have raised new specters of uneven development in a region where tourism is of enormous and still growing significance. Over the last decade, tourist arrivals and revenues in Southeast Asia have risen faster than in any other region in the world, almost twice the rates of industrialized countries, gaining an increased market share at the expense of America and Europe. Southeast Asia is now regarded as both a major generator and recipient of tourism, and a place for significant international tourism, regional tourism (within ASEAN) and domestic tourism. Governments have increasingly recognized, by desire and by default, that tourism is a powerful engine of growth and a creator of employment, prompting a greater ease of travel and the removal of complex (and sometime quite costly) visa requirements, part of a wider process of globalization. At the same time continued internationalization with the still growing presence of all the major global tourism chains has established tourism as a lead sector for economic growth. The natural environment (especially coasts), and its recreational possibilities, shopping, and 'exotic' and diverse cultures are the three major marketing themes and attractions.

Tourism growth has also been spurred by the birth of many regional low-cost airlines, taking mass tourism to a wider range of destinations. Lower operating costs, and cheaper airfares, have reduced the cost of travel, making air travel the dominant mode of travel in the region. New airlines have launched new routes to secondary destinations, especially in Thailand, Malaysia and Indonesia, to serve the emerging and growing tourism market in secondary cities and small islands, which has spread rather than reduced the metropolitan focus of tourism. Places such as Hua Hin (Thailand), Dalat (Vietnam), Pangkor and Langkawi (Malaysia) are growing national rather than international destinations, as tourism steadily incorporates new regions.

The devaluation of some regional currencies in the wake of the Asian Financial Crisis of the late 1990s belatedly sparked a rise in international tourism, through favorable exchange rates, while the subsequent Asian economic revival contributed to a parallel increase in national and regional tourism. In Indonesia, for example, tourism from elsewhere in Asia had surpassed tourist numbers from European countries by the end of the century (Cochrane 2009). A wealthy new middle class of Asians, and greater leisure time, has subsequently turned the region increasingly into one of intra-regional migration, especially if tourism from PR China is included. By 2012 China had become the most important country in the world for tourism expenditure, and more than 120 million tourists left China in 2015, especially to east Asia, with more than 4 million traveling to Thailand. The growth of Chinese tourism has been a massive factor in the expansion and restructuring of tourism in Southeast Asia, but not without some tensions.

Tourism is one of the most important sectors in ASEAN economies, and increasingly so. It is the most important sector and major source of foreign exchange earnings in Thailand, ranked second in Malaysia and the Philippines, third in Singapore and Indonesia and gathering pace and prominence elsewhere. The tourism sector in the Greater Mekong Sub-region (Cambodia, Laos, Myanmar, Vietnam and Thailand) was the main economic growth sector from 2006 to 2013, and is no less important in Malaysia and Indonesia. Between 2007 and 2014, tourism contributed more than 10 percent of the gross domestic product (GDP) of Thailand, Laos and Vietnam, and in Cambodia it contributed 23 percent (Nonthapot 2016). The tourism sector in Thailand supports over 1.5 million jobs. Each country promotes tourism widely, with a growing focus on particular niches, such as adventure tourism, shopping and medical tourism. The industry has acquired an unusual economic prominence compared with other world regions, with growing internationalization and the steady rise of national tourism and travel. The bulk of all tourism is regional (Table 13.1) while domestic tourism has been encouraged at least since the 1980s as means of fostering national integration (especially in Indonesia), of redistributing income from rural to urban areas and, in a form of 'staycation,' encouraging the domestic expenditure of disposable income (Cochrane 2009; Erb 2009).

In 2014 just over 100 million tourists visited the ASEAN countries, and almost half of those traveled within ASEAN, a threefold growth in the twenty-first century (Table 13.2). That total excludes the many millions of national tourists who are probably at least as numerous as international tourists. Increasingly, tourists in Southeast Asia are from Asia (Table 13.1), about half traveling between ASEAN states and more than a quarter coming from other Asian states (notably China, followed by Korea, Japan, India and Taiwan). With other tourists traveling from the Gulf and elsewhere that has meant that the numbers traveling from Europe, Australasia and North America are fewer than a quarter of all tourists, despite the slow growth of tourism from Russia. While Malaysia, Thailand and Singapore dominate destinations, the Singaporean and Malaysian numbers are magnified by many people traveling briefly and frequently across the international border. Thailand is the main economic beneficiary in the region (and is the seventh most visited country in the world) followed by Malaysia. Chinese tourists are the mainstay of Thai tourism, shifting its dependence and focus away from the west. Indonesian economic benefits are highly concentrated in Bali, as are Cambodian interests in Siem Reap. Singapore has tourist numbers some three times the national population, the highest ratio in the region.

Vietnam has grown rapidly in this century, partly as state controls have been relaxed, despite different perceptions of what a more market-oriented economy and foreign investment entail, but also because corrupt practices allow bypassing and thwarting of state bureaucracy (Lloyd

Table 13.1 Regional sources of visitors to ASEAN

Country of origin	2012		Country of origin	2014	
	Number of tourists (thousands)	Percentage of total		Number of tourists (thousands)	Percentage of total
ASEAN	39,845.5	44.7	ASEAN	49,223.0	46.8
China	9,283.2	10.4	China	13,059.9	12.4
European Union 28	8079.1	9.1	European Union 28	9,275.2	8.8
Japan	4,275.3	4.8	Japan	5,018.4	4.8
Australia	4,059.6	4.5	Australia	4,634.2	4.4
Republic of Korea	4,011.4	4.5	Republic of Korea	4,383.6	4.2
USA	2,984.2	3.3	USA	3,254.3	3.1
India	2,839.6	3.2	India	3,071.0	2.9
Taiwan (ROC)	1,846.0	2.1	Taiwan (ROC)	2,377.5	2.3
Russian Federation	1,834.6	2.1	Russian Federation	1,920.4	1.8
Top ten country/ regional sources	79,058.4	88.6	Top ten country/ regional sources	96,217.0	91.6
Rest of the world	10,166.8	11.4	Rest of the world	8,866.7	8.4
Total tourist arrivals in ASEAN	**89,225.2**	**100.0**	**Total tourist arrivals in ASEAN**	**105,083.8**	**100.0**

Source: ASEAN Tourism Statistics

Table 13.2 Tourist arrivals by country in Southeast Asia

Country	in thousand arrivals		
	1999	2012	2014
Brunei Darussalam	636.6	209.1	252
Cambodia	262.9	3,584.3	4,500
Indonesia	4,727.5	8044.5	9,440
Lao PDR	614.3	3,330.1	4,160
Malaysia	7,931.1	25,032.7	28,000
Myanmar	198.2	1,059.0	3,080
Philippines	2,170.5	4,272.8	4,830
Singapore	6,958.2	14,491.2	15,100
Thailand	8,651.3	22,353.9	24,780
Viet Nam	1,781.8	6,847.7	7,870
ASEAN	**33,932.4**	**89,225.2**	**102,012**

Source: ASEAN Tourism Statistics

2004; Bennett 2009). More than most other Southeast Asian countries Vietnam has also experienced the 'roots tourism' of overseas Vietnamese, Viet-Kieu, returning to former homes (Nguyen and King 1998). Laos has followed a somewhat similar trajectory on a much smaller scale, but in other respects represents a northern extension of Thailand, as it opens up to tourism and tourists seek new experiences and places. Cambodia too has grown steadily in this century in the wake of peace – much as in Vietnam – but mainly focused on Angkor Wat. That has given it similar numbers to the Philippines, which has never managed to convey a distinct culture and is more isolated from airline connections. Tourism is growing exceptionally fast in Myanmar, where only a decade ago ethical concerns were raised over any travel there (Henderson 2003). Isolated Timor-Leste has only a token tourist economy because of intervening opportunities, poor communications and infrastructure, and high costs. Brunei, marketing itself as 'The Green Heart of Borneo,' has been a little more successful.

At one level tourism is experiencing a greater geographical spread across the region; however, it has also tended to focus on certain prominent city nodes, especially as regional tourism numbers have grown. As international chains have played a greater role in the industry the benefits have thus spread rather less far and less evenly than hitherto, though they have increased substantially. The region has acquired a more metropolitan focus.

Niches

As tourism has grown in numbers it has gone substantially beyond shopping and beaches, although both – in the rapidly expanding shopping malls, iconic places such as Boracay and dozens of small islands – are crucially important and continue to be the key rationale for regional tourism. Indeed Singapore claims "The range of Singapore shopping malls is so vast that some visitors to this tiny island state book their plane tickets purely for one reason: to shop til they drop! In fact, Singapore has more high-end shopping malls per capita than anywhere else in Asia, and visitors are simply spoilt for choice in terms of both quality and quantity" (Singapore Guide, 2017; see Henderson et al. 2011). Bangkok makes similar claims and few significant tourist centers are without malls. Nonetheless expansion and marketing of ecotourism, adventure travel, cruises, golf tours, arts and entertainment, spas, food, health and medical tourism and other possibilities have given tourism increased diversity. Tourism marketing has focused on what are perceived to be 'high value' niches rather than on backpacking, perhaps the oldest niche of all, squeezed out by ideology and image, but a major contributor, especially in some nodes, as backpackers stay longer and spend more in country. Because Southeast Asia covers a great diversity of geographies, cultures, lifestyles and preferences, a wide range of niches has emerged.

The increasing sophistication of local, regional and international visitors has required the tourism industry to offer quality, comfort, convenience, relaxation, independence, adventure, excitement and learning: the whole gamut of visitor experiences. Convention centers and theme parks attract quite different tourist groups by family, age, nationality and socio-economic status, and other niches are as diverse. New niches, from birdwatching (especially in Indonesia) to white water rafting, mark a more sophisticated and specialized tourism. Film, food and yoga tourism have emerged. Cruise tourism has grown fast in the region, both of international cruise companies expanding their operations into Asia, and of regional cruises, many again with a significant Chinese market component. Cruises too have become more specialized. The Meridian company offers a cruise in Vietnam – 'Rockin' the Mekong with Elvis': an eight-day cruise with an Elvis impersonator performing nightly, that is touted as 'absorbing two cultures simultaneously.' By contrast dark tourism exists in Cambodia especially, centered on the Khmer Rouge killing fields and museums (Hughes 2008), and war tourism to the tunnels and battlefields of Vietnam and the 'Bridge on the River Kwai' in Thailand. Ironically parts of Laos are effectively closed to tourism since the artefacts of war remain dangerous. A number of these niches are examined below to emphasize the diversity that increasingly exists, though justice cannot be done to the multiple specialist regional and local tours that exist across a range of possibilities.

Backpackers

A generation ago, backpacking, where travelers preferred budget accommodation and transport, longer holidays, meeting other travelers and some local involvement, seemed to characterize tourism in Southeast Asia. Backpacker tourism has proved a rather vague economic and cultural phenomenon: cut-price, extended in time, of youthful self-identified 'travelers' rather than tourists, passing through something of a rite of passage and inhabiting some famous, even notorious, tourism enclaves. From the 1960s, backpackers paved the way for tourism in a range of Asian places, but the 'loss' of Afghanistan to Islamic insurgency brought more regional 'hippy trails' in this century. Traveler nodes were cut back from the 4Ks (Kathmandu, Kabul, Kuta, and Khao San Road, Bangkok) to just the latter two, but a Vietnam Trail stimulated the growth of such places as Hoi An and Sapa, now centers for national and international tourism.

Gap years have taken the place of hippies, banana pancakes replaced magic mushrooms, and full moon parties have been complemented by Irish pubs, as backpacking has become less anarchic and more regulated, marking a political economy of different times. Backpackers could often be impatient with local values and customs, and countries could be impatient with backpackers, who they wrongly saw as contributing little economically. Although backpackers are less likely to prefer to move off the beaten track, and enthuse over what is not in the guidebook (McGregor 2000), they have enabled some small and remote places to thrive, such as Malaysia and Thailand's small islands, or the Gili islands off Lombok (Indonesia). Party islands such as Koh Phangan and Koh Samui became more commercialized rather than matching the more laidback image of *The Beach*. Specialist operators and transport facilities have made backpacking more akin to flashpacking, being driven to particular nodes by commercial operators (Hampton 2013; Hampton and Hamzah 2016) but still relentlessly moving on.

Like other forms of tourism, backpacker tourism has also become national and regional, with the emergence of a 'new generation' of Chinese, Japanese and other Asian backpackers. Although Chinese backpackers, so-called 'donkey friends,' are not significantly different from their Western counterparts in terms of travel motivation, they tend to travel for rather shorter time periods, rely on the Internet and are more formally organized (Lim 2009; Chen et al. 2014). Not only do the places they visit change to meet new needs, but they change further as new 'mass' tourists replace the early backpackers. Thus at Sapa in highlands Vietnam, no more than a decade of tourism brought significant changes, at least to foreign backpackers as national tourism took over.

> Sa Pa town is perceived by the majority of backpackers as noisy, unsightly and ultimately an infringement on nature. Modernity shocks them, urban sprawl drives them away, karaoke excesses and rampant prostitution are judged sickening; they came here precisely to get away from it all ... There is already a tendency in their discourse to ... label it as increasingly 'worn out,' a 'spoilt' destination where the damages of 'bad tourism' have made interactions too 'commercial' and rendered a visit less appealing.
> (Michaud and Turner 2006, 799)

Tourism is always poised for a potential fall. Yet, what may here be unappealing to backpackers is admirable to the growing number of urban Vietnamese tourists; a new tourist cycle takes over as Sapa goes from small market town to mass tourism destination. The rapidly evolving tourism life cycle has been most evident in Bali, and particularly Kuta, in the 1960s a small village destination for backpackers and surfers on the Asian overland trail, where new tastes and desires were constantly linked into new tourist niches (from raves and honeymoons to whitewater rafting, camel safaris and butterfly parks) that accommodated the needs and whims of all kinds of tourists: it became 'whatever you want it to be' (Connell 1993). In the course of these changes the role of local people declined, as Javanese and international interests constructed hotels, restaurants and other facilities. Balinese were displaced from now valuable coastal land, and visual culture took on a variety of transformations, simultaneously being diminished, transformed, reinvented and globalized for the tourist gaze (Howe 2005). As Kuta changed, other parts of Bali evolved in quite different ways; some specialized in manufacturing particular tourist artefacts, or staging cultural events, as, under the tyranny of supposedly 'alternative' guide books (McGregor 2000), tourists followed conventional paths through the island. Kuta itself became notorious as a site of drunken dissoluteness, Nusa Dua became an enclave of expensive and elegant hotels, and in the village of Pengosekan, near Ubud, a group of workshops took up the manufacture of what had hitherto been Australian didjeridus for a European market (Gibson and Connell 2005). Mass tourism from Java, and a diversity of Asian tourists, maintained 'traditional' beach and new resort tourism. Backpackers moved away as environmental problems in Kuta especially accelerated. Theme parks arrived in Kuta, international literary festivals, film tourism, meditation and yoga reached Ubud, with greater cosmopolitanism and a blurring of distinctions between tourists and resident expatriates (MacRae 2016), and between Balinese culture and materialism. As one of the most important and famous destinations for regional and global tourism it is hardly surprising that Bali has changed in complex and comprehensive ways, and become a microcosm of a range of evolutionary changes elsewhere.

Natural places and ecotourism

As Southeast Asia has become more urban and tourism more national, national parks, natural places and wildlife have acquired greater prominence, but that has yet to extend to any significant focus on ecotourism. Geoparks, where geology is perceived as heritage, are an even newer phenomenon. Both culture and a particular perception of a more 'authentic' nature, lost elsewhere, have drawn tourists from China and Japan (Yamashita 2009). Tropical birds and wildlife, such as orangutans in Sumatra and Borneo, have drawn tourists to particular reserves and sanctuaries and played a very small part in discouraging environmental degradation of their habitats (e.g., Rajaratman et al. 2008). Yet sanctuaries and national parks have not been established without conflict and tension. Largely a by-product of the rise of national tourism, the growing pressure to preserve landscapes has caused conflicts between people who lived off the resources of parks and the interests of tourists, invariably to the disadvantage of former forest residents displaced beyond the parks (Wong 2008, Rugendyke and Son 2008). Yet 'natural' tourism remains in its infancy; additional disposable income has made relatively few seek out rural pleasures other than in regimented and sanitized form. Thus in highlands Vietnam, at Sapa, the new Vietnamese "affluent population is not that keen for prolonged contact with nature, and enjoying the town's amenities proves far more attractive than visiting unclean highlander villages" (Michaud and Turner 2006, 793). Conventional beach tourism dominates and ecotourism has proved rare, disappointing and not integrated into the wider commercial sphere of tourism, partly through lack of communication and cooperation between policymakers and other tourism stakeholders and partly because of inadequate marketing. While many adventure tours are tantamount to ecotourism, their relative closeness to nature is a function of remoteness rather than design. Thailand has had a National Ecotourism Policy since 1998 but ecotourism is caught up with adventure tourism and Thailand remains a mass tourism destination (Leksakundilok and Hirsch 2008). Other Southeast Asian countries have experienced even less ecotourism, despite the emergence of tourism in relatively remote areas, and in search of 'natural' and 'unique' experiences.

Culture

The particular realm of cultural heritage tourism constitutes one of the most significant and fastest growing segments of the tourism industry, with Asia said to be 'enriched by living cultures,' linked to agricultural lifestyles, arts and handicrafts, musical traditions, foods and cuisine, and spiritual and religious practices that constitute tangible and intangible heritages. Almost all have been reinvented in some way. Distinctive cultures have long been a drawcard for tourism to Southeast Asia, and travel brochures and websites, and even airlines, relentlessly emphasize idealized versions of cultural distinctiveness and the 'exotic,' often focusing on particularly distinctive ethnic groups. Several states, notably Thailand and Vietnam, emphasize the potential of tourism in distinctive hill tribe regions involving the commodification of tribal cultures and of such visually distinctive places as Bali and Toraja (Picard and Wood 1997). Ethnic people represent a form of exotic heritage, whose dress, longhouses and recent status as 'headhunters,' fossilized in time, are eminently marketable (Zeppel 1997; Yea 2002). Hill tribe girls, who make a living from being photographed with tourists on the Thai-Burma border, wear 'a contrived tribal costume' invented from elements of different tribal attire, but marketed as 'traditional' (Cohen 2001, 164). Food, from Singapore's World Gourmet Summit to every country's street food, and hawkers are deliberate drawcards (Chaney and Ryan 2012; Henderson et al. 2012; Henderson 2016). Somewhat more subtly floating (and other) markets have been touted as urban versions of creative heritage.

More broadly cultural tourism has been promoted in most countries both for national markets and through festivals of various kinds. That is especially so in Singapore where other tourism possibilities are more limited, heritage has been deliberately conserved and 'natural' environments and theme parks created. National parks have become more numerous, and monuments such as Angkor Wat, Borobudur and entire cities such as Hue and Luang Prabang have been restored, promoted and commercialized. Many have become simply 'bucket list' places to visit as much as places of critical engagement with cultural history (Winter 2007; Smith et al. 2012). Likewise festivals have been created and promoted, such as the annual Sarawak Rainforest World Music Festival held in the Sarawak Cultural Village, "a living museum where the traditional inhabitants of Sarawak's major ethnic groups have been lovingly reproduced" (quoted in Gibson and Connell 2005, 213). The majority of festivals cater for local tourists.

Film tourism has brought new cultural diversity, dominated by Elizabeth Gilbert's book *Eat, Pray, Love* (2006), and the film that followed, starring Julia Roberts, that massively boosted yoga and spirituality tourism to Ubud, and resulted in a local hotel construction boom. One upmarket travel chain linked Bhutan, Bangkok and Bali, where guests could choose from cleansing temple rituals, sessions with Ketut (Elizabeth's Gilbert's own teacher) or outings to locations where the movie was shot. In 2012 and 2013 Chiang Mai (Thailand) became a major destination for Chinese tourists, enthused by the comedy hit film *Lost in Thailand*, then the highest-grossing Chinese film ever. As shown in the comedy, Chiang Mai is a city steeped in Buddhist culture and bursting with an exotic atmosphere. In the movie, Thais are polite, nonchalant and fun-loving people who have learned to live life at a slow yet peaceful pace. Elsewhere in Thailand, after Alex Garland's *The Beach* (1997) was turned into a film starring Leonardo di Caprio, Maya Bay became a popular destination (Law et al. 2007). Following a spate of novels and films Bangkok may become a film noir tourism destination. In this niche, as elsewhere, diversity continues to increase.

Sport, adventure and beyond

Rising affluence has contributed to the expansion of sports tourism, both within the region, notably for golf, and from more distant countries for surfing, rock climbing and other activities.

Malaysia has hosted the Malaysian Grand Prix since 1999, as one stage in the Formula One World Championship, as has Singapore, but Southeast Asia has otherwise been unable to attract the global Olympics and World Cup football that its northern neighbors have attracted. Nonetheless sport is a substantial niche. Surfing has contributed to development in several small and remote places notably the Nias and Mentawai islands (Sumatra) and in better known Bali (Towner 2016). Diving tours are common as are rock climbing (Krabi, Thailand) and mountaineering (Mount Kinabalu, Borneo) that also involves volcano tours (as at Bromo and Lombok (Indonesia). Gunung Mulu National Park in Borneo (Malaysia) is home to the largest cave system in Southeast Asia, which continues to attract international caving expeditions.

Sex tourism has always been a part of tourism in Asia, and in an earlier age was even advertised in guide books (Connell 1993) with notorious nodes emerging in Pattaya and Bangkok (Thailand) and in most significant destinations. Sex tourism extends from temporary liaisons between tourists and local residents, with or without commercial relationships, that are loosely romantic, with both male and female tourists seeking short-term partners (Dahles 2009), and which may become long-term relationships (Chan 2009), into basic transactional sex, partly a function of anonymity for tourists and the regional incidence of poverty that draws more people into the industry. While sex tourism is conventionally regarded as being a phenomenon associated with Western tourists it is also a regional phenomenon associated with many border towns and crossing points, including the Thai borders with Laos and Malaysia (Lyttleton 2014), the Vietnamese border with China (Chan 2009) and Batam island (Indonesia) conveniently located for Singaporean mobility. Some of the same borders, and that between Myanmar and China, also account for trans-border tourism for gambling in casinos.

A particularly unpleasant component of sex tourism is child sex tourism, notably in Cambodia, where the trafficking of girls continues to be a serious problem, and is a destination for Asian and other foreign men. Recently, Indonesia islands like Bali and Batam have also become known for child sex tourism and sex trafficking, and Laos and Myanmar emerged as destinations, as Thailand has clamped down on its own sex industry, general tourism has increased in both countries and government regulation is weak or corrupt.

As in other forms of tourism sex tourism has contributed to distinct patterns of migration as tourism centers have developed. Both prostitution and gambling (in a formal casino sense) remain mainly urban phenomena, especially of borderlands, though often drawing workers from rural areas, sometimes through coercion and trafficking (e.g., Phongpaichit 1982). In some small but significant tourism destinations, such as Kuta and Koh Samui, massage parlors are particularly common, even in smaller towns (Garrick 2005). Within sex tourism niches have developed and regional involvement is significant.

Medical tourism

A specialized tourism niche is medical tourism that began in Southeast Asia in the 1990s and has expanded rapidly. It paralleled the wider rise of tourism and marked an early phase of privatization and corporatization, with many Asian hospitals seeking profits, new markets and new clientele in the wake of the Asian financial crisis (Henderson 2004; Connell 2011). That corresponded with the rise of a new middle class with new demands and an ability to pay for healthcare, both in Asia and in developed countries. Other boosters of tourism (assisted by television and the internet) were increased familiarity with distant places, waiting lists that could be bypassed, and a growing acceptance that most medical practices, technology and human resources, certainly in the best hospitals in 'less-developed' countries, were the equivalent of those in developed countries: 'First world care at third world prices.' A globally aging population created new demands. Medical tourism was adopted and promoted by governments, especially in Asia, anxious to see a modern creative biotech and health industry develop with enhanced visitor numbers (Ormond 2013). While doubts extend to whether it can be truly considered a form of tourism, it usually involves some pleasurable components, such as shopping and dining, and involves patient-travelers and their relatives and friends using the same infrastructure as other tourists and engaging in similar pursuits. Indeed, in Malaysia at least, medical tourists have contributed more financially to non-medical ends, such as accommodation and food, than to direct medical expenditure (Klijs et al. 2016). In Thailand, medical tourists spend rather more than non-medical tourists, while they also travel with companions (Noree 2015). Marketing healthcare to tourists and the focus of some hospitals and practitioners on distant people's wants have raised awkward ethical questions over the role of national health services.

Malaysia especially has focused on its ethnic diversity and cosmopolitanism both to attract medical tourists, and more generally, especially as it has oriented toward the Gulf region. Being a Muslim country has offered a competitive advantage, with halal food readily available, whereas other destinations, notably Thailand, have had to make more specific arrangements in particular floors of hospitals to cater for Muslim visitors. Captured by the image of cosmetic surgery, medical tourism was initially seen as being almost entirely of Western visitors into the region, especially to Thailand and Malaysia that have actively promoted themselves as medical tourism destinations, for 'tourists' from the Gulf and developed countries (Ormond 2013). However, as new facets of cosmetic surgery became more desirable, from whitening to bodily reconstruction, partly stimulated by new workplace preferences, especially in China, regional medical tourism has become more evident. More recently a significant part of medical tourism is between neighboring countries and across local borders, such as from Laos to Thailand and Indonesia to Malaysia, for minor medical procedures, contributing to the establishment of new border trading regions (Ormond and Sulianti 2017; Connell 2016). Medical tourism has thus followed the same broad trends of regional mobility evident elsewhere.

Voluntourism

Volunteer tourism has grown rapidly in this century, with the word 'voluntourism' only coined in 1998, as an intended version of ethical and responsible tourism. Extending from Earthwatch, volunteering on kibbutzim and national volunteer programs, such as the Peace Corps, it expanded rapidly in Southeast Asia, evolving from an alternative holiday option into what has become something of a rite of passage for many university students, and others. It largely involves tourists who volunteer to assist on seemingly worthwhile projects intended to contribute to development in poor countries, such as Cambodia, while usually also linked to an additional more 'standard' holiday experience.

Voluntourism has increasingly been criticized as a new form of colonialism and exploitation that reinforces the dominant paradigm that the poor of developing countries require the help of affluent Westerners, including students, to induce development (Vodopivec and Jaffe 2011). Individual sentimentality responds to, and deflects attention from, structural inequality (Mostafanezhad 2014). Companies market packages that reduce the complexity of development practice to a service that can be performed in a matter of days by the volunteer as part of an adventure experience. Private operators organize 'meaningful tours' such as the 14-day 'Teaching and Temples Tour of Cambodia,' which includes in its trip highlights 'the killing fields', Tuol Sleng Genocide Museum and 'Cambodia's people' (McGloin and Georgeou 2016). Tourists seek to both 'make a difference' through their contribution to teaching English, constructing houses, managing animal refuges and similar activities, while also obtaining, and recording on social media, an 'authentic' tourism experience through working in 'real' local community contexts (Sin and Minca 2014; Kontogeorgopoulos 2016) despite a considerable lack of relevant experience. Volunteer tourism has thus been heavily criticized for its inappropriate projects, failed housing schemes, corruption and fake Cambodian orphanages (Carpenter 2015) and for the self-indulgence of the tourists working on projects of doubtful or negative local value. While voluntourism 'gives,' it is also taking, as the experiences of the authentic 'other' and of 'doing development' become acts of consumption in the voluntourism paradigm. Again, like other forms of tourism, a limited regional component exists with volunteers traveling from Singapore to countries such as Cambodia, Vietnam and Thailand.

Employment, economics and equity

As a labor-intensive activity, where relatively few skills are required in many occupations, tourism makes an enormous contribution to employment. Regular, secure wages, away from the uncertainty of agricultural labor, are attractive and have even led to migration away from skilled employment. Success may so transform particular destinations that 'old' jobs in agriculture and other arduous activities are abandoned. However employment is usually repetitive, little training exists to enable career advancement, and limited skills mean low wages. Tourism has contributed to the economies of many villages but, with rare exceptions such as upland areas of Thailand and Sulawesi (Indonesia), mainly to urban areas, small islands and accessible coastal villages, hence the returns to tourism are sporadic and localized.

In most remote areas, rural people are distant from tourism and where tourism exists the benefits may be few. Hill tribes in Thailand are often involved merely through hawking of cheap trinkets, such as bracelets, while others without such goods resort to begging (Cohen 1996). Even the small income gained from that is concentrated among relatively few households who are already relatively better off (Dearden 1996). At Sapa, in the nominally socialist state of Vietnam, cultural minorities, despite representing about 85 percent of the local population, "are basically left to watch and hope for beneficial effects to trickle down, deprived as they are from access to economic success and political power in the state apparatus due to their cultural distinctiveness, their lack of formal education, and their limited economic capital" (Michaud and Turner 2006, 803). After 12 years of increasing tourism, in one Thailand Hmong hill tribe village the economic impact was 'astonishingly negligible', as trekking tours were organized in urban areas using Thai vehicles, drivers and guides, and trekkers ate Thai food brought with them (Michaud 1997). However in most similar contexts, and subsequently, benefits have been rather greater and have enabled livelihood diversification and a useful flexibility.

Here and elsewhere the benefits to local people have often been slight, as more powerful interests bypass them. At the Komodo National Park on Komodo Island (Indonesia), tourists, intent on seeing Komodo 'dragons,' stay in nearby urban centers on other islands and never stay overnight. A handful of carvers and boat crew are the only village beneficiaries, with local people again disadvantaged by their lack of relevant skills, education and knowledge (Hitchcock 1993; Walpole and Goodwin 2000; Borchers 2009). The benefits are less likely to accrue to local people than to the private sector or government agencies. In some contexts, such as at Ayutthaya (Thailand), local people are prohibited from active commercial participation, hence their economic benefits are negligible while they simultaneously experience tourism-related environmental degradation (Thanvisitthpon 2016). As resort enclaves have replaced small hotels local people have become distanced further.

Small business

Tourism has invariably stimulated small business development. Local people have long been pioneers in meeting new demands for tourist facilities and "start homestays, food stalls and transportation, or hire out snorkel gear, motorcycles or mountain bikes" (Kamsma and Bras 2000, 170). In Bali men and some women are engaged in handicraft production while women are engaged in the informal sector, as masseurs, hair-braiders, drink sellers, traders, homestay (losmen) workers and men are itinerant vendors and drivers. Women may benefit more from tourism than from many other service and manufacturing activities as informal traders, handicraft producers, cultural performers, market sellers and hotel workers, even if their incomes are low. But in most contexts, as in Cambodia, and especially in adventure tourism, men secure better jobs in tourism than women (Brickell 2008; Tirasatayapitak et al 2015). However over time, small-scale local accommodation has gradually been displaced in prestigious locations by large hotels and resorts. In the 1970s Batu Ferringhi beach on the northern coast of Penang island, Malaysia, was characterized by fishermen's cottages that were the main form of tourist accommodation; two decades later they had all been replaced by hotels, and international tourists, mainly backpackers, replaced by domestic tourists. The informal sector is constantly unwelcome and opposed for what is said to be the inappropriate imagery it offers but also for the 'threat' it may pose to formal sector economic activities, both national and international, that have gradually become the main beneficiaries.

As tourism grows local people are more likely to be displaced from their homes, from their land (required for hotel resorts, golf courses, etc.) and from jobs, as part of both a coastal squeeze, demand for higher standards of service and facilities, and government involvement. Often local workers are not hired, on the grounds that they are unreliable, to be replaced by cheaper migrant workers (Connell and Rugendyke 2008, 9–10). Even small-scale employment and income benefits can thus be brief, precarious and illusory, as corporate players take over. Entertainment becomes more 'sophisticated' and global rather than local, while handicrafts are imported into the region, usually from China (to where they often return). Local and regional multiplier effects can be limited. Large hotel chains are usually supported by global chains of food and drink supply that are independent of local producers, though resorts have sought to stimulate local production, to encourage good relations with local people and secure a convenient regular supply of fresh food, especially fish.

In most places the benefits from tourism are quite uneven, as certain local people and groups, with better connections and education, more land or entrepreneurial skills, take advantage. Just as tourist places go through a life cycle, so local employment experiences a cycle, initially growing only to fall when resorts take over. Tourism becomes one more means of local socio-economic differentiation. It may also contribute to marginalization and loss of local autonomy as distant outsiders take over the critical components of the industry, such as hotels, restaurants and car hire. Moreover in every context local people gain only a small proportion of the tourist expenditure of tourists, and intermediaries, or overseas companies, may often be the key beneficiaries. No country has promoted tourism as a poverty-alleviating strategy. After the 2004 tsunami, restructuring of the tourism industry around Phuket was undertaken in a more formal way that benefited the better-off who could make strong claims to land ownership and marginalized and excluded those who had worked in the informal sector. Corporations rather than communities were the beneficiaries, with every Southeast Asian country drawn by the allure of the potential windfalls from high-end tourism.

Environmental and cultural change

Tourism has had diverse environmental impacts. Change is inevitable as facilities are constructed and expanded but may be less visually intrusive and damaging to the environment than many forms of development as tourists seek relatively pristine circumstances. Yet even ecotourism is not necessarily without damaging environmental repercussions, as in northern Thailand, where 'ecotourism' projects often "reproduce the same contradictions as other forms of capital-intensive development . . . exacerbating economic and social disparities, diverting resources and alienating the majority of people from their resource base" (Pholpoke 1998, 262). Generally in Southeast Asia, where tourism has created additional income sources, the frequency of cultivation has increased through hired rural labor and/or the expansion of the cultivated area through land purchase, as in north Thailand (e.g., Forsyth 1995). By contrast, close to tourism itself, as around Sapa, as tourism developed and incomes increased, the rate of deforestation declined and marginal agricultural fields with low productivity were abandoned (Hoang et al. 2014). A similar process has occurred in Bali, Phuket, Koh Samui and elsewhere as land was too valuable to be 'wasted' on agriculture and small farmers.

Accentuating this trend is that the relationship between rapid tourism growth and environmental degradation has usually been close. By the end of the 1980s the rapid growth of tourism at 'hotspots' like Kuta and Koh Samui had outpaced the development of tourist infrastructure so that drainage, sanitation, traffic congestion and air, water and noise pollution had all become problematic (Wall and Long 1996, 43), alongside the visual pollution of poles, posters, neon lights and garbage. Inadequate environmental planning and management, alongside sand mining, land clearance and other deleterious activities, including golf courses, have resulted in the loss of coastal mangroves, fisheries habitats, coral reef damage, coastal erosion and pollution from solid and liquid waste (Williamson and Hirsch 1996). The environmental degradation (and social transformation) of Pattaya (Thailand) as it grew from a small village into an expanding city has contributed to the quintessential tourist directed dystopia: "the most extreme example of the touristic transition of a seaside resort" (Cohen 2001, 159). Vung Tau, in Vietnam, with its bars, karaoke parlors and casinos aimed at Chinese tourists, and Olongapo in the Philippines have acquired similar reputations. The environmental costs of tourism merge with social costs. Although in only a handful of places have tourists been accused of being "unpleasant guests" who are irresponsible and "loud, lecherous, drunken and rude" (Boissevain 1996, 5) fears exist that they provide demonstration effects for local youth.

Social changes have occurred because of tourism and independent of it. Tourism has kept some elements of culture 'alive' but often as empty practice rather than an ongoing and respected component of lives, and what was once in the course of being abandoned may be revitalized for tourist consumption, adapted, embellished and staged at particular times for the tourist gaze. Throughout Asia versions of traditional performances have been shortened and made more appropriate to the short attention spans and lack of linguistic and cultural affinity of tourists (Gibson and Connell 2005, 146–150). In Toraja funeral ceremonies were truncated to meet the needs of tourists, which resulted in community resentment, but communities were forced to relent in order to sell the souvenirs and accommodation on which they had become dependent (Adams 1990). Songkran, the Buddhist New Year festival prominent in Thailand and Laos, has gradually become a backpacker ritual of water throwing rather than something sacred (Porananond and Robinson 2008). Somewhat ironically, most of the financial benefits of tourism are directed into forms of 'modern' consumption, yet even at Kuta local people have retained local agency and can distinguish what is performed for tourists from what is truly sacred and valuable.

Conclusion

Throughout Southeast Asia tourism is of increasing importance as an international and increasingly regional and national phenomenon, while the region is poised to become a dominant center of global tourism – mainly because of the rise of Chinese tourism – that has overturned the former 'north-south' trend in the region. No longer is backpacking so significant, as it has become institutionalized, international investment has taken over, resorts have replaced rustic shacks and local workers been displaced by migrants. Exotic and diverse cultures, and the soft multiculturalism of their foods, have become rather more of a marketing backdrop. Based on accelerating mobility, tourism has acquired a greater diversity, differentiation and subtlety, as newer, more specific markets emerge, requiring the delicate task of simultaneously marketing cultural diversity and heritage alongside modernity, evident in the opening quotation from Singapore and, perhaps, in its earlier slogan 'Instant Asia' (see Chang and Lim 2004). Activities have become as important as destinations in tourism marketing. Boutique hotels, spreading from Singapore, have diversified accommodation, and Airbnb is widespread. Medical tourism emphasizes technological change in the region, matched by the ubiquity of cable cars, shopping malls, upmarket restaurants and shows with global performers: a gradual convergence with themes and trends in other parts of the world, and extensions of it, with the new resorts and casinos seen as "cultural laboratories for testing new protocols for neoliberal governance and capitalist production" (Simpson 2016), a far cry from the localized impacts of backpacking.

Some metropolitan nodes, such as Singapore and Bangkok, remain particularly well-placed to benefit from all forms of tourism, from cruises and conventions, to food and cultural heritage, and have developed complex stop-over packages to develop this further. Islamic tours, featuring halal food, are possible in most countries, as tourism has become more complex and cosmopolitan. Continued quests for new arenas and niches such as convention tourism are seen as the key to diversity and thus growth, notably in established situations such as Singapore and Thailand. Relatively new players like Myanmar and Timor-Leste simply search for success. At every scale tourism has an uneven impact.

Environmental management has becoming more critical, as numbers increase and values change, with enormous pressures on those parts of the region that are important destinations, such as Siem Reap (Angkor Wat), fragile Halong Bay and several parts of Bali. Climate change is unlikely to help. Economic changes have endangered wildlife tourism, despite tourists formally preferring relatively pristine environments. In many coastal areas and on the edge of national parks critical conflicts have arisen, while the desire for modernity clashes with urban heritage conservation in emerging centers like Yangon. The prominence of the informal sector almost everywhere, and undercurrents of tension in sex tourism and voluntourism, emphasize that tourism has not removed poverty but is often intimately associated with it. Likewise women may be useful marketing devices but are rarely the main beneficiaries of tourism. Greater emphasis is placed on sustainability and 'green tourism' in marketing than in actual practice. Ecotourism is no panacea and ethical and 'responsible' tourism are mainly academic concerns.

Recognition is evident that tourism is highly sensitive to violence and political disruption, as in Bali (Tarplee 2008; Hitchcock and Darma Putra 2007) and the southern Philippines, to biological problems, such as SARS and avian flu, and subject also to physical hazards (e.g., Calgaro and Lloyd 2008). Tourism in West Papua is unwelcome. Localized issues such as southern Thailand's political unrest and Malaysian Airlines' two flight disasters in 2014 have had limited geographical or temporal impacts, with Chinese arrivals into the region increasing again by the end of the year. By contrast, greater stability and democratization, as in Myanmar, can be a great boost for tourism. In a global climate marked by some uncertainty, Southeast Asia is likely to thrive, as the Chinese tourism market continues to expand, based on relative stability and diverse cultures and geographies. Its impact however will be increasingly uneven.

Note

1 Your Singapore: www.yoursingapore.com/en_au.html. Accessed 2 February 2017.

References

Adams, K. (1990). Cultural commoditization in Tana Toraja, Indonesia. *Cultural Survival Quarterly*, 14(1), pp. 31–34.
Bennett, J. (2009). The development of private tourism business activity in the transitional Vietnamese economy. In: M. Hitchcock, V. King and M. Parnwell, eds., *Tourism in Southeast Asia*. Honolulu: University of Hawai'i Press, 146–164.
Boissevain, J. (1996). Introduction. In: J. Boissevain, ed., *Coping with tourists*. Oxford: Berghahn, 1–26.
Borchers, H. (2009). Dragon tourism revisited: The sustainability of tourism development in Komodo National Park. In: M. Hitchcock, V. King and M. Parnwell, eds., *Tourism in Southeast Asia*. Honolulu: University of Hawai'i Press, 270–285.
Brickell, K. (2008). Tourism-generated employment and intra-household inequality in Cambodia. In: J. Cochrane, ed., *Asian tourism: Growth and change*. Oxford: Elsevier, 299–309.
Calgaro, E. and Lloyd, K. (2008). Sun, sea, sand and tsunami: Examining disaster vulnerability in the tourism community of Khao Lak, Thailand. *Singapore Journal of Tropical Geography*, 29(3), pp. 288–306.
Carpenter, K. (2015). Childhood studies and orphanage tourism in Cambodia. *Annals of Tourism Research*, 55, pp. 15–27.
Chan, Y. (2009). Cultural and gender politics in China-Vietnam border tourism. In: M. Hitchcock, V. King and M. Parnwell, eds., *Tourism in Southeast Asia*. Honolulu: University of Hawai'i Press, 206–221.
Chaney, S. and Ryan, C. (2012). Analyzing the evolution of Singapore's world gourmet summit: An example of gastronomic tourism. *International Journal of Hospitality Management*, 31(2), pp. 309–318.
Chang, T. and Lim, S. (2004). Geographical imaginations of 'New Asia-Singapore'. *Geografiska Annaler*, 86B, pp. 165–185.
Chen, G., Bao, J. and Huang, S. (2014). Segmenting Chinese backpackers by travel motivations. *International Journal of Tourism Research*, 16(4), pp. 355–367.
Cochrane, J. (2009). New directions in Indonesian ecotourism. In: M. Hitchcock, V. King and M. Parnwell, eds., *Tourism in Southeast Asia*. Honolulu, University of Hawai'i Press, 254–269.
Cohen, E. (1996). Hunter-gatherer tourism in Thailand. In: M. Hitchcock, V. King and M. Parnwell, eds., *Tourism in South-East Asia*. London: Routledge, 227–254.
Cohen, E. (2001). Thailand in 'Touristic Transition'. In: P. Teo, T. Chang and K. Ho, eds., *Interconnected worlds: Tourism in Southeast Asia*. Oxford: Elsevier, 155–175.
Connell, J. (1993). Bali revisited: Death, rejuvenation and the tourist cycle. *Environment and Planning D*, 11, pp. 641–661.
Connell, J. (2011). *Medical tourism*. Wallingford: CABI.
Connell, J. (2016). Reducing the scale? From global images to border crossings in medical tourism. *Global Networks*, 16, pp. 531–550.
Connell, J. and Rugendyke, B. (2008). Tourism and local people in the Asia-Pacific region. In: J. Connell and B. Rugendyke, eds., *Tourism at the grassroots: Villagers and visitors in the Asia-Pacific*. London and New York: Routledge, 1–40.
Dahles, H. (2009). Romance and sex tourism. In: M. Hitchcock, V. King and M. Parnwell, eds., *Tourism in Southeast Asia*. Honolulu: University of Hawai'i Press, 222–235.
Dearden, P. (1996). Trekking in northern Thailand: Impact distribution and evolution over time. In: M. Parnwell, ed., *Uneven development in Thailand*. Aldershot: Avebury, 204–225.
Erb, M. (2009). Tourism as glitter: Re-examining domestic tourism in Indonesia. In: T. Winter, P. Teo and T. Chang, eds., *Asia on tour: Exploring the rise of Asian tourism*. Abingdon: Routledge, 291–301.

Forsyth, T. (1995). Tourism and agricultural development in Thailand. *Annals of Tourism Research*, 22(4), pp. 877–900.

Garrick, D. (2005). Excuses, excuses: Rationalizations of western sex tourists in Thailand. *Current Issues in Tourism*, 8, pp. 497–509.

Gibson, C. and Connell, J. (2005). *Music and tourism: On the road again*. Clevedon: Channel View.

Hampton, M. (2013). *Backpacker tourism and economic development: Perspectives from the less-developed world*. London and New York: Routledge.

Hampton, M. and Hamzah, A. (2016). Change, choice and commercialization: Backpacker routes in Southeast Asia. *Growth and Change*, 47, pp. 556–571.

Henderson, J. (2003). The politics of tourism in Myanmar. *Current Issues in Tourism*, 6, pp. 97–118.

Henderson, J. (2004). Healthcare tourism in Southeast Asia. *Tourism Review International*, 7, pp. 111–121.

Henderson, J. (2016). Local and traditional or global and modern? Food and tourism in Singapore. *Journal of Gastronomy and Tourism*, 2(1), pp. 55–68.

Henderson, J., Chee, L., Mun, C. and Lee, C. (2011). Shopping, tourism and retailing in Singapore. *Managing Leisure*, 16(1), pp. 36–48.

Henderson, J., Yun, O., Poon, P. and Biwei, X. (2012). Hawker centres as tourist attractions. *International Journal of Hospitality Management*, 31, pp. 849–855.

Hitchcock, M. (1993). Dragon tourism in Komodo, eastern Indonesia. In: M. Hitchcock, V. King and M. Parnwell, eds., *Tourism in South-East Asia*. London: Routledge, 303–316.

Hitchcock, M. and Darma Putra, I. (2007). *Tourism, development and terrorism in Bali*. Aldershot: Ashgate.

Hoang, H., Vanacker, V., Van Rompaey, A., Vu, K. and Nguyen, A. (2014). Changing human-landscape interactions after development of tourism in the northern Vietnamese Highlands. *Anthropocene*, 5, pp. 42–51.

Howe, L (2005). *The changing world of Bali*. London: Routledge.

Hughes, R. (2008). Dutiful tourism: Encountering the Cambodian genocide. *Asia Pacific Viewpoint*, 49(3), pp. 318–330.

Kamsma, T. and Bras, K. (2000). Gili Trawangan - From desert island to 'marginal' paradise: Local participation, small-scale entrepreneurs and outside investors in an Indonesian tourist destination. In: D. Hill and G. Richards, eds., *Tourism and sustainable community development*. London: Routledge, 170–184.

Klijs, J., Ormond, M., Mainil, T., Peerlings, J. and Heijman, W. (2016). A state-level analysis of the economic impacts of medical tourism in Malaysia. *Asian-Pacific Economic Literature*, 30, pp. 3–29.

Kontogeorgopoulos, N. (2016). Forays into the backstage: Volunteer tourism and the pursuit of object authenticity. *Journal of Tourism and Cultural Change*. Available at: http://www.tandfonline.com/doi/full/10.1080/14766825.2016.1184673. Accessed 21 July 2017. doi:10.1080/14766825.2016.1184673

Law, L., Bunnell, T. and Ong, C. (2007). *The Beach*, the gaze and film tourism. *Tourist Studies*, 7(2), pp. 141–164.

Leksakundilok, A. and Hirsch, P. (2008). Community-based ecotourism in Thailand. In: J. Connell and B. Rugendyke, eds., *Tourism at the grassroots. Villagers and visitors in the Asia-Pacific*. London and New York: Routledge, 214–235.

Lim, F. (2009). 'Donkey friends' in China: The Internet, civil society and the emergence of the Chinese backpacking community. In: T. Winter, P. Teo and T. Chang, eds., *Asia on tour: Exploring the rise of Asian tourism*. Abingdon: Routledge, 291–301.

Lloyd, K. (2004). Tourism and transitional geographies: Mismatched expectations of tourism investment in Vietnam. *Asia Pacific Viewpoint*, 45(2), pp. 197–215.

Lyttleton, C. (2014). *Intimate economies of development: Mobility, sexuality and health in Asia*. London: Routledge.

Macrae, G. (2016). Community and cosmopolitanism in the new Ubud. *Annals of Tourism Research*, 59, pp. 16–29.

McGloin, C. and Georgeou, N. (2016). 'Looks good on your CV': The sociology of voluntourism recruitment in higher education. *Journal of Sociology*, 52(2), pp. 403–417.

McGregor, A. (2000). Dynamic texts and tourist Gaze: Death, bones and buffalo. *Annals of Tourism Research*, 27, pp. 27–50.

Michaud, J. (1997). A portrait of cultural resistance: The confinements of tourism in a Hmong village in Thailand. In: M. Picard and R. Wood, eds., *Tourism, ethnicity and the state in Asian and Pacific societies*. Honolulu: University of Hawai'i Press, 128–154.

Michaud, J. and Turner, S. (2006). Contending visions of a hill-station in Vietnam. *Annals of Tourism Research*, 33, pp. 785–808.

Mostafanezhad, M. (2014). *Volunteer tourism: Popular humanitarianism in Neoliberal times*. Farnham: Ashgate.

Nguyen, T. and King, B. (1998). Migrant homecomings: Viet Kieu attitudes towards travelling back to Vietnam. *Pacific Tourism Review*, 1, pp. 349–361.

Nonthapot, S. (2016). Mediation between tourism contribution and economic growth in the greater Mekong subregion. *Asia Pacific Journal of Tourism Research*, 21(2), pp. 157–171,

Noree, T. (2015). Medical tourism: A case study of Thailand. In: N. Lunt, D. Horsfall and J. Hanefeld, eds., *Handbook on medical tourism and patient mobility*. Cheltenham: Elgar, 268–277.

Ormond, M. (2013). *Neoliberal governance and international medical travel in Malaysia*. London: Routledge.

Ormond, M. and Sulianti, D. (2017). More than medical tourism: Lessons from Indonesia and Malaysia on South-South intra-regional medical travel. *Current Issues in Tourism*, 20, pp. 94–110.

Pholpoke, C. (1998). The Chiang Mai cable-car project: Local controversy over cultural and eco-tourism. in P. Hirsch and C. Warren, eds., *The politics of the environment in Southeast Asia*. London: Routledge, 262–277.

Phongpaichit, P. (1982). *From peasant girls to Bangkok masseuses*. Geneva: ILO.

Picard, M. and Wood, R. (eds.) (1997). *Tourism, ethnicity and the state in Asian and Pacific societies*. Honolulu: University of Hawai'i Press.

Porananond, P. and Robinson, M. (2008). Modernity and the evolution of a festive tourism tradition: the Songkran Festival in Chiang Mai, Thailand. In Cochrane, J. (ed.), *Asian Tourism: Growth and Change*. Oxford: Elsevier, 311–321.

Rajaratnam, R., Pang, C. and Lackman-Ancrenaz, I. (2008). Ecotourism and indigenous communities: The lower Kinabatangan experience. In: J. Connell and B. Rugendyke, eds., *Tourism at the grassroots: Villagers and visitors in the Asia-Pacific*. London and New York: Routledge, 236–255.

Rugendyke, B. and Son, N. (2008). Priorities, people and preservation: Nature-based tourism at Cuc Phuong National Park, Vietnam. In: J. Connell and B. Rugendyke, eds., *Tourism at the grassroots: Villagers and visitors in the Asia-Pacific*. London and New York: Routledge, 179–197.

Simpson, T. (ed.) (2016). *Offshore Islands, enclave spaces and mobile imaginaries*. Amsterdam: Amsterdam University Press.

Sin, H. and Minca, C. (2014). Touring responsibility: The trouble with 'going local' in community based tourism in Thailand. *Geoforum*, 51, pp. 96–106.

Singapore Guide. (2017). www.singapore-guide.com/singapore-shopping/top-10-shopping-malls. Accessed 2 February 2017.

Smith, L., Waterton, E. and Watson, S. (eds.) (2012). *The cultural moment in tourism*. London: Routledge.

Tarplee, S. (2008). After the bomb in a Balinese village. In: J. Connell and B. Rugendyke, eds., *Tourism at the grassroots: Villagers and visitors in the Asia-Pacific*. London and New York: Routledge, 148–163.

Thanvisitthpon, N. (2016). Urban environmental assessment and social impact assessment of tourism development policy: Thailand's Ayutthaya Historical Park. *Tourism Management Perspectives*, 18, pp. 1–5.

Tirasatayapitak, A., Chaiyasain, C. and Beeton, R. (2015). Can hybrid tourism be sustainable? White water rafting in Songpraek Village, Thailand. *Asia Pacific Journal of Tourism Research*, 20(2), pp. 210–222.

Towner, N. (2016). Searching for the perfect wave: Profiling surf tourists who visit the Mentawai Islands. *Journal of Hospitality and Tourism Management*, 26, pp. 63–71.

Vodopivec, B. and Jaffe, R. (2011). Save the world in a week: Volunteer tourism, development and difference. *The European Journal of Development Research*, 23(1), pp. 111–128.

Wall, G. and Long, V. (1996). Balinese homestays: An indigenous response to tourism opportunities. In: R. Butler and T. Hinch, eds., *Tourism and indigenous peoples*. London: International Thomson Business Press, 29–48.

Walpole, M. and Goodwin, H. (2000). Local economic impacts of dragon tourism in Indonesia. *Annals of Tourism Research*, 27, pp. 559–576.

Williamson, P. and Hirsch, P. (1996). Tourism development and social differentiation in Koh Samui. In: M. Parnwell, ed., *Uneven development in Thailand*. Aldershot: Avebury, 186–203.

Winter, T. (2007). *Post-conflict heritage, postcolonial tourism: Culture, politics and development at Angkor*. London: Routledge.

Wong, T. (2008). Communities on edge: Conflicts over community tourism in Thailand. In: J. Connell and B. Rugendyke, eds., *Tourism at the grassroots: Villagers and visitors in the Asia-Pacific*. London and New York: Routledge, 198–213.

Yamashita, S. (2009). Southeast Asian tourism from a Japanese perspective. In: M. Hitchcock, V. King and M. Parnwell, eds., *Tourism in Southeast Asia*. Honolulu: University of Hawai'i Press, 189–205.

Yea, S. (2002). On and off the ethnic tourism map in Southeast Asia: The case of Iban longhouse tourism, Sarawak, Malaysia. *Tourism Geographies*, 4, pp. 173–194.

Zeppel, H. (1997). Headhunters and longhouse adventure: Marketing of Iban culture in Sarawak, Borneo. In: M. Oppermann, ed., *Pacific Rim tourism*. Wallingford: CAB International, 82–93.

PART 3

People and development
Introduction

Lisa Law, Fiona Miller and Andrew McGregor

In this section of the Handbook we return to the idea of the success story raised in the introduction: Is development in Southeast Asia creating more equitable, sustainable and empowering contexts for social worlds? As discussed in the Institutions and Economies section, the past few decades have witnessed the regional acquiescence to free market economics, with policies to promote regional integration and economic, political, social and cultural cooperation across Southeast Asian nations. These initiatives have integrated the region into broader regional and global markets in profound ways, but do the free flow of goods, services, labor and capital provide new avenues for economic, social and political inclusion?

A common theme running through the following chapters is that economic development is an uneven and relational process, with opportunity and disparity unevenly spread across a diverse region of 620 million people. Poverty is decreasing in an overall sense, but, as Gartrell and Hak show in their analysis of disability, the most underprivileged groups remain left behind. Basic health entitlements are now integral to most government policy frameworks, but Ormond, Chan and Verghis contend these are variegated across national borders depending on citizenship and residency/visa status. In their analysis of who is *entitled to* and *responsible for* health, they show how neoliberal reforms polarize access: the health and well-being of certain high-income mobile individuals seeking personal healthcare is privileged in the current system, while migrant workers with health insurance receive fewer benefits and endure more examinations. Remittances now exceed flows of foreign aid and form an important component of national GDP, but Philip Kelly stresses that migration comes with its own costs. Indeed, regional integration has meant men and women are now more mobile than ever, and these mobilities reproduce and transform social identities and relations across many axes (gender, ability, age, ethnic status, and religion or immigration status). These identities/relations are articulated in the transforming spaces of the household, workplace, hospitals, urban fringes, rural areas, etc., creating new social spaces in the region. The authors collected here address and re-think their impact on common categories and concepts shaping our understanding of Southeast Asia: filial piety/responsibility, indigeneity, agrarian transition, religious affiliation, remittances, gender complementarity and citizenship to name a few.

The modernizing forces unleashed by economic integration and increased mobility have had a profound impact on the organization of the family/household. Households in lower income countries like the Philippines, Myanmar, Laos and Cambodia export workers to countries like

Singapore, Malaysia and Thailand who in turn rely on labor migration and dual income households to keep their economies competitive. Brenda S.A. Yeoh and Shirlena Huang discuss the increasingly complex web of gendered mobilities that feed into child- and elder-care deficits in the region. They suggest intra- and transnational mobility reproduces some gendered stereotypes (women migrating as domestic workers, wives and flying grandmothers, and men migrating as construction workers, seafarers) while transforming others (left-behind husbands), although on balance they detect a resilience of gender ideals surrounding motherhood and filial piety. Philip Kelly also discusses these gendered mobilities in his analysis of the migration/development nexus. Kelly's chapter mediates the optimistic and pessimistic understandings of the remittances migrant workers make – poverty reduction, crisis insurance and investment versus brain drain, exploitation and family separation – and argues that given all the evidence critical questions need to be asked about how migration truly enhances the well-being and livelihoods of sending areas.

Critical research challenges us to remember not just the success stories, but the stories of the differently mobile, the left behind, ethnic minorities, the multiply disadvantaged and indigenous people. Bernadette P. Resurrección and Ha Nguyen's understanding of development and change through the prism of feminist political ecology reveals the powerful connections between gender, livelihoods and environment in the region. Large-scale investments in disasters and climate change have linked environmental and gender-differentiated impacts, and Resurrección and Nguyen advocate the value of a feminist political ecology perspective in recognizing the wider drivers that sustain inequality and disadvantage in the region. In a similar way, Alexandra Gartrell and Panharath Hak argue that regional integration and neoliberal reform has exacerbated disability-based disadvantage. People with disabilities are most likely to be left behind in villages, to be uneducated, underemployed and in poverty – thus compounding socio-economic disadvantage. Cultural attitudes and stereotypes can also make disability particularly challenging in Theravada Buddhist nations, where disability is viewed through the prism of *karma*/past actions. Gartrell and Hak point out the key gaps in national policies and a complete lack of engagement with the United Nations Convention on the Rights of Persons with Disabilities in the region.

In a regional sense there is a thorny task of mediating the universals declared in United Nations Conventions, which can uphold Eurocentric constructions of economic and personal development, and the vast diversity of Southeast Asia that yields a wide spectrum of culturally based understandings of development. In the 1990s this tension was expressed as Asian Values, but as Beazley and Ball's paper on children/youth and development shows, Western constructions of childhood and development over-generalize some types of work in poorer Southeast Asian countries and downplay any notion of children's agency and how they acquire self-esteem (e.g., street children working in the informal sector). Beazley and Ball illustrate their case with examples of street-connected children, children affected by disasters and forced migrant labor, and identify the Millennium Development Goals (MDGs) as an opportunity to bring into focus the cultural contexts of children. Sarah Turner's broad analysis of the political 'stance' toward ethnic minorities and indigenous groups in Southeast Asia makes a similar call for different understandings of development – ones that recognize alternative modernities and approaches to development. In the case of indigenous people, however, Turner shows that transnational discourses have had an empowering effect in terms of claiming 'indigenous status' and therefore access to resources.

Ethnic and religious diversity remains the hallmark of Southeast Asia. While it is impossible to generalize such diversity and development impacts across the region, Sarah Turner's chapter shows how processes such as agrarian transition, resettlement policies and environmental destruction have particular impacts on ethnic minority and indigenous groups. The capitalization of agriculture has facilitated mass conversions to sedentary cultivation and cash cropping, with particular implications for upland ethnic minorities and customary land tenure arrangements in places like northern Vietnam and Borneo. Population pressure has led to state-sponsored resettlement – Indonesia's transmigration program being most renowned – that bring dominant, lowland ethnic groups into contact with upland minorities and indigenous people in less densely settled areas. Indigenous and ethnic groups populate mountainous and forested regions rich in natural resources, which are being deforested, dammed and mined at rapid rates. Non-governmental organizations (NGOs) are fighting for their rights and making some inroads into environmental agendas like REDD+, however, and development always holds the prospect of being 'indigenized' in creative ways.

The practice of Buddhism, Islam, Christianity, Hinduism and traditions of animism, ancestor worship and shamanism make Southeast Asia one of the most religiously diverse regions in the world. Orlando Woods contemplates the articulation between religion and the modernizing impulses of development suggesting that layers of tradition and modernity have combined in the region to create complex religious assemblages. Such intertwining has been apparent from the earliest Arab mercantilism that brought Islam to the region to the onset of European colonialism and Christian conversion. Woods discusses the role that development plays in shaping religion using examples such as the shift from Taoism to Christianity amongst the Chinese population in Singapore; the modernization of Islam in Malaysia; and the Khmer Rouge's attempts to eradicate Buddhism in Cambodia. He also examines the tensions that arise when religious groups shape development agendas with international funding. More investigations in this field are needed and Woods recommends more contextualized, interpretivist research.

The final chapter in this section, by Tubtim Tubtim and Philip Hirsch, explores what regional integration and neoliberal reform looks like from the vantage point of the 'village': perhaps the original focus point of development imaginations. They explore the impacts of the transition from agrarian livelihoods toward more complexly differentiated resource bases in a peri-urban village in northern Thailand, and the tensions that arise as villagers adjust to contrasting ideals and realities of rurality and rural life. The village, Nong Khwai, has partly urbanized and has a diversifying population: migrant Shan construction workers from Myanmar; dormitory workers from remote northern Thailand; and middle class newcomers from Bangkok (and abroad). The population must negotiate contrasting constructions of the Thai countryside, as well as differing notions of (for example) privacy and collectivity/shared labor. They extend their village analysis to changing rural spaces across Southeast Asia, examining how rural places have been incorporated into wider political, economic, social and environmental domains, and how rurality can only be understood in relation to its others: 'urbanity' and 'modernity.'

Collectively this section argues there have been some wins in terms of prioritizing social welfare (e.g., healthcare, primary education) and meeting MDGs, but these wins are geographically uneven and there remain important inequalities across economic divides, social groups and citizenship status both across and between nations that exacerbate inequalities and differences in the region. There has not been a straightforward acquiescence to regional integration and neoliberal reform, however, and many have adapted and resisted/subverted the status quo through NGO movements, cultural resilience and indigenizing development processes in innovative ways.

Lisa Law, Fiona Miller and Andrew McGregor

14
FAMILY, MIGRATION AND THE GENDER POLITICS OF CARE

Brenda S.A. Yeoh and Shirlena Huang

Introduction

By embodying notions of home, family, gender, masculinity, femininity and sexuality, Southeast Asian families are central sites for the cultural expression and reworking of ideas of the 'modern' as well as for the expression of anxieties around the costs (and benefits) of reproduction and development (Brickell and Yeoh 2014). Intra-familial relations represent a continuous, fluid process of negotiations, contracts and exchange – whether altruistic, reciprocal, unequal or oppressive – within the context of broader political, economic and social change. Relations of equality and complementarity between Southeast Asian men and women have long been thought to be a regional characteristic (Andaya 2007), but much has changed with the advent of neoliberal globalization in recent times. Within the family/household,[1] Southeast Asian women play a critical role in ensuring the physical and social reproduction across generations, even though the normalization of the gendered division of labor and feminized care work often leave women's sacrifices and resilience unrecognized and unrewarded. Inasmuch as women are expected to be the lynchpins of the household responsible for shoring up the reproductive sphere, they are sometimes also blamed for household negligence in cases of marital dissolution or breakdown, as vividly demonstrated by Cambodian women's struggles for legitimacy in the face of stigma and shame resulting from the physical division of the marital house into two after a divorce (or other forms of marital disruption) (Brickell 2014).

The ideological and practical placing of women within the domestic sphere, alongside the overall expectation that the household functions as a major provider of care and welfare, is reinforced by prevailing government policies and discourses in the region (Ochiai 2009). The neoliberal strategies of the state in minimizing institutional support for household reproduction as part of economic restructuring in both developing and more advanced nation-states in Southeast Asia have led to the privatization and commercialization of care work. Where citizen women are unable to fill care deficits in middle-class households, these gaps have been plugged by migrant women from less developed economies in the region. This has led to the formation of gendered circuits of labor migration linked to the globalization of care work, or what Hochschild (2000, 132) calls 'global care chains,' referring to "a series of personal links between people across the globe based on the paid or unpaid work of caring." While these global care chains transfer social capital from the poorer to the richer countries and receive economic capital in return, it is also important to note that the economic value of the labor declines as it moves down the care chain (Parreñas 2012, 269), thus resulting in diminishing returns for each subsequent woman. In this vein, Rhacel Parreñas' notion of the 'international division of reproductive labor' alerts us to the fact that by moving down the care chain as opposed to across the gender divide, a system of gender substitution of care labor dependent on exploitative practices of extracting cheap labor from migrant women is emplaced. These migrant women are increasingly inducted "the global world's newest proletariat, into the global capitalist activities of the North, all under severely diminished citizenship regimes" (Dobrowolsky and Tastsoglou 2006, 3).

In Southeast Asia, the development of global care chains has resulted in a 'feminization' of migration to meet the gender-differentiated demand for care labor in order to plug care deficits in the family. This chapter examines the gender politics of care in families and households at both ends of the global care chain, and across the class spectrum. It begins with exploring the dual class-differentiated care strategies in the more developed Southeast Asian economies of importing migrant domestic workers as paid care labor, as well as drawing in migrant wives as unpaid care labor. It then turns to discussing the gendered care strategies among households at the southern end of the care chain that are affected by a significant outflow of parents – particularly mothers – as labor migrants. The final substantive section gives attention to the care strategies of more privileged Southeast Asian families as they navigate transnational migration circuits.

Importing paid care labor: migrant domestic workers

In the case of the more developed economies in the region such as Singapore and Malaysia, the rapid decline in fertility rates, coupled with increasing life expectancy as well as higher proportions of delayed or non-marriage, has led to looming child- and elder-care deficits within families that have to be plugged by global householding strategies. Coined by Douglass (2006) to refer to the way households sustain themselves by adopting market- and non-market-based options predicated on the international movement of people in order to resolve care deficits, global householding strategies include, for middle-class households, the market-based option of bringing in women from less developed economies in the region to serve as low-paid, surrogate care for children, the elderly and the infirm as well as perform domestic work. While eldercare work may also be 'outsourced' to (mainly female) migrant healthcare workers laboring in the institutionalized space of the nursing home, the prevalence of gendered ideologies based on 'Asian familialism' means that families continue to prefer to relegate the duty of elder-care to the privatized family sector in order to conserve some semblance of filial piety (Yeoh and Huang 2009). In this context, the 'live-in foreign maid' emerges as an increasingly common substitute to provide the care labor needed to sustain the household.

By outsourcing domestic and care work to other Southeast Asian women from less developed economies in the region at a low cost, socially and economically privileged women trade in their class privilege for (partial) freedom from the burden of household reproductive labor. This has the simultaneous effects of subordinating other (migrant) women to work conditions governed by retrogressive employer-employee relations and minimal occupational mobility; devaluing, racializing and commodifying household labor as unskilled and lowly paid work; and further entrenching and normalizing domestic and care work as resolutely 'women's work.' This is further compounded by state policies that treat migrant domestic workers as transient labor with hardly any socio-political rights to participate in wider civil society. In Singapore, for example, exacting policies have been instituted to ensure the surveillance of migrant bodies and that they gain no permanent foothold in the geobody of the nation; these include tying the validity of the work permit to specific employers, preventing family formation and settlement by disallowing accompanying dependents, prohibiting marriage to Singapore citizens and permanent residents, and immediately repatriating workers found to be pregnant. Similarly, in Malaysia, medical surveillance is also used to assert control over transnational domestic worker bodies, commonly associated with notions of contamination, prostitution and such related health issues as unwanted pregnancies and venereal diseases.

For middle-class households, drawing in transnational domestic workers as substitute reproductive labor has become a major strategy in releasing citizen women's labor into the sphere of paid work in contexts where the prevailing household division of labor remains ossified in patriarchal norms. Citizen women do not necessarily relinquish every aspect of care work to migrant workers, but exercise discretion in holding on to some tasks while delegating others. Middle-class mothers, for example, often draw on the 'maid' to take the drudgery out of domestic work and childcare while conserving maternal 'quality time' with the children in order to enact their identities as good mothers, even if such identification is not achieved without constant negotiation and a degree of ambivalence (Yeoh and Huang 2010). While the availability of a global workforce of transnational domestic workers has created space for women employers to reconstitute and redefine their roles in the provision of household care labor, "they are not necessarily able to disregard the conventional ideology of womanhood and motherhood as their frame of reference" in their dealings with their transnational domestic workers (Cheng 2003, 3–4). Indeed, negotiating motherhood vis-à-vis a foreign other woman in the house is a 'fraught terrain' as it lies uncomfortably at "the crossroads of anxieties of sameness and difference" (Pratt 1997, 173). Mothers who rely on the transnational domestic worker as substitute caregiver to juggle the demands of home and work often continue to wrestle with pangs of maternal guilt that they are failing their children, as well as fears of being supplanted by the domestic worker in the children's affections. This dilemma is often resolved – at least in part – by dividing mothering work into physical tasks that can be relegated to the transnational domestic worker and those involving emotional and/or nurturing labor that are seen to be embodied in the personhood of the 'real' mother.

In a similar vein, where it comes to taking care of elderly parents and relatives, many families who can afford it have devolved the responsibility (or at least the physical aspects of caregiving) to foreign domestic workers, particularly in the case of dual-career households striving to maintain middle- and upper-class lifestyles (Yeoh and Huang 2009). Ochiai (2010, 233) calls this 'liberal familialism,' where the cost of purchasing care labor is borne by the family but where filial piety is outsourced to others whose services are bought from the market.

The gender politics of the home is thus negotiated between local and foreign women vis-à-vis a racialized grid of highly asymmetrical power relations, while men continue to abdicate their household responsibilities. This genderized mode of care work substitution confirms the prevailing global gender order as it reinforces the construction of such work as women's work, while allowing host-countrymen to continue to play truant from the manual aspects of reproductive work. The politics of household reproduction that develops in many middle-class homes in Southeast Asia hence features mainly women – migrant women struggling to present themselves as docile bodies amenable to the disciplinary gaze of local women on the one hand, while disengaging from the role of the deferential inferior on the other.

Importing unpaid care labor: migrant wives

Somewhat analogous to the practice of middle-class families recruiting migrant domestic workers for householding purposes in the cities of Southeast Asia, working-class families without the financial means may turn to drawing on unpaid care labor by recruiting 'foreign brides.' With globalization and expanding educational and career opportunities for women in the more developed economies in Southeast and East Asia, working class men from the lower socio-economic strata who feel positionally 'left behind' by local women's participation in the workforce may seek to fill the care deficit in their households through international marriage with women from the less developed economies in the region who are considered more 'traditional' and willing to take on procreation and caring roles in sustaining the household.

For example, the increasing proportion of Singaporean men of Chinese ethnicity seeking 'foreign brides' from the less developed parts of Southeast Asia reflects the growing mismatch in marriage expectations between the two largest groups of singles: on the one hand, independent-minded, financially well-resourced graduate women with sophisticated expectations of marriage partners, and on the other, Chinese-speaking blue-collar male workers with low levels of education with a preference for women willing to uphold traditional gender roles and values (Yeoh et al. 2012). In a context of rapidly rising educational levels among young women, cultural and social impediments to women's 'marrying down' and men's 'marrying up' (in terms of education, occupation and income) have remained stubbornly resistant to change. Singaporean men continue to hold traditional attitudes about wifely roles while Singaporean women are increasingly unwilling to play such roles. A weakening in family and community match-making networks and the loosening of family ties have also led to young single adults being relatively free and independent in marriage decisions. At one level, the resulting delay in marriage and rise in singlehood may be an outcome of individual preferences; at the same time, structural difficulties such as the limited marriage market and a lack of match-making mechanisms also present themselves as major obstacles, particularly to men at the lower social scales who wish to marry. While marriage to men at the lower rungs of the socio-economic hierarchy may appear unattractive to their women co-nationals, their prospects may be more appealing to women from countries that are on the whole poorer (Yeoh et al. 2014). The resultant pattern hence features men from more affluent countries marrying women from poorer countries, with the women migrating to the men's home countries upon a successfully match-made marriage, following the logic of the 'marriage gradient,' or what Constable (2005) calls 'global spatial hypergamy.'

In this context, the larger structural inequalities of gender, race, class, culture and citizenship operating across a transnational stage are integral to an understanding of familial politics and household reproduction in Southeast Asia. Focusing their work on the incorporation of Southeast Asian marriage migrants in Taiwan, Wang and Belanger (2008, 92) draw on Aihwa Ong's concept of 'partial citizenship' to show "how the operation of differential legal and social citizenship justifies the perpetuation of a hierarchy of immigrants and serves to prop up the notion of a superior national Taiwanese identity." While the immigrant wives are theoretically folded into the nation-state as 'new citizens,' they are "set in relationships with the state, family and community, which together constitute their identities, and at the same time produce and reproduce a racialized and genderized society in Taiwan" (Wang and Belanger 2008, 92–93). For example, cast in the role of "a good wife, a good mother and a good daughter-in-law" (Wang 2007), female marriage immigrants are expected to be only interested in integration courses that tie them to their families and that help them improve their roles as carers of "their husbands (cooking, hairstyling), children (parenthood, healthcare, women and children safety) and the elderly (medical care training). There are no choices like political participation, Southeast Asian language media offering, local community facilities information, and so on" (Wang and Belanger 2008, 98). The links between marriage migration and citizenship are thus not only based on but constrained by notions of the patriarchal family, and of women as domestic caregivers and biological and social reproducers.

At the same time, marriage migration as an increasingly common form of social and geographical mobility within Southeast and East Asia provides women from less developed economies a range of opportunities to re-construct their positions vis-à-vis their natal families across transnational topographies, albeit with differing degrees of success in altering their roles and identities. The question of "whether remittances have the effect of reaffirming or reconfiguring gender ideologies and relations across transnational space" (King et al. 2006, 429) remains a complex one in the case of marriage migrants. Belanger and Tran's (2009) study on Vietnamese women-as-foreign brides in Asian countries, for example, suggests that while marriage migrants who have the opportunity to engage in paid work have the ability to negotiate disposable income to send remittances to their natal families, the majority who do not work find themselves largely dependent on their relationships with their husbands and family members if they wish to raise funds to remit. An interesting finding from their work shows that women who have children tend to have a greater ability to remit, suggesting that child-bearing improves the status of the marriage migrants within their marital families, which in turn confers greater bargaining power on them in carving out funds to send home to their natal families. In the context of Vietnamese migrants who marry Singaporean men, Yeoh et al. (2012) show how sending remittances are significant to the women as 'acts of recognition' in the construction of gendered identities as filial daughters, and in the re-imagining of the transnational family through the 'connecting' and 'disconnecting' power of remittances. Indeed, the act of sending remittances back home is an integral part of the project of mobility, where remittances represent "a special kind of transnational family money. . . [which] is equated with or measured against filial care" (Singh 2006, 375). This, however, has to be balanced against making contributions in building up resources to support their marital families. Placed between their natal and marital families, marriage migrants often find themselves navigating two sets of expectations in performing their care roles and identities as 'dutiful daughters' and 'sacrificial sisters' vis-à-vis 'worthy wives' and 'devoted daughters-in-law.' As Thai (2008, 166) notes, "international marriages present complex social expectations and imbalanced reciprocity among [marital] couples especially in confronting issues of allocation of wages to family left behind in the community of origin."

Care provisioning for the left-behind at the end of the care chain

In the wake of the increasing feminization of labor migration in Southeast Asia, relationships of care at the southernmost end of the care chain are evolving, resulting in millions of left-behind children growing up for part or all of their young lives in the absence of a migrant father, a migrant mother or both, and under the care of a 'single' parent or other surrogate caregivers. The transfer of care work and domestic duties down a series of personal links between people stretched across a hierarchy of nation-states and sub-regions often results in the 'off-loading' of care work to the migrant woman's family and kin who supply unpaid care labor at the least developed end of the chain.

Until recently, research on how care deficits are dealt with in families and communities with migrant members in sending countries at the southernmost end of the global care chain has been rather limited. From the turn of the new millennium, however, a broader range of migration studies featuring Southeast Asian source countries points to both the flexibility and durability of notions of familyhood underpinning transnational migration. For example, while out-migration in rural Thailand has become so widespread that it has a 'demonstration and emulation' effect on aspirations among members of these communities (Jones and Kittisuksathit 2003, 528), Knodel and Saengtienchai (2006) show that far from being deserted by migrating children, left-behind parents actively participate in providing financial and emotional support for their children's mobility, leading to the emergence of a translocal extended family linked by multi-stranded, fluid relations between members across geographical space.

Similarly, studies on migrating mothers and their left-behind children highlight the (re)enactment of episodes of family interaction through distanced communication as the family takes on transnational dimensions (Parreñas 2001; Asis et al. 2004). New work has also called attention to the gender politics implicit in a range of mutually constitutive interactions between migrants and the left behind: in areas such as rural Indonesia and Vietnam where feminized out-migration streams predominate, established gender ideologies may either be challenged by changing social practices where women assume breadwinner roles, or continue to regulate traditionally scripted roles for men and women but in new ways (Elmhirst 2000; Resurrección and Van Khanh 2007).

A major strand in current research on migrant mothers emphasizes the resilience of gender ideals surrounding motherhood even under migration in the transnational context. While mothering at a distance reconstitutes 'good mothering' to incorporate breadwinning, it also continues maternal responsibility of nurturing by employing (tele)communications regularly to transmit transnational circuits of care and affection to their families and children left in source countries. Asis (2002) and Graham et al. (2012) observed that most migrant mothers actively worked to ensure a sense of connection across transnational spaces with their children through modern communication technologies, while Sobritchea (2007) argued that 'long-distant mothering' is an intensive emotional labor that involves activities of 'multiple burden and sacrifice,' spending 'quality time' during brief home visits, and reaffirming the 'other influence and presence' through surrogate figures and regular communication with children. Migrant mothers often make considerable efforts to ensure that care work and responsibilities are transferred to other family members in their absence, although the available evidence seems to suggest that the 'intangibles' of the mothering identity are less yielding and not so easily reassigned. For example, Asis et al.'s (2004, 208) work shows that migrating mothers who depend on long-distance mothering tend to leave the children "in an indeterminate state of being 'neither here nor there' – of having a mother, yet not being able to enjoy her daily involvement in their lives." At the same time, it is important to guard against undue focus on "a discourse of maternal loss and absence" as such a discourse "not only works through conventional, potentially conservative notions of the family, but can and does quickly turn to blame" (Pratt 2009, 7–8). As Parreñas (2005) notes, migrant mothers are often stigmatized for leaving their children behind, while fathers working overseas are not. Long-distance mothering, often performed for the sake of advancing the children's education opportunities as part of what 'sacrificial breadwinner mothers' do, thus has to contend with highly entrenched views of what constitutes a sanctioned gender division of labor in the household. Gender ideals, particularly those concerning motherhood, continue to remain resilient even under migration in the transnational context.

In addition, the research thus far suggests that the care vacuum resulting from the absence of migrant mothers is often filled by female relatives such as grandmothers and aunts (Gamburd 2000; Scalabrini Migration Centre 2004; Parreñas 2005; Save The Children 2006). The continued pressure to conform to gender norms with respect to caring and nurturing practices explains men's resistance to, and sometimes complete abdication of, parenting responsibilities involving physical care in their wives' absence. These studies conclude that the "delegation of the mother's nurturing and caring tasks to other women family members, and not the father, upholds normative gender behaviors in the domestic sphere and thereby keep the conventional gendered division of labor intact" (Hoang and Yeoh 2011, 722).

More in-depth studies combining quantitative and qualitative analyses, however, have begun to reveal a more complex picture of more flexible gender practices of care in sending countries. In place of the image of the delinquent left-behind man who is resistant to adjusting his family duties in the woman's absence, some Southeast Asian men strive to live up to masculine ideals of being both 'good fathers' and 'independent breadwinners' when their wives are working abroad by taking on at least some care functions that signified parental love and authority while holding on to paid work (even if monetary returns are low) for a semblance of economic autonomy (Hoang and Yeoh 2011).

In the Philippines, for example, families are said to be traditionally 'patriarchal in authority' with the husbands' breadwinning role taking precedence over women's position (Castro et al. 2008). In terms of parenting, Harper (2010, 67) found that the Filipino childrearing style has also gradually altered from that of "strict parental discipline and child obedience" to one that is "nurturing, affectionate, protective and at times indulgent." While some researchers observe that the Filipino father's role continues to adhere solely to the provider and disciplinarian model, Medina (cited in Harper 2010, 68) suggests that Filipino fathers are actually involved in "activities such as storytelling, playing with children, helping with homework, driving children to and from school, and going on walks with their children," contending that a new role for fathers – as warm and supportive yet authoritative – may be emerging.

Given the longstanding migration of Filipino women to become global care workers over the last three decades (Asis 2005; Asis et al. 2004), role reversal of the traditional gender ideology that men are the 'pillars' while women are the 'lights' of the home is increasingly common in Filipino society. While Castro et al. (2008) found that the majority of Filipino men still reject and denigrate the notion of becoming househusbands, they would set aside their pride and perform household tasks for the sake of their family's survival needs. This finding aligns with that of Pingol's (2001) study on Filipino migrant wives and househusbands where she found that male respondents project themselves as important providers of care, even if they perform care differently from their wives. Her study reveals that while Filipino men may "experience sudden shifting of gears that heavily disorients them," the maintenance of their productive self keeps them going (Pingol 2001, 220). Filipino men shared that their ability to perform the additional caring tasks well "enhances their sense of self worth . . . and gives them pride not only as they view themselves but also as they are looked upon by the community" (Pingol 2001, 221).

Transnational care strategies among privileged migrant families

As noted above, the literature on the provision of transnational care by migrants to their family members in Southeast Asia and elsewhere has given considerable weight to mid- to unskilled (female) migrants providing long distance care to their 'left-behind' (non-adult) children. In comparison, there has been limited scholarship on the nexus between migration, care and the elderly, especially in the families of skilled and/or relatively well-off migrants to and from Southeast Asia. This neglect is not only because parents are considered outside the realm of the nuclear family (often taken as the conventional family unit in the Anglo-American scholarship that dominates this literature) but also because in Western societies, the care of the elderly is more often than not conceived as necessitating proximate physical care because eldercare usually only begins when the latter become physically or mentally frail (Kofman 2004; Baldassar et al. 2007; Masselot 2011). In Southeast Asia, as with much of the rest of Asia, however, the notion of 'family' and familial obligations often encompasses extended family members, and providing both material and emotional care for one's parents begins with the "mere negotiated recognition of 'getting old'" (Liu and Kendig 2006, 6).

Although usually able to move as a nuclear family unit, the families of skilled migrants face the challenge of providing inter-generational care to aging parents. While they are usually able to tap into new technologies and frequent flights home to fulfil their filial duties, members of this more privileged group still face their own politics of intergenerational obligations in trying to negotiate transnational caregiving. Beyond the employment of migrant care workers (usually female) for eldercare in nursing homes and other old-age institutions, or as live-in domestic help (as discussed earlier; see also Huang et al. 2012), there are other ways in which care, the elderly and migration intersect for middle class and more privileged families. Existing research done internationally suggests that these 'diasporas of care' (Williams, cited in Lie 2010, 1435) include issues of how migrants facilitate transnational eldercare for left-/stay-behind parents especially as the latter grow old and frail; healthcare options for elderly migrants who move when (or because) they are old, either to join their children who are overseas or as retirement migrants on their own; the cultural negotiations of providing eldercare for those who migrated when they were younger but who have different expectations of eldercare from that of the society into which they have settled; and importantly, the reciprocity of care with the 'zero generation' – the parents of first generation migrants – often acting as transnational caregivers rather than care receivers (van der Geest et al. 2004; King et al. 2014). While very limited, available research and anecdotal evidence suggest that several of these forms are important and growing in Southeast Asia.

In many Southeast Asian societies, there are strong cultural obligations on the part of adult children to provide care for aging parents who remain in the country of origin, leading to various negotiations by migrating family members to continue to provide care from a distance. Migrant adult children may provide parents with financial remittances even if not needed (hence acting more as a symbolic representation of the obligation to care) and care advice as well as maintain constant communication about physical and emotional health issues. For example, Lam et al. (2002) found that both single and married Chinese professionals from Malaysia who had migrated to Singapore to work made it a point to maintain regular contact with their parents, and even grandparents, who remained in Singapore, and/or send remittances that they saw as a substitute for physical care. Notably, while sons spoke about maintaining commuting or providing financial support, it was the daughters(-in-law) who ended up shuttling between Malaysia and Singapore to fulfil their obligation to care for frail, ill or widowed parents(-in-law) on the one hand, and their own children and spouses on the other, if the relocation of their parents to be with them was not a possibility even when the adult children wanted their aging parents to join them in Singapore. Thus, while often male-led, the physical burden of providing transnational care still falls on the women in these more privileged families.

Although parents are not always able or prepared to migrate permanently to live with their children in the host country (as Lam et al. 2002 found), Southeast Asia has become the recipient of increasing numbers of retirement migrants from the global north. Many have been drawn by the emerging 'retirement industry' bolstered by the governments of countries such as Malaysia, the Philippines and Indonesia in promoting retirement migration as a national development policy, targeting 'quality foreigners' over 50 years old and with financial capability from neighboring Singapore, but also China, Japan, South Korea, Taiwan as well as the Middle East, Europe and North America. These later-life migrants are also drawn by the lower cost of living, the easy availability and affordability of live-in domestic workers – transforming them from "average pensioners in one country" to "high-power[ed] consumers in another" (Toyota and Xiang 2012, 716), and for some single, Western male retirement migrants, the possibility of marrying a younger local woman who would hopefully take care of them at a later stage (Green 2015b, 3). While research suggests that these retirement migrants often continue to depend on their families and/or healthcare systems in their home societies to deal with health issues in the initial years, they may present burdens on local care systems in the longer term. Toyota (2006) found that some Japanese men ended up renting small rooms in local Thai neighborhoods because they had no regular incomes, while Green's (2015a, 2015b) study of educated, middle-class later-life migrants from the UK, US and Australia living in Thailand and Indonesia revealed that many of them preferred to stay on in Southeast Asia even as they aged and faced concerns about how to manage their chronic health issues, with some ending up destitute. A case in point was a 2011 *Bangkok Post* report about 400 or so older foreigners who were unable to pay their bills after receiving treatment in a state-run hospital in Phuket, Thailand (cited in Green 2015a, 7).

But retirement migrants may also be active care providers, either to their grandchildren (for example, Green [2015b] highlighted how grandchildren would fly in from the retirement migrants' homeland to spend time and receive care from their grandparents) and even to their older parents (for example, in the course of their fieldwork, Toyota and Xiang [2012, 711] "came across a number of elderly who had migrated to Southeast Asia with their parents who were in their 90s"). More commonly, however, as recent literature has increasingly recognized, grandparents have remained a significant resource for their families especially as caregivers to their grandchildren (Thang et al. 2011) even in transnational families. In particular, the care provided by 'flying grandmothers' – as Baldassar and Wilding (2014, 241) termed them – is a crucial aspect of maintaining intergenerational relationships in the transnational families of skilled migrants.

For ethnic minority communities, grandmothers often cross international borders to meet family commitments by contributing to child-rearing not just in a physical sense, but also in terms of language maintenance, religious development, cultural identification, etc. (Lie 2010). As with the grandmothers of PRC Chinese households studied by Lie (2010) in the UK, Japanese grandmothers (Huang and Lin 2014) and Chinese-Singaporean grandmothers (Baldassar and Wilding 2014, 242) were found to be willing to spend extended periods of several months or more, visiting their relocated sons/daughters in Singapore and Australia respectively, to assist with childcare, cooking, housework or simply to fulfill 'grandmothering duties,' especially during the birth of a grandchild or times of crisis. Elsewhere, Da (2003) and Mujahid et al. (2011) also observed that young children from professional Chinese families may be sent back to the home country to be taken care of by their grandparents, or very young children are left with grandparents while parents migrate first.

Transnational grandmothering is usually tied to gender ideology and gender role practices of migrants' home societies. Ethnic Chinese grandmothers are among the most documented global grandparents. Not only is it deemed important for the modern Chinese woman to work, but also, there are high level of intra-family responsibilities within Chinese families where the "Confucian belief in filial piety obliges children to care for elderly parents, and in return grandparents, especially grandmothers, often help to care for grandchildren" (Mujahid et al. 2011, 193; see also Da 2003; Lie 2010; Lee et al. 2015). The engagement of the elderly in 'transnational back-and-forth mobility' that is 'care-driven' challenges the view of elderly as purely care receivers and instead highlights how transnational care within families – though still highly gendered – is not only often multidirectional and multigenerational (King et al. 2014) but can also be asymmetrical and dynamic, transforming over time as families evolve and travel and communication technologies become increasingly affordable and flexible (Baldassar et al. 2007; Baldassar and Wilding 2014).

The migration of the more privileged class highlights the 'knock on' effect of mobility. As observed above, the movement of one family member "tends to provoke the movement of another" (Baldassar and Wilding 2014, 244). For these more well-off families, distance is not necessarily a hindrance to providing care to elderly family members, or receiving care in return. However, much research has demonstrated that the specific practices and processes of long-distance care are mediated not only by each family and family member's commitments and capacity to exchange care, the strength of the family's transnational networks of solidarity, but also what is considered culturally acceptable within the prevailing norms of one's home society (Baldassar 2008; Merla 2015). For example, outside the context of Southeast Asia, van der Geest et al. (2004) found that while it was acceptable to hire a private carer for elderly parents left behind in Crete, it was unacceptable for a non-relative to care for an ailing old person in Ghana; additionally, the presence or absence of social support systems beyond the family are also crucial. Further, not only are "middle-class and privileged families ... relatively absent in the transnational migration literature" (Conradson and Latham 2005) in Southeast Asia where the focus has been on migrants lower on the skills and class ladder, but even fewer studies have addressed the issue of transnational care of/by the elderly in skilled migrants' families (Masselot 2011, 301). The complex cultural and societal contexts across Southeast Asia as well as the differing gender regimes call out for more attention to be paid to the gender politics of the roles of and relationships between aging parents and their adult migrant children. Such research will provide deeper and more nuanced understandings of the social and policy implications of transnational families as societies age in a globalizing world.

Conclusion

As migrations and mobilities in the region accelerate, Southeast Asian households across the class spectrum – and at both ends of the care chain – are reworking care provisioning in order to sustain the 'family' as a social morphology and a regime of power. As Southeast Asian women migrate – whether as domestic workers or foreign wives – to fill care deficits in households of the more developed economies, the families they leave behind in the source communities are impelled to (re)negotiate prevailing gender discourses and the household division of labor. Similarly, transnational care strategies among the more mobile, elite Southeast Asian families are experiencing considerable flux. In the process, gender identities that are scripted into the performance of everyday and generational care within the family are being reconstituted or reaffirmed across the transnational stage. Amidst the dynamism of change, the fact that Southeast Asian women are playing increasingly agentic and varied roles holds out hope that forces have been awakened in the unmaking of patriarchal worlds in a region undergoing rapid transition.

Acknowledgements

Brenda S.A. Yeoh would like to acknowledge funding support from Singapore Ministry of Education Academic Research Fund Tier 2 (Grant number MOE 2015-T2-1-008, PI: Brenda Yeoh) and the Social Science and Humanities Research Council of Canada Partnership Grant (File No: 895–2012–1021, PI: Ito Peng) for the research drawn upon in this chapter.

Note

1 The two terms 'family' and 'household' usually connote different meanings – the 'household' is concerned with activities such as production, consumption and reproduction directed toward the satisfaction of human needs while 'family' is seen as inhering symbols, values and meanings. However, as Croll (2000: 107, quoting Rapp) argues, it is important not to miss the "essential connections" between them for "it is through their commitment to the concept of the family that people are recruited to the material relations of the household."

References

Andaya, B.W. (2007). Studying women and gender in Southeast Asia. *International Journal of Asian Studies*, 4(1), pp. 113–136.
Asis, M. M. B. (2002). From the life stories of Filipino women: Personal and family agendas in migration. *Asia and Pacific Migration Journal*, 11, pp. 67–94.

Asis, M. M. B. (2005). Recent trends in international migration in Asia and the Pacific. *Asia Pacific Population Journal*, 2(3), pp. 15–38.

Asis, M. M. B., Huang, S. and Yeoh, B. S. A. (2004). When the light of the home is abroad: Unskilled female migration and the Filipino family. *Singapore Journal of Tropical Geography*, 25(2), pp. 198–215.

Baldassar, L. (2008). Missing kin and longing to be together: Emotions and the construction of co-presence in transnational relationships. *Journal of Intercultural Studies*, 29(3), pp. 47–266.

Baldassar, L., Baldock, C.V. and Wilding, R. (2007). *Families caring across borders: Migration, ageing and transnational caregiving*. Hampshire: Palgrave Macmillan.

Baldassar, L. and Wilding, R. (2014). Middle-class transnational caregiving: The circulation of care between family and extended kin networks in the global north. In: L. Baldassar and L. Merla, eds., *Transnational families, migration and the circulation of care: Understanding mobility and absence in family life*. New York: Routledge Transnationalism Series, 235–251.

Belanger, D. and Tran, G. L. (2009). *Cross-border marriages, women's emigration and social development in rural Vietnam*. Paper presented at the International Congress of the International Union for the Scientific Study of Population, Morocco.

Brickell, K. (2014). 'Plates in a basket will rattle': Marital dissolution and home 'unmaking' in contemporary Cambodia. *Geoforum*, 51, pp. 262–272.

Brickell, K. and Yeoh, B. S. A. (2014). Geographies of domestic life: 'Householding' in transition in East and Southeast Asia. *Geoforum*, 51, pp. 259–261.

Castro, J., Dado, F. R. and Tubesa, C. I. (2008). When dad becomes mom: Communication and househusbands with breadwinner wives. *Far Eastern University Communication Journal*, 2, pp. 1–11.

Cheng, A. (2003). *Reframing work and identity: Changing meanings of domesticity, womanhood, and motherhood in globalization*. Paper presented at the Annual Meeting of the American Sociological Association, Atlanta. Available at: http://citation.allacademic.com//meta/p_mla_apa_research_citation/1/0/7/8/8/pages107889/p107889-1.php. Accessed 23 October 2015.

Conradson, D. and Latham, A. (2005). Transnational urbanism: Attending to everyday practices and mobilities. *Journal of Ethnic and Migration Studies*, 31(2), pp. 227–233.

Constable, N. (2005). *Cross-border marriages: Gender and mobility in transnational Asia*. Philadelphia: University of Pennsylvania Press.

Croll, E. (2000). *Endangered daughters: Discrimination and development in Asia*. London: Routledge.

Da, W. W. (2003). Transnational grandparenting: Child care arrangements among migrants form the People's Republic of China to Australia. *Journal of International Migration and Integration*, 4(1), pp. 79–103.

Dobrowolsky, A. and Tastsoglou, E. (2006). Crossing boundaries and making connections. In: E. Tastsoglou and A. Dobrowolsky, eds., *Women, migration and citizenship: Making local, national and transnational connections*. Hampshire: Ashgate Publishers, 1–35.

Douglass, M. (2006). Global householding in Pacific Asia. *International Development Planning Review*, 28(4), pp. 421–445.

Elmhirst, R. (2000). A Javanese diaspora? Gender and identity politics in Indonesia's transmigration resettlement program. *Women's Studies International Forum*, 23(4), pp. 487–500.

Gamburd, M. R. (2000). *The kitchen spoon's handle: Transnationalism and Sri Lanka's migrant housemaids*. Ithaca, NY: Cornell University Press.

Graham, E., Jordan, L. P., Yeoh, B. S. A., Lam, T., Asis, M. and kamdi, S. (2012). Transnational families and the family nexus: Perspectives of Indonesian and Filipino children. *Environment and Planning A*, 44(4), pp. 793–815.

Green, P. (2015a). Mobility regimes in practice: Later-life westerners and visa runs in South-East Asia. *Mobilities*, 10(5), pp. 748–763.

Green, P. (2015b). Mobility, stasis and transnational kin: Western later-life migrants in Southeast Asia. *Asian Studies Review*, 39(4), pp. 669–685.

Harper, S. E. (2010). Exploring the role of Filipino fathers: Paternal behaviors and child outcomes. *Journal of Family Issues*, 31(1), pp. 66–89.

Hoang, L. A. and Yeoh, B. S. A. (2011). Breadwinning wives and 'left-behind' husbands: Men and masculinities in the Vietnamese transnational family. *Gender & Society*, 25(6), pp. 717–739.

Hochschild, A. R. (2000). Global care chains and emotional surplus value. In: W. Hutton and A. Giddens, eds., *On the edge: Living with global capitalism*. London: Sage, 130–146.

Huang, S. and Lin, Y. T. (2014). *Singaporean-Japanese families: Negotiating transnational identities through everyday practices*. Paper presented at the International Geographical Union Regional Conference, Krakow.

Huang, S., Yeoh, B. S. A. and Toyota, M. (2012). Caring for the elderly: The embodied labour of migrant are workers in Singapore. *Global Networks*, 12(2), pp. 195–215.

Jones, H. and Kittisuksathit, S. (2003). International labour migration and quality of life: Findings from rural Thailand. *International Journal of Population Geography*, 9(6), pp. 517–530.

King, R., Cela, E., Fokkema, T. and Vullnetari, J. (2014). The migration and wellbeing of the zero generation: Transnational care, grandparenting, and loneliness among Albanian older people. *Population, Space and Place*, 20(8), pp. 728–738.

King, R., Dalipaj, M. and Mai, N. (2006). Gendering migration and remittances: Evidence from London and northern Albania. *Population Space and Place*, 12(6), pp. 409–434.

Knodel, J. and Saengtienchai, C. (2006). Rural parents with urban children: Social and economic implications of migration for the rural elderly in Thailand. *Population, Space and Place*, 13(3), pp. 193–210.

Kofman, E. (2004). Family-related migration: A critical review of European studies. *Journal of Ethnic and Migration Studies*, 30(2), pp. 243–262.

Lam, T., Yeoh, B. S. A. and Law, L. (2002). Sustaining families transnationally: Chinese-Malaysians in Singapore. *Asian and Pacific Migration Journal*, 11(1), pp. 117–143.

Lee, Y. S., Chaudhuri, A. and Yoo, G. J. (2015). Caring from afar: Asian H1B migrant workers and aging parents. *Journal of Cross Cultural Gerontology*, 30, pp. 319–331.

Lie, M. L. S. (2010). Across the oceans: Childcare and grandparenting in UK Chinese and Bangladeshi households. *Journal of Ethnic and Migration Studies*, 36(9), pp. 1425–1443.

Liu, W. T. and Kendig, H. (eds.) (2006). *Who should care for the elderly? An East-West value divide*. Singapore: Singapore University Press.

Masselot, A. (2011). Highly skilled migrants and transnational care practices: Balancing work, life and crisis over large geographical distances. *Canterbury Law Review*, 17(2), pp. 299–315.

Merla, L. (2015). Salvadoran migrants in Australia: An analysis of transnational families' capability to care across borders. *International Migration*, 53(6), pp. 153–165. doi:10.1111/imig.12024

Mujahid, G., Kim, A. H. and Man, G. C. (2011). Transnational intergenerational support: Implications of aging in Mainland China for the Chinese in Canada. In: H. Cao and V. Poy, eds., *The China challenge: Sino-Canadian relations in the 21st century*. Ottawa: University of Ottawa Press, 183–204.

Ochiai, E. (2009). Care diamonds and welfare regimes in East and Southeast Asian societies: Bridging family and welfare sociology. *International Journal of Japanese Sociology*, 18(1), pp. 60–78.

Ochiai, E. (2010). Reconstruction of intimate and public spheres in Asian modernity: Familialism and beyond. *Journal of Intimate and Public Spheres*, Pilot Issue, 2–22.

Parreñas, R. S. (2001). *Servants of globalization: Women, migration and domestic work*. Stanford, CA: Stanford University Press.

Parreñas, R. S. (2005). *Children of global migration: Transnational families and gendered woes*. Stanford, CA: Stanford University Press.

Parreñas, R. S. (2012). The reproductive labour of migrant workers. *Global Networks*, 12(2), pp. 269–275.

Pingol, A. (2001). *Remaking masculinities: Identity, power, and gender dynamics in families with migrant wives and househusbands*. Philippines: University Center for Women's Studies.

Pratt, G. (1997). Stereotypes and ambivalences: The construction of domestic workers in Vancouver, British Columbia. *Gender, Place and Culture*, 4(2), pp. 159–177.

Pratt, G. (2009). Circulating sadness: Witnessing Filipina mothers' stories of family separation. *Gender, Place and Culture*, 16(1), pp. 3–22.

Resurrección, B. P. and Van Khanh, H. T. (2007). Able to come and go: Reproducing gender in female rural–urban migration in the Red River Delta. *Population, Space and Place*, 13(3), pp. 211–224.

Save the Children. (2006). *Left behind, left out: The impact on children and families of mothers migrating for work abroad*. Sri Lanka: Save the Children.

Scalabrini Migration Centre. (2004). *Hearts apart: Migration in the eyes of Filipino children*. Manila: SMC.

Singh, S. (2006). Towards a sociology of money and family in the Indian diaspora. *Contributions to Indian Sociology*, 40(3), pp. 375–398.

Sobritchea, C. (2007). Constructions of mothering: The experience of female Filipino overseas workers. In: T. W. Devasahayam and B. S. A. Yeoh, eds., *Working and mothering in Asia: Images, ideologies and identities*. Singapore: NUS Press, 173–194.

Thai, H. C. (2008). *For better or for worse: Vietnamese international marriages in the new global economy*. New Brunswick, NJ: Rutgers University Press.

Thang, L. L., Mehta, K., Usui, T. and Tsuruwaka, M. (2011). Being a good grandparent: Roles and expectations in intergenerational relationships in Japan and Singapore. *Marriage and Family Review*, 47(8), pp. 548–570.

Toyota, M. (2006). Ageing and transnational householding: Japanese retirees in Southeast Asia. *International Development Planning Review*, 28(4), pp. 515–531.

Toyota, M. and Xiang, B. (2012). The emerging transnational 'retirement industry' in Southeast Asia. *International Journal of Sociology and Social Policy*, 32(11/12), pp. 708–719.

van der Geest, S., Mul, A. and Vermeulen, H. (2004). Linkages between migration and the care of frail older people: Observations from Greece, Ghana and The Netherlands. *Ageing & Society*, 24, pp. 431–450.

Wang, H. (2007). Hidden spaces of resistance of the subordinated: Case studies from Vietnamese female migrant partners in Taiwan. *International Migration Review*, 41(3), pp. 706–727.

Wang, H. and Belanger, D. (2008). Taiwanizing female immigrant spouses and materializing differential citizenship. *Citizenship Studies*, 12(1), pp. 91–106.

Yeoh, B. S. A., Chee, H. L. and Vu, T. K. D. (2012). Commercially arranged marriage and the negotiation of citizenship rights among Vietnamese marriage migrants in multiracial Singapore. *Asian Ethnicity*, 14(2), pp. 139–156.

Yeoh, B. S. A., Chee, H. L. and Vu, T. K. D. (2014). Global householding and the negotiation of intimate labour in commercially-matched international marriages between Vietnamese women and Singaporean men. *Geoforum*, 51, pp. 284–293.

Yeoh, B. S. A., Chee, H. L., Vu, T. K. D. and Cheng, Y. (2012). Between two families: The social meaning of remittances for Vietnamese marriage migrants in Singapore. *Global Networks*, 13(4), pp. 441–458.

Yeoh, B. S. A. and Huang, S. (2009). Foreign domestic workers and home-based care for elders in Singapore. *Journal of Aging and Social Policy*, 22(1), pp. 69–88.

Yeoh, B. S. A. and Huang, S. (2010). Transnational domestic workers and the negotiation of mobility and work practices in Singapore's home-spaces. *Mobilities*, 5(2), pp. 219–236.

15
HEALTHCARE ENTITLEMENTS FOR CITIZENS AND TRANS-BORDER MOBILE PEOPLES IN SOUTHEAST ASIA

Meghann Ormond, Chan Chee Khoon and Sharuna Verghis

Introduction

The health of a population is shaped by a complex interplay of economic, political, social, cultural and biological factors that transcend national borders. In this chapter, we consider how these factors work together within the Association of Southeast Asian Nations (ASEAN), a supra-national regional political space encompassing a population of over 600 million across ten countries. Specifically, we query how economic liberalization and corollary increases in the movement of people, goods and services are challenging and transforming entrenched ideas about who is *entitled to* and *responsible for* health and social care within and across ASEAN member states' national borders.

Our discussion unfolds over three sections. In the first section, we examine regional-level economic and social policies/charters shaping contemporary national and regional healthcare discourses and practices within ASEAN. We show how ideological shifts from national toward regional and global health governance affect ASEAN member states' governments, citizen and non-citizen residents, welfare structures and industries. In so doing, we pay attention to how measures that understand access to healthcare as both a human right and an economic good work to privilege certain political, economic and social subjects over others and how these different actors comply with, resist and/or ignore them.

In the second section we consider how a range of ASEAN member states have worked to establish universal health coverage (UHC) for their (national) populations, and identify some of the challenges they face as healthcare access becomes increasingly multi-tiered. Not only do neoliberal reforms work to further polarize access between rich and poor through the privatization of health and social care, they also work to exacerbate the divide between citizens and non-citizens. Though many ASEAN member states host ever-larger populations of economic migrants, asylum seekers and stateless peoples (many of whom hail from other ASEAN member states), healthcare coverage and social protection within them continue to be linked to citizenship and thus exclude large vulnerable portions of these countries' populations. This predicament is especially ironic when contrasted with substantial regional promotion of and growth in trans-border flows of private healthcare investment, ownership, provision and travel.

Building on this distinction between citizens' and non-citizens' healthcare access and social entitlements, in the third section we depict the challenges and vulnerabilities faced by peoples rendered mobile both within and outside of their country of origin. Specifically, we demonstrate how inequalities generated through intersections of gender and politico-legal status shape migrants' health and well-being and their options for accessing care and caring for themselves. These examples reveal problems in how governments in both destination and source countries manage tensions between citizenship and human rights. They also clarify the limited reach and application of regional- and global-level social charters meant to harmonize practices in order to ensure greater protection for all.

From national to regional and global health and healthcare governance

ASEAN's stated aim is to "enhance the well-being and livelihood of the peoples of ASEAN by providing them with equitable access to opportunities for human development, social welfare and justice" (ASEAN 2012). The dominant means for accomplishing this objective have been the "deepen[ing] and broaden[ing of] internal economic integration and linkages with the world economy" (ASEAN 2004) to realize an ASEAN Economic Community (AEC). Regional integration is held to be strengthened "through liberalization and facilitation measures in the area of trade in goods, services and investments; and promot[ion of] private sector participation" (ASEAN 2009c). This involves measures such as the harmonization of goods and services standards and regulations, increasing transport of goods through the improvement of transport network infrastructure and services and the streamlining of customs procedures, public-private partnerships, liberalization of trade in services, mutual recognition agreements (MRAs), facilitated mobility of skilled labor, visa harmonization and the elimination of visa requirements for ASEAN nationals.

Yet to ensure peace, stability and shared prosperity in the region, ASEAN Community policies are geared not only at economic integration but also at improving the social welfare of those living within the member states. One of the main challenges to the latter objective resides not only in the significant differences between lower- and higher-income countries but also within them. To address these differences, the ASEAN Socio-Cultural Community (ASCC) Blueprint was developed. Its social welfare and protection clauses focus on seven fields: hunger and poverty alleviation; social safety net and protection from the negative impacts of integration and globalization; food security and safety; access to healthcare and promotion of healthy lifestyles; control of communicable diseases; a drug-free ASEAN; and, finally, disaster resiliency and safer communities (ASEAN 2009a).

Healthcare, earmarked as a priority for regional integration, is conceptualized in ASEAN policies both as a marketable industry to be developed *and* as a human right vulnerable to the effects of globalization. As a result, diverse policies – based on the stance that "accessible health services for citizens is no longer the sole responsibility of the state" (Hashim et al. 2012) – focus not only on improving health equity and public health outcomes but also on developing and opening up healthcare as an industry. This interest in fostering self-responsibility by "empower[ing] consumers to become active participants in health care" (ASEAN 2009b, 75) is explicit: "Access to health care and promotion of healthy lifestyles focuses on improving primary health care and public health education and coordination as well as strengthening capacity and competitiveness in health products and services" (ASEAN 2009a, 8–9).

As subjects of political, economic and social intervention, ASEAN member states' populations are distinctly envisioned as *national* citizens who are *regional* consumers. To date, ASEAN's approach to healthcare has assumed a largely conventional international health governance (IHG) perspective. Due to the "primacy of national sovereignty, the prevalence of national interests over the common good and the culture of rule by consensus" (Lamy and Phua 2012, 238), ASEAN's engagement in health issues can be conceptualized as a "useful mid-way organization relaying engagements of international agreements to the region" (Lamy and Phua 2012, 248). Challenged by member states' reticence to politically and financially cooperate on the basis of a shared vision of health's status as a global public good, health remains little more than a trans-border security issue in times of emergency. Indeed, while member states' health ministers may appear to cooperate through coordination and dialogue on public health concerns (e.g., communicable diseases, tobacco regulation, air pollution, natural disasters, etc.), healthcare policy and regulation of access to healthcare are still primarily deemed national concerns. For example, the ASCC health and development agenda – which calls attention to the need for improved access to adequate and affordable healthcare, medical services and medicines, and promotion of healthy lifestyles – has so far mainly privileged control over communicable diseases and food quality (Acuin et al. 2011; Pocock 2015). By contrast, joint action regarding the management of cross-border population and health-related flows, the sharing of health resources and the development of trans-border universal health coverage remains limited. The national versus regional sovereignty question thus indicates an impasse to enacting meaningful region-wide improvements to healthcare.

A global health governance (GHG) perspective on ASEAN's approach to health and social welfare reveals a more nuanced picture. It entails paying attention to more than government-to-government commitments (as with the IHG perspective) by also examining multi-sector engagements (e.g., with actors from industry and civil society). The strategic objective to "ensure that all ASEAN peoples are provided with social welfare and protection from the possible negative impacts of globalization and integration by improving the quality, coverage and sustainability of social protection and increasing the capacity of social risk management" (ASEAN 2009a, 6) does not stipulate *individual member states'* responsibilities but rather indicates a broader post-national *regional* and *market*-led responsibility to accomplish the objective. With the ASEAN Free Trade Agreement's (AFTA) liberalization of trade in services within the region, for example, healthcare has constituted one of the first priority services sectors – alongside air transport, information technology and tourism – to be liberalized (ASEAN 2009c, 13).

One key aspect of this liberalization was the reduction of visa requirements for intra-regional travel by ASEAN nationals in order to facilitate temporary cross-border travel for individuals seeking to access market-based healthcare options to satisfy personal healthcare needs (a.k.a. 'medical tourism') (ASEAN 2009b, 26). This has been considered a concrete step toward "realising an ASEAN Community that is people-centred and socially responsible with a view to achieving enduring solidarity and unity among the nations and peoples of ASEAN by forging a common identity and building a *caring and sharing society* which is inclusive and harmonious where the well-being, livelihood, and welfare of the peoples are enhanced" (ASEAN 2009a, 1, our emphasis). Yet, as we observe in the following sections in greater detail, this regional effort privileges the health and well-being of certain socio-economic, political and legal categories of mobile *individuals* while rendering the domestic and cross-border healthcare pursuits of others (e.g., low-income nationals, documented and undocumented migrants and refugees, etc.) more suspect and prone to criticism, abuse and exploitation as well as posing risks to the public health systems in both source and destination countries (Ormond 2014).

Another key aspect of this region-wide harmonization-via-liberalization approach has been the ASEAN Framework Agreement on Services' (AFAS) push to enable the intra-ASEAN mobility of much-needed doctors, nurses and dentists to mitigate member states' often severe domestic health worker shortages (Kittrakulrat et al. 2014). This move has been welcomed by the Philippines, a major global player in health worker outsourcing and, now, potent regional supplier (Rodriguez 2010). Yet, several potential destination member states are resistant because their national professional associations and smaller healthcare players wish to both protect the quality of care provision and limit competition to ensure the stability of their practices. Such protectionism has led Vietnam to require foreign doctors wishing to practice there to pass a language test, for example, thus significantly limiting numbers (Luistro 2015). Significantly, regional liberalization of the healthcare 'industry' promises the greatest benefits to both higher-income individual 'free agents' and large corporate players that can more easily move between and invest in other countries (e.g., major multi-national hospital conglomerates, pharmaceutical and device producers, etc.) (Center for International and Strategic Studies 2009).

Divergent healthcare access and entitlements to citizens and non-citizens

The divergent perspectives outlined above vis-à-vis trans-border healthcare commoditization versus regionally negotiated social entitlements to healthcare are starkly evident within ASEAN, where politico-legal and socio-economic statuses play a significant role in determining healthcare access at the national level. National health systems are undergoing major transformations in order to extend coverage to more of their citizens all the while adhering to neoliberal reforms to cut care costs and to make citizens more financially responsible for their own health. At the same time, some member states (e.g., Malaysia, Singapore and Thailand) have emerged both as major players in the privatized regional health market and as hosts to sizeable migrant worker populations from other ASEAN (e.g., Indonesia, Myanmar and the Philippines) and South Asian (e.g., Bangladesh and Nepal) source countries, leading to new national and trans-border regional challenges to guaranteeing and managing access to quality care for non-nationals.

In recent years, World Health Organization (WHO) member states have been urged to hasten national-level reforms to ensure timely, universal access to quality health services as and when needed, without healthcare users falling into financial hardship (World Health Assembly 2005, 2011). Notwithstanding recent moves to corporatize some public healthcare institutions, Malaysian and Singaporean citizens have long benefited from widely accessible tax-funded government healthcare, while Brunei nationals (who do not pay personal income tax) enjoy wide-ranging health and social benefits at public expense. In Thailand, the 2002 National Health Security Act extended healthcare coverage beyond civil servants and their dependents, and employees in the formal (private) sector, to the vast bulk of those who hitherto had limited access to necessary healthcare. This new initiative (Universal Coverage Scheme, or UCS), covering about 75 percent of the population, is financed by general tax revenues and offers a comprehensive package of services encompassing curative and preventive care. Public healthcare facilities are the main providers of care for more than 95 percent of UCS beneficiaries (Health Systems Research Institute 2012). In Indonesia, ten years after the National Social Security System Law was enacted, the Social Security Management Agency was established in 2014 to implement a national health insurance scheme. Building on the experience with the Jakarta Health Card scheme he introduced for the city's underserved when he was Jakarta's governor (2012–2014), President Joko Widodo launched the National Health Insurance Scheme on 3 November 2014 with the ambitious targets of enrolling 121.6 million citizens in the first year and achieving universal coverage for a projected 250 million citizens by 2019 (Sciortino 2014).

In a national context, however, universal health coverage (UHC) often translates into *citizenship*-based entitlements, leaving (documented and undocumented) migrant workers, refugees and asylum-seekers to fall through the cracks. Considering that Southeast Asia is home to major labor-exporting and -receiving countries, this constitutes an urgent priority for the ASEAN regional agenda. Indeed, of the 14 million international migrant workers originating from ASEAN member states, about 6 million are intra-ASEAN migrants, with 90 percent of them hosted by Malaysia,[1] Singapore and Thailand[2] (Baruah 2012). The presence of sizeable populations of undocumented migrants in particular presents distinctive public health challenges. Recall that the SARS pandemic erupted, and subsided, over an eight-month period in 2002–2003 in the absence of therapeutics, clinically useful diagnostics and vaccines. One of the key control measures that helped break the chains of transmission and extinguish the pandemic – quarantine and meticulous contact-tracing – would be difficult to implement when large populations of undocumented migrants have a strong incentive to avoid contact with government agencies. At the end of his country visit to Malaysia (19 November–2 December 2014), UN Special Rapporteur on the Right to Health Dainius Pūras noted that undocumented migrants are

> considered illegal in the country [Malaysia] and face criminal penalties for being undocumented, ranging from fines to imprisonment and caning. During my visit, I learned about the establishment of immigration counters inside public hospitals to facilitate the referrals of undocumented migrants and asylum seekers to the police when they come seeking medical attention. I consider that this practice goes against public health interests and the code of ethics of doctors. The establishment of these counters will deter undocumented migrants from seeking health care for fear of being reported, which among other things could cause the spread of communicable diseases.
> (in OHCHR, 2014; see also Hospital Kuala Lumpur 2014)

The pandemic potential of the MERS coronavirus, a more lethal but less transmissible relative of SARS, is amplified for the region by the annual flows of Hajj pilgrims traveling between Southeast Asia[3] and the Saudi epicenter. When contemplating unsettling but quite plausible scenarios (e.g., if SARS or MERS were to spread into large undocumented migrant populations), it is all too easy to slip into a counter-productive health policing mind-set that reinforces xenophobic sentiments toward 'the diseased and threatening other.' A pragmatic (rights-based) public health approach at both national and supra-national regional levels, therefore, coupled with deterrent penalties for human trafficking and illicit employment of vulnerable and compliant undocumented migrants, would be more effective in the internal and trans-border control of communicable diseases and the protection of migrants' health and well-being (Mann et al. 1994).

While ASEAN member states continue to struggle with commitments on migrants' rights and benefits, governments in Malaysia, the Philippines, Singapore and Thailand are decidedly keen to attract a different category of health-seeking foreigner: 'medical tourists' (Ormond and Mainil 2015). The Malaysian federal government, for instance, which controls the world's second largest listed healthcare provider, IHH Healthcare, is more preoccupied with developing an integrated regional health market than with developing an interventionist social charter that regionalizes UHC on a multilateral basis (e.g., portability of benefits, tax options and social entitlements for foreign workers, etc.) (ASEAN 2009c; Chan 2012; PEMANDU 2015). Indeed, unlike Thailand, Malaysia and Singapore prefer to rely on private insurers for migrants' health coverage (Guinto et al. 2015). It is sobering to note that, just as the New Deal and the British welfare state emerged from the wreckage of the Great Depression and the Second World War, a regionally negotiated social contract for UHC and other entitlements ultimately also may require as midwife a (global) trauma on the scale of the apocalyptic scenarios envisaged by climate change catastrophists or uncontrolled pandemic outbreaks[4] (Trajano 2015).

Health rights for and vulnerabilities of internally and internationally mobile peoples

Within ASEAN, international migration has largely been driven by economic differentials between countries and political and ethnic conflict within certain countries. Internal migration, meanwhile, consisting mainly of rural-urban migration and environmental displacement (as witnessed, for example, in the aftermath of the 2004 tsunami and the 2008 Cyclone Nargis), has been propelled by uneven national economic development and the region's vulnerability to natural hazards (IOM 2010; Cole et al. 2015; Sugiyarto 2015).

However, internal and international migration are not discrete processes (Huguet and Chamratrithirong 2011; Soda 2014). The in- and out-migration flows within and from Indonesia's Riau Islands, for example, highlight the fluidity of migration and the ever-shifting identity of migrants as internal *and/or* international migrants traversing the continuum of geographical spaces and geo-political boundaries between Indonesia, Malaysia and Singapore (Lyons and Ford 2007). Migratory processes such as these have consequences for health equity, rights and justice for mobile populations because the healthcare entitlements that these migrants can claim vary as per their changing migratory status from internal to international migrants as they cross national borders. Equally, being documented or undocumented impacts their access to healthcare once they are international migrants who have crossed from Indonesia into Malaysia or Singapore. Simultaneously, even within the same geographical jurisdiction, the health needs and vulnerabilities of internal and international migrants and of different categories of international migrants differ. Thus, for example, the government of Indonesia would be challenged to develop distinctive policies to address the unique health(care) needs of internal migrants coming into the Riau Islands and of international migrants preparing to travel overseas or international migrants returning home after successfully completing their contract vis-à-vis deportees from Malaysia or Singapore.

Thus, it is within complex inter-relationships between intra- and trans-national mobility that multiple intersecting inequalities transpire. One of these relates to the intersection of gender, race and immigration status in the labor sector. For instance, both internal and international migration flows within ASEAN evidence a trend toward feminization. Women comprise the majority of migrants leaving source countries and engage in gendered occupations like domestic work, nursing and entertainment (Baruah 2012; UNDP 2015). However, domestic work is generally excluded from employment ordinances that protect labor and health rights, thus subjecting migrant women workers to social isolation, sexual abuse, and denial of labor and health rights (UN Women 2013). Furthermore, much intra-ASEAN migration is irregular in nature and eludes official statistics, which exacerbates the vulnerability of undocumented migrants to abuse, exploitation and exclusion from healthcare entitlements (CARAM Asia 2009; Sugiyarto 2015). The absence of durable solutions for the stateless and asylum-seekers, a factor exacerbating human smuggling and trafficking in the region, was patent in two important events in 2015: the discovery of mass graves of trafficked migrants in Thailand and Malaysia (Beh 2015) and the humanitarian crisis involving the stateless Rohingyas stranded at sea, whose boats were initially pushed back by Malaysian and Thai authorities (Associated Press 2015).

However, immigration documentation status and the distinction between documented and undocumented within ASEAN member states is often fluid and tenuous. Migrants have limited options to obtain and retain legal status owing to restrictive admission policies, border controls, immigration policies that discourage social integration, weak enforcement of labor laws, and few options for legal redress of grievances. Furthermore, being documented is a necessary but not sufficient condition for being able to access public healthcare because, as noted above, access to healthcare in most Southeast Asian destination countries is mediated by legal citizenship. In Malaysian public hospitals, for instance, not only do undocumented migrants risk arrest after obtaining treatment, but all non-citizens pay substantially higher user fees, are only eligible for a five-day supply of medication and must pay out-of-pocket for the treatment of infectious diseases. Thailand, too, in spite of progressive policies that allow even undocumented migrants to purchase health insurance, separates its Compulsory Migrant Health Insurance (CMHI) from UCS, the UHC scheme for its own citizens. Not only were the CMHI utilization rates found to be lower than UCS rates but, as the numbers of registered migrants decreased by 2006, CMHI's role in financing migrants' health needs was found to be eclipsed by out-of-pocket payments and hospital exemptions (IOM 2009).

The above-described binary opposition of national and non-national also exposes the tensions between citizenship rights and human rights with regard to healthcare within ASEAN member states' migration regimes. As much as they are contested within ASEAN countries,[5] citizenship and citizenship rights – as an exclusionary membership in a political community with concomitant privileged entitlements and protections – drive a wedge between citizen and non-citizen populations in accessing healthcare, as evidenced in the examples of Malaysia and Thailand above. Thus, as ASEAN member states strive for greater cohesion as a regional grouping, they are challenged to reconcile the interpretation of the concept of 'universality' in UHC with their regional and global endorsements, notably the ASEAN Socio-Cultural Community (ASCC) Blueprint 2009–15, the ASEAN Declaration on the Protection and Promotion of the Rights of Migrant Workers (ASEAN 2007), the ASEAN Declaration of Commitment: Getting to Zero New Infections, Zero Discrimination, Zero AIDS-Related Deaths, 2011 (ASEAN 2011), and, more importantly, the 2008 World Health Assembly (WHA) Resolution on the Health of Migrants (WHO 2008), which calls for migrant-sensitive health policies.

ASEAN member states also need to reconcile their disparate approaches – of promoting the development of a market-led, profit-focused health system with one premised on health as a human right. None of the region's top three destinations for migrants and refugees – Malaysia, Singapore and Thailand – have ratified the 1990 International Convention on the Protection of the Rights of All Migrant Workers and Members of their Families (UN General Assembly 1990), the 1951 Refugee Convention (UN General Assembly 1951) or the 1954 Convention relating to the Status of Stateless Persons (UN General Assembly 1954). Furthermore, Malaysia and Singapore, among the most affluent countries in ASEAN, are not party to the 1996 International Covenant on Economic Social and Cultural Rights (UN General Assembly 1966), which spells out the norms for the right to health. They have also not codified the right to health in their constitutions for their own citizens. Malaysia's two-tiered healthcare system, for example, has increasing space for affluent citizens to opt out of the public health system. This development has evaded debate on the long-term consequences for health equity and on the risks associated with decreased incentives for the affluent to cross-subsidize public healthcare.

Healthcare for migrants has also been instrumentalized for profit, with migrant workers now legally required to purchase private health insurance from companies assigned by the Malaysian government. While the revenue from this insurance scheme contributes to Malaysia's economic transformation, with the aim of elevating it to the level of a developed nation (PEMANDU 2010), migrant workers subscribing to the two private health insurance schemes are not covered for out-patient care, health promotion or prevention. Although Thailand also aligns with ASEAN's approach of attributing an instrumental value to health in achieving its broader economic goals, it has attempted to accommodate a rights-based approach to health and healthcare, even incorporating it as a constitutional guarantee, through an expanded social protection role for the state in the provision of healthcare, investment in healthcare infrastructure, healthcare financing and poverty alleviation (Sakunphanit and Suwanrada 2011). However, challenges remain with regard to the inclusion of migrant workers in the country's UHC.

Conclusion

In the last two decades, deregulated financial markets and their capital flows have led to financial volatility worldwide with spill-over damage at a range of scales. In a competitive race to the bottom, poorly (or arbitrarily) regulated labor inflows undermine social compacts by weakening the leverage of domestic labor and rendering undocumented migrants more vulnerable to abuse and exploitation. To secure equitable gains, regional economic development initiatives in which Southeast Asian countries are engaged (e.g., the ASEAN Free Trade Agreement [AFTA], the ASEAN Economic Community [AEC] and the Trans-Pacific Partnership Agreement [TPPA], essentially a bill of rights for investors and traders) need to be balanced by a regional social charter with binding commitments on the rights and entitlements not only of national citizens of the region's nation-states but also of the trans-border (labor) migrants residing within them.

Stand-alone, national-level government commitments continue to be patchy and prove insufficient in an era of global health governance. To reinforce this observation using an example discussed above, Thailand's introduction of a government-run Compulsory Migrant Health Insurance scheme (CMHI) in 2001 for documented and undocumented migrant workers alike – which can be considered a pioneering policy approach within the region – still offers differential benefits when compared with the Universal Coverage Scheme (UCS) reserved for Thai citizens. Other implementation issues such as the restricted portability of healthcare coverage, annual medical examinations requisite for CMHI policy renewal (chargeable to migrant workers themselves) and the reluctance of undocumented migrants to identify themselves as such have contributed, as of 2013, to a low CMHI registration rate of 60,000 out of a targeted 1 million registrants (Guinto et al. 2015).

Even if there were political will at the national and regional levels to tackle human trafficking, dubious migrant labor brokering practices and jurisdictional irregularities along the labor 'supply chain'(Szep and Grudgings 2013; Chao 2014; Al-Mahmood 2015), both sending and receiving countries within ASEAN remain wary about enforcing binding commitments fully compliant with international human rights and migrant rights conventions. An interim measure that perhaps could find traction is a multilateral binding agreement among ASEAN member states on taxation options for migrants and their dependents, which would entitle them to 'citizen-equivalent' healthcare and social benefits in their host country.[6] This would not be a one-size-fits-all solution, but would be customized to the taxation and social entitlement regimes of respective ASEAN member states, i.e., an expanded notion of ('adoptive') citizen rights and obligations with arguably universal appeal.

The journey to realizing the right to health for all – including migrants – in ASEAN member states is going to be a long one. A first step toward moving forward as a politically, socially, economically and culturally cohesive regional grouping in relation to health, healthcare and migrant rights includes achieving greater conceptual, normative and operational clarity of the avowed goals. Equally important would be the negotiation of the meaning of community and identity, and the manners in which member states manage and project their borders aside from geo-political boundaries.

Notes

1. Delegates attending the Issues and Challenges of Foreign Workers in Malaysia conference (16 December 2014, Kuala Lumpur) reported about 2.9 million legal foreign workers in Malaysia (out of an estimated 6 million), citing police sources; an earlier World Bank report estimated about 1.8 million registered foreign workers, and another 1 to 2 million unregistered workers in 2010 (Del Carpio et al. 2013).
2. The Bangkok office of the International Organization for Migration (IOM) estimated a total of 2.46 million low-skilled migrants from three neighboring countries (Cambodia, Laos and Myanmar), of whom 1.4 million were unregistered (Huguet and Chamratrithirong 2011).
3. Some 200,000 pilgrims from Indonesia and 30,000 from Malaysia, in addition to Thai, Bruneian and Filipino Muslim pilgrims travel annually to Saudi Arabia for the Hajj.
4. Recall that it was a regional initiative, led by the Indonesian government, which rallied health ministers of 18 Asia Pacific countries to issue the Jakarta Declaration (2007) calling upon the WHO "to convene the necessary meetings, initiate the critical processes and obtain the essential commitment of all stakeholders to establish the mechanisms for more open virus and information sharing and accessibility to avian influenza and other potential pandemic influenza vaccines for developing countries." These proposals were tabled at the 60th World Health Assembly in Geneva (14–23 May 2007) as part of a resolution calling for new mechanisms for virus sharing and for more equitable access to vaccines developed from these viral source materials, which were largely adopted (WHA 60.28).
5. Consider, for example, the contestation of citizenship with regard to highland people in Thailand (Vaddhanphuti et al. 2002) and of the Rohingya in Myanmar (de Chickera 2010) and the dichotomization of citizenship rights in Malaysia (Koh 2015).
6. In 2014, documented migrant workers in the plantation, construction and manufacturing sectors in Malaysia were already contributing de facto tax revenues of MYR 1611 (US$503) per worker per annum by paying for the annual levy, temporary visitor work pass, multiple entry visa, processing fee, Foreign Workers Compensation Scheme and annual medical check-up.

References

Acuin, J., Firestone, R., Htay, T. T., Khor, G. L., Thabrany, H., Saphonn, V. and Wibulpolprasert, S. (2011). Southeast Asia: An emerging focus for global health. *The Lancet*, 377(9765), pp. 534–535.

Al-Mahmood, S. Z. (2015). Palm-oil migrant workers tell of abuses on Malaysian plantations. *Wall Street Journal*, 26 July. Available at: www.wsj.com/articles/palm-oil-migrant-workers-tell-of-abuses-on-malaysian-plantations-1437933321. Accessed 26 July 2015.

ASEAN. (2004). ASEAN sectoral integration protocol for healthcare. *Vientiane*, 29 November. Available at: www.asean.org/news/item/asean-sectoral-integration-protocol-for-healthcare. Accessed 8 June 2015.

ASEAN. (2007). *ASEAN declaration on the protection and promotion of the rights of migrant workers*, 13 January. Available at: www.asean.org/communities/asean-political-security-community/item/asean-declaration-on-the-protection-and-promotion-of-the-rights-of-migrant-workers-3. Accessed 8 June 2015.

ASEAN. (2009a). *ASEAN socio-cultural community blueprint*. Jakarta: ASEAN Secretariat. Available at: www.asean.org/archive/5187-19.pdf. Accessed 8 June 2015.

ASEAN. (2009b). *Roadmap for an ASEAN community, 2009–2015*. Jakarta: ASEAN Secretariat. Available at: www.meti.go.jp/policy/trade_policy/asean/dl/ASEANblueprint.pdf. Accessed 8 June 2015.

ASEAN. (2009c). *Appendix I – Roadmap for integration of healthcare sector*. Jakarta: ASEAN Secretariat. Available at: www.asean.org/archive/19429.pdf. Accessed 8 June 2015.

ASEAN. (2011). *ASEAN declaration of commitment: Getting to zero new infections, zero discrimination, zero AIDS-related deaths, which was adopted by all 10 ASEAN member states in 2011*. Available at: www.asean.org/archive/documents/19th%20summit/ASEAN_Declaration_of_Commitment.pdf. Accessed 8 June 2015.

ASEAN. (2012). *ASEAN strategic framework on social welfare and development*. Available at: www.asean.org/resources/publications/asean-publications/item/asean-strategic-framework-for-social-welfare-and-development. Accessed 8 June 2015.

Associated Press. (2015). Malaysia and Thailand turn away hundreds on migrant boats. *The Guardian*, 14 May. Available at: www.theguardian.com/world/2015/may/14/malaysia-turns-back-migrant-boat-with-more-than-500-aboard. Accessed 8 June 2015.

Baruah, N. (2012). *Reflection and sharing on implementation of the Hanoi and Bali recommendations*. Paper presented at the 5th ASEAN Forum on Migrant Labour, 9–10 October, Siem Reap, Cambodia.

Beh, L.Y. (2015). Malaysia mass graves: Villagers tell of migrants emerging from secret jungle camps. *The Guardian*, 26 May. Available at: www.theguardian.com/world/2015/may/26/malaysia-mass-graves-villagers-tell-of-desperate-migrants-emerging-from-jungle-camps. Accessed 8 June 2015.

CARAM Asia. (2009). *Regional consultation on migrant worker's access to health care and services*, 30–31 October, Kuala Lumpur.

Center for International and Strategic Studies. (2009). *ASEAN roadmap for integration of the healthcare sector*, Jakarta. Available at: http://pdf.usaid.gov/pdf_docs/Pnadj869.pdf. Accessed 8 June 2015.

Chan, C. K. (2012). Healthcare policy in Malaysia: Universalism, targeting and privatization. In: E. T. Gomez and J. Saravanamuttu, eds., *The new economic policy in Malaysia: Affirmative action, ethnic inequalities and social justice*. Singapore: NUS Press. Available at: https://nuspress.nus.edu.sg/products/the-new-economic-policy-in-malaysia.

Chao, S. (2014). Malaysia's unwanted. *Al Jazeera*, 21 November. Available at: www.youtube.com/watch?v=qdV3sj76vnA. Accessed 8 June 2015.

Cole, R., Wong, G. and Brockhaus, M. (2015). *Reworking the land: A review of literature on the role of migration and remittances in the rural livelihoods of Southeast Asia, Bogor*. Indonesia: Center for International Forestry Research (CIFOR).

de Chickera, A. (2010). *Unravelling anomaly: Detention, discrimination and the protection needs of stateless persons*. London: Equal Rights Trust.

Del Carpio, X., Karupiah, R., Marouani, M. A., Ozden, C., Testaverde, M. and Wagner, M. (2013). *Immigration in Malaysia: Assessment of its economic effects, and a review of the policy and system, Social Protection and Labor Unit, East Asia and Pacific Region*. Kuala Lumpur: World Bank and Ministry of Human Resources of Malaysia.

Guinto, R. L. L. R., Curran, U. Z., Suphanchaimat, R. and Pocock, N. S. (2015). Universal health coverage in 'One ASEAN': Are migrants included? *Global Health Action*, 8(25749). doi: 10.3402/gha.v8.25749

Hashim, J., Chongsuvivatwong, V., Phua, K. H., Pocock, N., Yap, M. T., Chhem, R. K., Wilopo, S. A. and Lopez, A. (2012). *Health and healthcare systems in Southeast Asia*. United Nations University. Available at: http://unu.edu/publications/articles/health-and-healthcare-systems-in-southeast-asia.html. Accessed 8 June 2015.

Health Systems Research Institute. (2012). *Thailand's universal coverage scheme: Achievements and challenges. An independent assessment of the first 10 years (2001–2010)*. Nonthaburi and Thailand: Health Insurance System Research Office.

Hospital Kuala Lumpur. (2014). *Minit mesyuarat interaksi membincangkan isu-isu pesakit warga asing di hospital Kaual Lumnpur bersama wakil/agensi & pihak HKL pada 30 Mei 2014 / 3.00 petang*. Kuala Lumpur: Unit Perhubungan Awam. Pejabat Pengarah. Hospital Kuala Lumpur.

Huguet, J. W. and Chamratrithirong, A. (2011). *Thailand migration report 2011*. Bangkok: International Organization for Migration.

International Organization for Migration (IOM). (2009). *Financing healthcare for migrants: A case study from Thailand*. Bangkok: International Organization for Migration.

International Organization for Migration (IOM). (2010). *World migration report 2010: The future of migration: Building capacities for change*. Geneva: International Organization for Migration.

Kittrakulrat, J., Jongjatuporn, W., Jurjai, R., Jarupanich, N. and Pongpirul, K. (2014). The ASEAN economic community and medical qualification. *Global Health Action*, 7(24535). doi:10.3402/gha.v7.24535

Koh, S. Y. (2015). Unpacking 'Malaysia' and 'Malaysian citizenship': Perspectives of Malaysian Chinese skilled diasporas. In: A. Christou and E. Mavroudi, eds., *Dismantling diasporas: Rethinking the geographies of diasporic identity, connection and development*. London: Ashgate, 129–143.

Lamy, M. and Phua, K. H. (2012). Southeast Asian cooperation in health: A comparative perspective on regional health governance in ASEAN and the EU. *Asia Europe Journal*, 10, pp. 233–250.

Luistro, M. A. S. (2015). ASEAN medical sector feels birth pains of regional integration. *The Filipino Connection*, 30 June. Available at: http://thefilipinoconnection.net/asean-medical-sector-feels-birth-pains-of-regional-integration/. Accessed 30 June 2015.

Lyons, L. and Ford, M. (2007). Where Internal and international migration intersect: Mobility and the formation of multi-ethnic communities in the Riau Islands Transit Zone. *International Journal on Multicultural Societies*, 9(2), pp. 236–263.

Mann, M., Gostin, L., Gruskin, S., Brennan, T., Lazzarini, Z. and Fineberg, H.V. (1994). Health and human rights. *Health and Human Rights*, 1(1), pp. 6–23.

Office of the United Nations High Commissioner for Human Rights (OHCHR). (2014). *Preliminary observations and recommendations by the UN Special Rapporteur on the right of everyone to the enjoyment of the highest attainable standard of physical and mental health – Mr. Dainius Pūras, Country Visit to Malaysia, 19 November–2 December. 2014*, Kuala Lumpur, 2 December. Available at: www.ohchr.org/EN/NewsEvents/Pages/DisplayNews.aspx?NewsID=15370&LangID=E. Accessed 8 June 2015.

Ormond, M. (2014). Medical tourism. In: A. A. Lew, C. M. Hall and A. M. Williams, eds., *The Wiley Blackwell companion to tourism*. London: John Wiley and Sons, 425–434.

Ormond, M. and Mainil, T. (2015). Government and governance strategies in medical tourism. In: N. Lunt, J. Hanefeld and D. Horsfall, eds., *Handbook on medical tourism and patient mobility*. London: Edward Elgar, 154–163.

PEMANDU. (2010). *Chapter 16: Creating wealth through excellence in healthcare*. Available at: www.moh.gov.my/images/gallery/ETP/NKEA%20Penjagaan%20Kesihatan.pdf. Accessed 8 June 2015.

PEMANDU. (2015). *Economic transformation programme*. Annual Report 2014: National Key Economic Area (Healthcare), Putrajaya, Malaysia: Prime Minister's Department, 236–253.

Pocock, N. S. (2015). Double Movement: Health Professionals and Patients in Southeast Asia. In: J. D. Rainhorn and S. El Boudamoussi, eds., *New Cannibal markets: Globalization and commodification of the human body*. Paris: Editions Maison des Sciences de l'Homme, 163–182.

Rodriguez, R. M. (2010). *Migrants for export: How the Philippine state brokers labor to the world*. Minneapolis: Minnesota University Press.

Sakunphanit, T. and Suwanrada, W. (2011). The Universal Coverage Scheme. In: UNDP, SU/SSC, ILO (ed) Sharing innovative experiences: successful social protection floor experiences. UNDP, New York. Available at: www.ilo.org/secsoc/information-resources/publications-and-tools/books-and-reports/WCMS_SECSOC_20840/lang-en/index.htm. Accessed 8 June 2015.

Sciortino, R. (2014). Indonesia health cards and the 'missing middle'. *The Jakarta Post*, 29 November. Available at: www.thejakartapost.com/news/2014/11/29/indonesia-health-cards-and-missing-middle.html. Accessed 8 June 2015.

Soda, R. (2014). Approaches to rethinking rural-urban migration in Southeast Asia: The case of the Iban in Sarawak, Malaysia. In: K. Husa, A. Trupp and H. Wohlschlägl, eds., *Southeast Asian mobility transitions: Issues and trends in migration and tourism*. Vienna: Department of Geography and Regional Research, University of Vienna, 100–121.

Sugiyarto, G. (2015). Internal and international migration in Southeast Asia. In: I. Coxhead, ed., *Routledge Handbook of Southeast Asian economics*. New York: Routledge, 270–299.

Szep, J. and Grudgings, S. (2013). Authorities implicated in Rohingya smuggling networks. *Reuters*, 17 July. Available at: www.pulitzer.org/files/2014/international-reporting/reuters/01reuters2014.pdf. Accessed 8 June 2015.

Trajano, J. C. I. (2015). *An ASEAN nuclear crisis centre: Preparing for a technological disaster in Southeast Asia*, RSIS Commentary No. 135, 9 June, Nanyang Technological University, Singapore.

UN General Assembly. (1951). *United nations convention relating to the status of refugees*. Available at: www.unhcr.org/pages/49da0e466.html. Accessed 8 June 2015.

UN General Assembly. (1954). *Convention relating to the status of stateless persons*. Available at: www.unhcr.org/3bbb25729.pdf. Accessed 8 June 2015.

UN General Assembly. (1966). *International covenant on economic, social and cultural rights*. Available at: www.ohchr.org/EN/ProfessionalInterest/Pages/CESCR.aspx. Accessed 8 June 2015.

UN General Assembly. (1990). *International convention on the protection of the rights of all migrant workers and members of their families*. Available at: www2.ohchr.org/english/bodies/cmw/cmw.htm. Accessed 8 June.

UN Women. (2013). *Managing labour Migration in ASEAN: Concerns for women migrant workers*. Bangkok: UN Women.

UNDP. (2015). *The right to health: Right to health for low-skilled labour migrants in ASEAN countries*. Bangkok: United Nations Development Programme Bangkok Regional Hub.

Vaddhanphuti, C., Wittayapak, C., Buadaeng, K. and Laungaramsri, P. (2002). *State-making, contested spaces and identifications*. Thai Group – Southeast Asian Collaborative Research Network and Yale University.

Available at: http://ibrarian.net/navon/paper/State_Making__Contested_Spaces_and_Identifications. pdf?paperid=860309. Accessed 8 June 2015.
World Health Assembly. (2005). Resolution 58.33 Sustainable health financing, universal coverage and social health insurance, 58th World Health Assembly, 25 May. Available at: http://apps.who.int/medicinedocs/documents/s21475en/s21475en.pdf. Accessed 8 June 2015.
World Health Assembly. (2011). WHA64.9 Sustainable health financing structures and universal coverage, 64th World Health Assembly, 24 May. Available at: http://apps.who.int/medicinedocs/documents/s21474en/s21474en.pdf. Accessed 8 June 2015.
World Health Organization (WHO). (2008). Health of migrants. 61st World Health Assembly. Agenda item 11.9. WHA 61.17, 24 May 2008. Available at: http://apps.who.int/gb/ebwha/pdf_files/A61/A61_R17-en.pdf. Accessed 8 June 2015.

16
MIGRATION, DEVELOPMENT AND REMITTANCES

Philip Kelly

In Southeast Asia, the connection between migration and development can mean different things in varied national settings. For some countries, most notably the Philippines, exporting workers and receiving the financial remittances they send back home has been a longstanding and well-established source of foreign exchange, government revenue and livelihood support. In other places, such as Vietnam, emigrations have occurred in waves and have created an overseas diaspora that is increasingly supporting livelihoods and investments in the homeland. In several cases, however, such as Thailand, Malaysia and Singapore, international labor migration is predominantly an in-bound flow, with large (and sometimes undocumented) populations of workers who have moved across borders within the region. Here, low-wage and sometimes highly exploited migrant laborers have been central to the competitiveness and development of their host economies.

The past 20 years have seen a period of growing optimism about the potential of international migration, and the remittances that it generates, as a development strategy for the Global South (Sharma 2011; de Haas 2012). This optimism has grown in direct proportion to the volume of remittances flowing to developing countries, which reached an estimated US$431 billion in 2014 (Ratha et al. 2016). Financial flows in the form of remittances now far exceed foreign aid flows and have even reached a similar order of magnitude to foreign direct investment. Their potential as a source of poverty alleviation and development financing has been widely trumpeted. A range of benefits are pointed out by proponents who see them as: an efficient means of addressing household needs directly; a source of investment in productive activities in a variety of sectors; a countercyclical insurance against economic crisis and natural disaster; a potential source of skilled human capital in the form of returning migrants; and a boon to government revenues, foreign exchange earnings and national sovereign debt ratings (Maimbo and Ratha 2005; Bakker 2015).

There is also, however, a pessimistic view on the role of migration and remittances in development (Gamlen 2014). This view emphasizes the costs associated with predicating development on the departure of migrant labor. These costs include: the social consequences of separated families, especially for children with absent parents; the selectivity of access to migration opportunities, meaning that the poorest are unable to leave and inequalities may actually be exacerbated in sending areas; the loss of talent in specific technical job fields; and the disenfranchisement and exploitation sometimes experienced by migrant workers in overseas contexts who live without the rights and protections of full citizenship (Ahsan 2015).

This chapter will examine the basis for migration-development optimism or pessimism in the context of Southeast Asia, drawing upon empirical studies conducted in the region. The first section outlines the main channels of international migration and remittance flows in Southeast Asia, including both cross-border migrant flows within the region, and international migrations further afield. The second and third sections tackle, in turn, the evidence for positive and negative outcomes from across the region that are attributable to migration as a development strategy. The chapter concludes that the positive-negative binary usefully sets up the issues that need to be addressed in questioning migration as a strategy for enhancement of individual well-being and macro-economic prosperity, but actual outcomes are highly contingent and rather resistant to generalizations.

Mapping migration and remittance flows in Southeast Asia

Reliable data on migration and remittance flows in Southeast Asia are not easily available. Migrants are undoubtedly undercounted in almost every context, and in some places more than others. In part this is because some migrants may be living and working without legal documentation, but it is also due to inconsistencies in definitions and counting systems. Remittance data are also rife with inaccuracies, as money moves in a variety of unrecorded channels. The real flow of resources along migration corridors is therefore likely much higher than formal accounting suggests. In the case of the Philippines, for example, Ducanes shows how widely varying totals for remittances can be achieved depending on whether the data used are from local surveys (US$6.7 billion in 2006), the Central Bank of the Philippines (US$12.8 billion in the same year) or the World Bank (US$15.3 billion) (Ducanes 2015).

While the exact magnitude of migration flows and stocks, and remittances, may be uncertain, it is not difficult to identify several distinct migration corridors in Southeast Asia. Thailand has by far the largest number of non-citizens living within its borders, mainly arriving from its neighbors in mainland Southeast Asia. Of the 3.9 million non-Thais recorded in 2015, 96 percent came from Myanmar (1,978,348 people), Laos (969,267) and Cambodia (805,272) (UNDESA 2015). Another major magnet for migrants is Singapore. Among a total population of approximately 5.5 million, the city included over 2.5 million migrants in 2015, of whom just over 1.5 million were classified as non-residents. While many migrants in Singapore are from South Asia and China, large contingents also come from Indonesia, Malaysia and the Philippines. Malaysia is also a significant migrant destination, especially in East Malaysia (Sabah and Sarawak) where Filipino and Indonesia migrants work in oil palm plantations and other resource industries. The Commission on Filipinos Overseas estimates that almost 800,000 Filipinos are living in Sabah, many of them undocumented (Commission on Filipinos Overseas 2014).

Source countries differ in the profile of their overseas populations. Among the 1.3 million Lao recorded as living overseas in 2015, almost 1 million are in Thailand, many having crossed the Mekong to work in low waged jobs. The majority of other overseas Lao are part of well-established immigrant communities in the United States (just over 200,000) and France (42,000), having fled wars, political unrest or persecution in the 1970s and 1980s. The profile of Cambodians overseas is very similar, and for the same set of reasons. Unlike Lao and Cambodians, Vietnam's diaspora is predominantly located in wealthier countries. From a total of about 2.5 million, just over 2 million overseas Vietnamese are living in North America, Europe and Australia (UNDESA 2015).

Although its overseas migration remains low as a proportion of total population, Indonesia has deployed increasing numbers of overseas workers in the last decade. In 2005, 2.7 million Indonesians were living abroad, but by 2015 this figure had risen to 3.9 million. Of these, just over 1 million are located in Malaysia, where linguistic and religious commonalities make integration into agricultural and resource industries relatively easy. A further 1.8 million are living as contract workers in the Middle East, with 1.3 million Indonesians in Saudi Arabia alone.

The Philippines is a unique case because of the long history, scale and geographical dispersal of its diasporic and contract worker populations. American colonial power (1898–1946) led to early twentieth century Filipino migrations to work in agricultural and resource industries in Hawaii and the West Coast of the United States. Enlistment in the US Navy also led to the emergence of large Filipino communities in California in particular (Espiritu 2003). Changes in immigration policies in the late 1960s in the United States created opportunities for large numbers of Filipinos to immigrate. Emigrants have also moved to other countries that have programs allowing permanent settlement, including the UK, Australia and Canada. The result, in 2015, was the presence of almost 2 million Filipino immigrants in the United States (in addition to a locally born Filipino-American population of about the same number), and over 500,000 in Canada.

In addition, the Philippines has an extensive and well-documented set of institutional structures for the deployment of temporary contract workers around the world (Rodriguez 2010; Tyner 2009). Well over 1 million temporary migrant workers are now deployed on new contracts every year. These deployments range from the plantation workers in Malaysia already mentioned, to domestic workers in Hong Kong and Singapore, to construction labor in the Middle East, to entertainers in hotels and cruise ships around the world, to crew members and officers aboard ocean-going freighters (Aguilar 2014; Fajardo 2011; McKay 2012). While there is a long history of nursing migration from the Philippines, Overseas Filipino Workers (OFWs) are also increasingly found in other professional, technical and managerial fields. In total, the UN recorded 5.3 million Filipino migrants worldwide in 2015, while the Philippine government estimates a total overseas population of 10.2 million (including, in this figure, permanently settled emigrants and, in some cases, their descendants too) (UNDESA 2015; Commission on Filipinos Overseas 2014).

These patterns of migration are reflected in the remittances returned to migrant-sending areas in Southeast Asia. Table 16.1 indicates the magnitude of remittances in 2014, and their importance relative to national economies as a whole. Because of the size of their diasporas in high income countries, the Philippines and Vietnam have by far the largest flows of remittances. In the case of the Philippines, the US$28 billion sent back in 2014 represents the third highest remittance flow globally after China and India, and accounted for one-tenth of the Philippine economy. Other countries, such as Laos, Myanmar and Cambodia have relatively little formal dependence on remittance incomes, although given the cross-border and often informal migrations occurring in those cases it is fair to assume that far more money crosses the border than is officially recorded.

It is important to note, however, that studies of the migration-development nexus can all too easily be reduced to the mapping of these kinds of flows. This can lead to a lack of reflection on who migrants are, why they move, the circumstances in which they live, and what development might mean for them and for their communities. Several points can be made here. First, migrants' experiences are importantly shaped by class, gender, ethnic and religious identities and other axes of difference. Gender, in particular, defines migrations, and can itself be redefined in the process of mobility. The jobs that migrants go to are usually deeply marked as masculine or feminine (domestic maid, nurse, construction worker, seafarer, etc.), but the very act of migration can also redefine the performance of gender roles, both abroad and back at home (Belanger and Linh 2011; Lukasiewicz 2011; Silvey 2006).

Table 16.1 Migration and remittances in Southeast Asia

	Migrants living overseas in 2015 (UNDESA figures)	Overseas population as % of domestic population	2014 migrant remittance inflows (US$millions)	Remittances as a share of GDP in 2014
Philippines	5,316,320	5.3	28,403	10.0%
Vietnam	2,558,678	2.7	12,000	6.4%
Timor-Leste	37,311	3.1	45	2.9%
Cambodia	1,187,142	7.6	382	1.8%
Thailand	854,327	1.3	5,655	1.5%
Indonesia	3,876,739	1.5	8,551	1.0%
Laos	1,345,075	19.8	60	0.5%
Malaysia	1,835,252	6.1	1,565	0.5%
Myanmar	2,881,797	5.3	105	0.2%
Brunei Darussalam	46,237	10.9		
Singapore	313,884	5.6		

Source: UNDESA 2015

A second point is that migrants move with varying degrees of coercion and live with differing kinds of (non-)citizenship status in their place of work or settlement. The circumstances of Burmese migrants working in slave-like conditions on Thai fishing boats has, for example, been the subject of recent media scrutiny (Hodal 2016). But equally, portrayals of human trafficking can sometimes be misleading. Parreñas, for example, notes that in 2004 the US State Department declared Filipina hostesses in Japan to be the world's largest group of sex-trafficked individuals, but this was based on assumptions about the type of work they were performing (Parreñas 2011). Parreñas argues that, in reality, such women live in a paradoxical position of 'indentured mobility' with elements of both personal freedom and subjugation.

Third, it is important to think about development in a broader sense than the financial flows that contribute to the mainstream, monetized and countable economy. For better or worse, the experience of migration is often transformational for both individual migrants and for their family members. These changes in social relations and cultural meanings often require a detailed and qualitative understanding of the implications of migration. Indeed, all of these considerations – from gender identities, to citizenship status and working conditions, to the non-economic dimensions of development – mean that it is important to pay attention to the lived experiences of migrants and not just the quantifiable dimensions of the phenomenon they represent.

Evidence for migration-development optimism in Southeast Asia

Optimism about the positive developmental role of migration is, not surprisingly, articulated primarily by international financial institutions and national governments. Nevertheless, there are also qualitative studies based on personal experiences and ethnographic research that speak to positive gains.

National economic growth

A key argument of migration-development optimists is that remittances boost government revenues, foreign exchange earnings and national sovereign debt ratings. Increased revenues (through visa and passport fees, clearance certificates, placement fees, taxes on incomes and expenditures. etc.) are then used as part of a government's resources to provide public goods and services. Remittance flows are also a source in increased aggregate demand in the domestic economy, thereby increasing overall wealth, and a source of foreign exchange earnings, which allow imports. Economic analyses have found these connections to hold true, but a given increase in remittances does not translate directly into a commensurate rate of growth in the national economy. Across a range of countries, the Asian Development Bank has found that a small but generally positive correlation exists between remittances and overall growth in the developing economies of Asia (Vargas-Silva et al. 2009). They suggest that a 10 percent increase in overseas workers' remittances translates into approximately 2 percent of real GDP growth per capita.

While these figures are aggregated at the national scale, there is also evidence of remittances driving growth even more intensively at smaller, urban scales. In the Philippines, for example, it is estimated that 30 percent of all remittances are spent on real estate, and almost half of all residential construction in recent years has been concentrated in and around Manila (Cardenas 2014). Remittance dollars have therefore driven a localized form of economic growth in the national capital.

Poverty reduction and welfare gains

Perhaps the strongest argument laid out by remittance optimists is that a direct flow of funds to households provides an efficient and incorruptible channel to reduce poverty and increase household incomes and well-being. The World Bank's lead economist on migration and development issues, Dilip Ratha, argues that remittances constitute "a powerful anti-poverty force in developing countries" and that "the presence of remittance income in a household strongly and significantly corresponds with positive health outcomes, especially for children" (Ratha 2013, 4, 5).

In Vietnam, Pfau and Giang used the Vietnam Household Living Standards Survey to show that by 2004, 7.3 percent of Vietnamese households were receiving international remittances, almost two-thirds of them coming from North America (Pfau and Giang 2009). Using data from four installments of the survey from 1992 to 2004, they found a significant role for remittances in reducing the incidence of poverty, and even a slight effect in reducing income inequality (perhaps because the increased income from remittances is offset by the greater likelihood that the recipient household head will not be participating in the waged labor force).

In the Philippines, Ducanes provides perhaps the most rigorous answer to the question of how remittances can impact poverty and inequality (Ducanes 2015). Using several detailed surveys, he shows that 24 percent of all households in 2006 had some contribution coming from abroad, and 7 percent of households had at least one Overseas Filipino Worker. Ducanes argues that households who send a migrant worker overseas rapidly increase their total incomes and their expenditures on food, clothing, education, real estate, medical care and recreation. He also finds a spillover effect among other households, as contributions and gifts are extended to those without a migrant worker. This does, however, still leave open the question of whether the family is uplifted economically in the long term, and whether there are significant economy-wide effects. Ducanes concludes that:

> At least in the short term and for overseas workers' own households, overseas labor migration confers many economic benefits (e.g., increased income and spending, reduced poverty), which appear to far outstrip its costs (e.g., possibly higher overall inequality). All else being equal, average household economic welfare in the country would almost certainly be lower in the absence of migration of labor overseas.
> (Ducanes 2015, 103)

Overall, Ducanes finds that a household with an overseas worker is three to six times more likely to escape poverty, depending on the statistical technique used. He also points out, however, that this should not be taken as an argument in favor of governments actively promoting migration as a development strategy, because overseas migrants generally come from households that were already better-off (a point we will return to later). Ducanes also notes that we should expect the poverty reduction impact of migration to be significantly higher in cases where the poor are more likely to be migrating – as in the case of cross-border migration to Thailand from Myanmar, Cambodia and Laos.

Capital for productive activities

The role of remittances in providing capital for investment in new productive activities has been a matter of debate for some time. Some research suggests that remittances are generally used for consumption items, although the expenditures on education and healthcare noted above can be seen as investment in human capital. There are, however, cases where remittance funds do provide capital for new or expanded productive enterprises at the household scale. In Vietnam, the Central Bank estimates that about 70 percent of remittance inflows to Ho Chi Minh City went into production and business investment, and 22 percent to the real estate sector (Ratha et al. 2016).

In the Philippines Lukasiewicz examines the ways in which female spouses deploy their husbands' remittances in the expansion of agricultural holdings (Lukasiewicz 2011). Using research from the town of Lucban, in Quezon Province, he shows how the purchase of expanded landholdings for coconut trees or the construction of a pig-raising enterprise represent both strategic investments and a renegotiation of women's roles as they become both managers of new farming enterprises and employers of agricultural laborers. In earlier research from the 1990s in the northern Philippines, Deirdre McKay described a situation in which women from her case study village in Ifugao Province were drawn into international contract labor migration (primarily as domestic workers in Hong Kong and Singapore), and so capital became available to purchase both land and crop inputs. This contributed to a shift away from rice cultivation and toward bean gardens (McKay 2005; McKay 2003).

Aside from investment in mainstream enterprises, migration and remittance flows have also opened up the possibility of alternative economic practices that might be financed through collective remittances. A pioneer in this effort has been Unlad Kabayan Migrant Services Foundation Inc. in the Philippines, a social economy NGO that originated with a group of domestic workers in Hong Kong. By collectively saving, they were able to provide access to a rotating capital fund for their own enterprises back home in the Philippines, or to facilitate social entrepreneurship and pro-poor local economic growth (Gibson et al. 2010). J.K. Gibson-Graham describes some of the examples that emerged, including food processing enterprises using local crops, a transportation cooperative, and a group making and renting out ceremonial garments for school graduations and weddings (Gibson-Graham 2005).

Insurance against economic crisis and natural disaster

Remittances have been widely promoted as a form of insurance against income loss or natural disaster in poor and rural areas of the Global South (Ratha 2007). Where possible, relatives overseas will often increase their support if family members back home are facing sudden difficulties because of unemployment, emergency medical expenses, crop losses or environmental disasters such as typhoons, drought or flooding. This will both help to maintain income levels for recipient households and will prevent them from having to sell off productive assets (such as land or livestock) that are important for future income. There is some evidence that remittances do play this kind of insurance role. In one study from the Philippines, Yang and Choi used localized meteorological and income data to examine whether remittances seem to replace household incomes when rainfall 'shocks' (that is, much more or much less than usual) cause a disruption (Yang and Choi 2007). They found that when household incomes decline, roughly 60 percent of the shortfall is replaced by remittance inflows in those households that have a relative overseas.

Environmentally induced migration may not, however, be just a diversification of risk in anticipation of future crises; it may also be a way of coping with present crises. In Cambodia, Bylander shows how rural households in Chanleas Dai commune, Siem Reap Province, responded to both the routinization of migrant labor opportunities in Thailand, and increasing incidence of floods and droughts after the mid-2000s (Bylander 2015). Migration became a necessary coping strategy and not just an insurance against environmental risks in agricultural production. In some cases, agricultural production was even abandoned or scaled back in order to focus on the more secure income available from migrant work across the border. As Bylander notes "rather than being used to *mediate* the risk of investing at home, in Chanleas Dai migration is primarily understood as an opportunity to *replace* investment at home" (2015, 144; emphasis in original).

Returnees with enhanced human capital

Migration and development optimists suggest that migrants may develop skills overseas that can then enhance productivity when they return home. Montefrio et al. (2014) provide an example in the form of Filipino workers from the island of Palawan who are employed in the oil palm plantations of Sabah (Malaysia). Located less than 500 km by ferry from Sabah, the Southern municipalities of Palawan have been a major source for labor in the plantations and a steady circular flow has been established (much of it undocumented). These workers have brought home knowledge of the sector and the global demand that it satisfies, as well as an understanding of the technical requirements of oil palm cultivation and processing. This knowledge transfer appears to be impacting land use decisions back in Palawan – not just in terms of practical know-how, but also in terms of 'social remittances' that affect attitudes toward what constitutes 'development' (Montefrio et al. 2014). Migrants to Sabah, and those who know them, link ideas of prosperity, progress, productivity, improvement and development to oil palm production, thereby laying the discursive groundwork for such developments in Palawan. While oil palm in Palawan is still on a small scale compared to other sites in Malaysia and Indonesia, it has grown rapidly. After the first seedlings were planted in 2007, around 3,600 hectares were harvested by 2011, with licenses issued for a further 15,500 hectares (Larsen et al. 2014). This represents around 99 percent of the cultivable agricultural land in the southern municipalities of the island, where the development is concentrated.

In many cases, however, the promise of a returning skilled workforce is not realized. This may be because there is simply no context in which to apply the skills back home. Learning how to use an advanced piece of medical machinery while working in a Saudi Arabian hospital doesn't help if no such machinery is installed in rural or health facilities in Indonesia. It is also important to remember that overseas migrants may be working in jobs that are below, or unrelated to, their level of professional training and so they are more likely to lose skills than to gain them while overseas – the engineering graduate who works as a laborer on a construction project in the Middle East, for example, may have even fewer updated skills when he returns home.

Migrant labor and host country development

Although seldom noted by migration and development optimists, the presence of migrant workers is undoubtedly good for profitability, competitiveness and conventional measures of economic growth in the destinations where they work. In all Southeast Asian contexts where migrant laborers are used in large numbers, they are paid less than locals would work for, are concentrated in undesirable and sometimes dangerous work, and are disenfranchised from some basic labor and citizenship rights due to their temporary status. The result is a cheap, malleable and dispensable workforce that is the basis for competitive production in a global capitalist system.

Sai Latt (2011) provides an example in the case of Northern Thailand, where very large numbers of Burmese migrant workers (especially from Shan ethnic minorities) provide labor for intensive agricultural operations in Royal Development Projects. These projects are designed to be showcases for enlightened agricultural practices and have been developed since the 1960s to replace opium cultivation with cash crops as a source of livelihood for villagers. Hmong villages in the area have, however, generally used Shan migrant workers as their farm labor. These workers have never been recognized legally as persons displaced by conflict and/or persecution in Burma/Myanmar and so have existed in a precarious state of non-citizenship. Such precariousness has left them with the threat of periodic fines or deportation by Thai authorities always hanging over them. Their employers discursively represent them as 'hardworking,' 'strong' and 'well-suited' to agricultural work, but it is precisely because of their precarious status that they are forced to become hardworking and compliant. As Latt notes, these are not innate characteristics, but "special qualities that the Shan *came to embody and perform* in order to survive" (Latt 2011, 542; emphasis in original). Without these disenfranchised and discursively marginalized migrant workers, cash crops could not be cultivated in Northern Thailand with the same degree of labor intensity and cost competitiveness.

This example from Thailand emphasizes the point that Southeast Asia is a complex mix of sources and destinations for labor migrants. While migration and development can be assessed in the context of sending countries and the remittances they receive, migration is also a fundamental part of the developmental model of host contexts as well. In many ways, the case for optimism with regard to migration as a development strategy for sending countries can only exist if a blind eye is turned toward the exploitation and disenfranchisement faced by many migrant workers. This brings us to the pessimistic perspective on the migration-development nexus.

The case for pessimism

In recent years a growing body of research has highlighted the negative dimensions of migration as a development strategy. In some cases negative outcomes are the product of individual circumstances. A 'failed migration' due to unscrupulous recruiters, corrupt government officials or abusive employers can lead to an early return or a loss of any financial gains made through overseas earnings. Nicole Constable describes these kinds of circumstances for Indonesian and Filipino workers in Hong Kong (Constable 2015). This can be disastrous for the migrant and their family. But there are also wider systemic factors that support a negative view of migration/remittances and their developmental effects.

Unequal access to migration, and exacerbated inequalities

While migration can lead to dramatic upward mobility for some, there remains the important question of who gets to take advantage of such an opportunity. In many contexts, migration is an expensive undertaking, with recruitment agency fees, air tickets and other costs adding up to many times the average annual income in sending areas. Thus, while cross border undocumented migrations (such as Cambodian, Lao and Burmese in Thailand) might be relatively accessible to those with few resources and low levels of education, the more formalized channels of migration are likely to be accessible to select individuals and households only.

Using the Vietnam Household Living Standards Survey cited earlier, Nguyen and Nguyen used 2006 and 2008 data to show that international remittances tend to accrue to the wealthiest segments of society (Nguyen and Nguyen 2015). While 12.7 percent of the richest quintile of Vietnamese households received international remittances in 2008, only 2.0 percent of the poorest quintile were recipients. Although there may be some circularity to this argument (remittance income may be what propelled households into the upper quintile in the first place), it is also generally the case that international migration, especially when it involves more than crossing a land border, is the preserve of those who already have access to resources.

Access to migration may also be selective in other ways. In most cases, it is undertaken by those in younger age cohorts. Roy Huijsmans, for example, shows that over 70 percent of Lao international migrants are between 15 and 25 years old (Huijsmans 2014). Access to opportunities may also be highly gendered, reflecting the segmentation of specific sectors in the global labor market, and in some cases there has been a distinctive feminization of migration flows over time. In the case of Indonesia, for example, around 48 percent of formally recorded international migrant workers were women in 1988, but by 2009 this had increased to 83 percent (Randolph 2015). Finally, migrants tend to be drawn unevenly from different parts of a source country. In the Philippines, for example, the Ilocos region of Northern Luzon accounted for just 5.5 percent the national population in 2010, but 9.5 percent of all overseas Filipino workers (POEA 2015). In each of these ways, then, migration is a pathway to upward mobility that is highly selective.

Brain drain and labor market impacts

Aside from the question of who has the opportunity to leave, migration on a large scale has an impact on the supply of labor in the country of origin. Given the size of its out-migration flows, this has been most clearly an issue in the Philippines. In 2014, over 1.8 million OFWs were deployed. Of the 487,176 within that number who were new deployments to land-based jobs (i.e., excluding seafarers), over 50,000 were working in professional and technical fields, including almost 20,000 nurses (POEA 2015). It is also important to note that many of those leaving have training and professional or technical skills that will not be used during their deployment. Filipino migrants generally represent the more educated segments of Philippine society. Among all overseas workers deployed in 2007, for example, 35 percent were college graduates, compared with just 14 percent of the domestic workforce (Ducanes 2015).

A key question has been whether the loss of talent through migration has affected the Philippine economy. This is difficult to answer because many of these graduates had trained specifically in fields such as nursing with the intention of working overseas. In turn, training institutions in fields with high global demand produce far more graduates than the Philippine labor market can absorb. Hence it is not so much their loss to the local labor market that is at stake, but rather their loss to other fields of study and employment that might have fostered productivity and innovation. An exception is nursing, where local shortages of skilled and experienced staff *have* been a significant issue. Numerous studies have pointed to the understaffing of Philippine hospitals and clinics, and the problems of having a relatively inexperienced medical staff, when the overwhelming propensity of healthcare workers is to migrate overseas (Kelly and D'Addario 2008; Lorenzo et al. 2007).

Migrant working conditions

In many contexts, migrant workers face abuse because of their disenfranchisement from the rights and protections that are, at least nominally, extended to citizens. The threat of deportation always hangs over a migrant worker, whether they are living in a host country legally or not. These vulnerabilities are especially accentuated when young women are working in domestic settings where contractual obligations are only vaguely observed, where privacy may be compromised, and where the possibilities of physical, sexual or emotional abuse are heightened.

Returning to the case of Indonesian domestic workers in Saudi Arabia, Nurchayati (2011) notes that isolation in the homes of employers can expose women to economic exploitation, physical abuse and sexual harassment, while at the same time making it hard to seek help from police or other authorities. At the same time, Nurchayati points out that the Indonesian women she interviewed were not simply passive victims. They were also, in some instances, able to resist abusive employers, negotiate the terms of their relationship and find ways of coping with the circumstances in which they found themselves. In other words, even in the most exploitative of situations, female migrant workers are not without some degree of agency (Nurchayati 2011).

The consequences of family separation

When migrants leave their families and communities there are emotional costs associated with separation, loneliness and homesickness. Parents in particular must somehow reconcile a contradiction in which they do their best to provide materially for their children and yet must be separated from them in order to satisfy that imperative. There are also potential long-term consequences for children in terms of their social and educational development – impacts that are often overlooked when migration is advocated as a means toward economic development (Cortes 2015; Graham and Jordan 2011; Hoang et al. 2015; Parreñas 2001, 2005).

In cases where children may eventually join a parent overseas, the problems created by separation may not be resolved through reunification. For example, Filipino migrants who arrive in Canada under the Live-In Caregiver Program can, after gaining permanent residency, sponsor their family members to join them. Reconstituting the family in a new place is, however, far from unproblematic. Children in particular have to deal with the process of reacquainting themselves with a parent they may only have known at a distance for many years, and the experience of downward mobility from a remittance-supported middle class lifestyle in the Philippines to a new family circumstance of precarious and low paid employment in Canada. Several studies have suggested that anomalously low achievement levels among Filipino youth in terms of post-secondary pathways is a product of this difficult transition (Farrales and Pratt 2012; Kelly 2014; Pratt 2012; Pratt 2008).

Migration can, nevertheless, be emancipatory. Whether separation from a family unit is painful or liberating depends on the context. For young women in particular, migration may carry with it the possibility of independent travel, escape from the strictures of gendered and generational discipline at home, and the chance to earn an independent income and enjoy the pleasures of independent consumption. As Rachel Silvey notes in the case of Indonesian women who migrate to work in Saudi Arabia, they may embody many contradictory positions, including: "global consumers, devoted mothers, victimized laborers, pious pilgrims and heroines of local and national development" (Silvey 2006). Similarly, studies from Vietnam point to the elevated status within families of daughters who migrate and send home remittances (Belanger and Linh 2011).

Conclusion

Clearly there is evidence for both positive and negative developmental outcomes from migration in Southeast Asia. What is clear, however, is that some critical questions need to be asked before migration is embraced as a strategy for enhancing well-being and livelihoods in sending areas. First, although migrant remittances might boost economic growth and aggregate welfare, it is important to ask who has access to migration opportunities and how widely its benefits are distributed. Second, although migrants may return with skills and capital to invest in new economic activities, we should also be thinking of the 'brain drain' that skilled migrants represent, the distortions that are created in 'home' labor markets, and the consequences of migration for the socialization and psychological well-being of left-behind children. Third, while migration can represent a journey of personal liberation and free agency, it can equally be an experience of abuse, exploitation and coercion.

References

Aguilar, F. V. (2014). *Migration revolution: Philippine nationhood and class relations in a globalized age*. Kyoto: CSEAS Series on Asian Studies.

Ahsan, A. (2015). *International migration and development in East Asia and the Pacific*. Washington, DC: World Bank Group.

Bakker, M. (2015). *Migrating into financial markets: How remittances became a development tool*. Oakland, CA: University of California Press.

Belanger, D. and Giang Linh, T. (2011). The impact of transnational migration on gender and marriage in sending communities of Vietnam. *Current Sociology*, 59(1), pp. 59–77.

Bylander, M. (2015). Depending on the sky: Environmental distress, migration, and coping in rural Cambodia. *International Migration*, 53(5), pp. 135–147.

Cardenas, K. (2014). Urban property development and the creative destruction of Filipino capitalism. In: W. Bello and J. Chavez, eds., *State of fragmentation: The Philippines in transition*. Bangkok: Focus on the Global South, 35–66.

Commission on Filipinos Overseas. (2014). *Stock estimate of Filipinos overseas*. Manila and Philippines: Commission on Filipinos Overseas. Available at: http://cfo.gov.ph/program-and-services/yearly-stock-estimation-of-overseas-filipinos.html. Accessed 18 October 2016.

Constable, N. (2015). Migrant motherhood, 'Failed Migration', and the gendered risks of precarious labor. *Trans – Trans-Regional and -National Studies of Southeast Asia*, 3(1), pp. 135–151.

Cortes, P. (2015). The feminization of international migration and its effects on the children left behind: Evidence from the Philippines. *World Development*, January(65), pp. 62–78.

de Haas, H. (2012). The migration and development pendulum: A critical view on research and policy. *International Migration*, 50(3), pp. 8–25.

Ducanes, G. (2015). The welfare impact of overseas migration on Philippine households: Analysis using panel data. *Asian and Pacific Migration Journal*, 24(1), pp. 79–106.

Espiritu, Y. L. (2003). *Home bound: Filipino American lives across cultures, communities, and countries*. Berkeley: University of California Press.

Fajardo, K. B. (2011). *Filipino crosscurrents: Oceanographies of seafaring, masculinities, and globalization*. Minneapolis: University of Minnesota Press.

Farrales, M. and Pratt, G. (2012). Stalled development of immigrant Filipino youths. *Metropolis British Columbia Working Paper Series*, 10, pp. 1–62. Available at: http://mbc.metropolis.net/assets/uploads/files/wp/2012/WP12-10.pdf. Accessed 18 October 2016.

Gamlen, A. (2014). The new migration-and-development pessimism. *Progress in Human Geography*, 38(4), pp. 581–597.

Gibson, K., Cahill, A. and McKay, D. (2010). Rethinking the dynamics of rural transformation: Performing different development pathways in a Philippine municipality. *Transactions of the Institute of British Geographers*, 35(2), pp. 237–255.

Gibson-Graham, J. K. (2005). Surplus possibilities: Postdevelopment and community economies. *Singapore Journal of Tropical Geography*, 26(1), pp. 4–26.

Graham, E. and Jordan, L. P. (2011). Migrant parents and the psychological well-being of left-behind children in Southeast Asia. *Journal of Marriage and Family*, 73(4), pp. 763–787.

Hoang, L. A., Lam, T., Yeoh, B. S. A. and Graham, E. (2015). Transnational migration, changing care arrangements and left-behind children's responses in South-East Asia. *Children's Geographies*, 13(3), pp. 263–277.

Hodal, K. (2016). Slavery and trafficking continue in Thai fishing industry, claim activists. *The Guardian*, 25 February. Available at: www.theguardian.com/global-development/2016/feb/25/slavery-trafficking-thai-fishing-industry-environmental-justice-foundation Accessed 18 October 2016.

Huijsmans, R. (2014). Becoming a young migrant or stayer seen through the lens of 'householding': Households 'in Flux' and the intersection of relations of gender and seniority. *Geoforum*, January, 51, pp. 294–304.

Kelly, P. F. (2014). *Understanding intergenerational social mobility: Filipino youth in Canada*. Montreal: Institute for Research on Public Policy. Available at: www.irpp.org/en/research/diversity-immigration-and-integration/filipino-youth/. Accessed 18 October 2016.

Kelly, P. F. and D'Addario, S. (2008). 'Filipinos are very strongly into medical stuff': Labor market segmentation in Toronto, Canada. In: J. Connell, ed., *The international migration of health workers*. London: Routledge, 77–98.

Larsen, R., Di Mano, F. and Pido, M. (2014). *The emerging oil palm agro-industry in Palawan*. The Philippines: Livelihoods, Environment and Corporate Accountability. 2014–3. Working Paper Series. Stockholm: Stockholm Environment Institute.

Latt, S. S. W. (2011). More than culture, gender, and class: Erasing Shan labor in the 'Success' of Thailand's royal development project. *Critical Asian Studies*, 43(4), pp. 531–550.

Lorenzo, F. M. E., Galvez-Tan, J., Icamina, K. and Javier, L. (2007). Nurse migration from a source country perspective: Philippine country case study. *Health Services Research*, 42(3 Pt2), pp. 1406–1418.

Lukasiewicz, A. (2011). Migration and gender identity in the rural Philippines. *Critical Asian Studies*, 43(4), pp. 577–593.

Maimbo, S. M. and Ratha, D. (2005). *Remittances: Development impact and future prospects*. Washington, DC: World Bank.

McKay, D. (2003). Cultivating new local futures: Remittance economies and land-use patterns in Ifugao, Philippines. *Journal of Southeast Asian Studies*, 34(2), pp. 285–306.

McKay, D. (2005). Reading remittance landscapes: Female migration and agricultural transition in the Philippines. *Geografisk Tidsskrift-Danish Journal of Geography*, 105(1), pp. 89–99.

McKay, D. (2012). *Global Filipinos: Migrants' lives in the virtual village*. Bloomington: Indiana University Press.

Montefrio, M. J. F., Ortiga, Y. Y. and Josol, M. R. C. B. (2014). Inducing development: Social remittances and the expansion of oil palm. *International Migration Review*, 48(1), pp. 216–242.

Nguyen, C. V. and Nguyen, H. Q. (2015). Do internal and international remittances matter to health, education and labor of children and adolescents? The case of Vietnam. *Children and Youth Services Review* November, 58, pp. 28–34.

Nurchayati. (2011). Bringing agency back in: Indonesian migrant domestic workers in Saudi Arabia. *Asian and Pacific Migration Journal*, 20(3–4), p. 479.

Parreñas, R. (2001). Mothering from a distance: Emotions, gender, and intergenerational relations in Filipino transnational families. *Feminist Studies*, 27(2), pp. 361–390.

Parreñas, R. (2005). *Children of global migration: Transnational families and gendered woes*. Stanford, CA: Stanford University Press.

Parreñas, R. (2011). *Illicit flirtations: Labor, migration, and sex trafficking in Tokyo*. Stanford, CA: Stanford University Press.

Pfau, W. D. and Giang, L. T. (2009). Determinants and impacts of international remittances on household welfare in Vietnam. *International Social Science Journal*, 60(197–198), pp. 431–443.

POEA. (2015). *2010–2014 overseas employment statistics*. Manila and Philippines: Philippine Overseas Employment Administration.

Pratt, G. (2008). *Deskilling across the generations reunification among transnational Filipino families in Vancouver*. CERIS Working Paper, Metropolis British Columbia, Centre of Excellence for Research on Immigration and Diversity.

Pratt, G. (2012). *Families apart: Migrant mothers and the conflicts of labor and love*. Minneapolis: University of Minnesota Press.

Randolph, G. (2015). *Labor migration and inclusive growth: Toward creating employment in origin communities*. Washington, DC: Solidarity Center.

Ratha, D. (2007). *Leveraging remittances for development*. MPI Policy Brief. Washington, DC: Migration Policy Institute.

Ratha, D. (2013). *The impact of remittances on economic growth and poverty reduction*. MPI Policy Brief 8. Washington, DC: Migration Policy Institute.

Ratha, D., Supriyo, D., Plaza, S., Schuettler, K. and Shaw, W. (2016). *Migration and remittances – Recent developments and outlook*. Migration and Development Brief 26. Washington, DC: World Bank.

Rodriguez, R. M. (2010). *Migrants for export: How the Philippine state brokers labor to the world*. Minneapolis: University of Minnesota Press.

Sharma, K. (2011). *Realizing the development potential of diasporas*. Tokyo: United Nations University Press.

Silvey, R. (2006). Consuming the transnational family: Indonesian migrant domestic workers to Saudi Arabia. *Global Networks*, 6(1), pp. 23–40.

Tyner, J. A. (2009). *The Philippines: Mobilities, identities, globalization (Global Realities)*. New York: Routledge.

UNDESA (United Nations Department of Economic and Social Affairs). (2015). *Trends in international migrant stock: Migrants by destination and origin*. POP/DB/MIG/Stock/Rev.2015. New York: United Nations.

Vargas-Silva, C., Jha, S. and Sugiyarto, G. (2009). *Remittances in Asia: Implications for the fight against poverty and the pursuit of economic growth*. ADB Economics Working Paper Series 182. Manila: Asian Development Bank.

Yang, D. and Choi, H. (2007). Are remittances insurance? Evidence from rainfall shocks in the Philippines. *The World Bank Economic Review*, 21(2), pp. 219–248.

17
CHILDREN, YOUTH AND DEVELOPMENT IN SOUTHEAST ASIA

Harriot Beazley and Jessica Ball

This chapter is a critical reflection on globalization and international development agendas that affect children and youth in Southeast Asia. Although most Southeast Asian nations have had sustained economic growth and some success in reducing relative and absolute poverty, accelerated globalization has exacerbated income inequality, and the ranks of the urban poor have been swelling rapidly (Murray and Overton 2014; OECD 2015; Asia Foundation 2016; Indonesia Investment 2014). A clear manifestation of growing inequality in the region is the increased marginalization of less powerful members of society – including children and youth. Children are particularly vulnerable to the effects of poverty, especially if they are: a member of a minority ethnic group, living in rural or remote areas, or among the urban poor. This chapter highlights key development indicators of child well-being in Southeast Asia, and then focuses on how poor children in the region are typically socially constructed as victims within development discourse, and how programs that have been designed to protect and 'save' them are often based on preconceived, patronizing Western ideas that reinscribe their victimhood. The chapter then explores the lives and aspirations of marginalized children, including street children in Indonesia and Cambodia, children in post-tsunami Aceh, and forced migrant children from Myanmar who are growing up on the borders in Thailand, China and Malaysia. Many of the children in these populations are also stateless (Ball et al. 2014).

It is widely recognized that 'childhood' is a culturally and historically specific construct (Aries 1962; Holloway and Valentine 2001; Ansell 2004). In addition, notions of childhood relate to the global capitalist economy and the subsequent ways in which the elite in many countries have been influenced by the 'global export of modern childhood' (Stephens 1995, 15). The importance of children and youth in international development agendas was validated by the almost universal ratification of the United Nations Convention on the Rights of the Child (UNCRC) (United Nations 1989). Following the ratification of the UNCRC by all Southeast Asian nations, a key concern of regional development agendas has been implementing those rights – in particular to improve the welfare of millions of children and youth living in poverty, including working children. In 1992 the International Labor Organization (ILO) created the International Program against Child Labor (IPEC) (Bessell 1998; ILO 2016a) and Convention 182 against the Worst Forms of Child Labor in 1999 (ILO 1999). In 2000, the United Nations Millennium Development Goals (MDGs) focused on the rights of children, emphasizing a child's right to education and survival, reduction of infant and child mortality rates and universal access to primary education (United Nations 2015a).

Within scholarly development discourse the concept of the 'global child' enshrined within the UNCRC immediately generated extensive critique (Stephens 1995; Burman 1996; Penn 2005; Pence and Hix-Small 2009). Critics have pointed to the Euro-centric construction of the necessary conditions for child development, the goals and ideal outcomes of childhood, and the Western imposition of presumed 'best practices' for ensuring family, community and public contributions to children's well-being, irrespective of country, community or cultural contexts (Stephens 1995; Nieuwenhuys 1996; Pence and Hix-Small 2009). These precepts are often at odds with the lived realities of many children in Southeast Asia, with its vast cultural diversity resulting in a huge spectrum of culturally based approaches to child-rearing and public services to support child survival, health and development. Euro-Western goals for child development, foundational theories of child development, and the goals of interventions informed by Euro-Western values and practices seem particularly at odds with the realities and aspirations of children and youth who are poor and marginalized (Penn 2002).

This chapter takes up these issues with a focus on the realities of children and youth living in Southeast Asian nations who exist in diverse communities, on the margins of society. Drawing on findings of participatory research with children in Indonesia, Cambodia and Myanmar, the chapter explores different sociocultural norms and styles of child-rearing that challenge Western conceptions of 'family' and 'childhood.' The discussion highlights the importance of the cultural role of community networks and the feelings of filial responsibility that sometimes motivate children to work to contribute to household incomes.

Emphasizing children's emotional geographies (Blazek and Kraftl 2015) and children's agency in seeking solutions for survival, the chapter challenges assumptions that children living in poverty and in vulnerable situations are passive and helpless victims in need of protection. Instead, the chapter contends that policy makers, practitioners and scholars can bring theories of child development into the twenty-first century, and increase the dignity and effectiveness of interventions by bringing children and youth into both the research process and decision-making about policies and programs that are intended to support them (Beazley 2015a). The chapter concludes by reflecting on what the UN Sustainable Development Goals (SDGs) for 2015–2030 (United Nations 2015b) may mean for children and youth in Southeast Asia in the next 15 years.

Key indicators of child and youth well-being in the region

Before a discussion of marginalized children's experiences in Southeast Asia, it is useful to highlight some indicators of children's well-being in the region, including infant and child mortality, access to primary and secondary school, and the numbers of children who are in the workforce. Drawing upon reports on the Millennium Development Goals (MDGs) (United Nations 2015a) and UNICEF's "State of the World's Children" (2015a), the overview provided here focuses on the status of children in Indonesia, Cambodia and Myanmar from 2000 to 2015 (United Nations 2015a). These indicators illustrate the significant variations among children in Southeast Asia on key dimensions of well-being.

While the proportion of children and youth within populations in other parts of the world is diminishing (United Nations Development Program 2016), children and youth under the age of 24 account for nearly half of the population in Southeast Asia (CIA 2015). For example, in Cambodia, with a population of 15.7 million, 51 percent are under 24 years old; in Indonesia, with a population of 256 million, 43 percent are under 24 years old, and in Myanmar, with a population of 56.3 million, 44 percent are under 24 years old (CIA 2015).

Socio-economic status

Many socio-economic factors underpin development issues relating to children's well-being. The Human Development Index (HDI) is an important indicator for comparative appraisal of development in the Southeast Asia region and is a tool developed by the United Nations Development Program (UNDP) to measure and rank countries' levels of social and economic development (UNDP 2016). HDI is a summary measure (on a scale between 0 and 1, where 0 indicates the lowest human development and 1 the highest), and based on four criteria: life expectancy at birth, mean years of schooling, expected years of schooling, and gross national income per capita. Singapore has the highest human development in the Southeast Asia region with an HDI of 0.91 and Malaysia also has a high HDI with 0.77 (UNDP 2016) (Table 17.1). Indonesia has a 'medium' HDI of 0.68. At the lower end of the spectrum, Myanmar has the lowest human development in the region with an HDI of 0.53. Cambodia and Laos also score low with 0.55 and 0.57 respectively (Table 17.1) (UNDP 2016).

Within Southeast Asia, the poorest nations are Cambodia, Myanmar and Laos. Income inequality is also a significant issue in the region and particularly conspicuous in Malaysia, Singapore, Philippines, Thailand and Indonesia (World Bank 2014). Rising inequality is associated with lower economic and social mobility for younger generations (World Bank 2014). Today, 7.7 percent of Southeast Asian youth are unemployed (ILO 2016b).

Poverty, hunger and health

Southeast Asian nations contributed to the global achievement of the MDG of reducing extreme poverty by half (United Nations 2015a). In Indonesia, between 2005 and 2013, the

Table 17.1 Positional ranking based on children's well-being

	Country	IMR 2013	CMR 2013	Underweight (%) in 2013	Stunting (%) in 2013	Expected years of schooling 2014	HDI 2014
1st	Singapore	2	3	3	4	15.4	0.91
2nd	Malaysia	7	9	13	17	12.7	0.77
3rd	Brunei	8	10	-	-	14.5	0.85
4th	Thailand	11	13	9	16	13.5	0.72
5th	Vietnam	19	24	36	47	11.9	0.66
6th	Indonesia	25	29	20	36	13.0	0.68
7th	Philippines	24	30	20	30	11.3	0.66
8th	Cambodia	33	38	29	41	10.9	0.55
9th	Myanmar	40	51	23	35	8.6	0.53
10th	Timor-Leste	46	55	45	58	11.7	0.59
11th	Laos	54	71	27	44	10.6	0.57

Sources: UNICEF "State of the World's Children" (2015a) and UNDP *Human Development Reports* (2016)

number of people living in poverty fell from 16 percent to 11.5 percent (29 million people) (United Nations 2015a). However, the World Bank (2014) reports that there are 68 million 'near-poor' Indonesians who are living just above the poverty line. In 2015, more than one-third of 15.3 million Cambodians still lived below the poverty line on less than US$1.25 a day (UNICEF 2016). In Myanmar 26 percent of the 51.4 million population are living in poverty (UNDP 2016).

The MDG goal of reducing extreme hunger by half was achieved by 2015 in Southeast Asia: the proportion of undernourished people has fallen from 31 percent in 1990 to 10 percent in 2015, and among children under 5 years old, malnutrition fell from 31 percent to 16 percent between 1990 and 2015 in the region as a whole (United Nations 2015a). Yet, malnutrition remains a salient concern for children in the region, again pointing to inequities in access to basic requirements for normative growth and development. Children in Cambodia are particularly affected by malnutrition: 41 percent of children under 5 have moderate to severe stunting (below normative height for age) and 29 percent are underweight (UNICEF 2015a). Stunting affects 36 percent of children under 5 in Indonesia while 20 percent of children below the age of 5 are underweight (UNICEF 2015a). In Myanmar 35 percent of children under 5 are stunted and 23 percent are underweight (UNICEF 2015a).

By 2015, Southeast Asia achieved the MDG target for reduced infant mortality (IMR) to 24 deaths per 1000 live births by 2015,[1] and the MDG target for reduced child mortality (CMR) to from 71 to 27 deaths per 1,000 live births (United Nations 2015a). However, significant disparities between and within Southeast Asian countries persist. Recent demographic data demonstrates that children are most able to thrive in Singapore and Malaysia (UNICEF 2015a). As shown in Table 17.1, Singapore has one of the lowest IMR (2 deaths per 1,000 live births) and the lowest CMR (3 deaths per 1,000 live births), followed by Malaysia (7/1,000 and 9/1,000) and Brunei (8/1,000 and 10/1,000). Countries with the lowest rates of child survival are Timor-Leste (IMR 45/1,000 and CMR 55/1,000), Myanmar (IMR 40/1,000 and CMR 51/1,000), and Cambodia (IMR 33/1,000 and CMR 38/1,000), where children continue to suffer avoidable debilitating health and nutritional problems and experience the highest infant and child mortality rates (UNICEF 2015a) (Table 17.1).

While Indonesia ranked sixth in the region for infant and child mortality rates (IMR 23/1,000 live births and CMR 29/1,000 live births), Eastern Indonesia had far higher rates of infant mortality with 58 per 1,000 live births (UNICEF 2016). Almost half of infant deaths in Indonesia were attributed to complications from premature birth, still births and severe infections including pneumonia, meningitis and septicemia (UNICEF 2016).

Education

As one of the MDGs, access to education in the region has improved considerably in the past decade (UNESCAP 2013). While a 91 percent literacy rate for youth globally was predicted for 2015, Southeast Asia appears to have surpassed this with a 98 percent literacy rate among youth aged under 24 years old (United Nations 2015a). However, specific populations of children who have low or no access to primary education have been identified in the region: "Future interventions will have to be tailored to the needs of specific groups of children – particularly girls, children belonging to minorities and nomadic communities, children engaged in child labor and children living with disabilities, in conflict situations or in urban slums" (United Nations 2015a, 27). Children with disabilities are also predominant among children out of school in the region (UNESCAP 2013). In terms of the number of years a child is expected to stay in school, in 2014 Singapore ranked the highest with an expected 15.4 years of schooling (Table 17.1).

Indonesia has an average of 13 years of expected schooling, Cambodia 10.9 and Myanmar 8.6 of expected schooling for each child (Table 17.1).

While the net primary school enrollment has increased in the region, expanding access to secondary education has been more challenging with a net secondary school enrolment rate of 77 percent Southeast Asia as a whole (World Bank 2016b). As well, persisting lower participation by girls compared to boys has been attributed to "negative social and cultural attitudes, lack of appreciation of the value of female education, the burden of household work and long journeys to school" (UNESCAP 2013, 2).

Dominant perceptions versus local realities

The first half of this chapter established some promising trends and outstanding challenges affecting the well-being of children and youth in the region. The second half of the chapter considers Western discourses of development, and how these dominant perceptions have influenced the international policies and practices concerning children, especially those in vulnerable situations and those involved in labor at a young age. The section begins with an overview of global discourses on child labor and vulnerable children, and then for illustrative purposes reflects on the authors' own participatory research with children in Southeast Asia. We focus on the experiences of children working on the streets in Indonesia and Cambodia; disaster management and development for children after the devastating tsunami in Indonesia; and forced migrant children from Myanmar. By contrasting the social and cultural complexity of these children's lived experiences against the dominant perspectives and representations instilled in development discourse and programming, the chapter raises questions about what we really know about the goals and trajectories of children's development in vulnerable situations in the region. The chapter also calls for an interrogation of the assumptions that inform prescriptions for interventions intended to close equity gaps and promote optimal developmental outcomes.

Global discourses on child labor

The ratification of the UNCRC in 1990 gave impetus to a surge of international interest in child labor in the region, especially relating to the international economy, globalization and urbanization. In Southeast Asia, 77.8 million children aged 5 to 17 years, or 9.3 percent of the region's children, are in the labor force (United States Department of Labor 2014). Most frequently, children work for very low wages in agriculture, factories, cottage industries and the urban informal sector, such as working in roadside stalls, as market 'coolies,' shoe-shining, scavenging, selling newspapers, parking cars, washing car windows and busking (ILO 2016a). As well, some children engage in unwaged domestic duties in exchange for food and shelter (ILO 2016a). Child labor is rooted in poverty, history, culture and global inequality (Lloyd-Evans 2014). Most children in the Southeast Asia region are engaged in forms of labor that violate their right to be protected from work that is dangerous, harmful to health or a barrier to education (ILO 2016a).

In 1992, the International Labor Organization (ILO) created the International Program against Child Labor (ILO 2016a; Bessell 1998) as a response to these violations. The IPEC program targeted elimination of the Worst Forms of Child Labor as expressed in IPEC's Convention 182 (ILO 1999). IPEC fostered significant improvements to the state of child labor in Southeast Asia. In 2014, five ASEAN countries were reported to have improved their legal frameworks for protecting children from hazardous work (USDL 2014). For example, in 2002, Indonesia, adopted the Child Protection Law (Law No 23), which protects children under 18 years old from a variety of abuses and prohibits the employment of children in the worst forms of child labor, such as forced or bonded labor, sexual exploitation and the trafficking of drugs (Hitzemann 2004; ILO 2016a). In Thailand, the minimum age requirements were raised to 15 years old for agricultural workers and 18 years old for working on fishing boats (USDL 2014).

While exploitive, unsafe and harmful child labor has obstructed healthy development for already vulnerable children throughout the region, a growing body of research has shown that not all forms of work are detrimental for children. Many children take pride in supporting themselves and their families financially (White 1994; Ennew et al. 2005; Beazley 2015a). Yet, interventions inspired by the UNCRC and other Euro-Western tools have tended to cast an over-generalized, negative pall over the concept of child labor and failed to engage children and youth in generating a more nuanced and contextualized perspective on what children's involvement in specific kinds of work means and achieves under various conditions and in varied cultural and economic contexts (White 1994; Lloyd-Evans 2014).

The imposition of Western constructions of childhood and the requirements for growth and development are also evident in the meanings attached to particular public 'spaces' and their judged appropriateness for children (Beazley 2003; Beazley 2016; Lloyd-Evans 2014). For example, within this discourse children working in industrial complexes or on the streets are particularly considered to be 'out of place' (Lloyd-Evans 2014). Most families in Southeast Asia subscribe to culturally based constructions of childhood that may draw upon some Western concepts and aspirations but that are grounded in local realities, values and goals. Given rising global economic insecurity and income inequalities, poor children and families are particularly likely to consider children's involvement in wage-earning as an opportunity to supplement family income (Beazley 2003; Beazley and Miller 2016). Children's agency must be considered: children may make an active choice to work, motivated by cultural expectations of filial duty and contribution to the household income, and children may garner new skills and an enhanced self-esteem and self-efficacy as a result (Beazley 2003; Bessell 2009; van Blerk 2006).

Example 1: street connected children

With the exception of Singapore, children living on the streets without supervision or protection of an adult are almost ubiquitous in cities across Southeast Asia. Nobody knows how many street children there are in Southeast Asia. This is in part because they are highly mobile and many have not been registered at birth and have no identity documents (Beazley 2003). As a result, they are effectively stateless. Most street connected children are boys aged 8 to 17 years, although the number of street girls is increasing in the region (Beazley 2008). Street children are most prevalent in poorer countries where there is a robust urban informal sector, including Indonesia, Timor-Leste, the Philippines and Myanmar, where they engage in a variety of income-generating activities. For example, in Indonesia street-dwelling boys often shine shoes, sell bottled water, cigarettes and other goods, wash car windows and busk or beg (Beazley 2003). In Cambodia they sell books, postcards, trinkets and bottled water to tourists (Beazley and Miller 2016). Factors influencing the prevalence of street children in Southeast Asian cities include rapid urbanization, forced migration, internal displacement, growing income inequality, poverty, consumerism, conflict, famine, natural disasters, family breakdowns and the increase of domestic violence.

Street children are commonly believed to have been orphaned or abandoned by neglectful parents, and they are often seen as belonging to groups organized by adults to beg or commit crime. However, it is helpful to unsettle these generic and stigmatizing perceptions of street connected children in light of a growing global discourse on child labor and children's rights. Although research is limited, available evidence points to many reasons why children start to work and live on the streets and their experiences and outcomes are diverse. For example, Beazley and Miller (2016) describe how, in Siem Reap, Cambodia, children attend school, work on the street for part of the day, live at home and have continuous family contact. Others are no longer in school and spend all their days on the streets but go home at night. Sometimes these children become involved with the subculture of street-entrenched children, eventually moving away from home altogether (Beazley 2003, 2016). These children are essentially homeless, as they live, work and spend the majority of their time on the streets, and have very little if any contact with their families (Beazley 2003).

In Indonesia, the Philippines and Cambodia, street children are often portrayed by the media and the state as undesirable social pariahs who are a blight on the city's landscape. Due to this social construction, there is an almost universal negative response to them in cities throughout the region: they are both spatially and socially oppressed through multiple forms of social control, marginalization and physical oppression by the state and society. This is often in the form of verbal abuse, evictions, arrests, beatings, abuse and torture while in police custody, and other infringements of children's basic human rights (Beazley 2003; Beazley and Miller 2016). There are numerous accounts of state 'cleansing operations' occurring in cities in the Philippines, Cambodia and Indonesia, when children have been physically cleared off the streets and dumped out of town, or arrested and beaten, and sometimes killed (Beazley 2003; Beazley and Miller 2016).

Many non-governmental organizations in the region have vigorously invested in programs seeking to 'rescue' children by taking them off the streets, often without their consent, and placing them in institutions or returning them home. Typically, these programs counter dominant media images of street-dwelling children as criminals by portraying them in funding campaigns as helpless and abandoned victims. Children's own preferences, agency or the meanings that they ascribe to their lives are afforded little if any regard (Wells 2010; Allerton 2014). In participatory child-centered research by Beazley (2016), children living on the streets in Yogyakarta reported that they had no desire to leave the street or to return home. This was in spite of an ILO-IPEC target to return 25 children to their family homes each month from the NGO they attended (Beazley 2016). In recent years more effective strategies have been developed for street frequenting children in some Southeast Asian nations, based on rights-based advocacy agendas, with respect for children's rights on the parts of governments and police, and the provision of appropriate services through drop-in centers and outreach programs (West 2003).

Example 2: children and disaster management

The second example is focused on children in disaster situations in the region. The role of children and youth as agents and drivers of change has been acknowledged in the global commitment for child-centered forms of Disaster Risk Reduction (DRR) (UNISDR 2015). Save the Children (2010, 4), argues that "children should not be seen as victims, but as actors in addressing the impacts of natural disasters and climate change on their lives and the life of their community." Strategies that embody this view include engaging children in risk and disaster impact assessments, disaster reduction and post-disaster planning and preparation, and evaluations of disaster risk reduction and management strategies (Save the Children 2010).

Disasters are prevalent in Southeast Asian nations: active tectonic plates give rise to earthquakes and tsunamis; cyclones and typhoons are generated by the Indian and Pacific oceans; and mountain ranges combined with deforestation yield frequent landslides and flash floods (UNESCAP 2015). It is widely recognized that children and youth suffer the most negative impacts from disaster including a third to half of disaster-related deaths (Save the Children 2010). Dangers include immediate injury and death, illness and malnutrition. Disasters also cause children psychological trauma and emotional distress through family separation and death, loss of home, abrupt changes to daily life including ability to attend school, and abuse or exploitation of children in vulnerable situations (Save the Children 2010). Interviews with children in Southeast Asia reveal that following a disaster they are primarily impacted by disruption to their education (Save the Children 2010). For example, following the 2011 tsunami in Thailand, 2,000 schools were temporarily closed due to structural damage, and in the 2012 typhoon in the Philippines, 551 schools were closed due to damage, affecting 100,000 children (Save the Children Australia 2014).

The 2004 earthquakes and tsunami in Aceh, Indonesia, severely impacted children: an estimated 200,000 people died and over 550,000 children were left homeless, including an estimated 50,000 'tsunami orphans' (Gunawan 2005; Moore 2005). The established response on the part of international relief agencies was to place separated and presumed orphaned children in residential institutions. A deluge of financial aid for 'tsunami orphans' led to the rapid construction of over 35 orphanages in Aceh where 2,600 children were placed (Martin and Sudrajat 2006). However, a majority of children who were thought to be 'tsunami orphans' by international aid agencies were in fact not orphaned. By 2006, it was determined that more than 85 percent of the children who were placed into orphanages or religious boarding schools after the tsunami had at least one parent alive, and 42 percent still had both parents (Martin and Sudrajat 2006; Save the Children 2006; Beazley 2015b; Riley, 2013). International organizations at the time did not recognize the significance and resilience of extended family systems, many of which could absorb children who had lost their parents (Beazley 2015b; Abebe 2009).

For many economically destitute and displaced families, institutional care for their children seemed the only option available because the bulk of relief funding was directed to building children's homes, rather than providing families with the support they needed to care for their children (Save the Children 2006; Beazley 2015b). Conversely, funding to these institutions was based on the number of children that they supported; hence, many parents were reportedly encouraged to place their children in institutional care (Martin and Sudrajat 2006). For organizations and administrators, there was an incentive to recruit children and to keep them as long as possible (Save the Children 2006; Beazley 2015b).

The need for humanitarian and international development policy to reconsider long-term residential care as a child protection strategy in disaster situations was revealed through participatory, child-led research with children living in three institutions in Aceh (Beazley et al. 2009; Beazley 2015b). Violations of children's rights were uncovered in all three orphanages, including systematic physical and verbal abuse by carers, teachers and other children; life threateningly poor nutrition, hygiene and living conditions, lack of healthcare and privacy, and strict regimentation and harsh discipline. Children also expressed their longing for family, friends and the opportunity to live in a family home, to enjoy the affection of a family member and siblings from whom many were separated through institutional placement (Save the Children 2006; Beazley 2015b).

The institutionalization of children in Aceh after the tsunami reflects the prevailing Western notion of best practices in humanitarian aid at the time. The decision disregarded the best interests of the children and their right to express an opinion in matters that affect them, enshrined in Article 12 of the UNCRC (1989). As Abebe (2009) and Riley (2013) have emphasized, Western donors and aid organizations' inaccurate, ethnocentric construction of disaster orphans have failed to understand culturally based family structures, practices, resilience and needs. They have also disregarded the child's right to live in a family home, with institutional care only as a last resort and not as a default response as seen in Aceh.

Participatory, rights-based research with children, youth and their communities can ensure that disaster planning and post-disaster responses are appropriate for the region and affected communities, and that children's rights are not violated in the name of protection.

Example 3: forced migrant children from Myanmar

The final example is from Myanmar, which is the most significant contributor to the world's 60 million forced migrants[2] and refugees (UNHCR 2015). Since the military took control of Myanmar in 1962, millions of children and families facing violent suppression of ethnic minorities have fled their homeland to neighboring Thailand, China and Malaysia. Thailand, which shares a 2,401 kilometer border with Myanmar, has been host to the largest influx with approximately 2.5 million Myanmar migrants and refugees, of whom about one-fifth are under 19 years old (Refugees International 2004; UNHCR 2014).

Many forced migrant children and youth have experienced psychosocial trauma associated with armed conflict and family separation. Many lack identity documentation and are effectively stateless (Lynch 2010), unable to access formal systems of social protection, healthcare and education (Myanmar Education Integration Initiative [MEII], 2013). In some instances, children are unaccompanied: Some are sent by their families in search of safety and education; some are separated from their family during conflict and displacement; some are kidnapped by drug traffickers and militia; and some are orphaned (Committee for the Protection and Promotion of Child Rights, Burma 2009).

Beginning in 2011, the Myanmar government began a political shift toward a participatory democracy. The new government has begun welcoming forced migrants to come home, and international aid to forced migrant children who continue to live outside of Myanmar is being withdrawn (MEII 2013). However, there are significant barriers to forced migrant children's reintegration, including: lack of documentation to prove that Myanmar is their country of origin; lack of education in the Myanmar language so that returning to Myanmar will significantly disrupt the continuity of any schooling they may have been able to access (often through non-formal learning programs) (Save the Children and World Education 2015); absence of a family or village to return to; and risks to their physical safety because of their ethnic and family heritage (Ball and Moselle, 2015). As well, some forced migrant children have spent most of their growing up years outside of Myanmar, and may not wish to return. These children's development narratives have not been systematically documented and their views and preferences about reintegration, assimilation into the country where they temporarily reside, or resettlement to a third country are not solicited in adult-led processes that determine their futures.

Perched on the precipice of multiple intersecting sources of uncertainty, these children's experiences and processes of development defy conventional understandings of childhood and identity formation. Global institutions, including international organizations, educational bodies and the media, tend to understand migrant children as passive agents who are dependent on parents, the state and international organizations to determine their well-being and future. They are often seen as victims, and their experiences are often understood as secondary products of their parents' primary narratives of displacement. By contrast, preliminary research by the second author suggests that forced migrant children and youth from Myanmar are eager to have their voices heard in decision-making about their futures – a right enshrined in Article 12 of the

UNCRC (1989). Although few have played active roles as instigators of their migration, their stories tell how, in order to survive and thrive, they must engage in dynamic meaning-making of their shifting circumstances and learn new skills and attitudes to understand and adapt. Suggesting a challenge to the binaries posited by foundational theories of children's development, the children's stories convey that they are both victims and agents, both vulnerable and resilient, and carriers of both their cultures of origin and of globalization. While some children and youth voice specific goals to return to Myanmar or to settle in another country, many are oriented to an open-ended future that they feel ready to actively create. This will be by drawing on internal resources they have generated as liminal children growing up in extremely difficult circumstances on the margins of mainstream societies in Southeast Asia.

Conclusion

Southeast Asia is a highly diverse region with thousands of linguistic, cultural and social groups and varied populations of children and youth. Generalizations across all communities and countries are likely to have as many exceptions as confirming cases. However, this chapter points to significant indicators of progress for some populations on some important dimensions, including infant and child mortality, access to primary school and quality of life. While it can be concluded that conditions for children and youth in the region are decidedly improving, income inequality is increasing and in every country there are some children whose lives are extremely difficult due to chronic conditions, such as poverty and discrimination, or acute conditions, such as natural and anthropogenic disasters and epidemics. Political insurgencies and economic instability in some countries pose ongoing threats to children and families, particularly those who are already living in vulnerable circumstances. With an already high proportion of the world's young people and an overall high fertility rate, sustained efforts to generate valid, contextualized, nuanced understandings of the experiences, needs and goals of children and youth in the region and to create effective strategies to support their wellness must be a local, national and regional priority.

In this chapter, the three case studies from Southeast Asia illustrate how an understanding of the local cultural circumstances of children's lives must be explored to assess the relevance of Western constructions of what is best for children in developing country contexts (Penn 2005). The case studies reveal how poor children in the region are frequently represented as only victims within development discourse and how programs designed to protect them are often based on Western perceptions that underline their vulnerability. The rights-based participatory research utilized in the case studies, however, demonstrates the imperative of accessing children's own self-reported views of their lived experiences in order to gain a clearer understanding of their lives and to better address their needs through appropriate and sustainable interventions.

In September 2015 the UN Sustainable Development Goals (SDGs) were launched, following expiry of the MDGS. What has become known as the 'post-2015 agenda' includes 17 new global goals for the next 15 years (2015–2030), many of which implicate or affect children and youth. In 2015 a UNICEF report (UNICEF 2015b) describes how the SDGs build on MDG successes in addressing maternal and child mortality, poverty, hunger, primary education and gender equality, while also elevating the importance of reducing child poverty and violence against children through enhanced social protection systems. While the SDGs hold tremendous promise, they do not specify the means by which targets are to be achieved. Looking at this open-endedness as an opportunity, scholars, policy decision-makers, child-serving international and local organizations, and practitioners have ample justification and scope to bring into sharper focus the cultural contexts of children's development and children's own views about how to improve the quality of life for children in Southeast Asia and around the world.

Notes

1 For comparison, in Australia the 2015 infant mortality rate was 3/1,000 and the child mortality rate was 4/1,000 (World Bank 2016a).
2 The International Organization for Migration defines forced migrants as individuals who "leave their countries to escape persecution, conflict, repression, natural and human-made disasters, ecological degradation, or other situations that endanger their lives, freedom or livelihood" (IOM 2000).

References

Abebe, T. (2009). Orphanhood, poverty and the care dilemma: Review of global policy trends. *Social Work and Society*, 7(1), pp. 70–85.
Allerton, C. (2014). Statelessness and the lives of the children of migrants in Sabah, East Malaysia. *Tilburg Law Review*, 19(1–2), pp. 26–34.
Ansell, N. (2004). *Children youth and development*. London: Routledge.
Aries, P. (1962). *Centuries of childhood: A social history of family life*. New York: Vintage Books.
Asia Foundation. (2016). Asia's biggest issues in 2016? Experts weigh in. *Weekly Insight and Analysis in Asia*, 6 January. Available at: http://asiafoundation.org/in-asia/2016/01/06/asias-biggest-issues-in-2016-experts-weigh-in/. Accessed 22 January 2016.
Ball, J., Butt, L., Beazley, H. and Fox, N. (2014). *Advancing research on "Stateless children": Family decision making and birth registration among transnational migrants in the Asia-Pacific region*, Centre for Asia Pacific Initiatives Working Paper Series on Migration and Mobility, MMP 2014–2 University of Victoria.
Ball, J. and Moselle, S. (2015). Living liminally: Migrant children living in the Myanmar-Thailand border region. *Global Studies of Childhood*, 5(4), pp. 425–436.
Beazley, H. (2003). Voices from the margins: Street children's subcultures in Indonesia. *Children's Geographies*, 1(2), pp. 181–200.
Beazley, H (2008). The Geographies and Identities of Street Girls in Indonesia. In M. Gutman and P. Kraftl, eds., *Designing Modern Childhoods: History, Space, and the Material Culture of Children. An International Reader*. New Jersey: Rutgers, 233–249.
Beazley, H. (2015a). Multiple identities, multiple realities: Children who migrate independently for work in Southeast Asia. *Children's Geographies*, 13(3), pp. 296–309.
Beazley, H. (2015b). Inappropriate aid: The experiences and emotions of tsunami 'Orphans' living in children's homes in Aceh, Indonesia. In: M. Blazek and P. Kraftl, eds., *Children's emotions in policy and practice*. London: Palgrave Macmillan, 34–51.
Beazley, H. (2016). Bus stops and toilet: Identifying spaces, spaces of identity. In: T. Skelton, C. Dwyer and N. Worth, eds., *Identities and subjectivities*, Volume 4 of T. Skelton, ed., *Geographies of children and young people*. Singapore: Springer, 1–28.
Beazley, H., Bessell, S., Ennew, J. and Waterson, R. (2009). The right to be properly researched: Research with children in a messy, real world. *Children's Geographies*, 7(4), pp. 365–378.
Beazley, H. and Miller, M. (2016). The art of not being governed: Street children and youth in Siem Reap, Cambodia. *Politics, Citizenship and Rights*, 7(1), pp. 263–289.
Bessell, S. (1998). Child labour and the rights of the child. *Development Bulletin*, 44, pp. 29–31.
Bessell, S. (2009). Indonesian children's views and experiences of work and poverty. *Social Policy & Society*, 8(4), pp. 527–540.
Blazek, M. and Kraftl, P. (2015). Introduction: Children's emotions in policy and practice. In: M. Blazek and P. Kraftl, eds., *Children's emotions in policy and practice: Mapping and making spaces of childhood*. London: Palgrave Macmillan, 1–13.
Burman, E. (1996). Local, global or globalised? Child development and international child rights legislation. *Childhood*, 3(1), pp. 45–66.
CIA. (2015). The world factbook Cambodia. *Central Intelligence Agency*. Available at: www.cia.gov/library/publications/the-world-factbook/geos/cb.html. Accessed 2 December 2015.

Committee for the Protection and Promotion of Child Rights (Burma). (2009). *Feeling small in another person's country: The situation of Burmese migrant children in Mae Sot, Thailand*. Unpublished report. Available at: www.burmalibrary.org/docs08/Feeling_Small.pdf

Ennew, J., Myers, W. E. and Plateau, D. P. (2005). Defining child labor as if human rights really matter. In: B. H. Weston, ed., *Child labor and human rights: Making children matter*. London: Lynne Rienner, 27–54.

Gunawan, A. (2005). Govt. to house tsunami orphans. *Jakarta Post*, 6 February. Available at: www.thejakartapost.com/news/2005/02/06/govt-house-tsunami-orphans.html. Accessed 19 January 2016.

Hitzemann, A. (2004). *Institution building & mainstreaming child protection in Indonesia*. United Nations Children's Fund. Available at: www.unicef.org/evaldatabase/files/Indonesia_2003_Mainstreaming_Child_Protection.pdf. Accessed 21 December 2015.

Holloway, S. L. and Valentine, G. (2001). A window in the wider world? Rural children's use of information and communication technologies. *Journal of Rural Studies*, 17(4), pp. 383–394.

Indonesia Investments. (2014). Despite poverty reduction in Indonesia, gap between rich and poor widens. *Indonesia Investments*. Available at: www.indonesia-investments.com/news/todays-headlines/despite-poverty-reduction-in-indonesia-gap-between-rich-and-poor-widens/item2258. Accessed 11 February 2016.

International Labour Organization. (1999). *Convention 182: Convention concerning the prohibition and immediate action for the elimination of the worst forms of child labour*. International Labor Organization. Available at: www.ilo.org/dyn/normlex/en/f?p=NORMLEXPUB:12100:0::NO::P12100_ILO_CODE C182. Accessed 11 February 2016.

International Labour Organization. (2016a). *International Programme on the Elimination of Child Labour (IPEC)*. International Labor Organization. Available at: www.ilo.org/ipec/lang–en/index.htm. Accessed 10 February 2016.

International Labour Organization. (2016b). *Global youth unemployment is on the rise again*. Available at www.ilo.org/global/about-the-ilo/newsroom/news/WCMS_513728/lang–en/index.htm. Accessed 22 July 2016.

International Organization for Migration (IOM). (2000). *World migration report 2000*. International Organization for Migration, United Nations, Geneva.

Lloyd-Evans, S. (2014). Child labour. In: V. Desai and R. B. Potter, eds., *The companion to development studies*. London: Routledge, 207–211.

Lynch, M. (2010). Without face or future: Stateless infants, children, and youth. In: M. O. Ensor and E. M. Gozdziak, eds., *Children and migration*. New York: Palgrave Macmillan, 117–140.

Martin, F. and Sudrajat, T. (2006). *A rapid assessment of children's homes in post-tsunami Aceh, Indonesia*, DEPSOS Jakarta and Save the Children, UK.

Moore, M. (2005). Help came for orphans but so few lived to see it. *Sydney Morning Herald*, 8 March. Available at: www.smh.com.au/news/Asia-Tsunami/So-few-tsunami-orphans-lived-to-get-help/2005/03/07/1110160760488.html. Accessed 19 January 2016.

Murray, W. E. and Overton, J. (2014). *Geographies of globalisation*. 2nd ed. Oxford: Routledge.

Myanmar Education Integration Initiative (MEII). (2013). Myanmar migrant education review summary: Review, strategies, recommendations, plans. *Chaing Mai*. Available at: http://meii.co/wp-content/uploads/2013/09/Migrant-Education-Review-Summary_to-Upload.pdf. Accessed 22 February 2016.

Nieuwenhuys, O. (1996). The paradox of childhood and anthropology. *Annual Review of Anthropology*, 25(1), pp. 237–251.

OECD. (2015). *In it together: Why less inequality benefits all*. Paris: OECD Publishing.

Pence, A. and Hix-Small, H. (2009). Global children in the shadow of the global child. *International Critical Childhood Policy Studies*, 2(1), pp. 75–91.

Penn, H. (2002). The world bank's view of early childhood. *Childhood*, 9(1), pp. 118–132.

Penn, H. (2005). *Unequal childhoods: Young children's lives in poor countries*. Oxon: Routledge.

Refugees International. (2004). *Stolen futures: The stateless children of Burmese asylum seekers*. Washington, DC: Refugees International.

Riley, L. (2013). Orphan geographies in Malawi. *Children's Geographies*, 11(4), pp. 409–421.

Save the Children. (2006). *Indonesian orphans on the increase as tsunami pushes parents into poverty and children into institutions*. Save the Children. Available at: http://reliefweb.int/report/indonesia/indonesian-orphans-increase-tsunami-pushes-parents-poverty-and-children. Accessed 16 February 2016.

Save the Children. (2010). *Living with disasters and changing climate: Children in Southeast Asia telling stories about disaster and climate change*. Thailand: Save the Children.

Save the Children Australia. (2014). *No child left behind: Education in crisis in the Asia-Pacific region*. Save the Children Australia. Available at: www.protectingeducation.org/sites/default/files/documents/save_the_children_australia-_no_child_left_behind.pdf. Accessed 13 December 2015.

Save the Children and World Education. (2015). *Pathways to a better future: A review of education for migrant children in Thailand*. Save the Children and World Education, UK.

Stephens, S. (1995). Children and the politics of culture in 'late capitalism.' In: S. Stephens, ed., *Children and the politics of culture*. Princeton, NJ: Princeton University Press, 3–48.

UNESCAP. (2013). *Regional overview: Youth in Asia and the Pacific*. United Nations Economic and Social Commission for Asia and the Pacific. Available at: www.un.org/esa/socdev/documents/youth/fact-sheets/youth-regional-escap.pdf. Accessed 3 December 2015.

UNESCAP. (2015). *Overview of natural disasters and their impacts in Asia and the Pacific, 1970–2014*. United Nations Economic and Social Commission for Asia and the Pacific. Available at: www.unescap.org/sites/default/files/Technical%20paper-Overview%20of%20natural%20hazards%20and%20their%20impacts_final.pdf. Accessed 13 December 2015.

UNICEF. (2015a). *State of the world's children*. Available at: http://sowc2015.unicef.org/. Accessed 26 January 2016.

UNICEF. (2015b). *A post-2015 world fit for children: A review of the open working group report on sustainable development goals from a child rights perspective*. United Nations Children's Fund. Available at: www.unicef.org/post2015/files/Post_2015_OWG_review_CR_FINAL.pdf. Accessed 19 December 2015.

UNICEF. (2016). *Children in Indonesia*. Available at: www.unicef.org/indonesia/children.html. Accessed 3 February 2016.

UNISDR. (2015). *Sendai framework for disaster risk reduction 2015–2030*. Available at: www.preventionweb.net/files/43291_sendaiframeworkfordrren.pdf. Accessed 24 February 2016.

United Nations. (1989). *Convention on the rights of the child*. 20 November, United Nations, Treaty Series, Vol. 1577. New York: UN General Assembly. Available at: www.unicef.org.au/Upload/UNICEF/Media/Our%20work/childfriendlycrc.pdf. Accessed 20 January 2016.

United Nations. (2015a). *The millennium development goals report 2015*. United Nations. Available at: www.un.org/millenniumgoals/2015_MDG_Report/pdf/MDG%202015%20rev%20(July%201).pdf. Accessed 16 December 2015.

United Nations. (2015b). *Sustainable development goals*. Available at: https://sustainabledevelopment.un.org/?menu=1300. Accessed 10 February 2016.

United Nations Development Program. (UNDP). (2016). *Human development reports*. Available at: http://hdr.undp.org/en/composite/HDI. Accessed 2 February 2016.

United Nations High Commissioner for Refugees (UNHCR). (2014). *Statistics and operational data*. Available at: www.unhcr.org/pages/49c3646c4d6.html. Accessed 12 February 2016.

United Nations High Commissioner for Refugees. (UNHCR). (2015). *Worldwide displacement hits all-time high as war and persecution increase*, June 18. Available at: www.unhcr.org/558193896.html. Accessed 23 January 2016.

United States Department of Labour (USDL). (2014). *2014 findings on the worst forms of child labour*, USDL. Available at: www.dol.gov/ilab/reports/child-labor/asia.htm. Accessed 13 December 2015.

Van Blerk, L. (2006). Working with children in development. In: V. Desai and R. B. Potter, eds., *Doing development research*. London: Sage, 52–61.

Wells, K. (2010). Child-saving and child rights: Depicting the suffering child in international NGO fundraising leaflets. *Journal of Children and Media*, 2(3), pp. 235–250.

West, A. (2003). At the margins: Street children in the Asia Pacific. *Poverty and Social Development Papers* No.8, Asian Development Bank. Available at: www.adb.org/sites/default/files/publication/29163/margins.pdf. Accessed 30 June 2016.

White, B. (1994). Children, work and 'child labour': Changing responses to the employment of children. *Development and Change*, 25(4), pp. 849–878.

World Bank. (2014). *New poverty frontier in Indonesia: Reduction slows, inequality rises*. Available at: www.worldbank.org/en/news/feature/2014/09/23/new-poverty-frontier-in-indonesia-reduction-slows-inequality-rises. Accessed 20 February 2016.

World Bank. (2016a). *Mortality rate, infant (per 1,000 live births)*. Available at: http://data.worldbank.org/indicator/SP.DYN.IMRT.IN. Accessed 12 January 2016.

World Bank. (2016b). *Net enrolment rate, secondary, both sexes (%)*. Available at: http://data.worldbank.org/indicator/SE.SEC.NENR. Accessed 12 January 2016.

18
ETHNIC MINORITIES, INDIGENOUS GROUPS AND DEVELOPMENT TENSIONS

Sarah Turner

The end of World War Two heralded the solidification of power in many newly independent Southeast Asian states as they became recognized internationally. In turn, numerous modern state borders were formalized with little concern for the spatial distribution of the area's ethnic, racial, linguistic and religious groups. While economic growth in parts of the region has since benefited some ethnic minorities and indigenous individuals, these groups have more frequently been discriminated against, sometimes violently. This discussion does not focus on every group in the region, which would be a major undertaking (Duncan 2004; Michaud et al. 2016). Instead, it starts with a brief overview of each country's main ethnic minority linguistic groups, populations and minority terminology, with countries grouped by their broad political 'stance' toward ethnic minorities. This is followed by an examination of specific 'development' processes that significantly impact minority and indigenous groups in the region. The entry concludes with a call to recognize these groups' alternative modernities and approaches to development.

Southeast Asia is an incredibly heterogeneous region in terms of ethnicity, with ethnic identities shaped by complex histories. Many of the categories widely taken for granted today are products of colonial racial classificatory practices inscribed onto geographic space. Colonial rulers often associated 'tribal peoples' with the forests and uplands, while 'villagers' were identified with lowland intensive agriculture. These classifications are now being remade by some as links with transnational discourses of indigeneity enable local people to demand access to resources by claiming 'indigenous status' (Dressler and McDermott 2010). Indeed, since the 1990s, some governments have demonstrated increased sensitivity to minority and indigenous rights, as well as to the impacts of economic 'development,' but this has been uneven across the region; indigenous and minority groups still lose out too often when it comes to mainstream economic policies.

Simply put, indigenous groups are autochthons of a specific territory, while ethnic minorities are frequently settler populations who, nonetheless, may have migrated to an area hundreds of years ago (Clarke 2001); the Hmong, for instance, traveled from southwest China into Vietnam, Laos, Burma and Thailand. It is important to remember that such categories are often highly controversial, subjective and wrapped up in unresolved political tensions. Moreover, defining ethnicity itself is often not straightforward, and ethnic identity can be fluid over time and space. One must also choose how far back in history to go to decide whether a group is defined as 'indigenous' or not.

Examples abound regarding the complexity of defining ethnicity in Southeast Asia. For hundreds of years, the term 'Dayak' was used as an insult in the Indonesian province of Kalimantan and the Malaysian state of Sarawak, both on the island of Borneo; individuals' primary identities were instead associated with specific tribes, such as Benuaq, Kayan, Kenyah and Tunjung. Today, however, Dayak is commonly used as an ethnonym to refer to the non-Muslim, non-Malay natives of Borneo and has become a symbol of pride and unity (Thung et al. 2004). In Vietnam, the government insists that there are 53 minority nationalities (*dân tộc thiểu số*) alongside the Vietnamese lowland majority (*Việt* or *Kinh*). In reality, there are far more than 53 minority groups, but the government has enshrined this authoritative number in its constitution (Michaud 2009).

Table 18.1 attempts to list the most populous minority and indigenous groups in the region, but by no means reflects the full diversity present. For instance, Indonesia, an archipelago of over 13,000 islands with a population of 253 million (2014 est., CIA 2015a), has over 300 ethnic groups. The most recent census data have been used where possible, but other sources are also drawn upon.

Table 18.1 Ethnic minorities and indigenous peoples in Southeast Asia

Country	Major ethnic group	Ethnic minorities/indigenous	Group population
Brunei Total Population: 406,200 (2013)[i]	Major Ethnic Group: Malay: 267,200	Chinese	41,600
		Others	97,400
Burma (Myanmar) Total Population: 51,486,253 (2014)[ii]	Major Ethnic Group: Bamars/Burmans/ Burmese (68%)[iii] *(Census data on ethnicity not available).*	Shan	9%
		Karen	7%
		Rakhine	4%
		Chinese	3%
		Indian	2%
		Mon	2%
		Other	5%
Cambodia Total population: 13,395,682 (2008)[iv]	Major Ethnic Group: Khmer: 12,901,381 (96.3%)	Cham	204,055
		Vietnamese	72,337
		Phnong	37,522
		Other	35,721
		Tumpoon	31,007
		Kuoy	28,630
		Chaaraay	26,331
		Krueng	20,007
		Lao	18,754
		Chinese	6,698
East Timor[v] Total population: 1,053,971 (2010)	Major Mother Tongue: Tetum Prasa: 385,269 (36.55%)	Mambai	131,361
		Makasai	101,854
		Tetum Terik	63,519
		Baikenu	62,201
		Kemak	61,969
		Bunak	55,837
		Tokodede	39,483
		Fataluku	37,779
		Waima'a	18,467
		Naueti	15,045
		Idate	13,512
		Galoli	13,066
		Midiki	9,586

(Continued)

Table 18.1 (Continued)

Country	Major ethnic group	Ethnic minorities/indigenous	Group population
Indonesia Total Population: 236,728,379 (2010)[vi]	Major Ethnic Group: Javanese: 95,217,022 (40.22%)	Sundanese	36,701,670
		Batak	8,466,969
		Madurese	7,179,356
		Betawi	6,807,968
		Minangkabau	6,462,713
		Bugis	6,359,700
		Melayu	5,365,399
		Bantenese	4,657,784
		Banjarese	4,127,124
		Acehnese	4,091,451
		Balinese	3,946,416
		Sasak	3,173,127
		Dayak	3,009,494
		Chinese	2,832,510
		Papua ethnic origin	2,693,630
Laos Total Population: 5,621,982 (2005)[vii]	Major Ethnic Group: Lao: 3,069,602 (54.6%)	Khmou/Khmu	613,893 (10.9%)
		Mông (Hmong)	451,946 (8%)
		Tai	215,254 (3.8%)
		Phuthai	187,391 (3.3%)
		Leu	123,054 (2.2%)
		Katang	118,276 (2.1%)
		Makong	117,842 (2.1%)
		Akha	90,698 (1.6%)
Malaysia Total Population: 28,334,135 (2010)[viii]	Major Ethnic Group: Malay: 14,191,720	Chinese	6,392,636
		Other Bumiputera	3,331,788
		Indian	1,907,827
		Others	189,385
		Non-Malaysian Citizens	2,320,779
Philippines Total Population: 92,340,000 (May 2010)[ix]	Major Ethnic Group: Tagalog: 22,512,089	Bisaya/Binisaya	10,539,816
		Cebuano	9,125,637
		Ilocano	8,074,536
		Hiligaynon/Ilonggo	7,773,655
		Bikol/Bicol	6,299,283
		Waray	3,660,645
		Other local languages/dialects	24,029,005
Singapore Total Population: 5,469,700, with 3,870,739 residents (June 2014)[x]	Major Ethnic Group: Chinese: 2,874,329 residents	Malay	516,657 residents
		Indian	353,021 residents
Thailand Total Population: 65,981,659 (2010)[xi] [data is 'usual language spoken at home']	Major Ethnic Group: Thai (Census category = 'only Thai language'): 59,866,190	Malay/Yawi	1,467,369
		'Local languages'	958,251
		Burmeae [Burmese]	827,713
		Karen	441,114

Country	Major ethnic group	Ethnic minorities/indigenous	Group population
Vietnam[xiii] Total Population: 85,846,997 (2009)	Major Ethnic Group: Kinh (Việt): 73,594,427	Tay	1,626,392
		Thái	1,550,423
		Mường	1,268,963
		Khmer	1,260,640
		Mông (Hmong)	1,068,189
		Nùng	968,800
		Hoa (overseas Chinese)	823,071
		Dao (Yao)	751,067
		Gia Rai	411,275

i   Department of Economic Planning and Development (Brunei) (2014). Total Population by Racial Groups and Districts 2001–2013. Available online: www.data.gov.bn/Lists/dataset/mdisplay.aspx?ID=231. Accessed 10 June 2015.
ii The Republic of the Union of Myanmar, Ministry of Immigration and Population (2014). The 2014 Myanmar Population and Housing Census: The Union Report, Census Report Volume 2, p. 14. Available online: http://countryoffice.unfpa.org/myanmar/2015/05/25/12157/myanmar_census_2014/. Accessed 10 June 2015. These figures include both the enumerated population and the estimated population not counted during the census.
iii No ethnicity data have been released from the 2014 census to date. These percentages are from: Central Intelligence Agency (CIA) 2015. The World Factbook Country Profile: Indonesia. Available online: www.cia.gov/library/publications/the-world-factbook/geos/id.html. Accessed 10 June 2015.
iv National Institute of Statistics, Ministry of Planning, Phnom Penh, Cambodia (2009). *General Population Census of Cambodia 2008: National Report on Final Census Results*. Available online at: http://camnut.weebly.com/uploads/2/0/3/8/20389289/2009_census_2008.pdf. Accessed 15 June 2015. Numbers calculated from percentages given in tables 2.12 & 2.13 on pp. 29–30.
v   National Statistics Directorate and United Nations Population Fund (2011). *Population and Housing Census of Timor-Leste: 2010. Volume 2: Population Distribution by Administrative Areas*. pp. 203–204. Available online: http://dne.mof.gov.tl/published/2010%20and%202011%20Publications/Pub%202%20English%20web/Publication%202%20FINAL%20%20English%20Fina_Website.pdf. Accessed 21 August 2015. The Mother Tongue/ethnic affiliation category refers to "the language a person identifies with or speaks most of the time. A list of all the 32 languages spoken in the country was prepared and used. For non-Timorese people, the person's main language was recorded." Timor-Leste Ministry of Finance (2010) *Social and Economic Characteristics: Census 2010 Volume 3* (p11). Available online: www.mof.gov.tl/wp-content/uploads/2011/06/Publication-3-English-Web.pdf. Accessed 21 August 2015.
vi Badan Pusat Statistik (2010). *Kewarganegaraan, Suku Bangsak, Agama, Dan Bahasa Sehari-Hari Penduduk Indonesia*. Available online: http://sp2010.bps.go.id/files/ebook/kewarganegaraan%20penduduk%20indonesia/index.html. Accessed 21 June 2015.
vii Lao Statistic Bureau (2005). *Lao Population Census*. Available online: www.nsc.gov.la/en/PDF/update%20Population%20%202005.pdf. Accessed 10 June 2015.
viii Department of Statistics, Malaysia (2010). *Population distribution and basic demographic characteristics*. Available online at: http://web.archive.org/web/20110830200524/www.statistics.gov.my/portal/download_Population/files/census2010/Taburan_Penduduk_dan_Ciri-ciri_Asas_Demografi.pdf. Accessed 21 June 2015. Table 2.1 on p. 15: Total population by age group, sex, ethnic group, stratum and state, Malaysia 2010.
ix Philippine Statistics Authority (2015). *The Philippines in Figures*. Available online at: http://web0.psa.gov.ph/sites/default/files/2015%20PIF.pdf. Accessed 21 June 2015.
x   Department of Statistics Singapore (2014). *Population Trends*. Available online: www.singstat.gov.sg/docs/default-source/default-document-library/publications/publications_and_papers/population_and_population_structure/population2014.pdf. Accessed 10 June 2015.
    Note from page 3: "Total population comprises Singapore residents and non-residents. Resident population comprises Singapore citizens and permanent residents. Non-resident population comprises foreigners who were working, studying or living in Singapore but not granted permanent residence, excluding tourists and short-term visitors."
xi National Statistics Office (2011). *Thailand Population Census*. Available online: http://web.nso.go.th/en/census/poph/data/090913_StatisticalTables_10.pdf. Accessed 2 August 2015.
xii General Statistics Office of Vietnam (2010). *The 2009 Vietnam Population and Housing Census: Completed Results*. Available online: www.gso.gov.vn/default_en.aspx?tabid=515&idmid=5&ItemID=10799. Accessed 10 June 2015.

A 'post-socialist' take on selective cultural preservation: Vietnam and Laos

The 2005 census of the Lao People's Democratic Republic lists 49 ethnic groups (*sonphao*), with the ethnic Lao majority representing 55 percent (2.5 million) of the country's population of 5.6 million. Members of this Lao majority tend to live in the lowlands along with related minority groups that speak a variety of Tai languages. Groups belonging to the Mon-Khmer linguistic family reside in both the uplands and lowlands. Later arrivals include Tibeto-Burman speakers from southwest China and Hmong-Mien (Miao-Yao) groups from southern China, who are likely to reside in upland areas in the northern provinces. Khmu, a Mon-Khmer speaking group (10.9 percent), and Hmong (8 percent) form the two largest non-Tai-speaking ethnic minority groups (Lao Statistic Bureau 2005; Pholsena n.d.).

In the Socialist Republic of Vietnam, the lowland Vietnamese majority account for approximately 87 percent of the population (General Statistics Office of Vietnam 2010). The remaining 13 percent belong to one of the 53 national minorities listed in the Constitution and the 2009 census. Yet, three of these groups, namely the Hoa (Han Chinese), Khmer and Lao can be thought of as 'spill-overs' from the majorities of neighboring countries, while the Cham are actually remnants of an ancient feudal state. Vietnam's ethnic minorities are officially recognized as belonging to five linguistic families, namely the Austroasiatic (including the Viet-Muong and Mon-Khmer branches), Austronesian, Sino-Tibetan, Tai-Kadai and Miao-Yao (Michaud et al. 2016).

Following Marxist rhetoric, ethnic minorities in Laos and Vietnam have often been depicted as being positioned at the lowest stage of economic development, necessitating assistance to become enlightened 'socialist man.' Especially since the mid-1970s, numerous extensive sedentarization, collectivization and industrialization projects have been implemented based on an ideology that prioritizes the unity of the nation and actively promotes the majority Lao and Kinh cultures (McElwee 2004; Turner et al. 2015).

In 1986, the governments of both Vietnam and Laos began to open up their economies to the market while keeping socialist regimes firmly in charge; this pattern closely mirrored the process China started in 1978. In Vietnam, these changes have supported a more measured attitude regarding ethnic minority trade, education and cultural expression; however, government officials still remain highly sensitive to upland security issues such as Christian agitation in the Central Highlands and northern border regions, allegedly encouraged by outside agents. The state's selective cultural preservation policy decides unilaterally which aspects of an ethnic group's culture are sufficiently valuable and politically acceptable for preservation and showcasing (Michaud 2009). Although the Vietnamese government officially acknowledges freedom of religion, animist beliefs labeled as superstitious are forbidden, and accompanying rituals and feasting are strongly discouraged as being wasteful (McElwee 2004).

Laos initiated a relocation and resettlement policy in the 1960s, officially to improve the living conditions of upland groups. Most observers, however, see this policy as having far more to do with reducing support for communist insurgents and bringing upland ethnic minority groups closer to roads and under the watchful eye of authorities. Along with a general agenda of national 'progress' and economic modernization, relocation still forms the core of the state's policy toward upland non-Lao populations (Évrard and Goudineau 2004).

Following economic reforms in both countries, collective property has been replaced with private land-use rights, cash cropping has been encouraged, and shifting cultivators have been strongly advised to become permanent, settled farmers. New infrastructure such as highways, airports, hydroelectricity schemes and communications networks all support this modernization drive commonly labeled 'development.' In response, ethnic minorities on the geographic, cultural and economic fringes of these two states frequently face tough choices as to how to maintain viable and culturally appropriate livelihoods (Turner et al. 2015).

Assimilate or be labeled 'mountain people': Thailand and Cambodia

Unlike the socialist governments of Vietnam and Laos, which acknowledge a large number of ethnic groups within national census data, the Thai national census has not collected such information since 1937, instead having a fairly unhelpful 'language spoken at home' census category.[1] The country's total population at the time of the last census in 2010 was 65.9 million (National Statistics Office, Thailand 2011). Nonetheless, 'hill tribes' were estimated to include 922,957 people in 2003 (Mukdawan 2013, 214), and nine upland minority groups, known as *Chao Khao* ('mountain people,' 'hill tribes' or 'hill people') are recognized in official discourse: the Kariang/Karen, Hmong/Miao, Lahu, Ikaw/Ahka, Yao/Mien, H'tin, Lisaw/Lisu, Lua/Lawa and Khmu/Khamu (Tribal Research Institute 2002).

The term 'hill tribes' came into official use in 1959 when the Thai government established a special committee to consider the "hill tribe problem." During the Cold War, upland groups living in the country's border areas were considered a national security threat due to fears that members of these groups could easily be persuaded to become communist insurgents (Vaddhanaphuti 2005). The term *Chao Khao* reflects embedded social meanings and values, and carries a pejorative connotation. Vaddhanaphuti (2005, 157) has noted: "In the Thai context, 'mountain' means forested, remote, inaccessible, wild, and uncivilized." Pholsena (n.d.) adds that members of the nine officially recognized groups have been granted legal status but are often not recognized as Thai citizens. The majority of people belonging to these groups reside in northern Thailand's upland areas, with others in the western hills bordering Burma. There also exist lowland minority groups such as Thai-speaking Muslims in the south, some of whom identify themselves as Malay (and with whom open armed conflict has occurred over the last twenty years), and those of Chinese origin, who often identify as Haw or Sino-Thai.

In comparison to Thailand, Cambodia's census treats ethnic composition in a more straightforward manner. Over 96 percent of Cambodia's population of 13,395,682 (2008 census) is ethnic Khmer (National Institute of Statistics, Cambodia 2009). Those living in the uplands and considered 'Khmer Loei,' or 'mountain Khmer,' account for about 100,000 individuals, the majority of whom are Mon-Khmer and Austronesian-speakers. In addition, about 204,000 Cham individuals, whose ancestors ruled over the region's ancient Cham Kingdom, live in central Cambodia. As in Thailand, post-colonial regimes in Cambodia have devised strategies to attempt to assimilate these groups into mainstream Khmer culture and society. As noted by Ovesen and Trankell (2003, 194): "The ways in which governments, officials and elites in post-colonial Cambodia have perceived and treated the country's non-Khmer ethnic groups reflect an attitude of Khmer supremacy ... manifesting a profound ethno-centrism, a conviction that Khmer culture is superior to others."

Intense ethnic conflict: the cases of Burma and the Philippines

Burma (Myanmar), as a physical entity, is a colonial creation "rife with internal contradictions and divisions" (Lintner 2003, 174). Since 1982, the Burmese government has recognized eight 'major national ethnic races,' namely the majority Burman and seven minority groups: Karen, Karreni, Kachin, Chin, Mon, Arakan and Shan. These eight categories incorporate 135 ethnic groups (Holliday 2010). A national census was completed in 2014 (Republic of the Union of Myanmar 2014), the first since 1983, but the ethnicity data collected have not been released due to political sensitivities. Although these data are contested, the United States Central Intelligence Agency estimates Burma's ethnic makeup as 68 percent Burman, 9 percent Shan, 7 percent Karen, 4 percent Rakhine, 3 percent Chinese, 2 percent Indian, 2 percent Mon and 5 percent other (CIA 2015b). It should be added that the term Rohingya is not recognized by the government as one of the 135 ethnic groups, so for the 2014 census Rohingya had to identify as Bengali or 'other' (Yen Snaing 2014). Holliday (2010, 122) adds that Rohingya rights "to property, marriage, travel, education, employment and so on are largely nonexistent. This is one of the clearest cases of ethnic persecution in the world today."

While Chin, Kachin, Karen (Kayin), Kayah, Mon and Shan states were formed in the first half of the twentieth century to give ethnic minorities some degree of autonomy, this was turned on its head when General Ne Win and the Burmese military came to power in the 1962 coup (Sadan 2013). An unofficial 'Burmanization' policy began shortly after, that continued well into the 1990s (Clarke 2001). In rural areas this has involved entire communities being forced to move to fenced 'strategic hamlets' subject to military control, while in urban centers ethnic minority communities have been split up and resettled. Clarke (2001, 422) adds that "the government's 'Burmanization' drive sparked intense military conflict, especially in the 1980s, as insurgent armies organized along ethnic lines fought the government in attempts to secure greater autonomy," such as the United Wa State Army. Tens of thousands of civilians have been killed in the fighting, while hundreds of thousands of refugees have been internally displaced or have fled the country.

The military's aim to unite Burma around a common language, religion and identity has met with limited success. In the mid-1990s, the Junta-controlled State Law and Order Council (SLOC) negotiated ceasefires that continue to hold with many insurgent groups (Holliday 2010). Yet human rights abuses against ethnic minorities have continued: forced labor – including as military porters and clearing minefields – as well as child labor and rape are still all too common. 'Burmanization' and the resultant military conflict have had significant adverse effects on living standards in ethnic minority areas. Moreover, 'development' projects like road construction and hydroelectric plants are not discussed with local communities, while the benefits of natural resource extraction fall into the hands of lowland Burmans, particularly the military (Clarke 2001; Holliday 2010).

A similar pattern of ethnic conflict has occurred in the Philippines, with a population of 92 million according to the 2010 census, spread over 7,100 islands (Philippine Statistics Authority 2012). The main Tagalog ethnic group makes up just under 25 percent of the population, and Tagalog is the official language. In addition, the Philippines contains over 180 other indigenous ethnic groups. A wide range of ethnic tensions and conflicts have arisen, with Ferrer (2005) arguing that a corrupt, weak state and notable political interference from powerful vested interests are to blame. The centralizing goals of both the colonial and post-colonial states have in turn increased the center-periphery dichotomy and extended poverty in peripheral areas. Perhaps the most prominent ethnic tensions surround the Moro resistance in the far south of the country; in the main island grouping of Mindanao there are approximately 3 to 7 million Philippine Muslims from 13 ethno-linguistic groups (Ferrer 2005). "From 1972, following the declaration of martial law, the government of President Ferdinand Marcos tried to suppress Muslim Moro claims for autonomy led by the Moro National Liberation Front (MNLF). Throughout the 14 years of the Marcos dictatorship, the Armed Forces of the Philippines waged a bitter counter-insurgency campaign against the MNLF in the Muslim provinces of western Mindanao" (Clarke 2001, 423). As in Burma, thousands have been killed and tens of thousands internally displaced. Armed resistance along ethnic lines has also occurred in the Cordillera of the northern Philippines, where a struggle for autonomy began in the late 1970s (Ferrer 2005).

Unity amid diversity? Indonesia, East Timor and Malaysia

In both Indonesia and Malaysia, post-independence governments have considered ethnic and cultural diversity threats to national unity and social cohesion. They have thus poured time and energy into creating standardized national identities.

Indonesia's population stands at over 236 million people, with over 300 ethnic groups. Indonesia's 2000 population census was the first in the country's history to ask a question on ethnicity, and raw data from the more recent 2010 census provide "1,331 categories consisting of ethnic, subethnic, and sub-sub-ethnic groups, making a very rich and complicated data set" (Ananta et al. 2013, 2). The Javanese are the most populous ethnic group, comprising 40 percent of the population, followed by the Sundanese, Batak, Madurese and Betawi. Significant tensions have existed in recent years between the government and the Acehnese (population 4 million) and Papuan (2.7 million) ethnic groups (Badan Pusat Statistik 2010).

The sheer number of ethnic conflicts and their diverse causes and effects complicates any discussion of ethnicity in post-Suharto Indonesia. Since independence in 1949, Indonesia's postcolonial governments have been committed to incorporating the country's distinct ethnic groups into one homogenous nation-state. Despite the seemingly benign slogan of 'unity amid diversity,' invasions of West Papua/Irian Jaya in 1962 and East Timor in 1975 took place, and conflict arose in Aceh. In each case, resistance was violently suppressed. Finally, in 1999, amid growing international pressure, President Habibie allowed East Timor to hold a referendum on independence, and after further violence – widely believed to have been instigated by the Indonesian state – East Timor became a new state in 2002. Independence for East Timor stoked demands from Aceh and West Papua for greater autonomy. Aceh Province, at the northern tip of the island of Sumatra, has a majority Acehnese population that has pushed for independence since Dutch colonial times. The December 2004 Indian Ocean earthquake and tsunami, in which approximately 94,000 people died and 132,000 went missing on Aceh alone (United Nations 2005, 172), prompted a peace agreement in 2005 between the Indonesian government and the main separatist organization, *Gerakan Aceh Merdeka* (GAM, Free Aceh Movement) that has ended the conflict for the time being (Aspinall 2007). In West Papua, where Papuans are related ethnically and culturally to those in Papua New Guinea and other Melanesian islands, several thousand people have died in conflicts with the Indonesian military, while others have faced torture and alienation from their land (May et al. 2013). The local insurgency has been restrained by the Indonesian government, yet East Timor's independence keeps Papuan hopes for secession alive.

Chinese Indonesians have also faced discrimination in both colonial and post-colonial Indonesia. Anti-Chinese sentiment increased throughout the 1990s, culminating in riots targeting Chinese Indonesian property and businesses just before the fall of President Suharto in 1998. Since then, symbolic government reforms, such as the reversal of a ban on public displays of Chinese culture, have attempted to rectify such discrimination (Purdey 2006).

In Malaysia, with a population of 28.3 million according to the 2010 census, 92 percent are classified as Malaysian citizens and 8 percent as non-citizens. Malaysian citizens are classified as *Bumiputera* ('sons of the soil,' or ethnic Malay, 50 percent), other *Bumiputera* (43 groups plus 'other Sarawak *Bumiputera*,' 11.7 percent), Chinese (ten groups plus 'other Chinese,' 22.5 percent), Indians (eight groups plus 'other Indians,' 6.7 percent) and Others (23 groups) (Department of Statistics Malaysia 2011). Although not used as census categories, the government recognizes 18 *Orang Asli* (original people) groups, clustered in three categories: Proto-Malay, Senoi and Semang. Many of these indigenous groups would prefer to be classified in the census according to their own identities, rather than being lumped together for statistical purposes as 3.3 million 'other *Bumiputera*' (Ahmad and Kadir 2005).

In the 1970s, the Malaysian government implemented policies that favored *Bumiputera* through affirmative action in public education and employment opportunities; the state also aimed to soothe inter-ethnic tensions following extended violence against Chinese Malaysians in 1969. These policies have succeeded in creating a significant urban Malay middle class but have been less successful in eradicating rural poverty. Chinese and Indian Malaysian minorities also largely resent this situation. "Ethnic identity, ethnic compartmentalization, and ethnic political mobilization are, as such, facts of life and so far have not been conducive to a 'melting pot' situation in Malaysian society. By virtue of this as well as other factors, the interaction between ethnic groups is limited" (Ahmad and Kadir 2005, 44). *Orang Asli* remain among the poorest in Malaysian society, and mining and logging companies continually infringe on their land (Endicott and Dentan 2004).

Development tensions

Instead of making broad and possibly misguided generalizations regarding the impacts of economic growth and government 'development' policies on ethnic minority and indigenous groups across Southeast Asia, this section briefly outlines some common elements of change that hold implications for minority groups, namely agrarian transition, resettlement policies and environmental destruction.

But first, a brief word on tourism. Perhaps the most benign approach some states have taken regarding indigenous and minority groups has been selective cultural preservation, which plays nicely into government support for tourism. Material aspects of culture such as textiles and carvings, along with dances and music, are often considered nonthreatening to the state project and are encouraged to attract tourist dollars. This sector of the economy draws heavily on representations of ethnic minorities and indigenous groups for foreign income, for instance Balinese temples in Indonesia, the 'hill tribes' surrounding Chiang Mai in Thailand and the Banaue rice terraces in the Philippine Cordillera Mountains in Luzon. Yet vigorous debate continues as to the economic and broader social and cultural benefits that indigenous and ethnic minority groups reap from tourism. Outsiders – whether ethnic majorities, states or multi-national corporations – frequently operate the region's large tourism enterprises and hotels. Ecotourism and alternative tourism projects can positively impact indigenous and minority groups if planned carefully, but these remain comparatively small-scale within the region's tourism sector.

Agrarian transition

Swiddening, often known as 'slash-and-burn' in the past, is an ancient technique that comprises the main agricultural approach for numerous upland ethnic minority groups in the region. In the 1970s, Southeast Asian governments began to blame swiddening rather simplistically for soil erosion and the destruction of forests and in the years since have initiated policies strongly promoting or obliging its replacement with fixed agriculture (Michaud et al. 2006). Conversions to sedentary cultivation and cash cropping are occurring rapidly across the region's uplands, with choices of crops as well as cropping systems increasingly dictated by the agribusiness industry. The capitalization of agriculture and rapid growth of agri-food systems and agribusiness have made local farmers ever more dependent on market relations, eroding farmers' traditional independence (Peluso and Harwell 2001). In upland northern Vietnam, for instance, the introduction of hybrid rice and maize seeds, strongly encouraged by the government, is forcing Hmong and Yao ethnic minority farmers to find new sources of cash to an unprecedented degree in order to purchase hybrid seeds, fertilizers and pesticides (Turner et al. 2015).

In many parts of Southeast Asia, land grabbing to create large plantations for cash crops such as sugarcane, oil palm, rubber and jatropha has put tremendous pressure on ethnic minority communities as well as majority small-holder communities. Land grabbing "leads to dispossession and/or to 'adverse incorporation' of people into the emerging enclaves of the global agro-food-feed-fuel complex" (Borras and Franco 2011, 5). One well-known example has occurred in West Kalimantan, Indonesia, where oil-palm plantations, financed by national and overseas investors, have overrun the customary land tenure patterns of indigenous Dayak and Punan communities, expropriating their land through state-approved corporate land deals (White and White 2012).

Resettlement policies

Population pressure, often in low-lying intensive agricultural zones, has led to state-led resettlement programs bringing lowlanders, often from dominant ethnic groups, into direct contact and conflict with upland ethnic minorities and indigenous groups. In Indonesia, "populating the empty areas in the outer islands through colonial emigration policy and post-colonial transmigration policy has been a major demographic engineering initiative for almost a century" (Tirtosudarmo 2003, 214). This program has involved the permanent movement, with government support and incentives, of households from the islands of Java (and to a lesser extent Bali and Madura) to less densely populated areas including West Papua (until 2015), Kalimantan, Sulawesi and Sumatra. The majority of transmigrants moving to outer islands are Javanese, making it fairly obvious that 'Javanization' has been a central tenet of this program; Javanese norms and customs have become dominant in many former ethnic minority or indigenous areas. Moreover, day-to-day necessities such as farmland and water are allocated to or often over-utilized by new migrants. Other natural resources are also increasingly exploited by newcomers, including logging in Kalimantan supported by Javanese business interests (Clarke 2001).

Although the Indonesian government has pursued the region's most ambitious resettlement policies, numerous other state-driven migration strategies – in Burma, Laos, Vietnam and Malaysia, for instance – have also resulted in ethnic conflicts and the asymmetric acculturation of ethnic minority groups into mainstream societies. Despite these policies, rural poverty remains widespread and inequalities are deepening across the region.

Environmental destruction and community conflicts

Many indigenous and ethnic minority groups live in mountainous and forested areas that are rich in natural resources, including minerals, hydroelectricity opportunities and forest products. State-endorsed 'development goals' – including forest clearance for cash crops – along with rising lowland and urban demand for resources are increasing pressure to exploit these resources. Throughout Southeast Asia, forests are being cleared for food or agro-fuel crops as well as felled for timber even within so-called protected parks. The Food and Agriculture Organization (FAO 2006) has estimated insular Southeast Asia's average yearly deforestation rate as 2 percent from 1990 to 2000, while Miettinen et al. (2011) note an overall yearly decline of 1 percent for 2000–2010. The economic benefits of such exploitation commonly go to outsiders. Even if logging and clearing activities bring short-term employment opportunities to locals, deforestation often results in biodiversity loss and jeopardizes local livelihoods dependent on non-timber forest products – not to mention concerns with increased carbon emissions and pollution.

Hydropower dam projects and gas pipelines feeding the demands of industry and urban consumers hundreds of kilometers away often encroach on ethnic minority and indigenous lands and livelihoods. Seldom are thorough social impact assessments completed, and local communities near hydropower dams stand to lose access to fishing resources, clean water and irrigation. Moreover, energy-importing countries such as Thailand benefit from international infrastructure projects without importing the environmental and social concerns that neighboring energy-exporting countries including Laos and Burma must deal with. For instance, the Yadana gas pipeline runs from a well site in the Andaman Sea, across parts of southeast Burma, and then 260 kilometers into Thailand. Simpson (2007, 545) notes that Burma's Junta-controlled State Peace and Development Council (SPDC) "used the pipeline project as an excuse to conduct military offensives against ethnic minorities in the region, such as the Karen and Mon, who have been in conflict with the Burmese military for over four decades." These groups were also forced to become laborers on the pipeline project, suffering systematic sexual and physical abuse (ibid.).

Not all is negative: social movements

Despite the range of adverse impacts modernization and neoliberal projects and policies have wrought on ethnic minorities and indigenous groups, the regional story is not all negative. Non-governmental organizations (NGOs), many concerned with the plight of ethnic minorities and indigenous peoples, have multiplied in the past 15 to 25 years. In the Philippines, the social movements supporting indigenous groups that emerged in the 1970s have multiplied and expanded since the mid-1980s, often protesting against natural resource extraction on indigenous lands. Other countries slowly undergoing political liberalization have permitted the expansion of civil society to varying degrees. Since 1998, a variety of NGOs have begun supporting the rights of indigenous groups in Indonesia such as those on Borneo and ethnic minorities in Aceh and West Papua. Vietnam, Laos and Burma retain tight government checks on such organizations, meaning that reformers often find it easier to fly 'under the radar.' Thailand has a rocky history regarding social movements; when democratically elected governments are in place, civil society often flourishes, but NGOs and other reformist groups can find themselves under heightened pressure to conform to rigid codes of behavior in the wake of military coups. In some instances, indigenous and ethnic minority communities in the region are gaining visibility due to global challenges such as climate change and environmental agendas like REDD+, in turn helping local social movements to gain traction and support from overseas actors.

Concluding thoughts

The brief summaries above reveal how nationalism has often been at the forefront of government strategies in Southeast Asia, "pursued with particular intensity in the early post-colonial years as governments sought to project an image of unity and control" (Holliday 2010, 114). Many state policies toward indigenous and ethnic minority groups have sought to control space, resources and people, with ethnic problems seen as direct threats to the ruling regime. Ethnic minority and indigenous groups have rarely become better off in such circumstances.

While some processes linked to modernization, nationalism and globalization can standardize on-the-ground practices, other challenges – including the agrarian transition, resettlement policies and environmental destruction – result in complex and diverse reactions and outcomes across time, space and cultures. While these challenges can trigger compliance, they can also lead to debate, contestation and struggle that requires further scholarly attention (Edelman 2001; Hollander and Einwohner 2004; Kerkvliet 2009; Turner et al. 2015). Any consideration of ethnic minority and indigenous reactions to and involvement with the 'development project' necessitates a nuanced understanding of these groups' cultural norms, social connections and livelihood adaptations, along with their embedding in local histories, customs and economic exchanges. Indeed, as Gaonkar (1999, 16) notes, we must consider the ways in which alternative modernities are produced "at different national and cultural sites. In short, modernity is not one, but many." Most ethnic minority and indigenous household adaptations do not signal straightforward acquiescence to modernization projects and state power. Many of these individuals, households and communities have proven resilient in the face of adversity and domination. While these groups are indeed changed by outside forces, they also creatively use what power they have to interpret, adapt and even subvert these pressures, sometimes with the support of broader social movements (Turner et al. 2015). The agency of members of these groups must not be understated: globalization and numerous regional 'development' schemes are frequently adapted or 'indigenized' locally (Sahlins 1999). Hence while ethnic minority and indigenous groups throughout the region face numerous struggles, it is important to remember that alternative modernities exist.

Note

1 While this is clearly frustrating in terms of gaining a broad understanding of ethnicity in the country, it is interesting to also note that a question on ethnicity was only added to the United Kingdom's census in 1991, having been rejected earlier due to political security concerns (Ananta et al. 2013).

References

Ahmad, Z. H. and Kadir, S. (2005). Ethnic conflict, prevention and management: The Malaysian case. In: K. Snitwongse and W. S. Thompson, eds., *Ethnic conflicts in Southeast Asia*. Singapore: Institute of Southeast Asian Studies, 42–64.

Ananta, A., Arifin, E. N., Hasbullah, M. S., Handayani, N. B. and Pramono, A. (2013). Changing ethnic composition: Indonesia, 2000–2010. *XXVII IUSSP International Population Conference*, 26–31. Available at: http://iussp.org/sites/default/files/event_call_for_papers/IUSSP%20Ethnicity%20Indonesia%20Poster%20Section%20G%202708%202013%20revised.pdf. Accessed 10 June 2015.

Aspinall, E. (2007). From Islamism to nationalism in Aceh, Indonesia. *Nations and Nationalism*, 13(2), pp. 245–263.

Badan Pusat Statistik. (2010). *Kewarganegaraan, Suku Bangsak, Agama, Dan Bahasa Sehari-Hari Penduduk Indonesia*. Available at: http://sp2010.bps.go.id/files/ebook/kewarganegaraan%20penduduk%20indonesia/index.html. Accessed 21 June 2015.

Borras Jr, S. B. and Franco, J. C. (2011). *Political dynamics of land-grabbing in Southeast Asia: Understanding Europe's role*. The Netherlands: Transnational Institute.

Central Intelligence Agency (CIA). (2015a). *The world factbook country profile: Burma*. Available at: www.cia.gov/library/publications/the-world-factbook/geos/bm.html. Accessed 18 June 2015.

Central Intelligence Agency (CIA). (2015b). *The world factbook country profile: Indonesia*. Available at: www.cia.gov/library/publications/the-world-factbook/geos/id.html. Accessed 10 June 2015.

Clarke, G. (2001). From ethnocide to ethnodevelopment? Ethnic minorities and indigenous peoples in Southeast Asia. *Third World Quarterly*, 22(3), pp. 413–436.

Department of Statistics, Malaysia. (2011). *Population distribution and basic demographic characteristic report 2010*, Updated 5 August 2011. Available at: www.statistics.gov.my/index.php?r=column/cthemeByCat&cat=117&bul_id=MDMxdHZjWTk1SjFzTzNkRXYzcVZjdz09&menu_id=L0pheU43NWJwRWVSZklWdzQ4TlhUUT09. Accessed 17 June 2015.

Dressler, W. and McDermott, M. H. (2010). Indigenous peoples and migrants: Social categories, rights, and policies for protected areas in the Philippine uplands. *Journal of Sustainable Forestry*, 29(2–4), pp. 328–361.

Duncan, C. (ed.) (2004). *Civilizing the margins: Southeast Asian government policies for the development of minorities*. Ithaca, NY: Cornell University Press.

Edelman, M. (2001). Social movements: Changing paradigms and forms of politics. *Annual Review of Anthropology*, 30, pp. 285–317.

Endicott, K. and Dentan, R. K. (2004). Into the mainstream or into the backwater? Malaysian assimilation of Orang Asli. In: C. Duncan, ed., *Civilizing the margins: Southeast Asian government policies for the development of minorities*. Ithaca, NY: Cornell University Press, 24–55.

Évrard, O. and Goudineau, Y. (2004). Planned resettlement, unexpected migrations and cultural trauma in Laos. *Development and Change*, 35(5), pp. 937–962.

FAO. (2006). *Global forest assessment 2005*. FAO Forestry Paper 147, Food and Agriculture Organization (FAO) of the United Nations (UN), Rome.

Ferrer, M. C. (2005). The Moro and the Cordillera conflicts in the Philippines and the struggle for autonomy. In: K. Snitwongse and W. S. Thompson, eds., *Ethnic conflicts in Southeast Asia*. Singapore: Institute of Southeast Asian Studies, 109–150.

Gaonkar, D. P. (1999). On alternative modernities. *Public Culture*, 11(1), pp. 1–18.

General Statistics Office of Vietnam. (2010). *The 2009 Vietnam population and housing census: Completed results*. Available at: www.gso.gov.vn/default_en.aspx?tabid=515&idmid=5&ItemID=10799. Accessed 10 June 2015.

Hollander, J. A. and Einwohner, R. L. (2004). Conceptualizing resistance. *Sociological Forum*, 19(4), pp. 533–554.

Holliday, I. (2010). Ethnicity and democratization in Myanmar. *Asian Journal of Political Science*, 18(2), pp. 111–128.

Kerkvliet, B. J. T. (2009). Everyday politics in peasant societies (and ours). *Journal of Peasant Studies*, 36(1), pp. 227–243.

Lao Statistic Bureau. (2005). *Lao population census*. Available at: www.nsc.gov.la/en/PDF/update%20Population%20%202005.pdf. Accessed 10 June 2015.

Lintner, B. (2003). Myanmar/Burma. In: C. Mackerras, ed., *Ethnicity in Asia*. London: Routledge, 174–193.

May, R. J., Matbob, P. and Papoutsaki, E. (2013). New Guinea. In G. Baldacchino, ed., *The political economy of divided islands: Unified geographies, multiple polities*. Baskingstoke: Palgrave Macmillan, 34–57.

McElwee, P. (2004). Becoming socialist or becoming Kinh? Government policies for ethnic minorities in the Socialist Republic of Vietnam. In: C. Duncan, ed., *Civilizing the margins: Southeast Asian government policies for the development of minorities*. Ithaca, NY: Cornell University Press, 182–213.

Michaud, J. (2009). Handling mountain minorities in China, Vietnam and Laos: From history to current issues. *Asian Ethnicity*, 10(1), pp. 25–49.

Michaud, J., Ruscheweyh, M. B. and Swain, M. B. (2016). *Historical dictionary of the peoples of the South-East Asian Massif*. 2nd ed. Lanham, MD: Scarecrow Press.

Miettinen, J., Shi, C. and Chin Liew, S. (2011). Deforestation rates in insular Southeast Asia between 2000 and 2010. *Global Change Biology*, 17, pp. 2261–2270.

Mukdawan, S. (2013). Controlling bad drugs, creating good citizens: Citizenship and social immobility for Thailand's highland ethnic minorities. In: C. Barry, ed., *Rights to culture: Heritage, language, and community in Thailand*. Bangkok: Princess Maha Chakri Sirindhorn Anthropology Centre; Chiang Mai, Thailand: Silkworm Books, 213–237.

National Institute of Statistics, Ministry of Planning, Phnom Penh, Cambodia. (2009). *General population census of Cambodia 2009: National report on final census results*, 3–13 March 2008. Available at: http://camnut.weebly.com/uploads/2/0/3/8/20389289/2009_census_2008.pdf. Accessed 14 June 2015.

National Statistics Office, Thailand. (2011). *Thailand population census: Preliminary report*. Available at: http://popcensus.nso.go.th/upload/popcensus-08-08-55-E.pdf. Accessed 10 June 2015.

Ovesen, J. and Trankell, I. B. (2003). Cambodia. In: C. Mackerras, ed., *Ethnicity in Asia*. London: Routledge, 194–209.

Peluso, N. L. and Harwell, E. (2001). Territory, custom, and the cultural politics of ethnic war in West Kalimantan, Indonesia. In: N. L. Peluso and M. Watts, eds., *Violent Environments*. Ithaca, NY: Cornell University Press, 83–116.

Philippine Statistics Authority, National Statistics Office. (2012). *The 2010 census of population and housing reveals the Philippine population at 92.34 million*. Available at: http://web0.psa.gov.ph/content/2010-census-population-and-housing-reveals-philippine-population-9234-million. Accessed 10 June 2015.

Pholsena, V. (n.d.). *Ethnic minorities, the state and beyond: Focus on Mainland Southeast Asia*. SEATIDE: Integration in Southeast Asia: Trajectories of Inclusion, Dynamics of Exclusion, Working Paper.

Purdey, J. (2006). *Anti-Chinese violence in Indonesia, 1996–1999*. Singapore: Singapore University Press.

Republic of the Union of Myanmar, Ministry of Immigration and Population. (2014). *The 2014 Myanmar population and housing census: The union report, census report volume 2*. Available at: http://countryoffice.unfpa.org/myanmar/2015/05/25/12157/myanmar_census_2014/. Accessed 10 June 2015.

Sadan, M. (2013). *Being and becoming Kachin: Histories beyond the state in the borderworlds of Burma.* Oxford: Oxford University Press.

Sahlins, M. (1999). What is anthropological enlightenment? Some lessons of the twentieth century. *Annual Review of Anthropology*, 28, pp. i–xxiii.

Simpson, A. (2007). The environment – Energy security nexus: Critical analysis of an energy 'love triangle' in Southeast Asia. *Third World Quarterly*, 28(3), pp. 539–554.

Thung, J. L., Manuati, Y. and Mulok Kedit, P. (2004). *The (Re)construction of the "Pan Dayak" identity in Kalimantan and Sarawak: A study on minority's identity, ethnicity, and nationality.* Jakarta: Pusat Penelitian Kemasyarakatan dan Kebudyaan, LIPI.

Tirtosudarmo, R. (2003). Population mobility and social conflict: The aftermath of the economic crisis in Indonesia. In: A. Ananta, ed., *The Indonesian crisis: A human development perspective.* Singapore: Institute of Southeast Asian Studies, 213–244.

Tribal Research Institute. (2002). *Tribal population summary in Thailand 2002.* Chiang Mai and Thailand: TRI.

Turner, S., Bonnin, C. and Michaud, J. (2015). *Frontier livelihoods: Hmong in the Sino-Vietnamese Borderlands.* Seattle: University of Washington Press.

United Nations. (2005). *Economic and social survey of Asia and the Pacific 2005. Dealing with Shocks.* Thailand: United Nations ESCAP Office.

Vaddhanaphuti, C. (2005). The Thai state and ethnic minorities: From assimilation to selective integration. In: K. Snitwongse and W. S. Thompson, eds., *Ethnic conflicts in Southeast Asia.* Singapore: Institute of Southeast Asian Studies, 151–166.

White, J. and White, B. (2012). Gendered experiences of dispossession: Oil palm expansion in a Dayak Hibun community in West Kalimantan. *The Journal of Peasant Studies*, 39(3–4), pp. 995–1016.

Yen Snaing. (2014). Group urges 'Sensitivity' in release of Burma's census data. *The Irrawaddy*, 16 May. Available at: www.irrawaddy.org/burma/group-urges-sensitivity-release-burmas-census-data.html. Accessed 15 June 2015.

19
GLOBALIZATION, REGIONAL INTEGRATION AND DISABILITY INCLUSION
Insights from rural Cambodia

Alexandra Gartrell and Panharath Hak

People with disabilities are the largest, most marginalized and resource-poor social group in the world, and the majority live in rural areas of the Asia Pacific region (WHO and WB 2011). Despite this reality, people with disabilities continue to be invisible within Southeast Asian discourses of poverty, regional integration and theories of development. This chapter examines how the economic integration of the Association of Southeast Asian Nations (ASEAN) into a single regional market – and into global markets, networks of production and trade – intersects with specific forms of disability-based disadvantage in the region. These globalizing and neoliberal economic transformations are characterized by insecurity and material hardship for many workers (Wilton and Schuer 2006, 188). In the global South there is little knowledge of the impacts of these changes, particularly on people with disabilities (Eide and Ingstad 2011). We do know that industrial capitalism and contemporary economies are geared toward an able-body-mind norm, however, and that this has resulted in the systematic exclusion of people with disabilities (Gleeson 1999). Critical geographies must examine how ASEAN economic integration, as an example of globalized and neoliberal regional development, can provide people with disabilities with a living wage, job security, dignity and meaningful social inclusion.

In this paper we argue that new forms of regional and global integration exacerbate pre-existing disability-based inequalities and differences among people with disabilities themselves. Regional integration exposes people with disabilities to new sources of opportunity that they struggle to access, let alone benefit from. The poorest men and women with disabilities, often women, those with sensory, intellectual and complex, multiple impairments are the least likely to secure wage-based employment – including in globalized industries. They are also the most likely to be left in rural villages with minimal support while other household members migrate to work in globalized industries to earn income to support their households financially (Gartrell and Hoban 2015). The establishment of a single ASEAN market may foster economic forms of inclusion and notions of citizenship, but the socio-cultural relations and hierarchical structures of power that shape social lives and often exclude the poor, including those with disabilities, at best continue to be neglected and at worst are magnified. We illustrate our argument by drawing on research in rural Cambodia (see also Gartrell 2010; Gartrell and Hoban 2013; Gartrell and Hoban 2015).

Background: globalization and regional development

Globalization has brought unprecedented opportunities for cooperation and competition at every scale. But greater integration and increasingly complex networks of trade and production worldwide have not diminished the significance of 'regions' at sub-national and transnational scales (Coe et al. 2004). Regionalism has grown rapidly in the last 20 years, with more than 200 trade groupings reported to the World Trade Organization globally, most of these established after 1995 (Plummer 2006). Regional economic blocks aim to strengthen the economic prospects of all member countries and are part of a broader international trend to produce greater economies of scale. In the ASEAN region, numerous moves have been made to promote greater ASEAN intraregional integration and interaction with regional neighbors including the ASEAN Free Trade Area and the ASEAN Economic Community.

In this paper, we use Coe et al.'s definition of regional development as "a dynamic outcome of the complex interaction between territorialized relational networks and global production networks within the context of changing regional governance structures" (2004, 469). This conceptual framework acknowledges regional development as a relational process, with regions as permeable territorial formations informed by what occurs within and beyond their boundaries through relations of markets, competition, control and dependency (Coe et al. 2004, 469). Regions are increasingly shaped by international as well as intra- and inter-regional scale relations and networks. Regional economic development is thus configured by the synergistic and dynamic interaction between these multi-scalar factors and processes.

In the ASEAN region, substantive moves toward greater regional integration followed the Asian Financial Crisis (AFC). The crisis highlighted the region's vulnerability to external economic and political forces as well as the need to promote Southeast Asian countries economic cooperation and recovery (Anwar et al. 2009). Within ASEAN there were fears that global trends toward regionalism could have negative effects on ASEAN (Plummer 2006) and progress toward an ASEAN economic community hastened. Since this time the 'Asian Way' to problem-solving and development gained traction, and greater integration into global supply chains has gathered pace.

The AFC prompted the adoption of expanded neoliberal reforms (Springer 2009). Neoliberal reform concentrates on three main areas: free trade on goods and services, the free circulation of capital and freedom of investment (Power 2003). To date, neoliberal economic reform in Southeast Asia has reinforced the socio-political status quo where a small elite have consolidated their wealth and privilege whilst the benefits for lower classes have not materialized (Springer 2009; Lilja 2013). This has particularly been the case in post-transitional Cambodia. Neoliberal economic reform has facilitated rising inequality (Springer 2009; Harvey 2005). Critical research can illuminate the globalized politics of neoliberal reform, which often (re)produces the deprivation, disadvantage and poverty that underpin the production of normative body-mind differences. These differences are the grounds on which disability as an embodied social-spatial process of discrimination is produced (Gartrell et al. 2016; Soldatic 2013). Our conceptual and analytical lens must incorporate the production of impairment for poverty reduction and inclusive development strategies to be effective.

An ASEAN economic community: dynamic growth, deprivation and inequality

In December 2015, the ten ASEAN member countries (Malaysia, Thailand, Singapore, Indonesia, Brunei, The Philippines, Vietnam, Laos, Cambodia and Myanmar) officially linked as an ASEAN Economic Community (AEC) (ASEAN 2014; Reuters/AFP 2015). ASEAN sees the potential of an ASEAN Economic Community (AEC) to further boost the region's dynamic growth through gains in trade and investment, expanding consumer markets, spreading networks of infrastructure and trade connections, and a workforce of 300 million (ADB and ILO 2014). The AEC will institute a range of neoliberal reforms including the freer flow of goods, services, labor, investment and capital, reduced internal trade barriers amongst ASEAN members, the promotion of members as sites for investment and trade partnerships, and strengthened institutions (ADB and ILO 2014; El Achkar Hilal 2014). These regional level changes, particularly the reduction in intra-regional transaction costs, will enable ASEAN to promote the region to multi-national corporations as a vertically integrated market (Plummer 2006, 438–439). Although ASEAN member nations are closely integrated with international markets through current trade and multinational networks (Plummer 2006), improved integration into global supply chains is a policy focus and ongoing goal of ASEAN member governments (El Achkar Hilal 2014).

The AEC aims to achieve development with equity alongside greater global integration (El Achkar Hilal 2014). Whilst many agree that equitable and inclusive development is of particular importance in ASEAN (ADB and ILO 2014; El Achkar Hilal 2014; Zhuang and Ali 2010), its achievement will be a key challenge. Inclusion itself is poorly defined in ASEAN development discourse and tends to refer to spatial and not social inclusion – let alone disability inclusion (Zhuang and Ali 2010). Furthermore, ASEAN is very diverse with "dynamic Asian economies" that range from less-developed to middle-income countries (Plummer 2006, 438). The 620 million people in the region also have very different education levels, employment productivity and live in diverse political and economic environments. Workers in Thailand, for example, are four times more 'productive' than those in Cambodia, and literacy rates in Cambodia and Laos are less than 75 percent while other nations have much higher rates (ADB and ILO 2014). Such diversity will adversely affect the speed and feasibility of collective efforts to achieve full integration that is inclusive of all members let alone social groups in individual member nations.

In addition to being a dynamic region with some of the world's fastest growing economies, the Asia Pacific is home to the largest number of extremely poor people (Balakrishnan et al. 2013), including 400 million or two-thirds of people with disabilities worldwide (Edmonds 2005). Inequality across the region is rising and ubiquitous, and most evident in Cambodia (Zhuang and Ali 2010). The gap between rich and poor has expanded in terms of income and across a range of measures including underweight children, lack of access to sanitation, lack of access to safe drinking water and deaths of children under 5 years of age (ADB 2013). Hunger remains a challenge in most countries with nearly 26 percent of children under 5 being underweight (Brooks et al. 2014). These deprivations in basic healthcare, sanitation and nutrition greatly contribute to the production of impairments. While almost universal enrollment in primary education has been achieved in most ASEAN nations, nearly 3 million children of primary school age are not in school – having never enrolled or dropped out early (El Achkar Hilal 2014, 55). It is likely that girls and boys with disabilities are among these. Although poverty headcounts have been dropping annually they are still relatively high, and those who have emerged out of poverty face the constant risk of falling back into poverty (WB 2014). Inequality poses a significant challenge to sustainable and equitable development, and as Zhuang and Ali (2010) put it, there is a "suffering" side to the "shinning" face of Asia's rapid economic growth because not everyone has benefit.

Employment opportunities are projected to increase with ASEAN economic integration. In terms of occupational demand, the greatest growth is projected in unskilled and semi-skilled occupations associated with the informal sector: agricultural, forestry and fishery laborers, street and market sales people, mining and construction workers (El Achkar Hilal 2014). Growth in vulnerable and informal sector employment is a concern as many workers are already trapped in poor quality jobs where they earn too little to escape poverty.[1] There is a high risk of exploitation and entrenched poverty in the absence of decent work opportunities, weak commitments to labor standards, social protection and safety nets in ASEAN. ASEAN economic integration may well add to the current lack of decent employment, particularly for women and young people (ADB and ILO 2014), thus increasing inequality (El Achkar Hilal 2014). Indeed, more critical analysis of the impact of the regional integration on the poorest people in ASEAN nations, young men and women with disabilities, is urgently required.

Disability-based inequality in ASEAN

Disability is an axis along which multiple compound disadvantages and discriminations stick, and there is widespread recognition of the intractable links between disability and poverty worldwide and in ASEAN nations (UNESCAP 2012; WHO and WB 2011; Palmer 2011; Gartrell 2010; Graham et al. 2014; Ngo et al. 2013). Challenges faced by people with disabilities are pervasive and deep-rooted in negative stereotypes that result in discrimination. As a consequence, people with disabilities have poorer access to education, employment, healthcare and transport than their able-bodied counterparts (UNESCAP 2012; WHO and WB 2011). These inequalities are compounded by violations of dignity through violence, abuse and prejudice, and denial of their autonomy; women and girls with disabilities, especially those with sensory, intellectual and developmental impairments, are particularly vulnerable.

In the ASEAN region, disability-based discrimination is rooted in negative cultural attitudes and social stereotypes. In Malaysia, for example, people with disabilities are seen as "sick, not normal, and without abilities," and hence constantly need help (Islam 2015, 172). Parents in Thai society often refuse to have their children with disabilities exposed to community activities because they fear their children might become targets of pity, if not mockery, from others (Naemiratch and Manderson 2009; UNESCAP 2012). Across the Theravada Buddhist nations of Thailand, Cambodia, Laos and Myanmar disability is understood to be a deserved consequence of bad actions or *karma* from a previous life, including unpleasant attitudes toward people with disabilities (Naemiratch and Manderson 2009). In Cambodia, mental disabilities are thought to be curses from spiritual beings; therefore, patronizing and belittling stigmas are common (Connelly 2009; AusAID 2012) and treatment is not usually sought (Kalyanpur 2011). Derogatory terminology is used to refer to people with disabilities as "inferior, weak, stupid, or all of the above" (Connelly 2009, 127). Family members with a disability are often perceived to be a disgrace to the family (UNESCAP 2012), objects to be pitied and hidden at home (Gartrell and Hoban 2015). These cultural beliefs and attitudes underpin low social status and the institutionalized barriers people with disabilities face to socio-economic and political inclusion, including the denial of legal protection.

Institutionalized discrimination

The United Nations' Convention on the Rights of Persons with Disabilities (CRPD) (2006) gives unprecedented recognition of the human rights of people with disabilities and provides a political and legislative framework to address the disability-poverty link (Eide and Ingstad 2011, 2). There are significant gaps between the domestic laws of Asian nations and international law as reflected in the region's often non-existent implementation of the CRPD, (Perlin 2012). No comprehensive disability law mandates principles of non-discrimination in many Asian countries, and few nations have well-documented or extensive experience in the operation of disability discrimination legislation (ibid:13). Of all ASEAN nations, only the Philippines reported to the UNESCAP that they have anti-discrimination laws, and few nations actually define what constitutes discrimination. Furthermore, within ASEAN only Malaysia and Thailand define disability from a social model perspective in alignment with the CRPD (ibid:14).

Cambodia does not have a national definition of disability. The draft Law on the Protection and the Promotion of the Rights of Persons with Disabilities included a definition derived from the conceptual framework outlined in the CRPD (Connelly 2009). However, when the Law was promulgated in 2009, it adopted the following definition of disability:

> Any persons who lack, lose, or damage any physical or mental functions, which result in a disturbance to their daily life or activities, such as physical, visual, hearing, intellectual impairments, mental disorders and any other types of disabilities toward the insurmountable end of the scale.
>
> (RGC 2009)

This definition of disability conflates and normalizes impairment (body-mind difference) with restricted activities without explicit identification of the social and environmental dimensions that interact with an impairment to preclude participation and produce disability. In practice, the Cambodian government defines disability in relation to eight "vague" conditions: difficulties in seeing, hearing, speaking, moving, feeling, behaving, learning and staying fit (Connelly 2009, 137). Other Asian nations, such as China, continue to define disability as "abnormality" and use medical approaches that attribute disability to impairment in clear violation of the CRPD's social model approach (Perlin 2012, 13). The lack of an internationally agreed upon definition of disability (WHO and WB 2011), together with a weak and inconsistent understanding of disability, poses problems for the measurement of disability prevalence and the development of strategies to address disability-specific disadvantages.

In Cambodia, for example, disability statistics are "notoriously unreliable or frustratingly incomplete" (Zook 2010, 151) and differ significantly depending upon the sources (Kalyanpur 2011). The Cambodia Socio-Economic Survey in 2004 estimated disability prevalence to be just under 5 percent; the Ministry of Social Affairs in 2008 reported 4 percent and the National Census in 2009 reported 1.5 percent (Lindsay 2009). Some estimates go as high as 15 percent, while the United Nations once reported that 30 percent of Cambodian families have at least one person with disability (Zook 2010; AusAID 2012). The most up-to-date Cambodian Demographic and Health Survey (NIS et al. 2015) stated that 10 percent of people at the age of 5 and older have one kind of disability or more.[2] The immediate implication of inconsistent and vague definitions of disability are inaccurate data collection, which typically results in the underestimation of the number of people with impairments in the global South (Meekosha 2011). This renders disability invisible in official statistics, political dialogue and social policy (Gartrell et al. 2016). In this context, the implementation of disability laws and regulations is difficult, irrespective of entrenched socio-cultural attitudes.

Despite all ten ASEAN nations being signatories to the UN CRPD, they do not comply with the Articles of the CRPD and are not taking steps to create enabling environments through the provision of explicit and targeted support for people with disabilities. Even in ASEAN nations that have domestic disability laws, such as Cambodia, there is no course of action that would allow a person with a disability to file a claim or resolve a grievance (Perlin 2012, 15). The Asia and the Pacific is the only region in the world that does not have a regional human rights court or commission and its absence is a major impediment to the movement to enforce disability rights in Asia (Perlin 2012). The lack of a regional level redress mechanism and rights tribunal is typically justified by the supposed conflict between "'Asian Values' and universal human rights. Asian nations also argue that economic rights are prioritized over political rights because at different stages of development it is necessary to focus on different rights (ibid:17). Furthermore, ASEAN nations lack the political, economic and social foundations and institutions necessary to sustain a commitment to human rights culture (Narine 2012). These cultural and institutional barriers to progress toward rights-based economic and social development, including disability rights, are amplified by neoliberal reforms that further undermine social democratic concerns for equality, democracy, social solidarity and accountability (Springer 2009).

Cambodia's neoliberalist reform

Cambodia has undergone rapid socio-economic transformation since the institution of a liberal democratic/free market economy in the early 1990s (Springer 2009). Cambodia has transitioned from nearly three decades of civil war, impoverishment and instability to a rapidly growing economy tightly integrated into regional and global markets, networks of trade and production (Hughes and Un 2011). Mirroring ASEAN trends, Cambodia's economic growth has been impressive, but inequality has significantly increased (RGC 2014; WB 2014; ILO/NIS 2010, 11). Poverty has decreased from 53 percent in 2004 to 20 percent in 2013. In line with regional trends, however, many are just above the poverty line and vulnerable to falling back into poverty (WB 2014).

The most recent Cambodian Government national development blueprint – The Rectangular Strategy – identified the attraction of more domestic and foreign investment, greater regional and global integration, increased and expanded exports and export markets as a strategy to address the anticipated budget shortfall as development assistance diminishes (RGC 2013). There is substantial potential for growth in these areas through Cambodia's active role in ASEAN (AusAID 2012). However, the relationship between the AEC and the labor market will determine the potential impacts of regional integration (ADB and ILO 2014). At the regional scale Cambodia's role is labor supply. Cheap, unskilled labor is one of the few comparative advantages the Cambodian government can offer to foreign investors (Lilja 2013).

Labor migration and disability

Findings from ongoing research in Cambodia suggest that people with disabilities encounter physical and attitudinal barriers to wage-based employment, particularly in globalized garment factories in Phnom Penh and/or Thailand. Men and women with disabilities who are able to move and communicate independently have migrated and gained employed in garment and other factories. However, those who require support to move and travel are less likely to be able to negotiate access to household level resources to support them to seek such employment. Without household level emotional, practical and financial support it is very difficult – if not impossible – for a member with a disability to actively seek employment beyond their village, particularly if they do not earn their own income. Men and women with disabilities that require communication or mobility support from others are thus effectively excluded from access to new and emerging forms of wage-based employment and remain engaged in poorly paid informal sector work at the village level, if they earn an income at all.

If those with mild impairments or impairments that can be hidden do manage to get employed, they usually find themselves disadvantaged at their workplaces (NIS et al. 2015; Connelly 2009; RGC 2010a). They have a higher chance of losing their jobs because of discrimination and common (mis)perceptions that people with disabilities are unproductive and unable to do anything (Gartrell 2010). This is even though they often "have appropriate skills, strong loyalty and low rates of absenteeism" (WHO and WB 2011, 236). When non-disabled household members migrate for employment to financially support their village-based household, those with disabilities are often left at home with less available care and support. Although grandparents, siblings and other relatives may provide some assistance, people with disabilities can experience even greater difficulties in their daily lives. The household has lost the daily physical presence of the member who is likely in their productive prime and most able to attend to a member with disability.

Employers in garment factories and in the construction industry often reject potential employees with disabilities because they believe they are unable to perform required work tasks. Employers have informed wheelchair users that it is not safe to employ them in a factory because the physical space is cramped and potentially dangerous, particularly in the event of a fire. Employers are concerned that if they employ someone with a vision impairment, they may not be able to find the toilet and/or get lost on the way back to their work space. Employers also fear being unable to instruct men and women with hearing impairments on their work tasks. Employers are concerned they would be looked upon poorly if they employ people with disabilities as such action would be seen to create difficulties for the person with disability themselves and thus be unsupportive of them. The better course of action would be to care for and protect people with disabilities at home and, by doing so, remove any potential difficulties they may face if they were to work (Gartrell and Hoban 2013).

In Cambodia, as is the case across ASEAN nations, people with disabilities have poor educational attainment, which restricts them to unskilled and low-skilled segments of the labor market. Rates of school enrollment of children with disabilities are half of that of non-disabled children, which in the long term means they are less competitive as employees (Kalyanpur 2011). Parents can be reticent to send their children with disabilities to schools that are not disability-friendly either socially or pedagogically; they typically lack appropriately skilled staff and relevant materials to support children with disabilities to learn. A lack of education undermines economic participation and greatly contributes to the economic marginalization of men and women with and without disabilities alike (AusAID 2012; Connelly 2009; NIS et al. 2015). In education and employment, there is lack of understanding that the provision of appropriate supports for people with disabilities can ensure their positive and valuable contribution. In their absence, entrenched patterns of disadvantage can result in inter-generational poverty (UNESCAP 2012).

Cambodia as a regional supplier of cheap, unskilled labor

At the regional level there is no research that examines the impact of disability on international migration. But current evidence suggests that if people with disabilities are able to migrate for work alongside increasingly large numbers of short-term migrants, they are likely to be employed at the lowest level and face significant risks to their health, safety and well-being. Within the region, Cambodia is a relatively large net-exporter of low-skilled labor migrants and the International Organisation of Migration estimate that by 2018 the number of Cambodian migrants to Thailand may reach 316,000 (Maltoni 2010).[3] The majority of these migrants work in agriculture, construction and fisheries (Paitoonpong 2011; Chan 2009) and are irregular migrants (ILO 2006). Labor migration is part of the Cambodian Government's employment generation strategy (RGC 2010b), as the economy cannot absorb the 250,000 new workers added to the economy each year (Heng 2013). With widespread poverty, low market access and environmental degradation, migrants themselves look at migration to Thailand as a "livelihood strategy" that averts poverty through greater employment opportunities and higher wages (Meyer et al. 2014, 200; CDRI 2007).

Domestic factors in Thailand also drive the growing numbers of Cambodian migrants. Thailand's declining fertility rate alongside the increasing number of senior citizens has meant a shrinking working population and higher dependency ratio (Chalamwong 2008). At the same time, basic educational requirements have gone from a minimum of nine to 12 years. This means fewer low-educated and hence fewer low-skilled workers – especially for three D jobs (dirty, dangerous and demeaning) – which in turn produces higher demand from abroad. The Thai government has requested low-skilled labor from Laos and Cambodia, but expensive, time-consuming and complicated official migration procedures resulted in a short-fall (Chantavanich 2008). Thailand has therefore "opted to employ illegal" migrant workers from Cambodia, and tolerance of their illegality will encourage greater inflows in the future (Heng 2013, 97). The majority of Cambodians who migrate to Thailand are thus irregular migrants who are exposed to the mental and physical risks associated with the daily fear of capture and expulsion (Zimmerman et al. 2011; ILO 2006). Discrimination, harassment, violence, communication barriers, underpay and restriction of all sorts are far from being uncommon. But worse – on top of their usually low literacy and education (ILO 2005) – their irregular status discourages them from seeking help and accessing protective services (Heng 2013).

In general, the poor and poorest in Cambodia are most likely to migrate for work and remittances reduce the "level, depth and severity of poverty" at home (Tong 2010 cited in Heng 2013, 101). Tong asserts that remittances from international migration have a greater impact on the living standards of migrants and their families than from internal migration. Remittances earned domestically are mostly used, in a prioritized order, for food, debt payment, medical treatment, agriculture, schooling, home improvements, savings and consumer goods (Dahlberg 2005, 6). Whilst Cambodian migrants to Thailand can improve the visible status of their household through such consumption, it has a negative impact on young people's education as they are drawn to migrate and earn money rather than continue their studies (Heng 2013). Disincentives for young people to continue at school will have long-term implications for the Cambodian labor market, particularly the supply of semi-skilled and skilled labor.

When and if men and women with disabilities do secure employment in Thailand, dangerous work exposes men and women with (and without) disabilities to injury and accidents, and little compensation is ever made. If an accident does occur, workers can be left with lifelong disabilities that fundamentally undermine their employability; this impacts the entire household. When a family member acquires a disability, others are bound to provide some sort of care, at least for a short term. Caregivers are therefore unable to engage in paid work as they may have previously, which has multiple negative impacts on the household. Furthermore, young unmarried women with disabilities who migrate live away from family and are vulnerable to rape and sexual violence; they have poor knowledge of how to protect their sexual and reproductive health and are inexperienced in negotiating relationships with men. Whilst employment in such contexts does provide an income, it is accompanied by significant risks that have long-term impacts across the life cycle. Current migration patterns, the demand for and supply of low-skilled laborers may well not serve Cambodia's longer-term development aspirations for equitable and inclusive growth.

Conclusion

Despite ASEAN member countries being signatories to the UN CRPD and committed to inclusive growth, at the national and ASEAN regional level legal, institutional and administrative absences undermine disability rights protection and inclusive development. ASEAN has indicated the desire to protect vulnerable groups – including women, children, people with disabilities and migrant workers – and to achieve democracy in the region.[4] However, the political and institutional development needed to ensure employment rights and protections for the most vulnerable are absent. The achievement of equitable, socially and spatially inclusive growth also requires further critical examination of neoliberal reform agendas.

The first step to ensure the visibility of disability in national statistics, political discourse and social policy, as well as in a regional human rights commission, is the adoption of a regionally agreed upon definition of disability based upon the social model. Limited understanding of disability and employment rights, non-discrimination and equal opportunity continue to result in multiple disability-specific barriers to employment. The physical barriers involved with traveling in inaccessible environments together with socio-cultural attitudes to migration and employment means that men and women with disabilities are often locked in feminized, domestic home spaces (Gartrell and Hoban 2015). Working age Cambodian men and women with disabilities are thus more likely to be unemployed, working in low status, poorly paid and informal sector work than their abled-bodied counterparts (Gartrell 2010). Improved enrollment in primary school, retention in secondary and high school, and clear pathways from high school to tertiary education/vocational training are key prerequisites to improve the skill level of the workforce and access to decent, skilled employment. ASEAN and individual member states need to adopt specific policies and strategies to support employers to provide enabling environments for employees with disabilities, and to provide inclusive learning environments in schools, higher education institutions and in vocation training.

Neoliberal conceptions of development increasingly underpin globalization and development (Norman 2011). While neoliberal marketization creates new opportunities for economic advancement for some, inequality and exclusion is created for others. Villages are seasonally de-populated as those in their productive prime migrate in the dry season and leave the elderly, children and those with disabilities at home. The very poorest are increasingly marginalized: too poor and without appropriate financial, practical and emotional support to migrate and access more lucrative forms of employment. They are trapped in poverty and poorly paid informal sector work.

Although the Cambodian Government has identified that access to secure work remains the most viable strategy for poverty reduction (NIS et al. 2015), specific strategies and supports are required to achieve disability inclusion and equity both within ASEAN nations and at the regional level. The key development challenge for ASEAN is to sustain growth whilst reducing poverty and inequality and ensuring all enjoy the benefits of economic growth (ILO/NIS 2010, 11). But the ASEAN economic community is further embedding neoliberal economic changes and inequality across the region. For the ASEAN economic community to achieve both increased economic and social prosperity, stronger commitment to labor standards, social protection and the expansion of decent work opportunities is needed. A continued singular economistic focus on intra-regional trade and investment will not achieve disability inclusion and other ASEAN social objectives.

Notes

1 At present approximately three in five (179 million) workers are trapped in vulnerable employment and 92 million earn too little to escape poverty (ILO and ADB 2014).
2 The Cambodia Demographic and Health Survey (NIS, DGH and ICF International 2015) disaggregated disability data based on types: 5 percent with difficulties seeing, 4 percent walking or climbing stairs,

4 percent remembering and concentrating, 3 percent hearing, 2 percent communicating, and 1 percent self-care.
3 Thailand is the largest importer of Cambodian labor migrants in ASEAN; Malaysia, South Korea and Japan are other popular destinations (Heng 2013).
4 See ASEAN Plan of Action for a Security Community (2004).

References

ADB. (2013). *Framework for inclusion growth indicators*. Manila: Asian Development Bank (ADB).
ADB and ILO. (2014). *ASEAN community 2015: Managing integration for better jobs and shared prosperity*. Bangkok: Asian Development Bank (ADB) and International Labour Organization (ILO).
Anwar, S., Doran, C. and Sam, C.Y. (2009). Committing to regional cooperation: ASEAN, globalisation and the Shin Corporation – Temasek Holdings deal. *Asia Pacific Viewpoint*, 50, pp. 307–321.
ASEAN. (2014). *ASEAN economic community*. Available at: www.asean.org/communities/asean-economic-community
AusAID. (2012). *Australia-Cambodia joint aid program strategy 2010–2015*. Canberra: Australian Government.
Balakrishnan, R., Steinberg, C. and Syed, M. (2013). *The elusive quest for inclusive growth: Growth, poverty, and inequality in Asia*. IMF Working Paper, Asia and Pacific Department, International Monetary Fund (IMF).
Brooks, D. H., Joshi, K., McArthur, J. W. et al. (2014). A ZEN approach to post-2015 development goals for Asia and the Pacific. *Ecological Economics*, 107, pp. 392–401.
CDRI. (2007). *Youth migration and urbanization in Cambodia*. Phnom Penh: Cambodian Development Research Institute (CDRI).
Chalamwong, Y. (2008). *Demographic change and international labor mobility in Thailand*. Draft paper presented the PECC-ABAC Conference on Demographic Change and International Labor Mobility in the Asia Pacific Region: Implications for Business and Cooperation, 25–26.
Chan, S. (2009). *Costs and benefits of cross-country labour migration in the GMS: Cambodia country study*. Working Paper Series No. 44. Phnom Penh: Cambodia Development Resource Institute.
Chantavanich, S. (2008). *The Mekong challenge: An honest broker-Improving cross-border recruitment practices for the benefit of government, workers and employers*. Bangkok: International Labour Organization (ILO).
Coe, N. M., Hess, M., Yeung, H. W. C. et al. (2004) 'Globalizing' regional development: A global production networks perspective. *Transactions of the Institute of British Geographers*, 29, pp. 468–484.
Connelly, U. B. (2009). Disability rights in Cambodia: Using the convention on the rights of people with disabilities to expose human rights violations. *Pacific Rim Law & Policy Journal Association*, 18, pp. 123–153.
Dahlberg, E. (2005). *Insights into migration and spending patterns based on a small-scale study of garment workers in Phnom Penh*. Stockholm: The European Institute of Japanese Studies.
Edmonds, L. (2005). *Disabled people and development*. Philippines: Asian Development Bank (ADB).
Eide, A. H. and Ingstad, B. (2011). *Disability and poverty: A global challenge*. Bristol: Policy Press.
El Achkar Hilal, S. (2014). *The impact of ASEAN economic integration on occupational outlooks and skills demand*. Bangkok: Regional Office for Asia and the Pacific, International Labour Organization (ILO).
Gartrell, A. (2010). 'A frog in a well': The exclusion of disabled people from work in Cambodia. *Disability & Society*, 25, pp. 289–301.
Gartrell, A. and Hoban, E. (2013). Structural vulnerability, disability, and access to nongovernmental organization services in rural Cambodia. *Journal of Social Work in Disability & Rehabilitation*, 12, pp. 194–212.
Gartrell, A. and Hoban, E. (2015). 'Locked in space': Rurality and the politics of location. In: S. Grech and K. Soldatic, eds., *Disability in the global South: Mapping terrains*. New York, NY: Springer, pp. 337–350.
Gartrell, A., Jennaway, M. G., Manderson, L. et al. (2016) Making the invisible visible: Disability inclusive development in Solomon Islands. *The Journal of Development Studies*, 52(10), pp. 1389–1400.
Gleeson, B. (1999). *Geographies of disability*. London: Routledge.
Graham, L., Moodley, J., Ismail, Z. et al. (2014). *Poverty and disability in South Africa: Research report*. Johannesburg: Centre for Social Development in South Africa.
Harvey, D. (2005). *A brief history of neoliberalism*. New York: Oxford University Press.
Heng, M. (2013). Revisiting poverty-migration nexus: Causes and effects of Cambodia-Thailand cross-border migration. In: *Graduate school of international development*, PhD thesis, Nagoya: Nagoya University.
Hughes, C. and Un, K. (2011). *Cambodia's economic transformation*. Copenhagen: Nordic Institute of Asian Studies Press.

ILO. (2005). *The Mekong challenge – destination Thailand – a cross-border labour migration survey in Banteay Meanchey province, Cambodia*. Bangkok: International Labour Organization (ILO).

ILO. (2006). *The Mekong challenge: Underpaid, overworked, and overlooked*. Bangkok: International Labour Organization (ILO).

ILO/NIS. (2010). *Labour and social trends in Cambodia 2010*. Phnom Penh: National Institute of Statistic (NIS), Ministry of Planning, International Labour Organization (ILO).

Islam, M. R. (2015). Rights of the people with disabilities and social exclusion in Malaysia. *International Journal of Social Science and Humanity*, 5, pp. 171–177.

Kalyanpur, M. (2011). Paradigm and paradox: Education for all and the inclusion of children with disabilities in Cambodia. *International Journal of Inclusive Education*, 15, pp. 1053–1071.

Lilja, M. (2013). *Resisting gendered norms: Civil society, the juridical and political space in Cambodia*. Surrey: Ashgate Publishing, Ltd.

Lindsay, B. (2009). *New definition of 'disabled' for census raises concerns*. Available at: www.cambodiadaily.com/archives/new-definition-of-disabled-for-census-raises-concerns-64263/

Maltoni, B. (2010). *Analyzing the impact of remittances from Cambodian migrant workers in Thailand on local communities in Cambodia*. Phnom Penh: International Organization for Migration.

Meekosha, H. (2011). Decolonising disability: Thinking and acting globally. *Disability & Society*, 26, pp. 667–682.

Meyer, S. R., Robinson, W. C., Chhim, S. et al. (2014). Labor migration and mental health in Cambodia: A qualitative study. *The Journal of Nervous and Mental Disease*, 202, pp. 200–208.

Naemiratch, B. and Manderson, L. (2009). Pity and pragmatism: Understandings of disability in northeast Thailand. *Disability & Society*, 24, pp. 475–488.

Narine, S. (2012). Human rights norms and the evolution of ASEAN: Moving without moving in a changing regional environment. *Contemporary Southeast Asia: A Journal of International and Strategic Affairs*, 34, pp. 365–388.

Ngo, A. D., Brolan, C., Fitzgerald, L. et al. (2013). Voices from Vietnam: Experiences of children and youth with disabilities, and their families, from an Agent Orange affected rural region. *Disability & Society*, 28, pp. 955–969.

NIS, DGH and ICF International. (2015). *Cambodia demographic and health survey 2014*. Phnom Penh, Cambodia, Rockville and Maryland: National Institute of Statistics (NIS), Directorate General for Health (DGH), and ICF International.

Norman, D. (2011). Neoliberal strategies of poverty reduction in Cambodia: The case of microfinance. In: C. Hughes and K. Un, eds., *Cambodia's economic transformation*. Copenhagen: Nordic Institute of Asian Studies Press, 161–181.

Paitoonpong, S. (2011). *Different stream, different needs, and impact: Managing international labor migration in ASEAN, Thailand (immigration)*. Makati: Philippine Institute for Development Studies.

Palmer, M. (2011). Disability and poverty: A conceptual review. *Journal of Disability Policy Studies*, 21, pp. 210–218.

Perlin, M. L. (2012). Promoting social change in Asia and the Pacific: The need for a disability rights tribunal to give life to the UN Convention on the Rights of Persons with Disabilities. *The George Washington International Law Review*, 44, pp. 1–37.

Plummer, M. G. (2006). ASEAN – EU economic relationship: Integration and lessons for the ASEAN economic community. *Journal of Asian Economics*, 17, pp. 427–447.

Power, M. (2003). *Rethinking critical geographies*. London: Routledge.

Reuters/AFP. (2015). *ASEAN leaders agree to economic 'Community' at summit shadowed by terrorism worries*. Available at: www.abc.net.au/news/2015-11-22/asean-creates-economic-community-at-summit/6962286

RGC. (2009). *Law on the protection and the promotion of the rights of persons with disabilities*, ed., The National Assembly. Phnom Penh: The Royal Government of Cambodia (RGC).

RGC. (2010a). *National social protection strategy for the poor and vulnerable*. Written in consultation with the Council for Agricultural and Rural Development (CARD) and Interim Working Group on Social Safety Nets (IWG-SSN). Phnom Penh: The Royal Government of Cambodia (RGC).

RGC. (2010b). *Policy on labour migration for Cambodia*. Written in consultation with the Ministry of Labour and Vocational Training. Phnom Penh: The Royal Government of Cambodia (RGC).

RGC. (2013). *Rectangular strategy: Phase III*. Phnom Penh: The Royal Government of Cambodia (RGC).

RGC. (2014). *National strategic development plan 2014–2018*. Phnom Penh: The Royal Governmental of Cambodia (RGC).

Soldatic, K. (2013). The transnational sphere of justice: Disability praxis and the politics of impairment. *Disability & Society*, 28, pp. 744–755.

Springer, S. (2009). Renewed authoritarianism in Southeast Asia: Undermining democracy through neo-liberal reform. *Asia Pacific Viewpoint*, 50, pp. 271–276.

UNESCAP. (2012). *Disability, livelihood and poverty in Asia and the Pacific: An executive summary of research findings*. Thailand: United Nations Economic and Social Commission for Asia and the Pacific.

WB. (2014). *Where have all the poor gone? Cambodia poverty assessment 2013*. Washington, DC: The World Bank (WB).

WHO and WB. (2011). *World report on disability*. Malta: World Health Organization (WHO).

Wilton, R. and Schuer, S. (2006). Towards socio-spatial inclusion? Disabled people, neoliberalism and the contemporary labour market. *Area*, 38, pp. 186–195.

Zhuang, J. and Ali, I. (2010). Poverty, inequality, and inclusive growth in Asia. In: J. Zhuang, ed., *Poverty, inequality, and inclusive growth in Asia: Measurement, policy, issues, and country studies*. New York and Manila: Anthem Press and Asian Development Bank, 1–32.

Zimmerman, C., Kiss, L. and Hossain, M. (2011). Migration and health: A framework for 21st century policy-making. *PLoS Medicine*, 8(5), e1001034.

Zook, D. C. (2010). Disability and democracy in Cambodia: An integrative approach to community building and civic engagement. *Disability & Society*, 25, pp. 149–161.

20
RELIGION AND DEVELOPMENT IN SOUTHEAST ASIA

Orlando Woods

"Money had come to the tropical land of forest and river and villages; and money created new frenzies and frustrations."

V.S. Naipaul, *Among the Believers* (2001)

Introduction

Southeast Asia comprises a mosaic of religions operating within a highly variegated development landscape. These characteristics define Southeast Asia and distinguish it from anywhere else in the world. It is a region of religious heterogeneity, with dominant encampments of nearly all major world religions – Buddhism, Islam, Catholicism and pockets of Protestantism and Hinduism – plus a range of minority groups as well. It is also a region of great socio-economic and political disparity. At the national level, the Human Development Index[1] for 2013 ranked Singapore ninth in the world, and Myanmar 150th (UNDP 2014); whilst throughout the region's modern history, both communism and capitalism have at various times been a scourge for some countries, but an apparent boon for others. At the sub-national level there remain stark divisions between the region's urban/industrial centers and rural/agricultural peripheries, and between ruling elites and the working classes. Throughout the region, practices of development serve to bridge, divide and corrupt individuals and communities, and often exacerbate the tensions between traditional and modern ways of life as well.

This chapter suggests that in Southeast Asia religion can both enable and stifle development, whilst development can influence processes of religious continuity and change. Development seeks to disrupt past practices in order to make way for new ways of being. Religious change is similar to development insomuch as it seeks a departure from old mind-sets and practices; religious continuity often resists such disruptions. There is, therefore, a fundamental tension between the modernizing impulses of development processes (especially when imparted by foreign religious groups) and more traditional ways of life. Such tensions constitute the fault lines that emerge from the coalescence of Western ideas of 'progress' and non-Western traditional (often religious) values. These tensions have evolved in line with the region's developmental and religious changes over time, and embody the 'frenzies and frustrations' of which Naipaul speaks.

Complicating the dialectic of tradition and modernity is the fact that some religions are better aligned with the modernizing impulses of development than others (see Woods 2012, 2013). This complication is felt particularly strongly in Southeast Asia, as all of the region's dominant religious groups – Islam in Indonesia, Malaysia and Brunei; Catholicism and (to a lesser extent) Protestantism in the Philippines; and Theravada Buddhism in Thailand, Myanmar, Cambodia and Laos – are imports from Arabia, Europe and South Asia respectively. They are imported ideas and traditions that have overlain, distorted and merged with what came before; they are religions into which people have converted over time. This has created layers of religious modernization and change, which overlay deep-rooted traditions of animism, folk religion and shamanism. Given that such layers also coincide with periods of socio-economic advancement – from Arab mercantilism to European colonialism – it becomes clear that religion in Southeast Asia is intimately, but often uncomfortably, entwined with the practices and processes of development. Indeed, because religion and development so often exist in tandem, it is necessary for each term to be defined from the outset.

The terms 'religion' and 'development' are both nebulous and multi-faceted (or "vague yet predictive" according to Haynes 2007, 3), and are therefore imbued with a sense of definitional complexity. The purpose of this chapter is not to explore such complexity in detail (see Asad 1983; So 1990; and Fountain 2013), and both shall be treated in a broad sense. That said, the problem with religious discourse is that "when people around the world use the same category of religion, they actually mean very different things... [meaning] can only be elucidated in the context of their particular discursive practices" (Casanova 2009, 9). This has particular relevance in Southeast Asia, where layers of religious tradition and modernity intertwine and create complex and often syncretic religious assemblages. Thus whilst any discussion of 'religion' is undermined by its definitional opacity, I take it to mean a framework of belief that unites a community of believers, and which is regulated and controlled by figures of religious authority. Throughout this chapter, references to religious groups will encompass a range of actors, including religious organizations, religious NGOs (non-governmental organizations) and FBOs (faith-based organizations).

'Development,' on the other hand, is taken to mean a series of processes that attempt to bring about some sort of social and/or economic advancement[2] (although not in unison; often the advancement of one is at the expense of the other). The focus on advancement is important, as development is often associated with (assumedly positive) socio-economic change and modernization. Such modernization implicitly refers to the 'Western developmental model' (Haynes 2007, 6), which draws on capitalist ideals to bring about an improvement in material well-being. The communist model provides an alternative development pathway (focused on reducing inequality through the equitable distribution of the means of production) and gained considerable traction in parts of Southeast Asia throughout the 1960s and 1970s. Both models fundamentally contrast with more traditional ways of life, and both are based on the assumption that the social, political and economic standing of religion must be reduced or eliminated in order to make way for development and modernization[3] (see Glahe and Vorhies 1989). Importantly, I take the socio-economic advancement associated with development to also represent a departure from previous mind-sets, behaviors and other ways of being.

Whilst recent decades have yielded a great variety of ideas and discourses surrounding each of these terms, it is only relatively recently that each has been explored in conjunction with the other. Thus despite the latent religious aspect of many development initiatives, religion remains 'a taboo subject' (McGregor 2010, 729; after Ver Beek 2000) that has proven to be of little interest to the economists and political scientists that have long claimed development studies as their own.[4] Whilst "development has long been haunted by the spectre of religion," the fact remains that "one of the most important ways in which religion has left its mark has been in the repeated and sustained attempts by mainstream actors to expunge it from the work of development" (Fountain 2012, 144). Things are, however, changing. Religion is starting to be heralded as an inextricable aspect of development processes (see Tiongco and White 1997), which is coupled with the realization that there is no secular-religious dichotomy in development studies (Fountain 2013; Fountain forthcoming). Indeed, the marginalization of religion is squarely at odds with the on-the-ground realities of development interventions, not least in regions like Southeast Asia, where most societies remain profoundly numinous in spite of socio-economic advancement.

This chapter sketches a broad overview of the dynamics and tensions that exist at the nexus of religion and development in Southeast Asia. It is not an exhaustive survey of each country or religion within the region, but an overview of some of the discourses and debates at play. The chapter is split into three sections. Sections one and two reflect Haynes's (2007) dialectic of religion and development: the first explores how development shapes religion, whilst the second explores how religion shapes development. The third explores outcomes at the intersection of religion and development, how the coalescence of socio-economic need and religious provision can bring about a number of questionable outcomes.

Development shaping religion in Southeast Asia

Some parts of Southeast Asia have developed to such an extent that they have experienced a number of social shifts, which in turn have contributed to processes of religious transformation. Bornstein (2002, 9) observes that "as economic development combines with the religious transformation of cultures, it is a process that entails, and enables, spiritual conversion," with one of the most widespread side effects of development processes being a "change [in] belief systems and practices." Such changes are a necessary corollary to development, but can also be a source of tension within and between religions, and between religious and secular agencies. This section focuses on two ways in which socio-economic shifts in Southeast Asia have impacted religion. The first examines how they have encouraged a transition from traditional folk religions to more 'rational' patterns of belief in Singapore. The second explores how modernization has been a source of religious change and upheaval in Malaysia and Cambodia.

In the first instance, it has long been noted that socio-economic modernization correlates with conversion out of superstition-based traditional religions (such as animism, ancestral worship and other folk religions) and into more 'rational' world religions (such as Christianity, Buddhism and Islam). Two inter-related factors drive such conversions. One pertains to the reorientation of societies toward more rational thought, with world religions being better able to address the ethical, emotional and intellectual challenges of modern life through the proclamation of "the existence of a transcendental realm vastly superior to that of everyday reality" (Hefner 1993, 3; see also Haynes 1993). The second pertains to the reality that one of the most pervasive outcomes of development is greater access to material wealth, which is often at the expense of widespread social alienation and anomie. The relative deprivation thesis, for example, posits that the contemporary growth of world religions (especially Christianity) is due to the fact that they are better able to meet the social needs of a modernizing society (see Woods 2012, 442–443 for a review). Thus the modernization of Southeast Asia – and associated processes of urbanization, industrialization and technological transformation – has facilitated a shift toward world religions. These processes have been ongoing for centuries, but their contemporary forms are clearly evinced in Singapore.

Over the past few decades, Singapore – the region's economic frontrunner – has witnessed a religious shift amongst its Chinese population, from Taoism to Christianity. According to Tong (2007), Singapore's education system has developed in a way that encourages systematic and rational thought processes, which in turn has contributed to the intellectualization of Singapore society. This has facilitated a widespread questioning of inherited beliefs and a search for more intellectually engaging religious alternatives. Increasingly, religion is no longer accepted as something to be passively followed based on family tradition and superstition, but something to be actively sought out based on its relevance to individual needs and circumstances (see Sng and You 1982; Clammer 1985). Compounding this is the fact that traditional religions like Taoism are increasingly regarded as low-status. World religions, on the other hand, are associated with higher levels of education, the English language and the burgeoning professional classes (Tong 2007). These factors have caused a transition from religions like Taoism – popularly perceived in Singapore to be a Chinese folk religion – to religions like Christianity, which tend to be associated with 'rational' thought and a broader idea of progress (see Woods 2012).

In the second instance, modernization can be a source of tension amongst existing, often dominant religious groups. This tension finds meaning and application in the dialectic of economic modernization and religious traditionalism. Most notably, amongst Southeast Asia's Muslim communities there have been concerted efforts – on behalf of both the state and society – to forge a new form of Islam that is relevant and compatible with a modernizing society. In Malaysia, for example, the *Bumiputera*[5] policies of the 1970s were designed to counteract the advancement of non-Malay (and non-Muslim) communities by creating a new Malay middle class through affirmative action in education, housing, company ownership and employment (see Shamsul 1986; Alatas 1996). Such policies have helped to advance the socio-economic status of Malays and to modernize Islam through the development of new types of Islamic institutions and lifestyles. These include new practices (such as conducting religious discussions, seminars and ceremonies in non-mosque settings, such as hotels), new fashions (such as the *jilbab* for women and *baju koko* for men[6]) and new Islamic educational institutions. In particular, the Malaysian government has worked closely with private foundations to establish various Islamic higher educational institutions (notably, the International Islamic University Malaysia and Universiti Sains Islam Malaysia), which offer both broad-based education and more specific Islamic instruction. Such institutions have contributed to the modernization of Malaysia's Muslim society and have helped it keep pace with broader processes of socio-economic change.

Finally, in Southeast Asia's modern history, communism has played as much of a role in shaping religion in some countries than more pervasive, capitalist systems of economic modernization. In Cambodia, for example, over a period of five years in the 1970s, the Khmer Rouge attempted to eradicate Buddhism; the religion of more than 90 percent of the population (see Kamm 1998; Poethig 2002; Haynes 2009). This involved the systematic execution of religious leaders and adherents, and the destruction of religious buildings and other artifacts. Doing so was part of a broader program of destroying traditional systems of power and control, which typically claim legitimacy from various aspects of Cambodian culture. Buddhism was identified as playing a prominent – and unifying – cultural role, whilst Buddhist monks and nuns claimed considerable social authority and were, amongst other things, responsible for education and social welfare at the village level (Haynes 2007). In this instance, therefore, fear that religion did not fit with the state's communist ideology of progress resulted in attempts to eradicate Buddhism altogether.

The Cambodia example highlights the role of religious institutions in traditional practices of socio-economic advancement (in this case, education and social welfare). Whilst such practices may be traditional, they remain a prominent part of Southeast Asia's development landscape.

Thus not only does development shape religion in Southeast Asia, but religion plays a role in shaping development pathways and outcomes as well.

Religion shaping development in Southeast Asia

Across the multiple religious traditions that define Southeast Asia, the one thing that unites them is their ameliorative role in addressing development-related needs. Such a role has become more pronounced as the drivers of development have shifted toward private sector agencies and NGOs, itself a function of the widespread corruption in the region's public sectors (Haynes 2007). In many instances, religion is not just a channel through which the tangible drivers of development (such as money, aid and food) can flow, but also an initiator of developmental processes, a source of advocacy and influence, and the glue that binds communities together during processes of development-related upheaval and change (Deneulin and Rakodi 2011). More often than not, religious groups act as bridges and mediators, connecting state agencies with communities, donors with beneficiaries and elites with the impoverished. In some cases, religious groups have even helped to shape the political landscape of development. In the Philippines, for example, Protestant churches played an influential role in the reformist movement to oust President Ferdinand Marcos in the mid-1980s (Youngblood 1990; Hunt 1992). In this instance, Marcos's authoritarian and corrupt regime was believed to be hindering economic development and miring the country in debt; his overthrow was a move to establish a more equitable (re)distribution of power and wealth throughout the country.

Throughout the region, therefore, religion is recognized as being 'indexed' to development (Aragon 2000). This means that religious groups – whether advertently or not – are development actors, and have the power to effect both socio-economic and religious changes in society. This relationship is clearly shown in a report by the Mennonite Central Committee (MCC – a US-based Anabaptist group that provides relief, development and peace-building services) on their activities in West Kalimantan, Indonesia in 1977. In the report, it was stated that:

> The Dayak[7] people in our focus area are still quite primitive compared to other areas, for this reason the nature of our ministry in this area is very different from our ministry in Java and Sumatra. It is comprehensive. The aspects of ministry which we want to carry on together are spiritual, educational, agricultural, health and others *with spiritual being first and foremost*.
>
> (cited in Fountain 2012, 150, emphasis added)

This excerpt highlights three relevant issues. The first relates to the fact that the MCC is a foreign religious group attempting to shape development outcomes in Southeast Asia. The second relates to the clear conflation of developmental and religious objectives – something that enables religious ministry (or 'mission') to be concealed behind the development imperative. And the third relates to the fact that much development work is religion-led ("spiritual being first and foremost"), which in turn can both help and hinder development outcomes. The rest of this section, and the next, shall explore these issues in more detail; first by examining how domestic religious groups shape development outcomes in Southeast Asia, and second by exploring how foreign[8] religious groups shape – and have shaped – development outcomes since the colonial epoch.

In the first instance, throughout Southeast Asia religious organizations are deeply embedded within development processes; in many instances they are the origins of development. The region's religious infrastructures – its temples, mosques and churches – have always been important nodes in development processes. They are gathering points for communities, with the strength of a localities' religious infrastructure often being synchronous with the strength of its community development programs (Weller 2001). More practically speaking, however, religious groups and buildings also serve as distribution and engagement channels through which alms and donations are collected, processed and redistributed amongst the community. Indeed, research by UBS and INSEAD (2011) demonstrates that wealthy philanthropists in Thailand prefer to channel their donations through the country's Buddhist temples, as they are believed to be a more trustworthy and respectable source of community development than NGOs. As such, religious organizations play a vital role in shaping development processes and outcomes throughout the region.

Above all other development processes, religious organizations in Southeast Asia play a prominent role in education. The need to socialize children into faith traditions is best done within the classroom of the *madrassah*, mission or temple school, as is the impartation of a unified worldview amongst communities. The provision of education by religious organizations serves not just an ideological need, but a practical one as well. Education enables religious groups to grow and spread, and to help close the welfare gaps that cannot be met by government provision, especially in rural areas. Yet whilst religious groups go a long way to providing more equitable access to education, the outcomes of such development processes can have negligible benefit to the individual. Such outcomes are discussed in more detail in relation to Indonesia's *pesantren* education system below.

In the second instance, the import of foreign religions into Southeast Asia has coincided with various waves of socio-economic modernization. The relationship between foreign religions and development was brought into sharp focus during the colonial epoch – the most recent wave of large-scale religious importation that coincided with a period of significant socio-economic change. Under colonial rule, most of Southeast Asia (the exception being Thailand) was annexed by European powers, which facilitated the large-scale introduction of Christianity and Catholicism to the region. Churches, FBOs and the contingents of missionaries that accompanied them played a formative role in advancing social welfare within the colonies. Western-style schools and healthcare facilities were established and, in many instances, served as catalysts for the promotion of Christianity. The conflation of religious and developmental goals did, however, give rise to the pejorative term 'Rice Christians' to describe individuals who converted to Christianity in order to align themselves more closely with colonial administrations in the expectation that doing so would yield social and material benefits. Whilst such a conflation of goals has since been the subject of widespread condemnation (see Heelas 1998), it does not detract from the fact that as much as converts were converting into Christianity, they were also converting into a more 'modern' way of life.

Converting individuals into more 'modern' ways of life has proven to be a key characteristic of foreign religious groups operating in Southeast Asia (see van der Veer 1995). It sets them apart from incumbent religious groups and can be a source of tension within and between communities. Indeed, tension is an inherent outcome of any development program in which "foreign religious values and beliefs come into contact with local religious landscapes" (McGregor 2010, 729) and is often exacerbated by the practices of foreign religious groups. Such groups tend to be deeply embedded within international aid chains – systems of linkages that connect beneficiaries with donors from around the world – and in many instances use socio-economic modernization as a pretext for religious modernization and change. In many instances, the problem lies in the fact that "the policies underlying transnational development networks reify a western approach, transferring secular principles and religious-political distinctions to nonsecular spaces where they are contested, debated, and sometimes reviled" (McGregor 2010, 743; see also De

Cordier 2009). Such a dynamic is pervasive and is most acutely felt in situations when immediate, short-term developmental assistance is required (such as in the aftermath of the 2004 tsunami, which devastated large parts of the coastal regions of Indonesia and Thailand). Such situations produce a number of outcomes, two of which will now be explored.

Outcomes at the intersection of religion and development in Southeast Asia

The involvement of religious groups in development processes is often a political act; it can either bind communities together or pull them apart. The widespread and systemic failure of postcolonial development agendas in most countries throughout the region – and the corresponding search for meaning and identity in nationalism – has caused religion to become politicized by the "crisis of modernity" (Thomas 2000, 49). Compounding this is the fact that the coalescence of religion and development often leads to the "conflation of practices and goals that muddies the water between what are legitimate or illegitimate grounds for proselytization" (Woods 2012, 449). Indeed, the most contentious questions surrounding any discussion of religion and development are those of religious influencing, propagation and conversion. These are the most divisive outcomes of religious involvement in development practices, outcomes that are brought into sharp focus in Southeast Asia given the chronic forms of inequality that persist, the vulnerability to natural disasters and the fact that Christianity – an aggressively proselytizing and well-funded religion – remains a minority player in most local contexts, and yet is often an influential catalyst of socio-economic modernization. In Malaysia, for example, Muslim agitators continue to see Chinese conversion to Christianity as a symbol of the foreign exploitation that has kept Malays in relative poverty (Jenkins 2007). As this example suggests, Christian-initiated development programs often pose a challenge to the dominance of incumbent religious groups and the state, and can be a source of deep-rooted tensions within and between communities (Bautista and Lim 2009; see also McGregor 2008b).

This final section briefly introduces two outcomes at the intersection of religion and development. The first involves religious proselytization under the pretext of development, with a specific focus on foreign Christian groups operating in Indonesia after the 2004 tsunami. The second examines the Islamic *pesantren* education system in Indonesia, showing how as much as religion can be an enabler of development, it can be a barrier to it as well.

In the first instance, there are many examples of Christian groups using the pretext of development to manipulate and impose their religious worldviews on their beneficiaries, the aim being to convert or to weaken previous religious ties. Such groups often operate as NGOs or FBOs, using such organizational structures as a strategy of concealment that can enable religious influencing to begin (see Woods 2013). For example, in the aftermath of the 2004 tsunami, Muslim communities became suspicious of Christian NGOs operating in Aceh, Indonesia, and ended up condemning their actions. Disaster relief was believed to be a pretext for the propagation of Christianity in Indonesia's most conservative Islamic province. In the capital, Banda Aceh, the construction of an office building by one such organization – Aceh Relief – angered the community to such an extent that they chased away the construction workers, razed their makeshift huts and forced Aceh Relief to leave the area; only the half-finished shell of their building remains. Whilst the Aceh Relief example is an extreme case, it is also a symptom of the fact that "relationships between local communities and aid teams in Aceh are often strained" (McGregor 2010, 729), and the more wide-ranging reality that "the work of faith institutions, however effective, is primarily motivated by a desire to gain converts or to serve a limited segment of the community" (Marshall 2005, 7). To counteract this tension, many Christian NGOs ended up partnering with Islamic organizations and working through Islamic networks, the aim being to generate acceptance and build close proximity to local communities, and in doing so to wield a softer form of power.

This example highlights two issues. The first is that religious presence – whether as a dedicated religious organization or as a faith-based developmental organization – is often viewed as an affront to incumbent religious groups and can stimulate inter-religious competition and conflict. The second is subtler and pertains to the fact that the conflation of religious and developmental goals is inherently coercive and can result in the exploitation of the most vulnerable segments of society (Flanigan 2010; see also Haynes 2007; Mahadev 2014). It was for this reason that in 2010, a family of American charity workers based in Aceh attracted the ire of the local community and was subsequently deported for attempting to convert Muslims to Christianity (Woods 2012). This situation is clearly exacerbated when religious development actors are a minority religious group (e.g., Christian) and when they rely on international sources of funding and support, thus enabling them to wield disproportionate economic influence over local communities. Given the vast differences in material wealth throughout Southeast Asia – and in Southeast Asia relative to other parts of the world – it is clear that foreign religious involvement in local development initiatives is often as rife with opportunities as it is challenges and politics.

In the second instance, the fact that religious groups can provide more equitable access to the processes of development does not mean that they also provide equitable access to the outcomes of development. To the extent that religion can shape development in Southeast Asia, it can also stand in the way of it. In Indonesia, for example, Islamic boarding schools called *pesantren* have long served to inculcate their pupils with an Islamic worldview that helps them to reimagine the environments and contexts within which they live. Originally the preserve of impoverished rural Javanese communities, they have since spread throughout the country and have expanded in scope to become not just places of religious instruction, but of social development (through vocational training) and empowerment as well. Nonetheless, the efficacy of the *pesantren* in achieving development outcomes should not be assumed, just because the means of creating such outcomes are provided. In his ethnography of Islam in Indonesia, Naipaul (2001, 378) provides one of the most scathing indictments of the *pesantren* system of education:

> It [the *pesantren*] was a breaking away from the Indonesian past; it was Islamization; it was stupefaction, greater than any that could have come with a western-style curriculum. And yet it was attractive to the people concerned, because, twisted up with it, was the old monkish idea of the celebration of poverty: an idea which, applied to a school in Java in 1979, came out as little more than the poor teaching the poor to be poor.

Naipaul's view may be extreme, but it brings to light the tension between religious tradition and modernity ("It was Islamization; it was stupefaction, greater than any that could have come with a western-style curriculum"), and the effect of such tension on development processes and outcomes. In the *pesantren* case, religion is seen to hamper development outcomes ("the poor teaching the poor to be poor"), even if it enables the processes of development to reach more people. In this case, development is used to serve the religion more than it is the communities from which religious groups draw strength and support. It shows how religion can corrupt development, privileging social and economic stagnation over improvement. In doing so, it also reifies the fact that more often than not development is associated with improvements in material well-being, and is often based on the assumption of a Western idea of 'progress.' Whilst such an idea may have become normative, it does not accurately reflect the kaleidoscopic range of development goals, pathways and outcomes that are at play within Southeast Asia.

Conclusions

The promise of development is that it creates pathways and opportunities for socio-economic advancement; the problem is that such pathways can be politically driven and can result in the strengthening of one (religious) group at the expense of others. Such problems become pronounced when it is religious groups shaping the development agenda, and even more pronounced when such groups draw upon international sources of funding, support and ideas. Indeed, as outside investment continues to flow into the economies of Southeast Asia, issues of charitable governance and transparency become increasingly important and deserving of closer consideration. To the extent that religious groups play an integral role in addressing the region's developmental needs, their involvement can easily create divisions within and between communities, and competition between religious and state agencies.

That said, scholarly understandings of the relationships between religion and development – especially in contexts as complex and multi-layered as those of Southeast Asia – remain embryonic. Religion is a prism through which social, economic and political realities are viewed, understood and often distorted. To fully and sensitively appreciate the types of development that are at play in Southeast Asia, more rigorously contextualized, interpretivist research is needed. Doing so will provide a departure from normative, positivist approaches to development studies (see Deneulin and Rakodi 2011), and will hopefully create new possibilities for community engagement and upliftment in time to come.

Notes

1. The Human Development Index takes into consideration life expectancy, education and income in order to assess the overall level of human development. The resulting index ranges from 0–1. The highest level of human development is 1; 0 is the lowest.
2. This reflects the expansion of the term to include more socially focused, 'quality of life' indicators that go beyond economic advancement only (see Haynes 2007). Increasingly, issues of environmental protection and sustainability have been included under the development remit, although these considerations are outside the focus of this chapter.
3. Despite this, Protestantism is often viewed as an avatar of economic globalization in the developing world.
4. The marginalization of religion in/and development has been attributed to the legacies of modernization theory in the 1950s, secularization theory in the 1960s, and the overarching reality that many postcolonial governments pursued development pathways defined – and funded – by Western donors (see Haynes 2007; Deneulin and Rakodi 2011).
5. *Bumiputera* is a term used to describe the Malay race and other indigenous peoples of Southeast Asia.
6. These refer to Muslim headscarves and Muslim-style shirts respectively.
7. The Dayaks are an indigenous tribe native to the island of Borneo, Indonesia.
8. In many instances the 'foreign' label is artificial, not least because many (if not all) religious groups operating in Southeast Asia were once considered 'foreign.' Nonetheless, the categorization brings to light a number of issues and legacies related to the perceived character of religions along either local (and therefore accepted) or foreign (and therefore different) lines.

References

Alatas, S. H. (1996). *The new Malay: His role and future*. Singapore: Association of Muslim Professionals.
Aragon, L. V. (2000). *Fields of the lord: Animism, Christian minorities, and state development in Indonesia*. Honolulu: University of Hawai'i Press.
Asad, T. (1983). Anthropological conceptions of religion: Reflections on Geertz. *Man*, 18(2), pp. 237–259.
Bautista, J. and Lim, F. K. G. (eds.) (2009). *Christianity and the state in Asia: Complicity and conflict*. Oxford: Routledge.

Bornstein, E. (2002). Religious nongovernmental organisations: An exploratory analysis. *Voluntas: International Journal of Voluntary and Non-Profit Organisations*, 14(1), pp. 15–39.

Casanova, J. (2009). Religion, politics and gender equality: Public religion revisited. In: S. Razavi, ed., *A debate on the public role of religion and its social and gender implications*. Geneva: UN Research Institute for Social Development, Gender and Development Programme Paper No. 5, 5–33.

Clammer, J. (1985). *Singapore: Ideology, society, culture*. Singapore: Chopmen Publishers.

De Cordier, B. (2009). The 'humanitarian frontline', development and relief, and religion: What context, which threats and which opportunities? *Third World Quarterly*, 30, pp. 663–684.

Deneulin, S. and Rakodi, C. (2011). Revisiting religion: Development studies thirty years on. *World Development*, 39(1), pp. 45–54.

Flanigan, S. T. (2010). *For the love of god: NGOs and religious identity in a violent world*. Stirling, VA: Kumarian Press.

Fountain, P. M. (2012). Blurring mission and development in the Mennonite central committee. In: M. Clarke, ed., *Mission and development*. London: Continuum, 143–166.

Fountain, P. M. (2013). The myth of religious NGOs: Development studies and the return of religion. *International Development Policy: Religion and Development*, 4, pp. 9–30.

Fountain, P. M. (forthcoming). Proselytising development. In: E. Tomalin, ed., *The Routledge handbook of religions and development*. London and New York: Routledge, 80–97.

Glahe, F. and Vorhies, F. (1989). Religion, liberty and economic development: An empirical investigation. *Public Choice*, 62(3), pp. 201–215.

Haynes, J. (1993). *Religion in third world politics*. Buckingham: Open University Press.

Haynes, J. (2007). *Religion and development: Conflict or cooperation?* Basingstoke: Palgrave Macmillan.

Haynes, J. (2009). Conflict, conflict-resolution and peace-building: The role of religion in Mozambique, Nigeria and Cambodia. *Commonwealth & Comparative Politics*, 47(1), pp. 52–75.

Heelas, P. (1998). *Religion, modernity and post-modernity*. London: Blackwell.

Hefner, R. (1993). World building and the rationality of conversion. In: R. Hefner, ed., *Conversion to Christianity: Historical and anthropological perspectives on a great transformation*. Berkeley: University of California Press, 3–45.

Hunt, C. L. (1992). Pentecostal churches and political action. *Philippine Sociological Review*, 40(1–4), pp. 24–31.

Jenkins, P. (2007). *The next Christendom: The coming of global Christianity*. Oxford: Oxford University Press.

Kamm, H. (1998). *Cambodia: Report from a stricken land*. London: Arcade.

Mahadev, N. (2014). Conversion and anti-conversion in contemporary Sri Lanka: Pentecostal Christian evangelism and Theravada Buddhist views on the ethics of religious attraction. In: J. Finucane and M. Feener, eds., *Proselytizing and the limits of religious pluralism in contemporary Asia*. Singapore: Springer, 211–235.

Marshall, K. (2005). *Faith and development: Rethinking development debates*. Washington, DC: World Bank.

McGregor, A. (2008b). Religious NGOs: Opportunities for post-development? In: A. Thornton and A. McGregor, eds., *Southern perspectives on development: Dialogue or division?* Proceedings of the Fifth Aotearoa New Zealand International Development Studies Network. Dunedin: Otago University Print, 165–183.

McGregor, A. (2010). Geographies of religion and development: rebuilding sacred spaces in Aceh, Indonesia, after the tsunami. *Environment and Planning A*, 42, pp. 729–746.

Naipaul, V. S. (2001). *Among the believers*. London: Picador.

Poethig, K. (2002). Movable peace: Engaging the transnational in Cambodia's Dhammayietra. *Journal for the Scientific Study of Religion*, 41(1), pp. 19–28.

Shamsul, A. B. (1986). *From British to Bumiputera rule: Local politics and rural development in Peninsular Malaysia*. Singapore: Institute of Southeast Asian Studies.

Sng, B. and You, P. S. (1982). *Religious trends in Singapore – with special reference to Christianity*. Singapore: Graduates Christian Fellowship and Fellowship of Evangelical Students.

So, A. Y. (1990). *Social change and development*. London: Sage.

Thomas, S. (2000). Religious resurgence, postmodernism and world politics. In: J. Esposito and M. Watson, eds., *Religion and global order*. Cardiff: University of Wales Press, 38–60.

Tiongco, R. and White, S. (1997). *Doing theology and development: Meeting the challenge of poverty*. Edinburgh: Saint Andrew Press.

Tong, C. K. (2007). *Rationalizing religion: Religious conversion, revivalism and competition in Singapore society*. Leiden: Brill.

UBS and INSEAD. (2011). *UBS-INSEAD study on Family Philanthropy in Asia*, UBS. Available at: www.insead.edu/facultyresearch/centres/social_entrepreneurship/documents/inseadstudy_family_philantropy_asia.pdf. Accessed 2 April 2015.

UNDP. (2014). *Human development report 2014*. United Nations Development Programme. Available at: http://hdr.undp.org/sites/default/files/hdr14-report-en-1.pdf. Accessed 1 April 2015.

Van der Veer, P. (ed.) (1995). *Conversion to modernities: The globalization of Christianity*. London: Routledge.

Ver Beek, K. (2000). Spirituality: A development Taboo. *Development in Practice*, 10(1), pp. 31–43.

Weller, R. P. (2001). *Alternate civilities: Democracy and culture in China and Taiwan*. Boulder, CO: Westview.

Woods, O. (2012). The geographies of religious conversion. *Progress in Human Geography*, 36(4), pp. 440–456.

Woods, O. (2013). The spatial modalities of evangelical Christian growth in Sri Lanka: Evangelism, social ministry and the structural mosaic. *Transactions of the Institute of British Geographers*, 38(4), pp. 652–664.

Youngblood, R. L. (1990). *Marcos against the Church: Economic development and political repression in the Philippines*. Ithaca, NY: Cornell University Press.

21
A FEMINIST POLITICAL ECOLOGY PRISM ON DEVELOPMENT AND CHANGE IN SOUTHEAST ASIA

Bernadette P. Resurrección and Ha Nguyen

Gender, livelihoods and environment are powerful prisms with which to view and unpack processes of development in Southeast Asia. Changes in people's lives and their identities are in part defined through the resources they use for daily living and livelihoods, and further mediated by their unequal gender, class and ethnic differences in society. Wider political and economic drivers also shape development and change in people's lives, which may create paths of well-being, or place them at greater disadvantage. More intense and frequent exposure to climate and disaster risks, meanwhile, make life generally more difficult, as this impinges on an already long list of uncertainty factors that characterize vulnerable living for poor women and men in the region.

Feminist political ecology (FPE) emerged out of a concern for social equity and social justice issues in environmental change, and draws from the intrinsic political character of feminist theory: power and difference. FPE offers a multi-scalar analysis of gendered rights and responsibilities, knowledge production and, more pointedly, the workings of power and politics in the use, access and distribution of resources in the context of contemporary neoliberal economic growth trajectories in Southeast Asia.

Through an FPE lens, this chapter will aim to discuss and explain people's experiences of two related but often treated as distinct drivers of development and change in Southeast Asia today: large-scale development investments, and climate change and disasters. The sections below will briefly discuss political ecology and its sub-field feminist political ecology, to be followed by a mix of original and secondary case studies conducted in parts of the region.

The view from feminist political ecology

No genealogical discussion of feminist political ecology is ever complete without meandering momentarily into political ecology itself, like a river artery to its source. Political ecology has fundamentally argued that environmental degradation is not an 'unfortunate accident' under advanced capitalism, but instead a part of the logic of that economic system, a consistent symptom of various logics and trajectories of accumulation (Peet et al. 2011, 26). Political ecology not only addresses degradation, but also current neoliberal efforts to 'green' the economy, governance and environment through conservation, clean technologies, carbon trade and offsets, and techno-managerial approaches to climate change mitigation and adaptation (Taylor 2014; Peet et al. 2011).

Like political ecology, the core defining feature of FPE is its critical positioning toward political economic drivers that appropriate resources and heighten people's gendered and social risks to development-induced disasters. In the landmark 1996 publication of *Feminist Political Ecology: Global Issues and Local Experiences*, Rocheleau, Thomas-Slayter and Wangari (1996, 4) described FPE as a sub-field of political ecology that recognizes gender as power relations that are a "critical variable in shaping resource access and control interacting with class, caste, race, culture, and ethnicity to shape processes of ecological change." FPE is a living, evolving platform of ideas that draws from the rich history of feminist theory. From its inception in the 1990s, when it aimed to highlight the materiality of women's political struggles around resources and rights (Moeckli and Braun 2001), FPE has of late shown strong poststructuralist leanings that question received wisdoms on the production of gender and other social identities. It also brings the staunchly critical reading of the workings of power – neoliberal, androcentric and environmental injustices – to constantly new levels of analyses, instilling a fresh advocacy for sustainable development (Buechler and Hanson 2015; Harcourt and Nelson 2015; Leach 2015; Rocheleau and Nirmal 2015; Elmhirst 2011; Hawkins and Ojeda 2011; Nightingale 2006; Harris 2006). More than 20 years after the publication of *Feminist Political Ecology*, Rocheleau (2015, 57) tells us that:

> FPE is more about a feminist perspective and an ongoing exploration and construction of a network of learners than a fixed approach for a single focus on women and/or gender. This constant circulation of theory, practice, policies and politics, and the mixing of various combinations of gender, class, race, ethnicity, sexuality, religion, ontologies and ecologies, with critique of colonial legacies and neoliberal designs, has characterized many feminist political ecologists. It is a work in process.

The intensification of environmental degradation and climate change (non-renewable energy markets and fossil fuel dependence, deforestation, desertification and urbanization in massive scales) has led to more risks to lives and livelihoods. In turn, solutions to mitigate these stresses – such as the emergence of the green economy (carbon trade, conservation enclosures, bio-energy development, payment for ecosystem services or PES) – pose difficult questions regarding trade-offs between environmental sustainability and social well-being. In these emerging contexts, FPE focuses on complex dimensions of gendered and social experiences of loss, disadvantage, dispossession and displacement within the multiple ecologies that human beings are intrinsically part of. This focus on multi-dimensional experiences marks out FPE's concern for first, an intersectional analysis on society-environment relations that does not disentangle gender from race, ethnicity, class and disability and other social categories. Second, FPE also recognizes the importance of conducting 'science from below' or examines people's embodied experiences of resource degradation, disasters, mobility and displacements, or dispossessions as these connect with other scales of power and decision-making (Harding 2008; Hanson 2015). And third, FPE also interrogates knowledge production, governance and policy making, as they herald new forms of intervention and environmental governance that may be inflected with assumptions that deepen differentiated and unjust life opportunities and exclusions. Some of these principles will be applied to contexts of development and climate change in the next sections.

Investment, dispossession, and clean and green investments

The Dawei Special Economic Zone lies on a large swathe of borderland between Thailand and Myanmar, occupying former agricultural lands of rice farmers. Better-paying jobs go to men while women living in the Myanmar side, whose former livelihoods were farm-based, now devote their time to small, irregular and informal enterprises and are facing more insecure futures as a result of the labor-shedding outcomes of the new industrial complex (Lin Aung 2012). As in Dawei, large-scale investments increasingly transcend territorial borders in many parts of Southeast Asia today, where decisions made in one territorial state often impact on the lives of people in another state. Transboundary investments in many parts of today's globalized world depart from the Westphalian view that the constitutional order of the modern territorial state determines patterns of advantage and disadvantage of its citizens (Fraser 2009). Land concentration, in this manner, can lead both to dispossession and exclusion, as well as unfavorable inclusion (Prugl et al. 2012).

State actors and local elites in Cambodia, Lao PDR, Myanmar, Thailand, Philippines, Indonesia and Vietnam have transacted the lease of huge land concessions with foreign and domestic public and private firms, similar to the Dawei case. The governments of Thailand and Vietnam have also transacted big projects such as hydropower and the lease of concessions for rubber plantations with Cambodia and Laos. Additionally, the agro-fuels boom in the region is being driven by rising global, regional and domestic demand for bio-energy and other uses, which in turn compete with traditional food crops, induce encroachment on so-called marginal lands where women's food and garden crops are usually cultivated, and threaten to pollute soil and water sources. The expansion of ethanol production in Thailand and oil palm in Indonesia, for instance, is expected to have a serious impact on water quality near processing facilities (Leonard 2011). Vast acreage of land has been converted to produce 'flex crops' or those crops with multiple uses across food, feed, fuel and industrial complexes (e.g., sugar cane, palm oil, maize and soya), other major commodities (e.g., rice, wheat and other cash crops), and industrial materials such as rubber. In turn, water and use rights of local farmers are placed at risk or re-configured especially when land and water are 'grabbed' by these concessions.

Large-scale investments show a diverse spectrum of outcomes ranging from short-term benefits and opportunities for local people, increased mobility in search for better opportunities due to declining assets, to outright displacements from dispossession and eviction. These outcomes are in large part shaped by the historical conditions of the investments themselves, past and existing land tenure and labor regimes where these investments occur, politico-legal regimes that are supportive of these investments, and persistent gender-unequal practices that shape how these investments translate into uneven benefits or threats to existing rights and livelihoods. Meanwhile, documentations of the grim impacts of dispossession and land grabbing are circulating widely. Dramatic environmental and politico-legal issues often occupy the analytical center stage in these documentations, which side-step important gender issues, as well as market-driven logics. To compound matters, ambivalence is building around some of these projects because they are also being framed as serving the ends of sustainable development and are seen as supporting the build-up to a green economy. For instance, palm oil and other biofuel plantations are being seen to reduce dependence on fossil fuels and also offer income-earning options to some groups within local populations; hydropower development is considered a source of clean energy, and in large part fuels the regional power trade as a potent engine of economic growth favored by nascent developing Lower Mekong countries. These 'green and clean' framings are gaining traction, and a growing number of studies from different parts of the world have recently drawn attention to their gender-differentiated implications that highlight the disturbing social justice implications of these projects (Behrman et al. 2012; Daley et al. 2013; Julia and White 2012; Koopman and Faye 2012; Lin Aung 2012).

A feminist political ecology perspective will highlight the embodied – or the grounded phenomenon – of women and men's lives, attending to: reconfigurations in women and men's labor; threats or enhancement of gendered rights to land, water and other resources; political actions and responses that traverse multiple scales of intimate relations, households, nation and the global; gender intersections with class, ethnicity and age that allow us to go beyond easy and simplified views of men as discrete winners and women as losers; and the effects of the implements of legal, discursive and normative gendered power that shape the trajectories and operations of large-scale investments. Briefly discussed below are recent studies on these investments that disclose gendered workings in commercial contract farming, urban real estate development and oil palm plantation expansion in parts of Southeast Asia that resonate with key aspects of FPE.

A recent FAO study on five sites of recent agricultural investments on rice, corn, cassava and bananas in Laos indicate that plantation-style investments and new contract farming arrangements provided more income opportunities for local residents, including women. But doubt remains whether these would lead to more secure livelihoods in the longer term, which is linked to the overall sustainability of the investing domestic and foreign-owned firms and their operations (Daley et al. 2013). Residents themselves are diverse, and intersections between class and gender define how groups of women may benefit from contract farming in very specific ways. For instance, married women who optimize household earnings by intensifying their time and labor in contract farms experience increased workloads because their husbands do not normally share reproductive and care obligations. Poorer women who head their households find their households more at risk because they incur debts to cover start-up costs for getting involved in contract farming. Despite the risks, they may have found it important to get involved in contract farming because not to do so creates greater social exclusions for them, as contract farming is becoming a source and index of wealth in these communities.

In Myanmar, apart from labor issues earlier described, land rights are being formalized dramatically after the end of the military regime. Under Myanmar's new 2012 Farmland Law, Land Use Certificate (LUC) titling has been found to increasingly erode women's customary land claims and rights. In the Dawei Special Economic Zone (SEZ) areas, both Karen and Tavoyan ethnic groups have bilinear inheritance traditions, however LUCs were instead granted to male household heads, effectively disenfranchising women from joint land claims that they had under their respective customary systems (Faxon 2015). While recognizing the adverse effects of the shift of customary to formal land rights on women in the Dawei SEZ context, FPE takes the view that neither customary or statutory regimes by themselves guarantee women's autonomy in exercising their land and tenurial rights. Instead, FPE recognizes that their identities and multiple positions as different and changing types of women over time (e.g., eldest or youngest daughter, wife, widow, mistress, sister) define and shape their relationships with land and, in turn, define their rights under any regulatory regime (Faxon 2015; Jackson 2003).

FPE also recognizes that people make livelihood decisions not only on rational terms, but also on calculations of their social standing within their communities and the symbolic value of newly-introduced commercial crops as they appear to link rural communities with global value chains, such as possibly in the earlier Lao cases. This may also in part explain the optimism with which women favored the new income sources from oil palm plantations, as in the case of a community living in a forest margin in Jambi, Sumatra. Puzzled researchers were compelled to abandon their former assumptions of women having more sustainable resource management practices than men (Villamor et al. 2013). In West Kalimantan, indigenous women have become a new class of plantation laborers attracted to receiving regular cash income, despite losing their customary land tenure rights in light of the new statutory recognition of (male) family heads in smallholder plot registration. This ambivalence can also be found in similar patterns found in many other cases of expanding commercial agricultural production and dispossession of common resources (Julia and White 2012). This case also chimes in well with another FAO study on biofuels in the Mekong region citing that the privatization and dispossession of common resources – or the new carbon sink enclosures – pose potential risks to women through biodiversity loss and reduced availability of forest resources. These in turn affect women's reproductive work, food security and household resilience to shocks. Despite the potential outcomes, the authors however suggest that the new technology within biofuel production might provide women with more employment opportunities than men, because women workers are often preferred in plantation agriculture (Rossi and Lambrou 2008).

All these demonstrate that outcomes of large-scale investments are complex: often multiple, varied and ambiguous in their gains and losses. By also employing a FPE optic, Elmhirst et al. (2015) present a multi-layered view with which to understand complex variations that unsettle easy dichotomies of discrete winners and losers, agents and victims, in their investigation of five cases of expanding oil palm plantations in East Kalimantan, Indonesia. Depending on historically and ecologically embedded gender norms, landscape and livelihood histories and the communities' modes of incorporation into a wider capitalist economic system, oil palm cultivation had varied outcomes. For some, oil palm cultivation augmented income returns from traditional livelihoods like rice swidden agriculture, and for others it was increasingly displacing swidden farming and non-timber forest product collection. These ramified the intensity of involvement of different groups of women in oil palm cultivation, where some placed more priority in swidden farming. In returning migrant communities, some women returnees from oil palm plantations in Malaysia invested in smallholdings to cultivate oil palm, mobilizing returnee kin networks from the same ethnic groups, thus blurring gender inequality but highlighting ethnic and kin elite formation. In a community with a large oil palm company driving an increase in wage labor opportunities, women began to sell processed food for a growing non-farming but wage-earning consumer market due to oil palm investments. While enterprises were heavily and traditionally feminized, formal land ownership was recognized as 'household-owned' tacitly referring to ownership of male heads typically found all over Indonesia. At the same time, persistent gender ideologies compel women to carry out their reproductive obligations and swidden farming tasks despite the increased diversity in their activities and heavier workloads. Overall, gender ideologies of women as finance and business managers of small enterprises and men as the public face of resource ownership remained persistent despite the class and ethnic variations in the people's relationship with oil palm.

A significant point in the study attributed 'agency' to oil palm itself, as it carried different meanings to different communities: on one hand, to some communities, it was the new portent of wealth and 'modernity,' and to others, it was seen as the exclusive crop of the rich man, the benefits of which were unattainable by common folk.

Other studies squarely look into the unrest wrought by dispossession from large-scale development, highlighting the grounded experiences of women activists and how their experiences blur the boundaries between the public and the private in land eviction. Brickell's study of 14 women activists' accounts of land eviction due to land concessioning near an urban wetland at the heart of Phnom Penh, the Cambodian capital, demonstrates how forced land evictions are an embodied, emotional and grounded phenomenon (Brickell 2011). A Chinese-backed private development company planned a satellite city with private villas and commercial establishments, contravening Cambodia's 2001 Land Law that enshrined local residents' formal possession rights. The company evicted thousands of residents in August 2008, as they poured mud and sand into the lake to pave the way for real estate construction. The study shows that women's activism against forced evictions were both 'intimate' and 'geopolitical,' linking their homes, bodies, the nation state and the geo-political, as their voices drew 'the whole world watching.' Their eviction and activism traverse multiple scales of engagement, which have intimate implications on their private domestic lives as some husbands threatened to put a stop to their activism, but simultaneously drew the attention of global audiences to their cause and plight.

Harvey (2007) refers to this as the new round of primitive accumulation through dispossession, or the acquisition of resources by firms for control and production, in many instances dispossessing poor communities of vital natural assets. As these studies draw attention to the gendered and social dynamics, ambivalences and variations of large-scale investments in parts of Southeast Asia, shocks and stresses impinge on people's resources and their daily lives. An FPE understanding of climate and disaster risks serves to unpack the 'marginal notes' – or the more textured stories of people – that are often lost in managerial planning that offer resilient solutions.

Increasing risks to climate change and disasters

Almost half of natural disasters between 2004 and 2013 occurred in the Asia Pacific, where Southeast Asia – predominantly Indonesia and the Philippines – were hardest hit, killing more than 350,000 in more than 500 incidents (UNESCAP 2015). The political ecology of climate change and disaster risks is increasingly drawing wider analytical attention. Hazards are not completely natural or inevitable, but that they have a history and may be co-produced by and include social and biophysical elements.

The feminist political ecology of disaster risk and climate change is still fairly nascent. In more recent gender, climate change and disaster studies, Hyndman (2008), Cupples (2007) and Arora-Jonsson (2011) challenge ideas of essentializing women's vulnerability to disasters. Along with others, they instead emphasize the need to recognize the historical contexts of women's (and men's) lives prior to a disaster, which could explain the differentiated vulnerable positions of women in the wake of a disaster that do not easily fit into the singular and undifferentiated category of 'disaster victim' (Enarson and Chakrabarti 2009; Enarson 2012; Bradshaw 2015; Resurrección and Sajor 2015). These ideas also chime with political ecology's growing concern with socio-natures (Castree and Braun 2001), where in particular, disasters are viewed as being socially, politically and biophysically produced and instantiated, but additionally, as Cupples (2007) argues, subjectivities are also performed, materialized and reworked through both extreme and slow-onset disasters. In what follows, I will discuss women's embodied experiences with climate-related water risks, culled from original research in Vietnam and the Philippines. The women's water practices in these research contexts also show us that there are other ways to understand how water is tied to power and inequality that go beyond differences in technical distribution and access rights (Truelove 2011).

Sta Rosa, Laguna, a municipality lying at the low-lying peri-urban fringes of Metro Manila in the Philippines has a pluralism of formal and informal water distribution institutions relying largely on groundwater extraction through deep wells. Exclusive gated residential communities, high-end shopping malls, theme parks and huge industrial complexes have their own independent deep wells and distribution systems. The top five of 45 companies account for 68 percent of industrial demand for water. The dense clustering of wells, especially among residential and industrial wells, may result in the lowering of the water table in the long term (WWF 2011). While the local government is able to monitor the extraction behavior of a private local water service provider, the massive extraction of water by independent commercial and private users having their own deep wells is, however, not monitored. The unregulated extraction of water by these privileged water users – exacerbated by longer dry spells due to a changing climate – is the major driver for water shortages now being felt by poorer communities living at the periphery of industrial and commercial complexes.

Women in an informal slum settlement, Barangay Sinalhan, manage the supply and distribution of water in their households. They heavily relied on pumping out water from their artesian wells to meet their daily domestic water supply requirements. But since two years ago, the water began drying up, and the little that was extracted emitted a foul smell. In addition to the problems with access to household water supply, most wells are poorly maintained, located in flood zones, beside canals, near toilets, piggery farms and other point sources of pollution. With more frequent flooding due to more severe cyclones and heavier precipitation occurring in this low-elevation zone, water quality is also steadily worsening. The public water supply provider, Laguna Water, is unable to offer legal connections to households in Barangay Sinalhan because the land these households occupy is privately owned by a local resident of Sta Rosa, and therefore residents are considered as illegal squatter dwellers. None of these women can show formal land titles. Recently, some women have been paying water sellers monthly for a supply of piped water through illegal connections with households outside the informal settlement. Poorer, younger women with young children have weaker or no access to these connections.

Another study in Central Vietnam showed that Ky Nam commune is susceptible to storms and droughts and localized chinooks (warm dry wind) in the dry season (January to August) and heavy rains and floods during the annual rainy season (Huynh and Resurrección 2014). According to climate change scenarios for Vietnam, this commune will be among the most vulnerable regions to water shortages due to increasingly high temperatures and dwindling rainfall (MoNRE 2011). To complicate the situation, Ky Anh district authorities recently confiscated nearly 100 hectares of which nearly 80 percent was cropland being cultivated by local households for establishing shrimp farming. Shrimp farming was a business that was run by small firms that were not from the locality. Local farm residents and firms began to increasingly compete for scarce water.

Irrigated rice cultivation partly reflects gender disparities in rights to water. In Vietnam, farmers usually exercise water use rights for irrigation regardless of their land tenure rights. Yet individual female farmers exercise their rights to water differently. In male-headed households, adult men had chief responsibility for irrigation, but both spouses flexibly manage sufficient time and labor to irrigate their rice fields. This, however, is a major task of women in female-headed households. As compared with farmer couples, female heads have de facto limited irrigation water rights especially during periods of water stress because (a) they are not able to compete physically with male farmers and their spouses at channeling water to their fields and/or (b) they have difficulty devoting time and labor for irrigating fields due to their heavy domestic workloads. A household survey also showed that households with female heads who have irrigated rice land experienced about 20 percent lower rice yield per hectare. They registered even lower yields particularly in 2008, the year of a severe drought, because of their inability to irrigate their fields. In short, in their attempts to adapt to water scarcity through irrigation management, female heads found themselves more disadvantaged than couples from male-headed households, and as a result, some of them had to altogether reduce their farming activities.

The cases in the Philippines and Vietnam demonstrate that inequalities in water access between women under conditions of rising temperatures and low precipitation combine with land issues of tenure and state appropriation, and water resource competition with industrial estates. The relationship that women have with water in these study sites is therefore fraught with tension and uncertainty due to growing water scarcity and pollution complicated by both climatic changes, insecure tenure arrangements, competition with industry, and state appropriation of both land and water. Privatization and commodification of the natural 'commons' such as water and land, Wichterich (2015) points out, indicate the decline of real and direct democracy, and compromise the public good, compelling women in the Philippine case to resort to unprotected means of accessing water. Embodied, everyday and emotional experiences in the risky use of water – due to its increasing scarcity and toxicity – runs the grain of favoring commercialization of the water commons for advancing neoliberal urban economic growth (Harris 2015).

Concluding remarks: connecting the drivers of development and change

This chapter highlights the agile capacities of feminist political ecology to attend to people's socially and politically grounded and gender-differentiated experiences of livelihoods, development and environmental change, and addresses the multi-scaled drivers that generate and maintain inequality, injustice, vulnerability and disadvantage in their lives. The chapter also demonstrates how changes in people's gendered lives and livelihoods in Southeast Asia are driven by large-scale development investments and climate change and disasters.

Development is evolving in many parts of rural, urban and peri-urban Southeast Asia where huge and medium-sized foreign and domestic capital investments reap profits at low wage rates utilizing semi- and low-skilled labor changing the physical and social landscape of resource use. Livelihoods have also become more multi-local, designed for a mobile rural-urban footloose workforce to address the need for optimizing livelihood opportunities and buffering income shortfalls from declining resources probably due to slow-onset climate changes. Incidences of dispossession create insecure living conditions for people who already have meager assets to live by. Simultaneously, more intense and frequent exposure to climate and disaster risks exacerbates livelihood survival strategies that place lives even more precariously on the edge, and embed them in conditions of insecurity and resource scarcity to the point of resorting to unprotected access and other precarious means.

Investments, climate and disasters are drivers of development and change often treated apart from each other, where respective solutions and responses however similarly undergo neat technocratic scrutiny and program design in boardrooms and conference halls.

Analyses of large-scale investments, climate risks and disasters have their own respective epistemic communities who converse rarely and maintain their own blind spots. On the one hand, the concern for climate change and disasters connects very little with agrarian and urban development, and the incorporation of livelihoods and markets into the wider neoliberal-driven economy. On the other hand, social analyses on large-scale investments hardly discuss the effects of climate change and disasters on the investments themselves, and the new labor and property regimes that support them. The empirical cases in this chapter remind us of the need to make these vital connections, as people experience these changes in often interrelated ways.

Feminist political ecology – through its ontological, grounded, embodied and intersectional lenses to the workings of gender embedded in ecological and agrarian contexts – can make the connections between economic growth through investments and climate and disaster risks. This begins with its critical positioning toward neoliberal projects and practices, unequal resource capitals, and differentiating the workings and effects of power among groups of poor, ethnic or racialized women and men.

FPE is deliberately conscious of how gender constantly intersects with other axes of power such as ethnicity and class, therefore accounting for variations of outcomes, and it follows, underscoring the need for nuanced responses to disadvantage from large-scale investments, climate change and disasters. Additionally, with its focus on embodied experience and grounded ontological workings of power, FPE can also creatively engage with other power frameworks that are critical of 'truth' and 'expert' claims of technocratic and managerialist discourses and practices, many of which offer one-size-fits-all solutions, which may stand at stark odds with people's complex and changing lives on the ground.

References

Arora-Jonsson, S. (2011). Virtue and vulnerability: Discourses on women, gender and climate change. *Global Environmental Change*, 21(2), pp. 744–751. Available at: https://doi.org/10.1016/j.gloenvcha.2011.01.005

Behrman, J., Meinzen-Dick, R. and Quisumbing, A. (2012). The gender implications of large-scale land deals. *The Journal of Peasant Studies*, 39(1), pp. 49–79. Available at: https://doi.org/10.1080/03066150.2011.652621

Bradshaw, S. (2015). Engendering development and disasters. *Disasters*, 39(Suppl 1) (s1), pp. S54–S75. Available at: https://doi.org/10.1111/disa.12111

Brickell, K. (2011). We don't forget the old rice pot when we get the new one: Discourses on Ideals and Practices of Women in Contemporary Cambodia. *Signs*, 36(2), pp. 437–462. Available at: https://doi.org/10.1086/655915

Buechler, S. and Hanson, A-M. S. (eds.) (2015). *A political ecology of women, water and global environmental change*. 1st ed. Abingdon, Oxon and New York: Routledge.

Castree, N. and Braun, B. (eds.) (2001). *Social nature: Theory, practice and politics*. Oxford: Blackwell Publishing.

Cupples, J. (2007). Gender and hurricane mitch: Reconstructing subjectivities after disaster. *Disasters*, 31(2), pp. 155–175. Available at: https://doi.org/10.1111/j.1467-7717.2007.01002.x

Daley, E., Osorio, M. and Park, C. M.Y. (2013). *The gender and equity implications of land-related investments on land access and labour and income-generating opportunities: A case study of selected agricultural investments in LAO PDR*. Bangkok, Thailand: Food and Agriculture Organization of the United Nations.

Elmhirst, R. (2011). Introducing new feminist political ecologies. *Geoforum*, 42(2), pp. 129–132. Available at: https://doi.org/10.1016/j.geoforum.2011.01.006

Elmhirst, R., Siscawati, M. and Basnett, B. S. (2015). *Navigating investment and dispossession: Gendered impacts of the oil palm "Land Rush" in East Kalimantan, Indonesia*. Presented at 'Land grabbing, conflict and agrarian-environmental transformations: perspectives from East and Southeast Asia', Chiang Mai University, June.

Enarson, E. and Chakrabarti, P. G. D. (eds.) (2009). *Women, gender and disaster: Global issues and initiatives*. Los Angeles: Sage.

Enarson, E. P. (2012). *Women confronting natural disaster: From vulnerability to resilience*. 1st ed. Boulder, CO: Lynne Rienner Publishers.

Faxon, H. O. (2015). *The Praxis of access: Gender in Myanmar's national land use policy*. Conference Paper No. 17. Chiang Mai University: BRICS Initiatives for Critical Agrarian Studies (BICAS). Available at: www.iss.nl/fileadmin/ASSETS/iss/Research_and_projects/Research_networks/LDPI/CMCP_17-_Faxon.pdf

Fraser, N. (2009). Feminism, capitalism and the cunning of history. *New Left Review*, 56(2), pp. 97–117.

Hanson. (2015). Shoes in the seaweed and bottles on the beach: Global garbage and women's oral histories of socio-environmental change in coastal Yucatan. In: S. Buechler and A-M. Hanson, eds., *A political ecology of women, water and global environmental change*. Oxon, UK: Routledge, 165–184.

Harcourt, W. and Nelson, I. L. (2015). *Practicing feminist political ecologies: Moving beyond the "Green Economy"*. London: Zed Books.

Harding, S. (2008). *Sciences from below: Feminisms, postcolonialities, and modernities*. Durham, NC: Duke University Press.

Harris, L. M. (2006). Irrigation, gender, and social geographies of the changing waterscapes of southeastern Anatolia. *Environment and Planning D: Society and Space*, 24(2), pp. 187–213. Available at: https://doi.org/10.1068/d03k

Harris, L. M. (2015). Hegemonic waters and rethinking natures otherwise. In: W. Harcourt and I. L. Nelson, eds., *Practicing feminist political ecologies: Moving beyond the "Green Economy."* London: Zed Books, 157–181.

Harvey, D. (2007). *A brief history of neoliberalism*. Oxford and New York: Oxford University Press.

Hawkins, R. and Ojeda, D. (2011). Gender and environment: Critical tradition and new challenges. *Environment and Planning D: Society and Space*, 29(2), pp. 237–253. Available at: https://doi.org/10.1068/d16810

Huynh, P. T. A. and Resurrección, B. P. (2014). Women's differentiated vulnerability and adaptations to climate-related agricultural water scarcity in rural Central Vietnam. *Climate and Development*, 6(3), pp. 226–237. Available at: https://doi.org/10.1080/17565529.2014.886989

Hyndman, J. (2008). Feminism, conflict and disasters in post-tsunami Sri Lanka. *Gender, Technology and Development*, 12(1), pp. 101–121. Available at: https://doi.org/10.1177/097185240701200107

Jackson, C. (2003). Gender analysis of land: Beyond land rights for women? *Journal of Agrarian Change*, 3(4), pp. 453–480. Available at: https://doi.org/10.1111/1471-0366.00062

Julia and White, B. (2012). Gendered experiences of dispossession: Oil palm expansion in a Dayak Hibun community in West Kalimantan. *Journal of Peasant Studies*, 39(3–4), pp. 995–1016.

Koopman, J. and Faye, I. M. (2012). *Land grabs, women's farming, and women's activism in Africa*. Paper presented at the International Conference on Global Land Grabbing II. Land Deals Political Initiative, Department of Development Sociology, Cornell University, 17–19 October, Ithaca, NY.

Leach, M. (2015). *Gender equality and sustainable development*. Oxon: Routledge.

Leonard, R. (2011). *Agrofuels: A boost of energy for the Mekong region? Occasional Paper 10 A Report for the Focus on the Global South*. Bangkok, Thailand: Focus on the Global South.

Lin Aung, S. (2012). *Women and gender in Dawei*. Available at: www.newmandala.org/women-and-gender-in-dawei/. Accessed 25 July 2017.

Ministry of Natural Resources and Environment (MoNRE). (2011). *Scenarios for climate change, sea level rise in Vietnam. Hanoi, Vietnam*. Hanoi: Vietnam.

Moeckli, J. and Braun, B. (2001). Gendered natures: Feminism, politics, and social nature. In: N. Castree and B. Braun, eds., *Social nature: Theory, practice and politics*. Oxford: Blackwell Publishing, 112–132.

Nightingale, A. (2006). The nature of gender: Work, gender, and environment. *Environment and Planning D: Society and Space*, 24(2), pp. 165–185. Available at: https://doi.org/10.1068/d01k

Peet, R., Robbins, P. and Watts, M. (eds.) (2011). *Global political ecology*. 1st ed. London and New York: Routledge.

Prugl, E., Razavi, S. and Reysoo, F. (2012). *Gender and agriculture after neoliberalism?* Geneva: UNRISD.

Resurrección, B. P. and Sajor, E. E. (2015). Gender, floods and mobile subjects: A post-disaster view. In: R. Lund, P. Doneys and B. P. Resurrección, eds., *Gendered entanglements: Revisiting gender in rapidly changing Asia*. Copenhagen: Nordic Institute of Asian Studies, 207–234.

Rocheleau, D. E. (2015). A situated view of feminist political ecology from my networks, roots and territories. In: W. Harcourt and I. L. Nelson, eds., *Practicing feminist political ecologies: Moving beyond the "Green Economy."* London: Zed Books, 29–66.

Rocheleau, D. E. and Nirmal, P. (2015). Feminist political ecologies: Grounded, networked and rooted on earth. In: R. Baksh and W. Harcourt, eds., *The Oxford Handbook of transnational feminist movements*. Oxford: Oxford University Press, 793–814.

Rocheleau, D. E., Thomas-Slayter, B. P. and Wangari, E. (1996). *Feminist political ecology: Global issues and local experiences*. London and New York: Routledge.

Rossi, A. and Lambrou, Y. (2008). *Gender and equity issues in liquid biofuels production: Minimising the risks to maximize the opportunities*. Rome: FAO.

Taylor, M. (2014). *The political ecology of climate change adaptation: Livelihoods, agrarian change and the conflicts of development*. New York: Taylor and Francis.

Truelove, Y. (2011). (Re-)Conceptualizing water inequality in Delhi, India through a feminist political ecology framework. *Geoforum*, 42(2), pp. 143–152. Available at: https://doi.org/10.1016/j.geoforum.2011.01.004

UN Economic and Social Commission for the Asia and Pacific (UNESCAP). (2015). *Overview of natural disasters and their impacts in Asia and the Pacific, 1970–2014*. Bangkok, Thailand: UNESCAP.

Villamor, G. B., Desrianti, F., Akiefnawati, R., Amaruzaman, S. and van Noordwijk, M. (2013). Gender influences decisions to change land use practices in the tropical forest margins of Jambi, Indonesia. *Mitig Adapt Strateg Glob Change*. Available at: https://doi.org/10.1007/s11027-013-9478-7

Wichterich, C. (2015). Contesting green growth, connecting care, commons and enough. In: W. Harcourt and I. L. Nelson, eds., *Practising feminist political ecologies: Moving beyond the "Green Economy"*. London: Zed Books, 67–100.

World Wide Fund (WWF)-Philippines.(2011). *Santa Rosa Watershed: Managing water resources in urbanizing landscapes*. Quezon City, Philipppines: WWF-Philippines.

22
RETHINKING RURAL SPACES
Decropping the Southeast Asian countryside

Tubtim Tubtim and Philip Hirsch

Introduction: decropping the countryside

In the popular imagination, and as a subject of study, the Southeast Asian countryside is commonly associated with particular forms of economic activity, types of social organization and characteristics of landscape. It has also long been considered a target of 'development,' in the sense of overcoming backwardness, isolation and deprivation. Rurality is generally understood to be characterized by mainly smallholder-based agricultural livelihoods, by villages that form the basis for residence and social organization amongst households, by cultivated fields surrounded by natural features including rivers and forests, and by poverty relative to urban areas.

Much recent study of the Southeast Asian countryside has focused on the rapid pace of change away from relatively uncomplicated peasant-based livelihoods toward more complex, differentiated and mobile occupations (Rigg 2001). The village as a neo-logistic construct was critiqued by Kemp a generation ago (Kemp 1987), and village as an uncomplicated overlap of locality and social space has similarly been problematized in its specific country context for some time (Hirsch 1993). Yet, conceptions of rurality among city dwellers have in many cases failed to keep up with the changing countryside. The emergence of a large urban middle class has also constructed certain images and expectations of a countryside that harks back to an idealized past, raising questions not dissimilar to those associated with post-productionist countryside studies in Europe and elsewhere (Halfacree 1995; Wilson and Rigg 2003) and putting a different perspective on the need for or desirability of certain aspects of development.

In this chapter we seek to delink rural space from exclusively agrarian livelihoods, communities and landscapes, and to challenge some of the more static and romanticized constructions of Southeast Asian rurality, based largely but not exclusively on the experience of Thailand. We seek to understand rurality as a relational term, in particular as that which is not urban yet which is also not wild, and as a discursive construct rather than as an essentialized phenomenon. We challenge persistent popular and – to some extent – academic notions of rural spaces, both as essentially agricultural in character and as being made up of uncomplicated and harmonious communities. We thus employ the term 'decropping' in two related senses. First, we show how, despite the continued overall expansion of agricultural croplands in an absolute sense, the Southeast Asian countryside has in the process of development become progressively delinked from farming as a defining part of rural landscapes and social organization. Second, we explore the 'cropped' views of the countryside that persist amongst many urban middle class groups and that are reinforced through media, helping to shape a new politics of urban-rural relations. We 'decrop' these views with examples of encounters that challenge outsiders' normative notions of the rural, which in turn lead to disappointment and tensions when rural spaces and the people who occupy them turn out to be something different.

We commence with a case study of a peri-urban village in northern Thailand that has seen rapid and extensive change in a single generation, away from a quintessentially rural landscape and community, with few markers of development as commonly understood, toward a complex and partly urbanized physical and social space. We reflect on the expectations of middle class newcomers to this village and their encounters with a reality different to imaginaries derived from persistent urban constructions of the Thai countryside. We use this case as a springboard to introduce themes that we then extend to a review of related literature on changing rural spaces in Southeast Asia, bringing in examples from several countries. Finally, we conclude by considering what this decropped picture of rurality means for ways of thinking about rurality as the development moniker fades.

Encountering and decropping a changing rural space

Until at least the 1980s, the village of Nong Khwai in Hang Dong District of Chiang Mai Province was unreservedly rural in most of its markers of landscape and community. The village adjoined the forested slopes of Doi Suthep mountain, the landscape surrounding the village was dominated by paddy fields and some dryland cultivation with forests as a backdrop, the houses were mainly traditional wooden constructions, and the roads were unsurfaced laterite. The overwhelming majority of residents were rice farmers, who supplemented their livelihoods with local forest resources both for subsistence (mushrooms and other non-timber forest products) and for household-level artisanal purposes, notably wood turning and a range of other handcraft products. Irrigation was organized and locally governed through traditional *muang-faai* (channel and weir) systems, and the temple was the main focus of ritual and social activity. Although the village is located less than 15 kilometers from the center of Chiang Mai, road access was poor and the majority of villagers had only infrequent interaction with the city. This remained a rural place subject to contemporary discourses of development, which produced ways of thinking about future prosperity and served as a means for the state to enact its programs (Hirsch 1990).

A generation later, development has more or less disappeared from the vocabulary through which people discuss or think about their futures. Nong Khwai is part of the outer peri-urban zone of Chiang Mai. It is linked to the city by a four-lane highway along the main irrigation canal. While the landscape continues to include significant areas of rice cultivation, most of the fields are owned by outsiders, and most of those working them from temporary encampments during the wet season are Hmong and other upland farmers from elsewhere who rent the land. Few people living in Nong Khwai consider themselves to be farmers any more. Nevertheless, they refer to themselves collectively, and in distinction from newcomers, as '*chaobaan*,' or villagers.

Whereas virtually all the residents of Nong Khwai in the 1980s had been born there, had married into the village or had arrived there to clear forest land for cultivation, the residential population of Nong Khwai is now very diverse. Other than descendants of the earlier village households and their spouses, Nong Khwai is home to many migrant workers. These include Shan construction workers from Myanmar who live in camps run by well-to-do villagers who have invested in small construction businesses that employ these migrants. They also include a large number of dormitory workers, most of whom come from more remote villages elsewhere in northern Thailand and who work either in Chiang Mai town or for more proximate employers including the Night Safari and Royal Flora Exhibition that are part of Chiang Mai's visitor-oriented economy. In turn, these two attractions brought with them an expanded road, which has catalyzed other changes in the area. The dormitories are also owned by better-off villagers, built within the residential space of the village and hence neighboring the houses of the original residents.

The diversity of Nong Khwai is further extended by a significant number of middle class and foreign residents who have bought land and built houses on the edge of the village, taking advantage of its green landscape, mountain views and relative proximity to urban and peri-urban services. This phenomenon of "moving to the [peri-urban] countryside" (Tubtim 2012) is characteristic of large parts of peri-urban Chiang Mai and extends to many other parts of Thailand. Many of the newcomers bring expectations of rural life that are in part shaped by idealized notions of community, peaceful and traditional aspects of rural environments, and a conscious desire to live away from the city and in a more 'organic' mixed setting than would be found in one of the many gated communities that also dot the peri-urban countryside around Chiang Mai.

Newcomers do not always find what they expect. Many encounters help us to illustrate the gap between the spaces of rurality found in this slice of Thai countryside, on the one hand, and the expectations of newcomers on the other, a gap that is further widened by certain expectations of local people toward the newcomers. Three of these are detailed below by way of illustration. Our contention is that the confounded expectations exhibited by these encounters are the product of a highly 'cropped,' selective notion of the rural amongst Thailand's urban middle class. As such, the case of Nong Khwai, while no more 'typical' of a Southeast Asian rural community than any other could be in such a variegated region within and between countries, nevertheless provides us an entry into a broader consideration of the dynamic rural spaces that are part of Southeast Asia's contemporary development landscape.

Encounter 1: property and privacy

While the residential section of the village of a generation ago was largely unfenced, today fencing between houses is very common. Longstanding residents may have hedges with barbed wire, and some of higher economic status have concrete fences. Nevertheless, front gates are left wide open during the daytime. In contrast, most of the middle class newcomers have taller fences for privacy and tend to fully or partially close their sliding gates most of the time. However, each of the middle class houses tends to have one or more families of locals with whom they have closer relations, often through hiring for domestic help and who therefore have access to the house and garden.

On one occasion, a local villager who worked in the garden of a middle class householder cut a bunch of ripe bananas from a tree in her garden and then invited another elderly local villager, who lived nearby, to cut the banana leaves for her to use in making fermented bean wrappings. The owner of the garden in which the banana tree was located found out and expressed surprise that the local villager had cut the banana leaves that still had decorative value without asking her for permission, given that it was inside the fence. In turn, the elderly lady wondered out loud, with amusement, what the point was of keeping leaves on a tree that produces only one crop of fruit and is then barren.

Apart from the aesthetics versus utilitarian values associated with banana plants, this encounter reflects different notions of privacy. For the middle class newcomer, the fence is a spatial marker of privacy, within which any act requires authorization or permission from the owner. Villagers respect such privacy in many respects, but this does not extend to use of some plants, nor to issues such as noise from village cottage industries during the daytime or music at nighttime, or to smoke pollution from burning leaves. Furthermore, for newcomers there is a reluctance to assert their ideals of privacy, for fear of giving offence. More often, space is shut off by physical measures such as closed gates. In a more fundamental sense, newcomers achieve privacy by building their houses on the available larger spaces at the edge of or even completely separate from the main residential sections of the village.

Encounter 2: trees and normative landscapes

Many newcomers are attracted to Nong Khwai for its proximity to the hills and its generally open and green environment, with many residual markers of rurality such as rice fields, trees and bamboo-hedged lanes. The association between environment and countryside is strong, the more so for those moving from Bangkok. There is a significant urbanite discourse about the value that villagers place on conserving their environment and managing their natural resources, and this is particularly strong in the case of northern Thailand with its forested mountains, traditional irrigation systems and so on.

Newcomers' assumptions are challenged in Nong Khwai in a number of ways, none so materially poignant as in the case of trees. Along public lanes and in a few cases of trees on private land adjacent to the lanes, there have been several cases of removal of ancient trees for fear of their threat to power lines in the event of toppling during a storm. In 2012 there was a case of a large tamarind tree in the community area of one of the hamlets of Nong Khwai that the village elders wished to cut down. They went to the length of selling the tree to a lopper, without consulting the rest of the community. Some of the middle class group expressed concern about the loss of this greenery, and in the end a compromise was reached whereby the tree was coppiced. But in many other cases, trees have been removed with little sentimentality on the part of villagers, to the dismay of newcomers who find their shade and their views compromised. This included the cutting of several trees in the community area for the expedience of not having to do maintenance of branch lopping to protect the community buildings and clearing leaves.

Similarly, more and more villagers are replacing hedges with concrete fences exposed to the road. This is for a combination of showing status, the convenience of not having to trim vegetation, as well as providing security for those who are away for work during the daytime. Meanwhile, newcomers often place a green façade in front of their iron-railed front fences.

Encounter 3: expectations of community and shared labor

Longstanding residents in the village maintain a practice of shared labor, albeit in a much reduced set of activities than previously. Reciprocal labor use in agriculture disappeared a generation ago. However, for certain public works, in particular trimming of roadside weeds and hedgerows, maintaining community buildings and grounds, rubbish collection and beautification, there are regular public service days on which each household is expected to provide one member to contribute her or his labor, usually for a morning. Announcements are made a few days ahead of time through the village loudspeaker. This labor is organized on a sub-village hamlet basis.

The urban newcomers have a complex engagement with this practice. On the one hand, shared labor is entirely in conformity with their expectations of collective action and tends to reinforce their idea that they are living within a 'community.' On the other the timing of labor days is often inconvenient, occurring irrespective of weekdays and weekend distinctions, and during working hours. The timing is decided with no consultation. Furthermore, the announcements over the loudspeaker are hardly audible, compounded by the fact that they are made in the local northern Thai dialect. Sometimes the newcomers learn of the day by word of mouth, or when they see villagers embarking with hoes and rakes on the day itself. Sometimes they join in, but more often not.

At the same time, many locals do not want to bother the middle class newcomers directly, feeling ambivalent about talking to them on this matter in cognizance of class difference. Many of the newcomers are '*achaans*' (university lecturers or others of educated status), making it awkward for villagers to invite them to help given that it involves a request to use manual labor. Partly too, there is a continuing sense among the villagers that community membership for these purposes does not extend to newcomers, neither the wealthier urbanites nor the poorer construction camp and dormitory residents, who in any case are too busy making a day-to-day living to join in. The option of paying money (100 baht) instead of sending labor on the day has been discussed among villagers but has not yet been enacted, partly because the villagers who do join in are unable to do the same to the significant number of villagers, particularly from younger families, who do not participate, in part because they also work outside. There is also reluctance to create a norm of people paying money instead of giving time and labor.

Challenging quintessentially rural spaces

The changing nature of the Southeast Asian countryside has been documented in numerous studies, sometimes explicitly in terms of development and its impacts, but also through productionist-oriented conceptual lenses such as that of agrarian change. Here, we emphasize those aspects that challenge the longstanding ideas of rurality in contraposition to urbanity and urban values. Our challenge to the quintessentially rural extends to a brief review of literature that shows the phenomenon to be much more generalized than the single case of a peri-urban community in northern Thailand.

To begin with, it is important to note the difference between absolute and relative trends away from rurality as commonly conceived. For example, the total area of cultivated land in Southeast Asia has continued to expand, at the same time that agriculture as a share of GDP, of employment and as a component of rural people's livelihoods has declined (De Koninck and Pham Thanh Hai 2017, Table 3). In this sense, the 'decropping' of the countryside denotes diversification rather than decline.

Despite the wealth of studies on agrarian change and other aspects of a changing rurality in Southeast Asia, popular and touristic images of the quintessential Southeast Asian countryside have nevertheless also been reinforced by certain scholarly works. Such studies need to be understood in the socio-political context of the times and places in which they were written. In Thailand, for example, we can trace some of the persistent understandings of time-honored rurality to writing dating from the post-leftist movements of the 1980s. For example, the political economist Chatthip Nartsupha's influential 1984 book *Setthakhid Muubaan Thai nai Adit* (The Thai Village Economy in the Past) presented a detailed analysis of traditional rural livelihoods and culture, based on oral histories carried out in Thailand's poor, and at that time still quite isolated, northeastern region of Isan (Nartsupha 1999, translated from 1984 original). Seri Phongphit (1986) of the so-called community culture school, presented a quite idealized purview of rural living (see also Phongphit and Hewison 1990). These works were written in the context of civil society and academic attempts at the time to challenge the negative ideas of rural life as backward and uncultured. They also sought to promote alternative paths to development triggered by disappointments with failures in the ability of fast-track mainstream growth strategies to improve the position of the rural poor, and also in the wake of the collapse of the earlier leftist challenges to capitalist development. Documenting, and sometimes idealizing, rural culture was prominent in such studies. The work of many NGOs has been built on ideas in these studies, often reflecting quite a static perspective on rural Thailand, and furthermore one that easily transforms into normative expectations that "the answer is in the village" (NGO Coordinating Committee, cited in Phatharathananunth 2002, 27) in a traditionalist sense.

More recently, ethnographic studies of rural change have emphasized change away from the rural past. Of particular note is Jonathan Rigg and Peter Vandergeest's (Rigg and Vandergeest 2012) collection of community restudies by senior authors one or two generations following their original (mainly PhD-level) accounts. Fourteen separate village re-studies carried out for this book between 2006 and 2009 document the fundamental changes in rural ways of life over the space of a generation or two. Five of these were based in different regions of Thailand. Michel Bruneau's work on agricultural and social change in villages of northern and north-central Thailand (Chiang Mai and Phitsanulok provinces respectively) since the 1960s, Philip Hirsch's analysis of generational change in a community of western Thailand (Uthaithani Province) since the mid-1980s, Jonathan Rigg's detailed analysis of household level change in northeastern Thailand (Udonthani Province) since the early 1980s, Peter Vandergeest's study of a virtual disappearance of the physical markers of village as recognizable rural community despite a seemingly re-agrarianized landscape in Songkhla Province, and Chusak Withayapak's study of ethnic identity and village change in northern Thailand (Nan Province) since the early 1990s all present the countryside as highly dynamic but in often surprising ways that reveal non-linear paths of development and change (Bruneau 2012; Hirsch 2012; Rigg and Salamanca 2012; Vandergeest 2012; Witayapak 2012). Each brings a unique lens to the understanding of rural change. Yet, findings common to all these studies are the relative decline of agriculture in rural livelihoods, the geographical widening of horizons and experiences by rural people, and hence the unpacking of common assumptions about rurality and trajectories of rural change. Other studies in this collection cover village-level change in Cambodia, Indonesia, Malaysia, the Philippines and Vietnam, and the generalized findings reveal broadly similar patterns.

The above studies seek to understand what has happened in particular local contexts as rural places are incorporated into the wider political, economic, social and environmental currents of development. There is, however, also a growing literature that specifically and critically targets the construction of rurality based on stereotypical images and imaginings. These often go on to deconstruct the manufactured landscape or portrayal of such. One of the most interesting and entertaining of these is Rigg and Ritchie's (2002) study of the Regent Hotel near Chiang Mai to show how rural landscape has been created for high-paying tourists in northern Thailand. They identify a post-productivist trend that sees "a shift from production *in* the countryside to consumption *of* the countryside" (ibid, 360).

The findings of these critical academic studies of Southeast Asian rurality are not always mirrored in popular imaginings. While it is somewhat speculative to trace the source of the popularly imagined countryside, a combination of media and educational influences no doubt plays an important part, showing that such imaginaries go well beyond the constructed tourist experience to embed themselves in domestic urban understandings of non-urban spaces and livelihoods. In particular, urban middle class views of the countryside have been shaped by a combination of textbooks together with enduring magazine, literature and television images. In Thailand, the popular 1960s novels on village headman Phu Yai Li have been serialized in numerous television versions. Thai textbooks inculcate 'traditional' values that are supposedly based on simple rural ways of life. Soap operas present the countryside and country people in the most bucolic way imaginable. A recent program has a well-known actor filming herself going 'back' to the land in a series titled, "*Chan ja pen chaonaa* [I'm gonna be a (rice) farmer]." And so on. Furthermore, the rural images that many middle class Thais have inculcated are influenced in no small part from their visits, or at least media exposure, to Europe, and this can be seen in the architectural and landscape design features of rural idylls that the wealthier amongst these construct in northern Thailand, around the edges of Khao Yai National Park northeast of Bangkok, in Kanchanaburi Province to the west of Bangkok and elsewhere. These are sometimes manifested as kitsch (for examples restaurants adorned by Dutch-style windmills on northern mountain roads) and sometimes achieve closer resemblances to the pastoral character of European landscapes.

There is also a political project in construction of the quintessentially rural, Buddhist way of life in Thailand based on limited wants and needs. Most starkly, the 'sufficiency economy [*setthakid phor phiang*]' ideal has found its way into official discourse since it was first publicly espoused by the king in response to the 1997 financial crisis. The ideal exhorts moral behavior and modest material expectations of the rural poor, with an implicit critique that debt and other ills are the product of their illicit and extravagant behavior. Indeed, so pervasive has this normative idea of appropriate ways of living become that it has post-facto been employed as an interpretive framework to laud and promote rural development programs that predate the sufficiency economy narrative (Singsuriya 2015).

Thailand is perhaps more exaggerated in the gaps between urbanite images and rural realities than other countries in Southeast Asia, given the combination of wealth gaps, urban primacy of Bangkok and the place of rural-urban dynamics in the country's recent and ongoing political imbroglio. However, we can see at least nascent parallels in Vietnam, the Philippines and Indonesia as those countries' mainly city-based middle classes at once distance themselves economically from their rural compatriots while building nostalgic images of rural authenticity as a basis for national culture.

The nostalgic imagery of the countryside from Bangkok and other urban centers in Southeast Asia contains a mix of a disappearing past and one that in many respects never existed. Many years ago, Jeremy Kemp wrote about the "seductive mirage" of the Southeast Asian village community, at a time when the point of critique was as much fellow anthropologists as it was a more superficial popular understanding of the stereotypical Southeast Asian village (Kemp 1987). Jan Bremen wrote in a companion volume of the "shattered image" in his historical account of the construction and deconstruction of village society in Southeast Asia (Bremen 1987), showing how European notions of village and community had influenced not only understandings of, but also colonial policy toward, rural social organization in the region. We find an interesting resonance between these 'outsiders' unwitting imposition of their own constructs of rurality and that of contemporary domestic urbanite expectations of their own rural landscapes and social behavior.

Of course, not all of the imagery in question is invented or imagined. Some of the nostalgia is based on partial realities. Rural production *was* previously more exclusively geared to farming, and rice farmers *did* often privilege cultivation of their own food before producing for the market. People *did* share their labor for planting and harvesting. Housing materials *were* derived more from wood and other local materials than they are at present. Buffalo *were* used to plough the fields. But these aspects of rural Southeast Asia hark back a generation or more in most places. To emphasize the ever-growing gaps between lingering imagery of rural Southeast Asia and present-day realities, we briefly review the work of three authors who have explored rural-urban interactions to demonstrate the decreasing relevance of dichotomized discussion of rurality and urbanity.

Singapore-based US anthropologist Eric Thompson shows the gap between nostalgic, yet derogatory, Malaysian urban elite representations of the village (*kampung*) and the realities of everyday life. In particular, he shows that despite superficial characteristics of housing and landscape, urban values are firmly rooted in village society. Most economic activity is based on commodity production and services in and outside of agriculture, unlike the bucolic and subsistence-based assumptions of Kuala Lumpur middle classes. Social interactions and other relations, along with occupational stratification, moreover, follow a more urban than rural pattern. In other words, the spaces of *kampung* have become "socially urban" (Thompson 2004).

Australian-based Vietnamese specialist in agrarian change and natural resource management To Xuan Phuc carried out a fascinating study of wealthy urbanites' (*dai gia*) second homes northwest of Hanoi at Ba Vi. Ba Vi is at the transition of the Red River Delta into the mountainous northwest of Vietnam that the French geographer Pierre Gourou described in stark terms in the 1930s (Gourou 1936), and the area has long been farmed by ethnic minority *Muong* people. Phuc shows that members of the rising Hanoi-based upper middle class have occupied this rural space through speculative purchase of land, based on a desire to situate themselves during their leisure time amidst scenic rural landscapes driven by a type of nostalgia for markers of traditional village life, but with little desire for interaction with social aspects of rurality. The huge gap between urban rich and rural poor in a country long committed politically to the ideal of economic equality is made stark by the contrast in respective house styles and mutual disdain between the newcomers and locals (To Xuan Phuc 2012).

Finally, and in the most encompassing, indeed almost encyclopedic way, the prolific Singapore-based British geographer Jonathan Rigg (1998) has written extensively over at least two decades of the "rural-urban interpenetration" that breaks down any notion of an isolated rurality separate from urban influence and experience. This is driven by the increased mobility, communication and economic interactions between countryside and city. He shows systematically that the basis for rural economy is much "more than the soil" (Rigg 2000), delinking poverty in Thailand from full-time farming (Rigg 2005). Tellingly, Rigg's prescient writing was counter-trend from the early 1990s in his skepticism about grassroots-based development as espoused by the community culture school of NGOs in the face of the rapid pace of rural change (Rigg 1991). More recently, Charles Keyes has written of 'cosmopolitan villagers' in recognition of the expanded world view and mobility of people with homes and/or origins in rural northeastern Thailand (Keyes 2012).

Beyond the constructions of rurality and the gap between imagined and actual realities of the Southeast Asian countryside, there remains an economic and political salience to rural imagery. At one level, portraying the countryside in its cropped form continues to attract tourists into the rural economy, albeit with limited trickle-down to people living in the sites visited. This extends increasingly to domestic tourism in search of getaways from the pressures of urban life. At another level, performance of rurality is part of the political scene, for example when farmers protest in 'typical' rural garb, such as the indigo *mor hom* shirts that northern and northeastern Thai farmers long ago ceased to produce and wear in daily life, but that immediately mark them as representing downtrodden rural interests. When less familiar markers of rural ways of life such as loud music, pickup trucks and other 'modern' trappings find their way into protests on city streets, such as at Ratchaprasong during the drawn out 2010 protests that were ultimately put down violently by the government and military, there is less sympathy from the powers that be and many urban residents. Andrew Walker (2012) has written of 'Thailand's Political Peasants' to show not only how rural people have become fully incorporated into the rough and tumble of parliamentary and extra-parliamentary politics, but also of how the peasant imagery and discourse is employed within it. Meanwhile, Charles Keyes has shown, from his studies in northeastern Thai communities since the early 1960s, how rural people have finally 'found their voice' in national affairs (Keyes 2014), presenting uncomfortable challenges to urban elites accustomed to deciding who rules the country.

Conclusion: developed or decropped Southeast Asian rural spaces?

The countryside in Southeast Asia is changing rapidly, perhaps even more so than urban spaces. Many markers of development exist in places they were absent a generation ago. But not all such changes follow the linear paths and patterns of modernization, commodification and de-agrarianization that come with the moniker of 'development.' Indeed, there is a reflexive aspect to understandings and associated expectations of rurality, on the one hand, and physical, social and economic manifestations of rurality on the other. In other words, outsiders' impositions and rural people's expressions both shape the nature of the 'rural' in Southeast Asia. But this rurality is always a relative notion, particularly in relation to urbanity and sometimes also to modernity.

The gaps between imagined and experienced rurality can lead to clashes. Many of these clashes are everyday, even mundane in character, such as the encounters described above in the case of Nong Khwai village. Others are clashes of an academic nature, both within the scholarly literature as critique and counter-critique that refine our understanding of rural Southeast Asian realities, and between academic and popular notions of the rural. Yet other clashes are of a political nature, as rural majorities express themselves vis-à-vis actual or perceived oppressive urban elite interests, and as political actors take advantage of, or otherwise employ, rural identities for particular ends.

Our intention in this chapter has been to 'decrop' the Southeast Asian countryside, not only by reminding of the ever-diminishing relative significance of farming in the economic and social lives of people who live there, and of the fading relevance of development as a way of thinking about change, but also by suggesting that 'cropped' versions of the rural lands are based on selective understandings or deliberate obfuscations of reality. We do not take issue with the deliberate and discursive deployment of such selectivity. Rather, we suggest that a fuller understanding of the Southeast Asian countryside needs to be based in research that explores such selective representations of rurality as a subject in its own right.

References

Bremen, J. (1987). *The shattered image: Construction and deconstruction of the village in colonial Asia.* Amsterdam: Centre for Asian Studies Amsterdam.
Bruneau, M. (2012). Agrarian transitions in northern Thailand: From peri-urban to mountain margins, 1966–2006. In: J. Rigg and P. Vandergeest, eds., *Revisiting rural places: Pathways to poverty and prosperity in Southeast Asia.* Honolulu: University of Hawaii Press, 38–51.
De Koninck, R. and Thanh Hai, P. (2017). Population growth and environmental degradation in Southeast Asia. In: P. Hirsch, ed., *Handbook of the environment in Southeast Asia.* London: Routledge, 46–68.
Gourou, P. (1936). *Les paysans du delta tonkinois.* Paris-La Haye: Mouton.
Halfacree, K. H. (1995). Talking about rurality: Social representations of the rural as expressed by residents of six English parishes. *Journal of Rural Studies*, 11(1), pp. 1–20.
Hirsch, P. (1990). *Development dilemmas in rural Thailand.* Singapore: Oxford University Press.
Hirsch, P. (1993). *The village in perspective: Community and locality in rural Thailand.* Chiang Mai: Social Research Institute, Chiang Mai University.
Hirsch, P. (2012). Nong Nae revisited: Continuity and change in a post-frontier community. In: J. Rigg and P. Vandergeest, eds., *Revisiting rural places: Pathways to poverty and prosperity in Southeast Asia.* Honolulu: University of Hawaii Press, 112–134.
Kemp, J. (1987). *The seductive mirage: The search for the village community in Southeast Asia.* Amsterdam: Centre for Asian Studies Amsterdam.
Keyes, C. (2012). "Cosmopolitan" villagers and populist democracy in Thailand. *South East Asia Research*, 20(3), pp. 343–360.
Keyes, C. (2014). *Finding their voice: Northeastern villagers and the Thai state.* Chiang Mai: Silkworm Books.
Nartsupha, C. (1999). *The Thai village economy in the past.* Chiang Mai: Silkworm Books.

Phatharathananunth, S. (2002). The politics of the NGO movement in Northeast Thailand. In: P. Phongpaichit, ed., *Asian review 2002: Popular movements*. Bangkok: Institute of Asian Studies, Chulalongkorn University, 21–36.

Phongphit, S. (1986). *Back to the roots: Village and self-reliance in a Thai context*. Bangkok: Rural Development Document Centre.

Phongphit, S. and Hewison, K. (1990). *Thai village life: Culture and transition in the Northeast*. Bangkok: Mooban Press.

Rigg, J. (1991). Grass-roots development in rural Thailand: A lost cause? *World Development*, 19(2), pp. 199–211.

Rigg, J. (1998). Rural – urban interactions, agriculture and wealth: A southeast Asian perspective. *Progress in Human Geography*, 22, pp. 497–522.

Rigg, J. (2000). *More than the soil: Rural changes in Southeast Asia*. Toronto: Pearson Education.

Rigg, J. (2001). *More than the soil: Rural change in Southeast Asia*. Harlow: Prentice Hall.

Rigg, J. (2005). Poverty and livelihoods after full-time farming: A South-east Asian view. *Asia Pacific Viewpoint*, 46(2), pp. 173–184. Available at: www.scopus.com/inward/record.url?eid=2-s2.0-29044434004&partnerID=40

Rigg, J. and Salamanca, A., (2012). Moving lives in northeast Thailand: Household mobility transformations and the village, 1982–2009. In: J. Rigg and P. Vandergeest, eds, *Revisiting rural places: Pathways to poverty and prosperity in Southeast Asia*. Honolulu: University of Hawaii Press, 88–111.

Rigg, J. and Ritchie, M. (2002). Production, consumption and imagination in rural Thailand. *Journal of Rural Studies*, 18(4), pp. 359–371.

Rigg, J. and Vandergeest, P. (2012). *Revisiting rural places: Pathways to poverty and prosperity in Southeast Asia*. Honolulu: University of Hawaii Press.

Singsuriya, P. (2015). Sufficiency economy and backdated claims of its application: Phooyai (Headman) Wiboon's agroforestry and self-narrative. *Journal of Asian and African studies*, 50(Published online (December 2015)), pp. 1–13.

Thompson, E. (2004). Rural villages as socially urban spaces in Malaysia. *Urban Studies*, 41(12), pp. 2357–2376.

To Xuan Phuc. (2012). When the Đa. i Gia (Urban Rich) Go to the Countryside: Impacts of the Urban-Fuelled Rural Land Market in the Uplands. In: V. Nguyen-Marshall, L. Drummond and D. Belanger, eds., *The reinvention of distinction: Modernity and the middle class in urban Vietnam*. Dordrecht: Springer, 143–156.

Tubtim, T. (2012). Migration to the countryside. *Critical Asian Studies*, 44(1), pp. 113–130. Available at: www.tandfonline.com/doi/abs/10.1080/14672715.2012.644890

Vandergeest, P. (2012). Deagrarianization and re-agrarianization: Multiple pathways of change on the Sathing Phra Peninsula. In: J. Rigg and P. Vandergeest, eds., *Revisiting rural places: Pathways to poverty and prosperity in Southeast Asia*. Honolulu: University of Hawaii Press, 135–156.

Walker, A. (2012). *Thailand's political peasants power in the modern rural economy*. Madison: University of Wisconsin Press.

Wilson, G. A. and Rigg, J. (2003). Post productivist agricultural regimes and the south: Discordant concepts? *Progress in Human Geography*, 27(6), pp. 181–207.

Witayapak, C. (2012). Who are the farmers? Livelihood trajectories in a northern Thai village. In: J. Rigg and P. Vandergeest, eds., *Revisiting rural places: Pathways to poverty and prosperity in Southeast Asia*. Honolulu: University of Hawaii Press, 211–228.

PART 4

Environment and development
Introduction

Fiona Miller, Andrew McGregor and Lisa Law

Continuing the themes that emerge from earlier parts of the Handbook, the chapters in this section highlight both the uneven nature of development across Southeast Asia as well as the costs associated with resource intensive development and rapid economic growth. One of the most cogent critiques of modernist development has been the impact of resource intensification and growth-oriented economic strategies on the environment. Indeed, the positioning of the environment as somehow external to the economic and social processes of development is an assumption that has been widely challenged with the rise of sustainability discourses. Yet beyond framing the environment as experiencing impacts from development and seeing it as a set of resources underpinning livelihoods and economic activities requiring regulation, conservation and protection, the authors in this section highlight the ways in which rethinking society-environment relations can contribute to more equitable and sustainable development.

The imagery associated with the region's iconic landscapes of tropical forests, terraced rice fields, colorful reefs and golden beaches sits awkwardly with that of the grey concrete, glass and steel of the sprawling urban and industrial estates and busy transport corridors that constitute the landscapes more commonly populated by people in the region. The environment of Southeast Asia is both a source of well-being for millions, and something valued for its beauty, biodiversity and complexity. Yet, due to the negative effects of development, the environment is also a source of harm that threatens the health, well-being and prosperity of current and future generations. The problems of air and water pollution, food contamination, species and habitat loss, resource scarcity, and environmental extremes, disasters and climate change are not novel. Rather these problems have long been recognized by communities, researchers and governments throughout the region, with much scholarship focused on documenting and trying to solve these problems. Yet there is growing recognition that these problems require urgent and concerted responses at local, national, regional and global scales by all development actors, including government, business, donors, NGOs, communities and individuals. These problems cannot be contained within the boundaries of particular localities or nations, but rather many are inherently transboundary in nature. As highlighted in the many case studies presented in this volume, the way in which people value, relate to, use and are affected by or benefit from the environment is highly differentiated across society, space and time. There is also considerable variation in the way governments throughout the region have prioritized and dealt with particular environmental problems; some have chosen more market-based approaches to environmental matters over stronger regulatory responses, with the extent to which community-based natural resource management institutions are recognized versus more centralized, top-down regulation of the environment forming another key point of distinction across the countries of the region.

Three strong themes emerge from the collection, including: the social and environmental consequences of neoliberal development; the highly interconnected and regional nature of many environmental challenges; and the uneven consequences of environmental change and resource scarcity. As highlighted elsewhere in this collection, neoliberal development has led to an increased reliance on market-based approaches to environmental regulation and conservation. At the same time, with globalization, the region is increasingly interconnected economically through trade and investment flows as well as physically through environmental processes and structures such as energy power grids and transport networks. Finally, the highly uneven nature of the consequences of environmental change and resource scarcity is prominent in many chapters and reveals the inequity associated with current approaches to development, raising important questions of responsibility and justice.

The first chapter in this section, by Miller, provides an overview of the changing position and value of water in the region. Through the presentation of a combined analysis of regional data and case studies, Miller documents the way the modernist development paradigm that underpinned intensive and large-scale water use developments throughout Southeast Asia has resulted in profound material changes in water systems. She argues a turn toward the diverse, extant community-based and context sensitive water relations, institutions and practices found throughout the region is likely to generate valuable approaches to address the rising competition over water between humans and non-humans.

Competition over land is a strong theme within Fujita Lagerqvist and Connell's chapter. They provide a comprehensive review of the role of agriculture and land in development throughout the region, and focus on how particular policies including the promotion of foreign direct investment and trade, and private property rights have facilitated the expansion of commercial agriculture. They highlight the enduring importance of land in rural livelihoods, despite the declining contribution of agriculture to national GDPs and household incomes. The transformation of livelihoods documented by Fujita Lagerqvist and Connell is radical and includes shifts from subsistence to commercial and from extensive to intensive production and the rise of more pluri-active, multi-local and interconnected livelihoods.

The increasingly interconnected nature of livelihoods in the region is a theme reinforced in the chapters by both Bush, Marschke and Belton and Majid Cooke et al. Bush, Marschke and Belton provide a detailed study of the value chain and political, economic and geographical context of the seafood industry in Southeast Asia, encompassing both capture fisheries and aquaculture. Whilst to date there has been considerable attention given to environmental sustainability issues associated with the sector, this chapter redresses the neglect of social sustainability issues by considering the rights and equity concerns of workers. They draw a link between the weak governance that enables the overexploitation of fishery resources as also enabling the exploitation of workers. As such, sustainability in the sector requires not just attention on environmental performance but also a consideration of human rights, well-being and gender equality.

Sustainability concerns have also been associated with the expansion of oil palm plantations throughout the region, as documented in the chapter by Majid Cooke et al. They demonstrate through a series of local level case studies how smallholder farmers through their relations with mills, NGOs, the government and certification authorities negotiate a series of trade-offs between diversification and intensification, market access and economic security, and food security. The experience of communities in East Malaysia is quite varied, with Majid Cooke et al.'s chapter demonstrating the value of nuanced, context-sensitive ethnographic research.

Introduction

The chapter by Thomalla, Boyland and Calgaro considers the complex interactions between disasters and development in Southeast Asia – one of the most disaster-prone regions of the world. The chapter provides a valuable regional perspective on the uneven spatial distribution of disaster risk, drawing upon analysis of a number of data sets, case studies and program documents. High levels of poverty and inequality, demographic change and urbanization, unsustainable natural resource use and weak governance structures are shown to all contribute to rising vulnerability to disasters. They suggest a number of pathways by which development can be transformed toward more equitable resilience.

One of the factors identified as likely to exacerbate disaster vulnerability in the region in the future is that of climate change. Similar to Thomalla, Boyland and Calgaro, Uddin and Nylander offer a strong regional level analysis of efforts to transition toward low-carbon development through the upscaling of climate change mitigation efforts. Drawing particularly on examples from the energy sector, they discuss the different regional cooperation schemes relevant to climate change mitigation, identifying opportunities whereby these could complement national and global level efforts. They argue that in order to realize greenhouse gas mitigation goals, regional cooperation efforts in energy, technology and climate require greater harmonization, and suggest that regional markets could potentially play a valuable role toward this goal as could an ASEAN climate policy target.

The chapter by Neef and Sangkapitux contributes to wider environmental governance debates concerning so-called 'Green Capitalism.' They consider efforts to conserve the environment in Southeast Asia, not through traditional command and control-type regulatory mechanisms such as taxation, but rather through the provision of incentives in the form of payments for so-called ecosystem services. They present a series of comparative case studies that interrogate the promise of such schemes but also reveal some of the limitations of market-based approaches to conservation, particularly in terms of strengthening customary land rights and local control of resources.

Finally, much work on environment and development tends to look at impacts on the environment due to development, and to some extent considers how environmental processes influence development pathways. Few consider the agency of the environment as a development actor as McGregor and Thomas do in their chapter on forests. By adopting a more-than-human approach to forests they recognize "forests not as inert objects within development processes but dynamic assemblages of interacting things, whose materiality influences the types of development that can and does occur." Through engagement with more-than-human theories, the chapter opens up clear opportunities to rethink development by foreshadowing human and non-human entanglements that resist the hyper-separation of humans and nature that lies at the core of modernist development and its contribution to widespread environmental degradation. As such, through their thoughtful engagement with more-than-human theory, McGregor and Thomas provide a hopeful perspective on development in the final chapter of the collection.

Fiona Miller, Andrew McGregor and Lisa Law

23
MATERIAL, DISCURSIVE AND CULTURAL FRAMINGS OF WATER IN SOUTHEAST ASIAN DEVELOPMENT

Fiona Miller

Introduction

Water is synonymous with life and culture in Southeast Asia. The tropical landscapes of the region are defined by the seasonal monsoon rains, floods and rivers that shape and connect ecologies, livelihoods, commerce, cultures and social activities. Total water consumption in the region has been steadily rising due to the intensification of agriculture and industry, as well as changing living standards, lifestyles and population increases. This has led to an overall decline of 31 percent in estimated total renewable water resources per capita across the region from an average of 23,640 m^3/inhab/yr in 1992 to 16,272.63 m^3/inhab/yr in 2014 (FAO 2015). In comparison to other regions around the world, however, water availability in Southeast Asia still remains theoretically fairly high despite the steady rise in water demand.

Considering the centrality of water to human and non-human life, livelihoods, agriculture, industry and transport, distinct approaches to address issues of competition have long been present at various scales. Different systems to address the allocation of water are apparent at the community scale, amongst users across large-scale irrigation schemes in the region's deltas and plains, within regional transboundary water sharing agreements such as for the Mekong Basin, or in the codification of water rights, payments and allocation arrangements at the city or national scale. Water relations vary in the degree to which they are formalized, cooperative and bureaucratic, as well as the extent to which they are considered equitable, legitimate and sustainable. The pace of change accompanying development has placed pressure on these relations and institutions to effectively and fairly resolve water-based competition and prevent conflict.

This chapter seeks to provide an overview of the changing position and value of water in Southeast Asia and its role in development. It does this by first outlining the different entry points for understanding water-development relations in the region, before providing a brief historical overview of water in development. The rise of modernist development and the influence of particular water discourses are discussed before the chapter addresses the main drivers for material change in water throughout the region. In doing so the awkward coexistence of diverse knowledges, values and relations with water apparent in contemporary Southeast Asian development is highlighted.

Approaching water and development issues in Southeast Asia

There are different entry points for understanding water and its relationship with development in Southeast Asia. There are discrete literatures, professional, academic, civil society and activist networks, and organizations associated with water's different dimensions. Water is a common issue of concern within many sectors and aspects of development, as captured in Table 23.1.

Table 23.1 Multiple dimensions of water and development

Water dimension	Networks and key development actors
Catchment and basin management	River basin organizations, e.g., Mekong River Commission Network of Asian River Basin Organisations Civil society networks organized at a basin scale, e.g., www.savethemekong.org; Living River Siam Association Global Water Partnership International Water Management Institute
Hydropower dams	Energy and dam building industry ICOLD (International Commission of Large Dams) Multilateral lenders, such as World Bank, ADB and financial institutions Export credit facilities Technical advisors and bilateral aid agencies Activist and concerned scholar networks Social movements for rivers/anti-dams, e.g., Save the Salween Network
Traditional and community water management systems	Community groups Community and development NGOs Researchers, e.g., Thai Baan citizen scientist initiative
Livelihoods, irrigation and agriculture	Agribusinesses Community and development NGOs Agricultural research organizations, e.g., IRRI Extension officers, agronomists Government agriculture and rural development departments International Water Management Institute
Drinking water supply for community and large-scale urban systems	UNICEF WASH Health practitioners Urban planners Water engineers Public, state-owned and private water utilities Transnational water companies, e.g., Vivendi, Veolia, Thames Water, Southeast Asian Water Utilities Association International Water Management Institute
Waste water and water pollution	International Water Management Institute Waste water treatment companies Water management authorities Environmental protection agencies
River ecology and restoration	Ecologists Community groups Environmental groups

Water dimension	Networks and key development actors
Water related risks and disasters	Disaster risk reduction community Climate change adaptation UN International Strategy for Disaster Reduction (ISDR) Asian Disaster Preparedness Centre

Source: Fiona Miller

Each of the above dimensions of water open up different development concerns and priorities, as well as specialized knowledge and networks. What has emerged from work on catchment management is a strong policy emphasis on so-called integrated approaches to water at the scale of the basin. This, in part, seeks to address the multiple dimensions of water in a coordinated way in water use, planning and management. Much attention has focused on the idea of integrated water resources management throughout Southeast Asia as well as the creation of new institutions such as river basin organizations (Miller and Hirsch 2003; Molle 2008). Often the emphasis on so-called integrated approaches to water has accompanied or been a precursor to accelerated water resources development, such as hydropower development, reflecting a technocratic approach to water that assumes management is value-neutral and rational, driven by appropriate inputs of data, stakeholder consultation processes and accurate modeling (Miller and Hirsch 2003; Molle 2008). Critiques of 'integration' at the basin scale highlight how it has often occurred in a top-down manner, with community level and environmental water needs and knowledges often marginalized by more powerful development interests (Miller and Hirsch 2003). In contrast, research on community-based water management has generated different understandings of water ethics, approaches to water, social relations and ecologies yielding particular insights on how to avoid and resolve water conflicts, methods of allocation during times of scarcity, and practices that ensure longevity and equity within the system. Yet, community-based approaches face their own limitations particularly in terms of addressing wider scale processes and conflicts.

Beyond catchment and livelihood focused research, work on drinking water supply has tended to focus on the operation and maintenance of different systems, appropriate technologies for water supply, and different funding and management models. The water supply literature has been quite polarized by debates on the equity dimensions of water privatization (Postel 1997; Bakker 2007; Bakker 2010). The themes of integration, sustainability and equity permeate the literature on water and development in Southeast Asia and emerge from the region's particular history of development that is characterized by tensions between community-level and top-down, centralized control of water associated with state-building, and the technologies and institutions of modernist development.

Historical shifts in water and development

Human settlement patterns throughout Southeast Asia were historically partly derived by the presence of water sources and river flows, with many of the region's major cities located on or beside major rivers, such as Bangkok on the Chao Phraya River, Hanoi on the Red River and Vientiane and Phnom Penh on the Mekong River. Throughout the region, complex and large-scale water control networks, for irrigation as well as ceremonial and religious purposes, can be traced back to ancient times. The degree to which such systems were centralized and under state control has been the subject of considerable debate (Wittfogel 1956; Wittfogel 1957). The vast Angkor system in present day Cambodia (Christie 1992; Fletcher, Penny et al. 2008) and subak systems in Bali (Geertz 1972; Geertz 1980; Lansig 1987) both date from around the ninth century AD and have been the subject of intensive inquiry regarding their purpose, nature of cooperation and hydro-ecological processes. In more recent times, the colonial era saw the hydrological landscapes of the region transformed by state-building of a different kind. Massive canal building projects occurred in the deltas of mainland Southeast Asia at this time, aided by the introduction of mechanical dredgers by the French in Indochina (Brocheux 1995; Biggs et al. 2009; Biggs 2010), and flood control and reclamation efforts in the Irrawaddy Delta in Myanmar (Adas 1974) and Chao Phraya delta in Thailand during the same period. Much of the expansion of rice cultivation at this time was for international export with the mantle for being the region's most productive 'rice bowl' shifting historically over time between these three regions.

The integrity and continuity of traditional community-level water institutions, ethics and rights, as exemplified in the Balinese Subak, Northern Thailand's Muang Fai and the Zanjera indigenous irrigation system of the Philippines (Coward 1979; Coward 1980; Geertz 1980; Lansig 1987; Christie 1992; Rigg 1992), have been directly challenged by the imposition of state water management bureaucracies and the expansion of institutions that seek to scientifically measure, codify, commodify and commercialize water. The neglect, denial or enclosure of traditional systems and associated knowledge systems of water, catchments and ecologies has seen some systems disintegrate or ossify whilst others continue, adapting to new environmental challenges, social expectations and development pressures – co-existing with different water values, knowledges and technologies. For example, the Muang Fai system of Northern Thailand may have changed materially over time, through the replacement of the wood and bamboo weir structures with concrete, yet the Muang Fai water user groups continue to actively deal with water allocation and management, and watershed protection matters (Witthayaphak 1995; Ricks 2015). New understandings of water ethics and relations continue to be generated from the study of long-existing community-based water management, as seen in Palmer's (2010) study in Timor-Leste, which calls for the recognition and valuing of customary water institutions and practices. Her work seeks to profoundly challenge the imposition of modernist development approaches to water (Palmer 2010). As such, the diverse relations with and understandings of water across the region both challenge and redefine development and demonstrate alternatives to the imposition of efforts that seek to commodify and separate water from its ecological source.

Water is now variously framed in development discourses throughout the region as a resource, "a socially vital economic good" (Asian Development Bank 2001), a source of life, a commodity, a finite resource, a source of conflict, an opportunity for cooperation and an essential element of nature. With modernist development and in particular the shift to neoliberalism, water has increasingly been framed as a commodity that is prioritized for economic purposes and subject to profit motives (Bakker 2007). This represents a departure from treatment of water as a commons, an element of nature inseparable from its physical context (Bakker 2007). Though neoliberal approaches to development are influential, they have been resisted in the region, with contestation over water a high profile political battleground. The global push to address the severe lack of universal access to clean and safe drinking water in the 1990s coincided with the Washington Consensus and the push to commercialize and in some cases privatize water, often justified as a means of ensuring water use efficiency and generating funding for much needed water supply infrastructure (Barlow 2001; Barlow 2008). An increase in the role of the private sector in water matters in Southeast Asia accompanied this discursive framing of water as a scarce commodity and was facilitated by development actors, such as the ADB, the

World Bank and other donors through key networks, and policy reforms, technical assistance and loan conditionalities. However, apart from a few high-profile and strongly critiqued cases of water privatization, such as Jakarta and Manila where there were steep rises in water service prices and inadequate attention given to the water needs of the poor (Bakker, Michelle et al. 2008; Pierce 2015), water privatization remains fairly limited across the region (See 2015). The improvement in access to clean and safe water throughout Southeast Asia in recent years has largely been achieved through government, donor and community investments, and public reforms.

Material changes in the distribution of water

Despite growing development pressures and expectations, water availability in Southeast Asia in general is high by international standards. Maritime Southeast Asia (Brunei, East Malaysia, Philippines, Indonesia, Singapore and Timor-Leste) has very high rates of internal renewable water resources (IRWR) per inhabitant of 10,644 m^3/annum, followed by Mainland Southeast Asia (Myanmar, Cambodia, Laos, Thailand, Peninsula Malaysia and Vietnam) with 8,499 m^3/inhabitant, whereas East Asia has just 2,068 m^3/inhabitant and South Asia has 1,228 m^3/inhabitant (FAO 2015). For some countries in Mainland Southeast Asia, when water originating outside a country is taken into account, the distribution of total actual renewable water resources (TARWR) is higher, such as for Cambodia where the IRWR are 8,626 m^3/inhabitant yet the TARWR are almost four times as high at 34,061 m^3/inhabitant and for Laos the corresponding figures are 31,155 m^3 and 54,573 m^3 (FAO 2015). Such measures highlight the interconnected nature of the region's water resources and associated ecological and climatic processes.

In a material sense the processes of modernist development have transformed water availability and distribution across the region. Dams and irrigation expansion in particular, as discussed in this section, have resulted in water becoming increasingly separated from its natural setting. This has been facilitated by the discursive framing of water as an economic resource and commodity for economic activities such as agriculture, industry, transport and urban growth. This framing has helped justify 'megaproject' proposals such as the 'water grid' in Thailand (Molle and Floch 2008) – a series of dams, pipelines and inter-basin transfer schemes designed to move water to areas of highest economic demand. Though water has long been central to people's livelihoods and economic activities, the extent that it has been materially transformed by development has intensified. Large-scale river modification schemes have seen waterways channelized, and river flows regulated through dams, large-scale irrigation and extraction schemes, flood protection works and inter-basin diversion schemes. Such interventions, and the associated rise in water withdrawals, raise questions as to the sustainability of current approaches to water and development.

Water scarcity

Overtime, water availability in the region has declined. Figure 23.1 shows recent declines in total renewable water resources per capita for the countries of Southeast Asia between 1992 and 2014. Based on this physical measure alone Singapore is the only country that would be categorized as water stressed (defined as water less than 1,700 m^3/inhab). Yet Singapore is a special case; previously this city-state was highly reliant on the import of water from Malaysia, but through efforts to manage water demand and increase supply efficiency and the production of

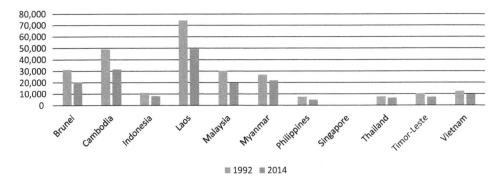

Figure 23.1 Total renewable water resources per capita (m^3/inhab/year) 1992–2014
Source: FAO AQUASTAT, 2015 <www.fao.org/nr/water/aquastat/data>

new water from recycling and desalination technologies, Singapore has successfully diversified its water sources and reduced its external water dependence (Tortajada 2006). Countries that face economic water scarcity,[1] include Cambodia, Indonesia, Malaysia and Myanmar (Marcotullio 2007).

Urbanization, agriculture and industry have placed significant pressure on water resources, as reflected in estimates of freshwater withdrawals as a percentage of total renewable water resources. The countries that use a high percentage of total available water resources include Singapore, Thailand, Vietnam and the Philippines, see Figure 23.2. These water withdrawals are primarily for agriculture and industry, as shown in Figure 23.3. Associated with the expansion in withdrawals is a rise in irretrievable water losses, due to consumption, irrigation, water pollution and other losses. It is estimated that irretrievable water losses for the region increased from 142 km^3/yr (1950) to 399 km^3/yr (1990) (Dudgeon 2000), threatening the sustainability of current uses. Despite the relatively high theoretical availability of water, due to rapid development, population growth and environmental degradation, water has in the space of a couple of generations transformed from being widely perceived as plentiful to being a hotly contested and relatively scarce resource (Marcotullio 2007). This scarcity has been socially constructed largely by the water-intensive nature of development and associated decline in the integrity and provisioning capacity of ecosystems and catchments; such scarcity, has been used to justify policies premised on increased efficiency, such as privatization.

Dams

Considering the tropical nature of the climate of Southeast Asia it is important to go beyond these aggregate figures to consider seasonal and spatial water variations. The seasonal variation in water has long presented a challenge to livelihoods and a constraint to development premised on intensive water use. The monsoons of Southeast Asia produce distinctive periods of drought in the dry season that alternate with periods of heavy rains and increased river discharge, spates and floodplain inundation in the wet season (Dudgeon 2000). The seasonal variation in rains and river flow, where river flow can vary by as much as ten times between the dry and wet seasons, has formed one of primary justifications for dam building. Human modification of river systems in Southeast Asia has long sought to even out this variation in water availability. The Asian region, including India and China, has been the site of some

Material, cultural framings of water

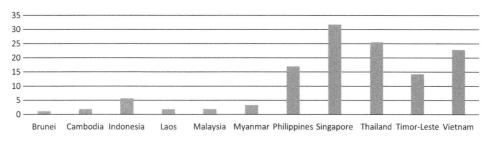

Figure 23.2 Total annual freshwater withdrawals (m³) as a percentage of total (2013)
Source: World Bank (2015)

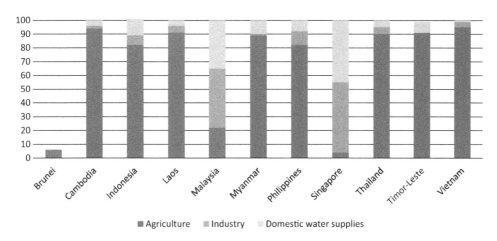

Figure 23.3 Annual freshwater withdrawals (m³) as a percentage of total (2013) for agriculture, industry and domestic
Source: World Bank (2015)

of the most intensive dam building on the planet, accounting for a high proportion of the global number of dams (Dudgeon 2000; McCully 2001). Dudgeon (2000, 793) reports that, "[i]n 1950 Asia had 1,541 large dams, accounting for 30 percent of the global total (van der Leeden et al. 1990); by 1982 that figure had grown to 22,701 (65 percent of the global total)." While most of these dams are in China and India, dam building in Southeast Asia has also been significant.

Indonesia, Vietnam, Thailand and Malaysia have built many large dams – 242, 100, 40 and 27, respectively, according to ICOLD data (FAO 2015). The total dam capacity (km³) for select countries, indicates the extent of investment in dam construction: Thailand leads the region at 68.3 km³, followed by Vietnam at 28 km³, Indonesia at 23 km³ and Malaysia at 22.5 km³(ibid). The intensity and scale of such modification has progressively increased with development pressures, with many dams premised on the need for expanded irrigation and hydropower production. There are now few large river systems that remain unregulated by dams, with the relatively unregulated Salaween River in Myanmar and the lower reaches of the Mekong River now the sites of rising tensions over vast dam building schemes.

Box 23.1 Mekong River dams water for energy at the cost of food security

The international transboundary river system of the Mekong has long been the site of conflict over dam building. Tensions have largely focused on water sharing arrangements between the six riparian nations, only four of which are party to any formal agreement. Yet, a major tension exists not just between upstream and downstream, rich and poor countries, but more fundamentally between development premised on control and modification of the highly variable water regime and the extant livelihoods adapted to these water conditions. The mean monthly water flow is highly variable; for example at Kratie in Cambodia it reaches its peak during the wet season in September at more than 36,000 m3/sec. yet is just 2,200 m3/sec. in April during the dry season (Commission 2009), with 75 percent of the flow occurring between July and October (Mekong River Commission 2010). Yet, dam building on the mainstream and major tributaries of the Mekong threatens this variability. Plans for dams on the lower Mekong can be traced back to the Cold War era when harnessing the development potential of water was strategic to America and its anti-communist allies' geopolitical interests. American engineers associated with the Tennessee Valley Authority sought to export a model of river basin development represented by the plan for a Mekong Cascade of dams (Nguyen Thi Dieu 1999). It is now, some 50 years after these large-scale mainstream dams plans were first proposed, that this vision is coming close to being realized.

China has already built seven dams on the upper Mekong River, with plans to build many more. The 11 proposed mainstream dams in the Lower Mekong Basin are premised on a more regular flow being provided by the Chinese dams, with the generation of hydropower in the dry season making such dams now more economically viable (Mekong River Commission 2010). Throughout the late 1990s and early 2000s Laos and Vietnam advanced their extensive plans to build dams on the major tributaries of the Mekong, whereas Thailand's dam building in the north and northeast of the country slowed toward the end of the 1990s, due to growing opposition to dams. Basin-wide development scenarios have considered the possibility of up to 30 tributary dams (Mekong River Commission 2011).

The Mekong system forms the largest inland fishery in the world (Ziv, Baran et al. 2012). Fish and other aquatic organisms are particularly sensitive to the construction of dams and other in-stream barriers as well as modification of the flood plains and decline in water quality and quantity. This is because of their migratory nature and the role floods and floodplains play in their diversity and productivity. Fish and other aquatic organisms form a significant source of protein in the diets of people throughout the lower Mekong Basin. As such, there is a direct conflict between the food and livelihood security served by the world's largest freshwater fisheries and the construction of hydropower dams for energy purposes (Ziv, Baran et al. 2012). Moreover, the largescale resettlement associated with dams is considerable, meaning the benefits dams generate come at the cost of widespread displacement and ecological harm. So, even though a lot more is now known about the harmful social and environmental impacts of the current and proposed dams, especially in terms of fisheries impacts, the central discursive position they play in framing and pursuing 'development' in the region has changed little since many of these dams were first proposed during the Cold War. One change, however, is that the discourse now alludes to 'green energy,' 'low-emissions energy' and 'sustainable development.'

Irrigation

Dams together with river control works have not only modified flow regimes but have also led to a reduction in flood plain and wetland ecology, loss of important estuaries and a reduction in aquatic species diversity. These water control works have enabled the expansion of irrigation throughout the region, with some of the most intensely cultivated areas on the planet found in Southeast Asia.[2] Some 11 percent of the world's irrigated crop area is found in Southeast Asia (FAO 2015), with irrigated areas now comprising more than a third of cultivated land across Asia (Mukherji, Facon et al. 2009). Significant expansion of irrigated agriculture in the region occurred from the 1960s onwards with the introduction of Green Revolution technologies accompanying modernist development, with the area under irrigation growing from 8.1 million hectares to 16.7 million hectares between 1961 and 2003 (ibid).

As can be seen in Table 23.2, the countries with the highest irrigation withdrawals in the region are Indonesia, the Philippines, Vietnam and Thailand – all major agricultural economies with more than 30 percent of their populations employed in agriculture. As a crude measure of water efficiency we can see, however, that the countries with higher water requirement ratios, such as Thailand, Brunei, Indonesia, Malaysia and the Philippines, tend to withdraw more water than is actually required, although all countries (other than Thailand) have water requirement ratios similar to or below the world average (of 56 percent).

Table 23.2 Irrigation water withdrawals

Country	Irrigation water withdrawal (km^3/yr)	Water requirement ratio
Brunei	0.005	51
Cambodia	1.928	48
Indonesia	92.76	51
Laos	3.193	48
Malaysia	2.505	51
Myanmar	29.57	42
Philippines	65.59	51
Singapore	–	–
Thailand	51.79	66
Timor-Leste	1.071	18
Vietnam	77.75	38
Region	326163.00	48–51

Source: FAO AQUASTAT 2015

Engineering landscapes for intensive agricultural development

There are few places in Southeast Asia where the scale of the human control of the landscape and the control of water for national development has been as intense as that of the Mekong Delta in Vietnam. The historical role of water control has been central to the development of this intensively cultivated and populated part of Southeast Asia (Brocheux 1995; Biggs, Miller et al. 2009; Biggs 2010). Yet, the dynamic ecology of the region continuously thwarts efforts to completely control water, and farmers have developed specific adaptive practices and relations with water in response to the annual floods, salt-water intrusion and variations in water availability (Huu Chiem 1994; Miller 2006a; Miller 2006b; Can, Duong et al. 2007; Miller 2014).

The intensification of water use in the delta has seen a rapid increase in rice exports from the delta, rising from 284,000 tonnes in 1880 during the French colonial period to more recent figures that estimate that the delta produces some 24 million tonnes of rice (General Statistics Office (GSO) 2012), of which close to 90 percent is for export (Can, Duong et al. 2007). The rapid rise in demand for water accompanying this increase in rice production has raised particular challenges in terms of how water use is coordinated at multiple scales as previously, under less competitive water conditions, there was little need for organization to coordinate water use (Miller 2014). There are rising concerns about the scarcity of water, especially in the agriculturally productive dry season period when competition over limited water is particularly fierce between users within and upstream of the delta. It is estimated that more than half of the water used in the irrigation sector in the Lower Mekong Basin is now consumed in the delta alone (Mekong River Commission 2010), making the development strategy of the delta highly risky, and indeed dependent upon the increased dry season flows promised by upstream storage reservoirs.

Whilst the state has long played a lead role in irrigation development and management throughout Southeast Asia, community built and operated irrigation systems often preceded bureaucratic control of water by the state. Community irrigation systems have been subject to significant change accompanying modernist development and the shift from subsistence- to market-oriented, and extensive to intensive agriculture threatening the sustainability and integrity of the collective organization underpinning such systems. These systems, Barker and Molle (2004) suggest, must accommodate more diversified economic activities, address the rising cost of maintenance and confront the rise of water competition from other users as well as the hydrological changes accompanying wider catchment processes, such as deforestation. Barker and Molle (2004) note that traditional rights and rules of water sharing have often not been acknowledged or respected by outsiders and that the imposition of water storages and infrastructure by the state has undermined these systems. They characterize the tension between community managed irrigation systems and state-imposed systems as a "conflict between flexibility and adaptation to microphysical and sociocultural contexts and top-down, capital-intensive, and large-scale macro-strategies of development" (ibid: 8). The value of these context-sensitive and adaptive water use strategies is becoming increasingly important to future livelihood and food security in light of the variability associated with climate change.

Drinking water supply

In addition to meeting the water needs of agriculture in a sustainable and fair manner, the challenge of drinking water supply, especially in urban areas, is considerable. Increased efforts to improve water supply and sanitation (WSS) accompanied the International Drinking Water Supply and Sanitation Decade (1981–90) and the subsequent Millennium Development Goals (MDG) (2000–2015), resulting in a steady rise in access to clean water and, to a lesser extent, sanitation in Southeast Asia. Access to 'improved water sources' in Southeast Asia increased from 71 percent in 1990 to 88 percent in 2010 (UNICEF and World Health Organisation 2012), see

Table 23.3. A note of caution though that behind such aggregate figures the situation of water access is likely to be worse. For instance, though Indonesia reportedly had 78 percent of the total population gaining access to improved water in 2002, only 17 percent had direct household connections (Marcotullio 2007).

The increased availability of clean water has nonetheless contributed to improvements in the quality of life of millions throughout the region, many of whom were previously subject to insecure, polluted, unsafe or difficult to access water, especially in the dry season. Table 23.4 captures key measures of access to improved water and sanitation across Southeast Asia, showing that there remains a continuing lag in the quality of access to water in rural areas (UNICEF and World Health Organisation 2012; Asian Development Bank 2013). Poor quality access to improved sanitation remains an ongoing development challenge. The figures on the incidence of diarrhea clearly indicate that the Least Developed Countries of Southeast Asia (excluding Timor-Leste) have a considerable health burden associated with poor water quality, with diarrhea continuing to contribute to more than 8.5 percent of deaths across the region (World Health Organisation 2012).

Table 23.3 Percentage of Southeast Asian population with access to improved drinking water sources

	Urban	Rural	National
1990	91	62	71
2000	92	72	80
2010	94	83	88

Source: UNICEF and WHO 2012

Table 23.4 Basic water, sanitation and health indicators

Country	Urban pop. with improved water access (%)	Rural pop. with improved water access (%)	Pop. with improved water access (%)	Pop. with access to improved sanitation (%)	Diarrheal incidence per 100,000 people
Brunei	100★	–	100★	80★	94
Cambodia	87	58	64	31	2,170
Indonesia	92	74	82	54	483
Laos	77	62	67	63	1,078
Malaysia	100	99	100	96	181
Myanmar	93	78	93	76	1,551
Philippines	93	92	92	74	528
Singapore	100	n/a	100	100	73
Thailand	97	95	96	96	584
Timor-Leste	91	60	69	47	556
Vietnam	99	93	95	76	296
Region	94	83	88	69	
Source:	UNICEF and WHO 2012; ★ADB 2013	UNICEF and WHO 2012;	UNICEF and WHO 2012; ★ADB 2013	ADB 2013 UNICEF and WHO 2012;	Based on DALY measure, ADB 2013

Source: World Bank (2015). World Development Indicators 2015

Improved water systems reflected in the data in Table 23.4 include, not only piped systems to people's homes, but also public standpipes, boreholes, protected dug wells, protected springs and rainwater collection systems. These systems have brought great health benefits and reduced the labor burden of water collection (especially on women and children). Yet, the ways in which improvements in water supply have occurred are not entirely uncontested and have not always resulted in adequate water services, with actual daily water consumption quite low (Marcotullio 2007). Choices around technology, financing, ownership and management are the subject of considerable debate in the region.

Despite the influence of neoliberal approaches to water in the region water remains largely in community and public hands. Phonm Penh's water agency, for instance, is one of the better performing water utilities in the region, and is publicly owned and operated (Biswas and Tortajada 2010) despite the push to privatize that occurred in the late 1990s–early 2000s. There has, subsequently, been a partial 'strategic retreat' of the private sector from water supply (Bakker 2013), and a number of the original contracts to private concessionaires have been canceled as in the cases of Manila and Jakarta (Pierce 2015). A large part of the controversy associated with water privatization schemes was the hasty manner in which they were implemented, often without trying alternative models of service delivery, without governments maintaining proper regulatory capacity and without considering more culturally appropriate approaches to water (Hirsch, Carrard et al. 2005).

Conclusion: everyday water cultures and relations with water

Cultural connections with water as essential to life and its celebration in water-related ceremonies are found throughout the region and are part of vibrant water-related traditions and everyday water practices. Prominent water festivals in Southeast Asia include the Songkran water festival for Buddhist New Year in Thailand and Laos and the Bon Om Tuok water festival in Cambodia, which celebrates the end of the rainy season and the turning of the Tonle Sap River. Yet in an increasingly urbanized Asia, reconnecting with water through such festivals and re-engaging with an ethics of care that seeks to place human needs in relation to ecological needs and cycles is becoming critical to future sustainable development.

The tropical cycles that define the climate and drive the ecology and hydrology of Southeast Asia are reflected in the everyday practices and relations people have with water. Overtime in response to the considerable seasonal variation in water availability development efforts have shifted from adaptation to control, resulting in the dramatic modification of rivers and catchments through dam construction, irrigation, flood control and coastal defenses. Although by international standards water continues to be a largely plentiful resource in Southeast Asia with industrialization, intensive agriculture, urbanization and decline in ecological systems it is increasingly subject to greater scarcity and competition.

The modernist development paradigm that underpinned intensive and large-scale water use developments throughout Southeast Asia has resulted in profound material changes in water systems and a rise in competition between users and uses. For a region that by international standards has high levels of water availability this rise in competition threatens the sustainability of current approaches. The primary policy response to managing rising scarcity has been to focus upon integration and efficiency. Yet, scarcity is largely socially produced and rethinking the nature of development, and the ecologically destructive and water intensive nature of particular industries and lifestyles, is required in order to reverse these processes that contribute to scarcity. This also provides an opportunity to identify how water use for both human and non-human needs can be met in more adaptive ways. Considering the challenges associated with existing water demands together with climate related risks, turning toward the diverse, extant community-based and context sensitive water relations, institutions and practices found throughout the region is likely to generate valuable approaches that are more adapted and suitable to changing social-ecological and hydrological conditions and processes.

Beyond the material changes associated with the intensification of water use and control accompanying modernist development in the region has been the discursive framing of water as a scarce resource and commodity requiring efficient use. This discourse has led to a reworking of the institutional architecture for water management by the state, private sector and communities, most notably in the area of drinking water provision. Yet this discourse is contested, reflecting the ongoing and awkward coexistence of diverse values and interests in water. Greater attention by policy makers, donors and development scholars on the diverse cultural connections that people in Southeast Asia have with water in everyday life and that support efforts to rekindle social-ecological connections and responsibilities is likely to generate the new understandings and approaches necessary to realize more equitable, adaptive and sustainable water futures.

Notes

1 Economic water scarcity differs from physical water scarcity in that sufficient water is available, but considerable investment in water infrastructure is required to access and use that water (Marcotullio 2007).
2 FAO estimates the cropping intensity for Mainland and Maritime Southeast Asia as 160 percent and 187 percent, respectively, with the world average estimated at 133 percent (FAO AQUASTAT 2015).

References

Adas, M. (1974). *The Burma delta: Economic development and social change on an Asian rice frontier 1852–1941*. Madison: The University of Wisconsin Press.
Asian Development Bank. (2001). *Water for all: The water policy of the Asian development bank*. Manila: Asian Development Bank.
Asian Development Bank. (2013). *Water development outlook*. Manila: Asian Development Bank.
Bakker, K. (2007). "The "Commons"Versus the "Commodity": Alter-globalization, Anti-privatization and the human right to water in the global South. *Antipode*, 39(3), pp. 430–455.
Bakker, K. (2010). *Privatizing water: Governance failure and the world's urban water crisis*. New York: Cornell University Press.
Bakker, K. (2013). Neoliberal versus postneoliberal water: Geographies of privatization and resistance. *Annals of the Association of the American Geographers*, 103(2), pp. 253–260.
Bakker, K., Kooy, M., Shofiani, N. E. and Martijn, E-J. (2008). Governance failure: Rethinking the institutional dimensions of urban water supply to poor households. *World Development*, 36(10), pp. 1891–1915.
Barker, R. and Molle, F. (2004). *Evolution of irrigation in South and Southeast Asia – comprehensive assessment research report 5*. Comprehensive Assessment of Water Management in Agriculture. Colombo, Comprehensive Assessment Secretariat.
Barlow, M. (2001). Water as commodity – The wrong prescription. *Backgrounder*, 7, pp. 1–8.
Barlow, M. (2008). *Blue covenant: The global water crisis and the coming battle for the right to water*. New York: New Press.
Biggs, D. (2010). *Quagmire: Nation building and nature in the Mekong delta*. Seattle: University of Washington Press.
Biggs, D., Miller, F., Hoanh, C.T. and Molle, F. (2009). The delta machine: Water management in the Vietnamese Mekong delta in historical and contemporary perspectives. In: F. Molle, T. Foran and M. Käkönen, eds., *Contested waterscapes in the Mekong region: Hydropower, livelihoods and governance*. London: Earthscan, 203–225.
Biswas, A. K. and Tortajada, C. (2010). Water supply of phnom penh: An example of good governance. *International Journal of Water Resources Development*, 26(2), pp. 157–172.
Brocheux, P. (1995). *The Mekong delta: Ecology, economy, and revolution, 1860–1960*. Madison: University of Wisconsin.

Can, N. D. et al. (2007). Livelihoods and resource use strategies in the Mekong delta. In: T. T. Be., B. T. Sinh and F. Miller, eds., *Challenges to sustainable development in the Mekong delta: Regional and national policy issues and research needs*. Bangkok: The Sustainable Mekong Research Network (Sumernet), 69–98.

Christie, J. W. (1992). Water from the Ancestors: Irrigation in early Java and Bali. In: J. Rigg, ed., *The gift of water: Water management, cosmology and the state in South East Asia*. London: School of Oriental and African Studies, 7–25.

Commission, M. R. (2009). *The flow of the Mekong*. MRC Management Information Booklet Series No. 2. Vientiane, Mekong River Commission.

Coward, E. W. (1980). *Irrigation and agricultural development in Asia: Perspectives from the social sciences*. Ithaca, NY: Cornell University Press.

Coward, E. W. J. (1979). Principles of social organization in an indigenous irrigation system. *Human Organization*, 38(1), pp. 28–36.

Dudgeon, D. (2000). Large-scale hydrological changes in tropical Asia: Prospects for riverine biodiversity. *BioScience*, 50(9), pp. 793–806.

FAO (Food and Agriculture Organisation of the United Nations). (2015). *FAOSTAT*. Available at: www.fao.org/nr/water/aquastat/countries_regions/profile_segments/asiaSE-WR_eng.stm. Water Resources, FAO. Rome, FAO.

Fletcher, R. et al. (2008). The water management network of Angkor, Cambodia. *Antiquity*, 82(317), pp. 658–670.

Geertz, C. (1972). The wet and the dry: Traditional irrigation in Bali and Morocco. *Human Ecology*, 1(1), pp. 23–39.

Geertz, C. (1980). Organization of the Balinese Subak. In: E. W. Coward, ed., *Irrigation and agricultural development in Asia: Perspectives from the social sciences*. Ithaca, NY: Cornell University Press, 70–90.

General Statistics Office (GSO). (2012). *Statistical yearbook*. Hanoi: Statistical Publishing House.

Hirsch, P. et al. (2005). *Water governance in context: Lessons for development assistance, report submitted to AusAID*. Sydney: University of Sydney.

Huu Chiem, N. (1994). Former and present cropping patterns in the Mekong delta. *Southeast Asian Studies*, 31(4), pp. 345–384.

Lansig, S. J. (1987). "Water Temples" and the management of irrigation. *American Anthropological Association*, 89(2), pp. 326–341.

Marcotullio, P. J. (2007). Urban water-related environmental transitions in Southeast Asia. *Sustainability Science*, 2(1), pp. 27–54.

McCully, P. (2001). *Silenced rivers: The ecology and politics of large dams*. London: Zed Books.

Mekong River Commission. (2010). *State of the Basin report 2010*. Vientiane: Mekong River Commission.

Mekong River Commission. (2011). *Assessment of basin-wide development scenarios: Basin development plan programme phase two*. Vientiane: Mekong River Commission.

Miller, F. (2006a). Environmental Risk in Water Resources Management in the Mekong Delta: A Multi-Scale Analysis. *A History of Water: Water Control and River Biographies*. T. Tvedt and E. Jakobsson. London: I.B. Tauris. 1: 172–193.

Miller, F. (2006b). "Seeing Water Blindness": The role of water control in agricultural intensification and responses to environmental change in the Mekong delta, Viet Nam. In: J. Connell and E. Waddell, eds., *Environment, development and change in rural Asia-Pacific: Between local and global*. London: Routledge, 196–207.

Miller, F. (2014). Constructing risk: multi-scale change, livelihoods and vulnerability in the Mekong Delta, Vietnam. *Australian Geographer*, 45(3), pp. 309–324.

Miller, F. and Hirsch, P. (2003). *Civil society and internationalized river Basin management*. Sydney: Australian Mekong Resource Centre, University of Sydney.

Molle, F. (2008). Nirvana concepts, narratives and policy models: Insights from the water sector. *Water Alternatives*, 1(1), pp. 131–156.

Molle, F. and Floch, P. (2008). Megaprojects and social and environmental changes: The case of the Thai "Water Grid." *Ambio*, 37(3), pp. 199–204.

Mukherji, A. et al. (2009). *Revitalizing Asia's irrigation: To sustainably meet tomorrow's food needs*. Colombo, Sri Lanka: International Water Management Institute.

Nguyen Thi Dieu. (1999). *The Mekong river and the struggle for Indochina: Water, war and peace*. Westport: Connecticut, Praeger.

Palmer, L. (2010). Enlivening development: Water management in post-conflict Baucau city, Timor-Leste. *Singapore Journal of Tropical Geography*, 31, pp. 357–370.

Pierce, G. (2015). Beyond the strategic retreat? Explaining urban water privatization's shallow expansion in low- and middle-income countries. *Journal of Planning Literature*, 30(2), pp. 119–131.

Postel, S. (1997). *Last Oasis: Facing water scarcity*. New York: W.W. Norton.

Ricks, J. I. (2015). Pockets of participation: Bureaucratic incentives and participatory irrigation management in Thailand. *Water Alternatives*, 8(2), pp. 193–214.

Rigg, J. (1992). *The gift of water: Water management, cosmology and the state in South East Asia*. London: School of Oriental and African Studies, University of London.

See, K. F. (2015). Exploring and analysing sources of technical efficiency in water supply services: some evidence from Southeast Asian public water utilities. *Water Resources and Economics*, 9(1), pp. 23–44.

Tortajada, C. (2006). Water management in Singapore. *Water Resources Development*, 22(2), pp. 227–240.

UNICEF and World Health Organisation. (2012). *Progress on drinking water and sanitation: 2012 Update*. Geneva: UNICEF.

Wittfogel, K. A. (1956). *The hydraulic civilizations: Man's role in changing the face of the earth*, ed., W. L. Thomas. Chicago: The University of Chicago Press, 152–164.

Wittfogel, K. A. (1957). *Oriental despotism: A comparative study of total power*. New Haven, CT: Yale University Press.

Witthayaphak, C. (1995). *Local institutions in common property resources: A study of community-based watershed management in Northern Thailand*. Paper presented at the Fifth Annual Common Property Conference of the International Association for the Study of Common Property. Bodo, Norway, 1–28.

World Bank. (2015). *World development indicators 2013*. Washington, DC: World Bank. doi 10.1596/978-0-8213-9824-1

World Health Organisation. (2012). *Water-related diseases*. Water Sanitation Health. 2015, Available at: www.who.int/water_sanitation_health/diseases/diarrhoea/en/

Ziv, G. et al. (2012). Trading-off fish biodiversity, food security, and hydropower in the Mekong River Basin. *Proceedings of the National Academy of Sciences of the United States of America (PNAS)*, 109(15), pp. 5609–5614.

24
AGRICULTURE AND LAND IN SOUTHEAST ASIA

Yayoi Fujita Lagerqvist and John Connell

Introduction

Referring to the historical process of agricultural development and industrialization that occurred in the West, the economist Theodore Schultz boldly claimed in 1951, in an article entitled "The Declining Economic Importance of Agricultural Land," that the role of agricultural income diminishes relative to the growth of national income and that, through the introduction of new technology and improved knowledge, agricultural productivity can be improved to overcome the diminishing returns to land. He wrote,

> Land is no longer the limitational factor it once was ... the economy has freed itself from the severe restrictions formerly imposed by land. This achievement is the result of new and better production possibilities and of the path of community choice in relation to these gains. This achievement has diminished greatly the economic dependency of people on land; it has reduced the income claims of this factor to an ever-smaller fraction of the national income; and it has given rise to profound changes in the existing forms of income-producing property. The underlying economic development has modified in an important way and relaxed substantially the earlier iron grip of the niggardliness of Nature.
>
> (Schultz 1951, 725)

Global agricultural productivity has improved since the 1960s as new agricultural technologies were developed and transferred from research stations to the field. Global yields of cereal crops nearly tripled between 1961 and 2014 from 1,422 to 3,886 kg per hectare (World Bank 2015). In the developing countries of East Asia and Latin America, crop yield quadrupled and tripled respectively, yet despite the marked improvement in agricultural productivity during the last five decades land has not been spared or released, but has simply become more valuable. Contrary to the claim made by Schultz more than half a century ago, soaring demand for food and energy has intensified competition for land and resources. This chapter examines contemporary land issues and problems to highlight the interconnectedness of agricultural transformation processes occurring in the region and looming land scarcity, which raises questions on environmental sustainability, equity and growing concern over land grabs.

The next section revisits agrarian transformation in Southeast Asia since the last quarter of the twentieth century till the present day. Key features of the transformation include institutional reforms aimed at establishing property rights, and expansion of commercial agricultural production and trade as Southeast Asia became increasingly integrated into the global market economy. Privatization of land and agricultural intensification have heightened pressures on land as various stakeholders, including the state, corporations, investors and smallholders, compete to legitimize their claims to control land. This has resulted in loss of communal land sharing practices, and collective and reciprocal work in previously marginal areas (e.g., Li 2014a, 2014b). Simultaneously food security has become a significant issue with urbanization, rising demands for food and increased food prices. This affects the most vulnerable population groups who are unable to easily adapt their livelihoods to economic transformations from beyond yet are still dependent on land and natural resources for their survival.

From agrarian society to industrialization: the changing role of agriculture and land

Following WWII, independent states across Southeast Asia adopted land reforms to distribute land to the peasantry and move away from more informal notions of land tenure to formal land ownership (Hayami and Kikuchi 1981; Feeny 1988). Governments directed institutional reforms in different ways, from recognition of private ownership of land to socialist reforms that collectivized ownership. Land reform in some countries, notably the Philippines, failed to remove a very unequal structure of land distribution, which has hampered national economic development and created considerable dissidence and radical social movements.

Since the 1980s, institutional reforms in several countries have promoted formalization of property rights – for both private and public lands – often aided by Western donors and international financial institutions (Table 24.1). These reforms involved cadastral surveys and registration of land parcels, establishing administrative rules that regulated land use and/or ownership rights (Jones 2010). Such reforms were promoted by institutions such as the World Bank to encourage private sector investment and re-allocation of land based on market principles (Deininger 2003). At the same time, a suite of land policies was introduced across the region to demarcate areas of public and private lands, and prescribe who had legitimate access to different types of land.

Alongside institutional reforms that defined people's access to land and resources, development organizations and international research agencies promoted modern agricultural technologies to boost production. The state played a central role in efforts to improve agriculture through the development of irrigation systems, the application of high-yielding varieties and technological innovation. State-led interventions of the 1960s and 1970s in conjunction with Green Revolution technologies contributed to expanding agricultural areas, particularly in the lowlands where staple crops such as rice were farmed. Between 1965 and 2013 the area of cereal crop production doubled in Thailand, while in Indonesia, Myanmar and Vietnam, the area of production increased 1.6 to 1.8 fold (Figure 24.1).

Green Revolution technologies boosted agricultural productivity across Southeast Asia. Annual land productivity between 1970 and 2009 in Southeast Asia increased by 2.2 percent, in comparison with 1.8 and 1.5 percent in Latin America and Sub-Saharan Africa respectively (Briones and Felipe 2013). In terms of cereal crop yields, Vietnam achieved the highest level, of more than 5,425 kg per hectare, followed by Indonesia's 5,085 kg in 2013 (Table 24.2). In

Table 24.1 Land reforms in selected Southeast Asian countries

Country	Year	Program
Indonesia	1960	Basic Agrarian Law, sporadic registration of land
	1995	Land Administration Project, sponsored by the World Bank and implemented by the National Land Agency
Malaysia	1965	National Land Code
Thailand	1954	Land Act, partial recognition of land rights
	1985	Systematic Land Titling, financed by the World Bank and implemented by the Department of Lands, Ministry of Interior
Cambodia	1992	Land Law, registration of land and recognition of use rights
	2001	Land Law, transfer of state public land to private land and allocation of state private land to economic land concessions
Laos	1994	Land Forest Allocation, registration of village land and recognition of communal land use rights
	1997	Land Law; Land Titling Project, sponsored by the World Bank and implemented by the Department of Land, introducing systematic land titling in urban and peri-urban areas and allocating land use rights
Myanmar	1948	Land Nationalization Act, state ownership of land and lease arrangements with farmers
	1953	Land Nationalization Act, state ownership of land and lease arrangements with farmers
	2012	Farmland Law and Vacant, Fallow, and Virgin Land Management Law
	2014	Draft National Land Use Policy
Vietnam	1988	Land Law, Resolution 10, de-collectivization of farms, registration of land and recognition of land use rights
Philippines	1963	Land Reform Program, abolishes share tenancy and establishes owner-operated family size farms
	1971–72	Land reform, allowing farmers to rent and in some cases own land
	1988	Comprehensive Agrarian Reform Program, land distribution and allocation to landless farmers
Timor-Leste	2009	Drafting of Transitional Land Law

Sources: Ledesma (1980); Fujita and Phanvilay (2008); Jones (2010); Prosterman and Vhugen (2012); Thu (2012); Neef et al. (2013); Woods (2013); Lau (2014)

Vietnam the yield particularly surged after the mid-1980s, as the government adopted a market economy, which encouraged active engagement of private sector investors and farmers (Figure 24.2). Laos also experienced a 4.7 fold improvement of cereal crop yield between 1961 and 2013 (Table 24.2, Figure 24.2). As will be discussed later in the chapter, improved accessibility to the market has enabled farmers across the region to increase their use of machinery, fertilizers and pesticides, and to diversify their agricultural activities. This includes planting more than one annual crop and experimenting with genetically modified (GM) seeds.

Improved productivity of agriculture in Southeast Asia has been accompanied by changes in the composition of agriculture between 1970 and 2010, notably the switch from production of cereal crops, such as rice, an essential staple across Southeast Asia, and traditional tropical export products, including coffee, tea, spices, sugar and nuts, to commodities highly sought after in the global market, including oil palm and rubber (Briones and Felipe 2013). Nonetheless rice has remained highly important throughout the region and, in both Thailand and Myanmar, overemphasizing rice production has prevented crop diversification and deprived these economies of

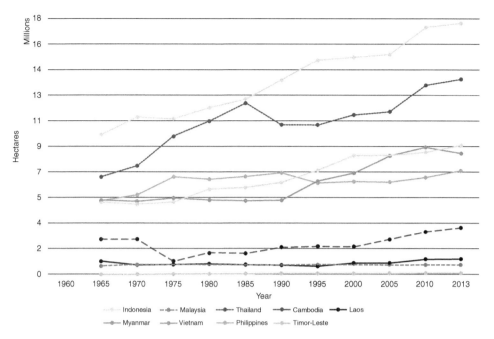

Figure 24.1 Land under cereal crop production for selected Southeast Asian countries
Source: World Bank Data 2015 (excluding Singapore, Brunei Darussalam)

Table 24.2 Yield of cereal crop in selected Southeast Asian countries (Unit: kg/ha)

Country	1961	1970	1980	1990	2000	2010	2013
Indonesia	1,542	2,001	2,866	3,800	4,026	4,878	5,085
Malaysia	2,097	2,383	2,836	2,740	3,040	3,660	3,889
Thailand	1,675	2,076	1,911	2,009	2,719	2,977	3,022
Cambodia	1,127	1,590	1,180	1,362	2,134	3,016	3,117
Laos	884	1,366	1,422	2,268	3,018	3,832	4,150
Myanmar	1,535	1,611	2,600	2,762	3,101	3,863	3,641
Vietnam	1,856	2,104	2,016	3,073	4,112	5,177	5,425
Philippines	996	1,350	1,606	2,065	2,581	3,232	3,532
Timor-Leste	1,404	1,277	1,270	1,608	1,937	2,451	1,880

Source: World Bank Data 2015 (excluding Singapore, Brunei Darussalam)

broad-based agricultural growth, while mismanagement and pricing policies have hampered the returns to agricultural labor. Meanwhile, Indonesia, Malaysia and the Philippines have developed a dependence on imported rice to ensure sufficient national stocks.

Shifts in crop composition occurred in response to the globalization of agricultural trade and investment, and the rapidly growing global demand for food and energy crops during the 1980s and 1990s (McMichael 2012; Cotula 2012). Choices of what crops to grow, how to grow them and where to market them are closely linked to the development of globalised webs of supply and market chains. In particular, oil palm rapidly expanded in Southeast Asia during the 1990s (Koh and Wilcove 2008; Borras and Franco 2010; UNEP 2011), especially in Indonesia and

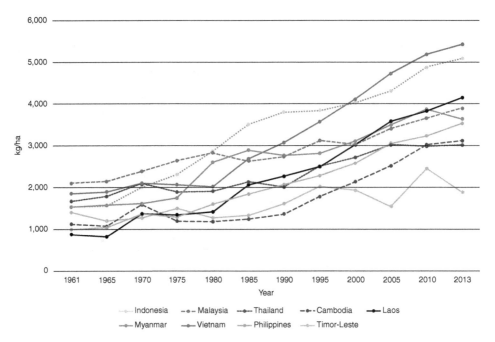

Figure 24.2 Yield of cereal crop production (kg/ha) for selected Southeast Asian countries

Source: World Bank Data 2015 (excluding Singapore, Brunei Darussalam)

Malaysia, where nearly 90 percent of the world's oil palm is produced (McCarthy 2010; Fold and Hansen 2007). Since the 1970s these two countries have lost 30 percent and 20 percent of forest areas respectively, mainly to oil palm, as the resource frontier has extended (Wicke et al. 2011). Demand for oil palm, a valuable biofuel, is projected to grow in the future, and the area of cultivation is expected to expand further in Indonesia and Malaysia, and extend into other Southeast Asian countries including the Philippines, Thailand and Cambodia. The rapid expansion of oil palm in Indonesia and Malaysia has also had effects on land use across the region as oil palm displaced other plantation crops including coconut and rubber. Rubber, crucial in Malaysia and an early component of settlement schemes in Thailand (Connell 1978), has extended further. In particular, fueled by demand in China, rubber has been rapidly expanding in traditionally non-rubber growing countries including Cambodia, Laos and Myanmar.

The expansion of oil palm especially, and of other large-scale commercial crops including coffee, cocoa and rubber, has resulted in massive deforestation, as secondary forests and swidden fields are turned into plantations, and diverse agricultural system transitioned toward monoculture (Padoch et al. 2007; Padoch and Sunderland 2014; Mertz et al. 2009; Hall 2011, Ziegler et al. 2009; Fox et al. 2009; Li and Fox 2012; Harrington 2015; McAllister 2015). Conversely, a general decline in swidden has occurred by choice and under pressure. Throughout the region, a dual household economy has been established where subsistence agriculture is combined with an increased engagement with the market economy (Dove 2011). That transition was most rapid in accessible lowland areas, but is quickly expanding into the upland areas. Processes of land conversion are extremely complex according to differences in topography, ethnicity, traditional ecological knowledge, market integration, government policies on development and land management and the evolution of cross border trade (Pham et al. 2015). Although there are pockets of remote areas where transitions are yet to occur, as in Myanmar's Shan State and northern Laos, where opium poppy cultivation remains significant (UNODC 2014), they too are at the forefront of the market economy (Sturgeon et al. 2013; Woods 2013). While in lowland areas of Thailand where change has been comprehensive, the 'traditional' peasantry have effectively disappeared (Dayley and Sattayanurak 2016).

More or less formal schemes have allocated land in the region, ranging from managed stakeholder schemes (such as FELDA in Malaysia), nucleus-estate smallholder schemes (Indonesia) and joint ventures with customary tenure systems (eastern Malaysia). Settlements have been relatively successful in Malaysia in incorporating the relatively poor but less so elsewhere, where impoverishment has occurred, government services are inadequate and cultural integrity has been lost, as in Indonesia and Laos (Li 2009, 2014a; Evrard and Goudineau 2004; High 2008). Land shortages and infertile land have been the fate of many resettled populations, exacerbating local tensions and disputes (Lagerqvist et al. 2014; Dao 2015, 2016) especially where ethnic minorities have been resettled elsewhere, or where land has been declared 'degraded' or 'state forest' and expropriated by state officials in the name of poverty alleviation with the unfulfilled promise that plantations would provide new wage labor opportunities for those dispossessed (McAllister 2015). Few schemes have met the needs of the poor (e.g., Sutton 2001), and the nucleus estate model has sometimes left customary landowners vulnerable to significant exploitation and losses (Cramb 2013). Although settlers were supposed to focus on agricultural activities, increasing numbers have engaged in non-farm livelihoods outside the settlements and, in Malaysia at least, some settlement land has been sold for urban expansion.

Agricultural change and development have been accompanied by institutional reforms on property rights, intensification of agricultural production through new technology and integration into national and global commodity markets. Alongside policies promoting foreign direct investment and trade, institutional reforms that recognized private property rights facilitated expansion of commercial agriculture. Development efforts have focused on strengthening priority value chains, for various crops from rice and maize to palm oil, coffee, tea and cocoa, to increase farmer productivity and profitability while reducing detrimental environmental effects. Such processes often enabled private investors and powerful elites to acquire land and accumulate wealth, while at the same time excluding some of the more vulnerable population groups. Where agriculture had been collectivized, as in Vietnam, peasants have fought strongly to return to family farming and individual ownership, but through an egalitarian structure (Gorman 2014). Integration into the global market economy, private investment and formalization of property rights have also opened new economic opportunities for smallholders across Southeast Asia and enabled them to stake their claim on land in various ways (Hall 2011; Suhardiman et al. 2015). Yet, transport inadequacies, poor access to credit and extension advice, new seeds, fertilizers and irrigation technology, and limited storage facilities, impede participation of some smallholders in national and international markets.

Institutional reforms and policies that aim to turn land into capital continue to expand agricultural horizons across Southeast Asia in several ways. Lambin and Meyfroidt (2011) point to a "rebound effect," where improved efficiency to produce highly sought after commodities has resulted in expansion rather than containment of agricultural land use. This is exemplified by the expansion of oil palm into marginal agricultural land across Southeast Asia, while oil palm expansion and the parallel expansion of rubber in traditionally non-rubber producing countries highlighted the need to pay attention to the transboundary "displacement effect" of land use transformation (Lambin and Meyfroidt 2011), as changes occurring in one location led to more rather than less cropland expansion in other parts of the region through various inter-connected mechanisms. Moreover, agricultural frontiers are not only being pushed further into the marginal lands by the expansion of large-scale plantations involving private investors and the state, but also through the expansion of smallholder agricultural activities, as households invested their income from non-farm activities into commercial farming: the "remittance effect" (Lambin and Meyfroidt 2011). Even where commercial agriculture exists remittances may constitute as much as 80 percent of rural incomes, especially where migration has resulted in agricultural labor shortages (Neilson and Shonk 2014; Manivong et al. 2014). To better understand this transformation, the next section further examines the changing context of rural livelihoods and the persisting importance of land.

Changing contexts of rural livelihoods and the unchanging importance of land

As countries achieved economic growth, agriculture's role in the national economies of Southeast Asia began to change. Across Southeast Asia, the proportion of rural population is declining rapidly (Table 24.3). Urbanization has been particularly prominent in Malaysia and Indonesia, and also in the Philippines and Thailand. In addition to the declining rural population, in Indonesia, Malaysia and Thailand, agriculture's share of GDP also declined sharply since the 1960s as these countries achieved industrial development and economic growth (de Koninck and Rousseau 2012; Briones and Felipe 2013). Figure 24.3 shows that a parallel decline in agriculture's contribution to the national economies in Cambodia, Laos and Vietnam occurred later, in the 1990s, as these countries adopted variants of market reform and economic growth was assisted by a new influx of private investment. The structural changes in agriculture influenced the context of rural livelihoods in a number of ways.

First, penetration of the market economy since the 1980s has allowed increasing numbers of rural households across Southeast Asia to engage in commodity production and move away from subsistence farming. Entrepreneurial farmers have harnessed agricultural technologies to actively participate in commercial crop production in both lowlands and uplands, and livestock rearing and sales, alongside the 'domestication' and sale of non-timber forest products. Among such farmers are ethnic minorities in mainly mountainous areas who have previously been blamed for resource degradation through their engagement in swidden agriculture in the uplands (Cramb et al. 2009; Cramb and Curry 2012; Fox et al. 2009; Hall 2011; Sturgeon et al. 2013; Li 2014a), but at the same time largely excluded from accessing other land and resources

Table 24.3 Rural population, percentage of total population

Country	1970	2010
Indonesia	83	50
Malaysia	67	29
Thailand	79	56
Cambodia	84	80
Laos	90	67
Myanmar	77	69
Vietnam	82	70
Philippines	67	55
Timor-Leste	87	70

Source: World Bank Data 2015 (excluding Singapore, Brunei Darussalam)

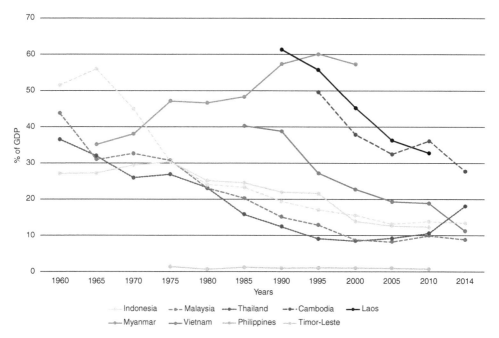

Figure 24.3 Share of agriculture, percentage of GDP for selected Southeast Asian countries
Source: World Bank Data 2015 (excluding Singapore, Brunei Darussalam)

(Peluso 1992; Peluso et al. 1995; McElwee 1999; Poffenberger 1999). Nonetheless upland farmers have increasingly become incorporated into regional and international commodity flows (e.g., Turner 2011; Harrington 2015). However, government pressure on upland farmers in several countries, including Vietnam, Thailand and Laos, to abandon shifting cultivation and to resettle has become a neo-colonial means of control (e.g., Dery 2000; Sikor 2007) alongside 'land grabs' that strengthen state sovereignty in marginal areas (Dao 2015). Remote and 'frontier' land in many regions has been taken over by commercial operations, and in some upland areas people are struggling to retain and formalize their claims to land by cultivating high value crops on what was communal land (Cramb et al. 2009; Cramb and Curry 2012; Hall 2011, 2013; Peluso and Lund 2011). Throughout the region, farmers have negotiated with other actors, both private and state, to legitimize their claims to land and resources, and their abilities to diversify and ingeniously adapt agricultural activities.

Second, transboundary land acquisition for production of food and energy crops, often referred to as international 'land grabs,' is raising new concerns as smallholders are often forcefully displaced from their land and excluded from accessing natural resources that are crucial to their livelihoods (Borras et al. 2011; Baird 2011; Li 2011; McMichael 2012; Neef et al. 2013; Woods 2013; McAllister 2015). A significant part of that dispossession can be attributed to large-scale land acquisition by investors and state authorities (Zoomers 2010; Deininger et al. 2011). 'Land grabs' produce social inequity by delimiting people's access to land, food resources and livelihood opportunities, particularly significant in this century with the growth of agribusinesses in Cambodia; China accounts for half of all foreign investment in land and agricultural projects in both Laos and Cambodia, with Thailand and Vietnam also playing a large role (Polack 2012). National elites play a further role in such 'investment,', and land grabbing is carried out by both domestic and transnational companies, often with encouragement and support from central governments. Most production – food, feed, fuel – is exported within the circuit and logic of the global agricultural-market complex. On occasion land grabs have been so excessive that land has been used highly inefficiently or not at all (Schönweger and Messerli 2015).

Land grabs usually involve large-scale plantations for the cultivation and processing of export commodities, but they also include mines, dams, special economic zones and tourist resorts. Conflicts have emerged when so-called vacant or unused land is in reality informally occupied by smallholder subsistence farmers who must be relocated so that the land can be 'improved' for large-scale commercial use (Lagerqvist et al. 2014). As in Cambodia and Malaysia, governments have often chosen not to recognize customary land tenure and ejected traditional owners, resulting in disputes and violent conflict, as landowners resist removal and relocation and are forcibly removed by government intermediaries, including national armies. Local groups may also come into conflict among themselves, as land acquires value and populations grow, while transactions may be manipulated by local bureaucrats without regard for the social consequences (McCarthy 2010; Li 2002). State intervention in land matters is often more harmful than beneficial (Lunkapis 2013; Urano 2014).

The growth of transnational agriculture-based businesses working in Southeast Asia is reflected in the land acquisitions. Malaysian conglomerates have a majority ownership in approximately two-thirds of companies acquiring land for oil palm in Indonesia. Sime Darby Berhad, a Malaysian company, is the world's largest agriculture-based transnational corporation, holding 633,000 hectares of land within Southeast Asia, mostly for growing palm oil and also rubber. Other ASEAN conglomerates include Thailand's Charoen Pokphand Food Public Company Ltd, which focuses on food processing, livestock and aquaculture. The Philippine company San Miguel Corporation has operations across ASEAN (Polack 2012). These companies, and their patrons, have been the main beneficiaries of the expansion of oil palm.

Third, complex agricultural changes and the acquisition and opening up of new areas have contributed to environmental degradation, especially where forests have been cleared and pesticides and fertilizers are commonly used, as in the Mekong Delta, where salinity has also threatened agricultural systems (Miller 2014). The environmental implications of land grabs, including the loss of biodiversity and regional air and water pollution, have attracted global concern, particularly because the environmental effects of land grabs are often transboundary and widespread, as demands for resources in one location trigger changes in resource use practices elsewhere and disrupt wider food production systems (Lambin and Meyfroidt 2011; Cotula 2012; Meyfroidt et al. 2013; Rulli et al. 2013; Davis et al. 2015; Cramb and Curry 2012; Varkkey 2012). Rice production systems have recently become increasingly threatened by the effects of climate change, a wild card in the region, as a large portion of the rice-growing areas are located in especially vulnerable regions: deltas and coastal lowlands. A number of countries have experienced a gradual stagnation in production levels. In the Philippines, farmers have found it more difficult to maintain rice production levels because they can no longer depend on seasonal rainfall to irrigate paddy fields and have to pump groundwater onto the fields, with negative financial implications. New dams have contributed to water shortages and soil degradation, so that food security becomes a moving target. More adequate land and water governance is crucial with greater demands on uncertain resources. The establishment of national parks has not always sequestered land from use and misuse as encroachment across their boundaries has been substantial. Management plans are often ineffective and ignored by farmers (Duangjai et al. 2015). Alongside forest depletion, this has also meant the loss of biodiversity and conservation areas, reduced potential for adequate watershed management and the increased significance of downstream floods.

Social and environmental implications of capitalist land relations

Amidst the changing context of rural livelihoods, displacement of rural households and proletarianization of labor in resource-rich developing countries has been problematized particularly with the expansion of land grabs across the region, which enable private investors and state authorities to expand agricultural horizons. This section therefore examines the implications of emerging capitalist land relations.

As smallholder farmers expand commercial agricultural activities, alongside the state authorities and large-scale investors, new capitalist land relations are emerging throughout Southeast Asia. Smallholders across Southeast Asia are devising various strategies to adapt their livelihoods simply to survive (Hall 2011; Suhardiman et al. 2015). Financial capital and political connections enable households to position themselves to benefit from the use of their land, whereas households without access to land, finance and connections are often forced off their land and marginalized, or forced to sell their labor to maintain their livelihood. The spread of capitalist land relations is far from allocating resources equitably and promoting efficient use of land, but rather producing land scarcity and entrenching already vulnerable population groups into greater destitution, while customary land sharing practices succumb to private ownership of land (Li 2014a, 2014b).

Strategies to survive intensive competition for land depend on smallholders' existing assets, including political and economic assets. In Laos, for example, households with access to larger agricultural land areas were more easily able to survive land grabs and participate in commercial agriculture, while poor farmers were exposed to market risks and pushed out of agriculture (Suhardiman et al. 2015). Both the relative decline of agricultural income and the improvement of communication networks across the region have triggered the greater mobility of rural populations to engage in off-farm activities both domestically and internationally in order to diversify livelihoods (Rigg 2001, 2006; Kelly 2011; Rigg and Vandergeest 2012). Larger and richer households have disproportionately benefited from this, contributing to new inequalities at village level (e.g., Rungmanee 2014; Thulstrup 2015; Martin and Lorenzen 2016). That has often been paralleled by urban expansion displacing some valuable agricultural land, and both necessitating and encouraging alternative livelihood choices, that older farmers especially are unable to take advantage of (e.g., Kelly 2000; Nguyen et al. 2016) and causing substantial conflict around cities like Ho Chi Minh City (Labbé 2015). With urbanization, the area available for rice paddies is decreasing. Some 50 percent of irrigated cropland in the Philippines has already been lost to urban development. Substantial annual losses have also occurred in Thailand and in Indonesia (Lorenzen 2015).

Although migration and engagement in non-agricultural livelihoods is an integral livelihood strategy for many rural households in Southeast Asia, most people are not entirely moving out of agriculture, but undertaking pluri-active and multi-local strategies (Elmhirst 2012; Rigg and Vandergeest 2012; Cole et al. 2015). In some places, as on the Thai-Lao and Thai-Cambodian borders and for Indonesians in Malaysia, migrants are obtaining higher incomes even in agricultural employment across borders, so contributing to the persistence of smallholder agriculture and the profitability of larger schemes (Estudillo et al. 2013; Rungmanee 2014; Rigg and Vandergeest 2012; Rigg et al. 2016). Income earned outside the village is often used to maintain and sometimes expand agricultural activities on ever scarcer land (Elmhirst 2012; Kelly 2011; Rigg et al. 2016), but also to engage in service industries.

Urbanization and the growth of a middle class have diversified diets away from staple cereals toward the meat, fruit, vegetable and dairy products that wealthier urban consumers prefer (e.g., Reardon et al. 2015). This has resulted in new, intensified specialization and the growth of trade in once minority foods, such as tomatoes, in areas with good infrastructure and access to large cities (Hernandez et al. 2015). Because food is often expensive in cities, and because of high demand for fresh agricultural produce, urban agriculture is a viable but marginal livelihood practice to supply urban food markets, through informal use of vacant sites and alongside urban waterways. Even Brunei and the city-state of Singapore, where over 80 percent of food is imported, have invested in intensive urban agriculture as a gesture toward food sovereignty. Nonetheless, food security is far from universal. As many as 20 million Indonesians are undernourished and many in other countries have limited access to food. Smallholders with less than 2 hectares of land, with limited access to technology, information and markets, form the bulk of the poor and food-insecure.

The future of agriculture, land and people in Southeast Asia

The problem of land raised by Schultz alludes to the increasing pressures on land and diminishing returns from land, and calls for the need to improve agricultural technologies and human capital to overcome the perceived threat of a Malthusian population trap. While diffusion of agricultural technology may have averted the problem of absolute food shortages, pressures on land and the problems of land degradation, distribution, access and scarcity remain of critical concern. More than 55 percent of the area of new agricultural land has come at the expense of intact forest areas in the tropical forest regions (Gibbs et al. 2010), and rapid clearance in Indonesia and Malaysia has brought a suite of negative environmental consequences, locally and regionally. Agricultural intensification and land use zoning in one country can trigger a series of changes elsewhere.

The agricultural frontier has continued to expand in Southeast Asia at the cost of forest, natural resources and the exclusion of the poor. Arable land expansion (and cropping intensity) are reaching their physical limits in many places, while urban growth, industrialization and the use of land for bio-fuels and livestock feed continue to take over lowland agricultural areas. The switch from rubber to oil palm in Malaysia and Indonesia has opened pathways for rural farmers, and a myriad of small and large investors have engaged in rubber plantations across the non-traditional rubber growing regions of mountainous mainland Southeast Asia. These changes have not only triggered a physical transformation of landscape, but have also intensified competition for land and produced problems of land degradation and scarcity across the region.

Emerging contestation over land in Southeast Asia occurs in the wider context of smallholder expansion of high-value crops, and their competition with other forms of land acquisition for food and energy crops, for dams and mines and growing urban areas (Rowcroft 2008; Hall 2011; Li and Fox 2012; Orr et al. 2012; Li 2014a; Leinenkugel et al. 2015; Friis et al. 2016). It can also be understood from a multi-scalar perspective indicating how processes of regionalization and globalization influence decisions to use and manage land, and whether the transformation of land and relationship with land is contributing to further social inequity.

Furthermore, pressures on land in Southeast Asia continue to intensify as urbanization and dams and mines expand, flooding, undermining and transforming agricultural land. Government efforts to promote industrialization in the upper Mekong Region have induced a cascade of effects on downstream farming and fisheries in Cambodia and the Mekong Delta region (Ziv et al. 2012; Orr et al. 2012). Intensified competition for land and water, particularly in mainland Southeast Asia, highlights the complexity of actors involved and the various actions taken to legitimize multiple claims to control resources (Suhardiman and Giordano 2014; Pittock et al. 2015). Intensified pressure on land has meant that rural areas no longer necessarily offer a safety net for migrants who return from the city (Li 2009).

Agriculture's relative importance to national and household economies has diminished in Southeast Asia as the countries industrialize and achieve economic growth. However, agriculture is very far from obsolete, continues to play an important role in national economies and provides the basis of livelihoods for the majority of the population. The critical agrarian question of land, particularly how land is utilized to accumulate wealth, is still relevant in the twenty-first century Southeast Asia (Rigg and Vandergeest 2012; Fairbairn et al. 2014), particularly as land becomes increasingly privatized and commercial agriculture expands. Although penetration of the market economy and improved road access across Southeast Asia have enabled smallholders to participate in commercial agriculture and other non-agricultural economic activities, access to land continues to define and shape household wealth.

Improving agriculture and food systems to combat hunger and poverty remain high on the global development agenda (Griggs et al. 2013; Townsend 2015; United Nations 2015). The new 15-year global commitment to achieve sustainable development, enshrined in the Sustainable Development Goals (https://sustainabledevelopment.un.org/sdgs) not only focuses on producing more food and energy for the growing world population, but also promotes sustainable agriculture, in the face of climatic uncertainties, and equitable access to food and resources. That target is intricately related to concerns over land: overcoming increasing demand for and pressures on land, averting land and other natural resource degradation, and recognizing land tenure (United Nations 2015). Such concerns are pertinent in Southeast Asia, one of the fastest growing economic regions in the world, but one that is experiencing unprecedented rates of resource degradation.

An increasingly wide range of actors, from subsistence farmers, smallholders, investors, agribusiness corporations, national elites and state authorities, have become involved in the structural transformation of agriculture, and in agricultural intensification and land acquisition. Individually and collectively, interactions between actors affect what happens on the land, and are shaped by cultural, economic, physical and political factors. And these are ever changing; as Dayley and Sattayanurak state for Thailand: "The rule-makers who once exploited peasants now court modern farmers with guaranteed subsidies and pledges of state support" (2016, 65). Yet, in this neoliberal age it is unlikely that any subsidy or pledge of state support can be guaranteed. Commercialization of land and agriculture, at the same time as population growth, urbanization and de-agrarianization, constantly restructures local economies, geographies and social relationships. Various local, regional and global factors operate simultaneously, outcomes vary, and no single approach can resolve issues of land grabs and land degradation, access and exclusion, the increasingly inequitable distribution of land and growing concerns over food security. Understanding the role and significance of land for development thus demands a nuanced understanding of social transformations and the agency of various actors at multiple levels as they reshape notions of authority, livelihood, identity and citizenship.

References

Baird, I. (2011). Turning land into capital, turning people into labor: Primitive accumulation and the arrival of large-scale economic land concessions in the Lao People's Democratic Republic. *New Proposals: Journal of Marxism and Interdisciplinary Inquiry*, 5(1), pp. 10–26.

Borras, S. and Franco, J. (2010). Contemporary discourses and contestations around pro-poor land policies and land governance. *Journal of Agrarian Change*, 10(1), pp. 1–32.

Borras, S. M., Hall, R., Scoones, I., White, B. and Wolford, W. (2011). Towards a better understanding of global land grabbing: An editorial introduction. *Journal of Peasant Studies*, 38(2), pp. 209–216.

Briones, R. and Felipe, J. (2013). *Agriculture and structural transformation in developing Asia: Review and outlook*. ADB Economics Working Paper Series, Manila, Philippines, Asian Development Bank.

Cole, R., Wong, G. and Brockhaus, M. (2015). *Reworking the land: A review of literature on the role of migration and remittances in the rural livelihoods of Southeast Asia*. Bogor, Indonesia: Center for International Forestry Research.

Connell, J. (1978). Thailand's Southern Land Settlement Scheme. *Asian Profile*, 6(6), pp. 577–586.

Cotula, L. (2012). The international political economy of the global land rush: A critical appraisal of trends, scale, geography and drivers. *Journal of Peasant Studies*, 39(3–4), pp. 649–680.

Cramb, R. (2013). Palmed off: Incentive problems with joint-venture schemes for oil palm development on customary land. *World Development*, 43, pp. 84–99.

Cramb, R., Colfer, C.J. P, Dressler, W., Laungaramsri, P., Quang, T.L., Mulyoutami, E., Peluso, N.L., and Wadley, R.L. (2009). Swidden Transformations and Rural Livelihoods in Southeast Asia. *Human Ecology*, 37, pp. 323–346.

Cramb, R. and Curry, G. (2012). Oil palm and livelihoods in the Asia-Pacific region: An overview. *Asia Pacific Viewpoint*, 53(3), pp. 223–239.

Dao, N. (2015). Rubber plantations in the Northwest: Rethinking the concept of land grabs in Vietnam. *Journal of Peasant Studies*, 42(2), pp. 347–369.

Dao, N. (2016). Political responses to dam-induced resettlement in Northern Uplands Vietnam. *Journal of Agrarian Studies*, 16(2), pp. 291–317.

Davis, K., Rulli, M. and D'Odorico, P. (2015). The global land rush and climate change. *Earth's Future*, 3, 298–311.

Dayley, R. and Sattayanurak, A. (2016). Thailand's Last Peasant. *Journal of Southeast Asian Studies*, 47(1), pp. 42–65.

De Koninck, R. and Rousseau, J-F. G. (2012). *Gambling with the land: The contemporary evolution of Southeast Asian agriculture*. Singapore: NUS Press.

Deininger, K. (2003). *Land policies for growth and poverty reduction*. Washington, DC: World Bank and Oxford University Press.

Deininger, K., Byerlee, D., Lindsay, J., Norton, A., Selod, H. and Stickler, M. (2011). *Rising global interest in farmland: Can it yield sustainable and equitable benefits?* Washington, DC: World Bank.

Dery, S. (2000). Agricultural colonisation in Lam Dong Province, Vietnam. *Asia Pacific Viewpoint*, 41(1), pp. 35–49.

Dove, M. (2011). *The Banana tree at the gate: A history of marginal people and global markets in Borneo*. Singapore: NUS Press.

Duangjai, W., Schmidt-Vogt, D. and Shresthra, R. (2015). Farmers' land use decision-making in the context of changing land and conservation polices: A case study of Doi Mae Salong in Chiang Rai Province, Northern Thailand. *Land Use Policy*, 48, pp. 179–189.

Elmhirst, R. (2012). Displacement, resettlement and rural change in Southeast Asia. *Critical Asian Studies*, 44(1), pp. 131–152.

Estudillo, J., Mano, Y. and Seng-Arloun, S. (2013). Job choice of three generations in rural Laos. *Journal of Development Studies*, 49(7), pp. 991–1009.

Evrard, O. and Goudineau, Y. (2004). Planned resettlement, unexpected migrations and cultural trauma in Laos. *Development and Change*, 35(5), pp. 937–962.

Fairbairn, M., Fox, J., Isakson, S. R., Levien, M., Peluso, N., Razavi, S., Scoones, I. and Sivaramakrishnan, K. (2014). Introduction: New directions in agrarian political economy. *Journal of Peasant Studies*, 41(5), pp. 653–666.

Feeny, D. (1988). The development of property rights in land: A comparative study. In: R. H. Bates, ed., *Towards a political economy of development: A rational choice perspective*. Berkeley: University of California Press, pp. 272–299.

Fold, N. and Hansen, T. (2010). Oil palm expansion in Sarawak: Lessons learned by a latecomer? In: J. Connell and E. Waddell, eds., *Environment development and change in rural Asia-Pacific*. London: Routledge, 147–166.

Fox, J., Fujita, Y., Ngidang, D., Peluso, N., Potter, L., Sakuntaladewi, N., Sturgeon, J. and Thomas, D. (2009). Policies, political-economy, and Swidden in Southeast Asia. *Human Ecology*, 37(3), pp. 305–322.

Friis, C., Reenberg, A., Heinimann, A. and Schönweger, O. (2016). Changing local land systems: Implications of a Chinese rubber plantation in Nambak District, Lao PDR. *Singapore Journal of Tropical Geography*, 37(1), pp. 25–42.

Fujita, Y. and Phanvilay, K. (2008). Land and forest allocation in Lao people's democratic republic: Comparison of case studies from community-based natural resource management research. *Society & Natural Resources*, 21(2), pp. 120–133.

Gibbs, H. K., Ruesch, A. S., Achard, F., Clayton, M. K., Holmgren, P., Ramankutty, N. and Foley, J. A. (2010). Tropical forests were the primary sources of new agricultural land in the 1980s and 1990s. *Proceedings of the National Academy of Sciences*, 107(38), pp. 16732–16737.

Gorman, T. (2014). Moral economy and the upper peasant: The dynamics of land privatization in the Mekong delta. *Journal of Agrarian Change*, 14(4), pp. 501–521.

Griggs, D., Stafford-Smith, M., Gaffney, O., Rockström, J., Öhman, M. C., Shyamsundar, P., Steffen, W., Glaser, G., Kanie, N. and Noble, I. (2013). Policy: Sustainable development goals for people and planet. *Nature*, 495(7441), pp. 305–307.

Hall, D. (2011). Land grabs, land control, and Southeast Asian crop booms. *Journal of Peasant Studies*, 38(4), pp. 837–857.

Hall, D. (2013). *Land*. Oxford: Wiley.

Harrington, M. (2015). 'Hanging by the Rubber': How cash threatens the agricultural systems of the Siang Dayak. *Asia Pacific Journal of Anthropology*, 16(5), pp. 481–495.

Hayami, Y. and Kikuchi, M. (1981). *Asian village economy at the crossroads: An economic approach to institutional change*. Tokyo: University of Tokyo Press.

Hernandez, R., Reardon, T., Natawidjaja, R. and Shetty, S. (2015). Tomato farmers and modernising value chains in Indonesia. *Bulletin of Indonesian Economic Studies*, 51(3), pp. 425–444.

High, H. (2008). The implications of aspirations: Reconsidering resettlement in Laos. *Critical Asian Studies*, 40(4), pp. 531–550.

Jones, D. S. (2010). Land registration and administrative reform in Southeast Asian states: Progress and constraints. *International Public Management Review*, 11(1), pp. 67–89.

Kelly, P. (2000). *Landscapes of globalisation, human geographies of economic change in the Philippines*. London: Routledge.

Kelly, P. F. (2011). Migration, agrarian transition, and rural change in Southeast Asia. *Critical Asian Studies*, 43(4), pp. 479–506.

Koh, L. P. and Wilcove, D. S. (2008). Is oil palm agriculture really destroying tropical biodiversity? *Conservation Letters*, 1(2), pp. 60–64.

Labbé, D. (2015). Media dissent and peri-urban land struggles in Vietnam: The case of the Van Giang Incident. *Critical Asian Studies*, 47(4), pp. 495–513.

Lagerqvist, Y., Woollacott, L., Phasouysaingam, A. and Souliyavong, S. (2014). Resource development and the perpetuation of poverty in rural Laos. *Australian Geographer*, 45(3), pp. 407–417.

Lambin, E. F. and Meyfroidt, P. (2011). Global land use change, economic globalization, and the looming land scarcity. *Proceedings of the National Academy of Sciences*, 108(9), pp. 3465–3472.

Lau, S. (2014). *Land reform in Myanmar and lessons learned from Southeast Asia*. Adelaide: Institute for International Development.

Ledesma, A. J. (1980). Land reform in East and Southeast Asia: A comparative approach. *Philippine Studies*, 28(4), pp. 451–481.

Leinenkugel, P., Wolters, M. L., Oppelt, N. and Kuenzer, C. (2015). Tree cover and forest cover dynamics in the Mekong Basin from 2001 to 2011. *Remote Sensing of Environment*, 158, pp. 376–392.

Li, T. (2002). Local histories, global markets: Cocoa and class in upland Sulawesi. *Development and Change*, 33(3), pp. 415–437.

Li, T. (2009). Exit from agriculture: A step forward or a step backward for the rural poor? *Journal of Peasant Studies*, 36(3), pp. 629–636.

Li, T. (2011). Centering labor in the land grab debate. *Journal of Peasant Studies*, 38(2), pp. 281–298.

Li, T. (2014a). *Land's end: Capitalist relations on an indigenous frontier*. Durham, NC: Duke University Press.

Li, T. (2014b). What is land? Assembling a resource for global investment. *Transactions of the Institute of British Geographers*, 39(4), 589–602.

Li, Z. and Fox, J. M. (2012). Mapping rubber tree growth in mainland Southeast Asia using time-series MODIS 250 m NDVI and statistical data. *Applied Geography*, 32(2), pp. 420–432.

Lorenzen, R. (2015). Disintegration, formalisation or reinvention? Contemplating the future of Balinese irrigated rice societies. *Asia Pacific Journal of Anthropology*, 16(2), pp. 176–193.

Lunkapis, J. (2013). Confusion over land rights and development opportunities through communal titles in Sabah, Malaysia. *Asia Pacific Viewpoint*, 54(2), pp. 198–205.

Manivong, V., Cramb, R. and Newby, J. (2014). Rice and remittances: Crop intensification versus labour migration in Southern Laos. *Human Ecology*, 42(3), pp. 367–379.

Martin, S. and Lorenzen, K. (2016). Livelihood diversification in rural Laos. *World Development*, 83, pp. 231–243.

McAllister, K. (2015). Rubber, rights and resistance: The evolution of local struggles against a Chinese rubber concession in Northern Laos. *Journal of Peasant Studies*, 42(3–4), pp. 817–837.

McCarthy, J. (2010). Processes of inclusion and adverse incorporation: Oil palm and agrarian change in Sumatra, Indonesia. *Journal of Peasant Studies*, 37(4), pp. 821–850.

McElwee, P. (1999). Policies of prejudice: Ethnicity and shifting cultivation in Vietnam. *Watershed*, 5, pp. 30–38.

McMichael, P. (2012). The land grab and corporate food regime restructuring. *Journal of Peasant Studies*, 39(3–4), pp. 681–701.

Mertz, O., Padoch, C., Fox, J., Cramb, R. A., Leisz, S. J., Nguyen, T. L. and Tran, D.V. (2009). Swidden change in Southeast Asia: Understanding causes and consequences. *Human Ecology*, 37, pp. 259–264.

Meyfroidt, P., Lambin, E. F., Erb, K-H. and Hertel, T. W. (2013). Globalization of land use: Distant drivers of land change and geographic displacement of land use. *Current Opinion in Environmental Sustainability*, 5(5), pp. 438–444.

Miller, F. (2014). Constructing risk: Multi-scale change, livelihoods and vulnerability in the Mekong Delta, Vietnam. *Australian Geographer*, 45(3), pp. 309–324.

Neef, A., Touch, S. and Chiengthong, J. (2013). The politics and ethics of land concessions in rural Cambodia. *Journal of Agricultural and Environmental Ethics*, 26(6), pp. 1085–1103.

Neilson, J. and Shonk, F. (2014). Chained to development? Livelihoods and global value chains in the coffee-producing Toraja region of Indonesia. *Australian Geographer*, 45(3), pp. 269–288.

Nguyen, T., Tran, V., Bui, Q., Man, Q. and Walter, T. (2016). Socio-economic effects of agricultural land conversion for urban development: Case study of Hanoi, Vietnam. *Land Use Policy*, 54, pp. 583–592.

Orr, S., Pittock, J., Chapagain, A. and Dumaresq, D. (2012). Dams on the Mekong river: Lost fish protein and the implications for land and water resources. *Global Environmental Change*, 22(4), pp. 925–932.

Padoch, C., Coffey, K., Mertz, O., Leisz, S., Fox, J. and Wadley, R. (2007). The demise of Swidden in Southeast Asia? Local realities and regional ambiguities. *Geografisk Tidsskrift, Danish Journal of Geography*, 107(1), pp. 29–41.

Padoch, C. and Sunderland, T. (2014). Managing landscapes for greater food security and improved livelihoods. *Unasylva*, 64(241), pp. 3–13.

Peluso, N. (1992). *Rich forests, poor people: Resource control and resistance in Java*. Berkeley: University of California Press.

Peluso, N. and Lund, C. (2011). New frontiers of land control: Introduction. *Journal of Peasant Studies*, 38(4), pp. 667–681.

Peluso, N. L., Vandergeest, P. and Potter, L. (1995). Social aspects of forestry in Southeast Asia: A review of postwar trends in the scholarly literature. *Journal of Southeast Asian Studies*, 26(1), pp. 196–218.

Pham, T., Turner, S. and Trincsi, K. (2015). Applying a systematic review to land cover change in Northern Upland Vietnam: The missing case of the Borderlands. *Geographical Research*, 53(4), pp. 419–435.

Pittock, J., Orr, S., Stevens, L., Aheeyar, M. and Smith, M. (2015). Tackling trade-offs in the nexus of water, energy and food. *Aquatic Procedia*, 5, pp. 58–68.

Poffenberger, M. (ed.) (1999). *Communities and forest Management in Southeast Asia*. Switzerland: Gland, Ford Foundation, DFID, Asia Forest Network and IUCN.

Polack, E. (2012). *Agricultural land acquisitions: A lens on Southeast Asia*. London: International Institute for Environment and Development.

Prosterman, R. and Vhugen, D. (2012). Land to the tillers of Myanmar. *New York Times*, 13 June, 57–68.

Reardon, T., Stringer, R., Timmer, C., Minot, N. and Daryanto, A. (2015). Transformation of the Indonesian agrifood system and the future beyond rice. *Bulletin of Indonesian Economic Studies*, 51(3), pp. 369–373.

Rigg, J. (2001). *More than the soil: Rural change in Southeast Asia*. Essex: Pearson Education Limited.

Rigg, J. (2006). Land, farming, livelihoods and poverty: Rethinking the links in the Rural South. *World Development*, 34, pp. 180–202.

Rigg, J., Salamanca, A. and Thompson, E. (2016). The puzzle of East and Southeast Asia's persistent smallholder. *Journal of Rural Studies*, 43, pp. 118–133.

Rigg, J. and Vandergeest, P. (eds.) (2012). *Revisiting rural places: Pathways to poverty and prosperity in Southeast Asia*. Singapore: NUS Press.

Rowcroft, P. (2008). Frontiers of change: The reasons behind land-use change in the Mekong Basin. *Ambio*, 37(3), pp. 213–218.

Rulli, M., Saviori, A. and Odorico, P. D. (2013). Global land and water grabbing. *Proceedings of the National Academy of Sciences*, 110(3), pp. 892–897.

Rungmanee, S. (2014). The dynamic pathways of agrarian transformation in the Northeastern Thai-Lao Borderlands. *Australian Geographer*, 45(3), pp. 341–354.

Schönweger, O. and Messerli, P. (2015). Land acquisition, investment and development in the Lao coffee sector: Successes and failures. *Critical Asian Studies*, 47(1), pp. 94–122.

Schultz, T. W. (1951). The declining economic importance of agricultural land. *The Economic Journal*, 61(244), pp. 725–740.

Sikor, T. (2007). Land reform and the state in Vietnam's northwestern mountains. In: J. Connell and E. Waddell, eds., *Environment development and change in rural Asia-Pacific*. London: Routledge, 90–107.

Sturgeon, J., Menzies, N., Fujita Lagerqvist, Y., Thomas, D., Ekasingh, B., Lebel, L., Phanvilay, K. and Thongmanivong, S. (2013). Enclosing Ethnic Minorities and Forests in the Golden Economic Quadrangle. *Development and Change*, 44(1), pp. 53–79.

Suhardiman, D. and Giordano, M. (2014). Legal plurality: An analysis of power interplay in Mekong hydropower. *Annals of the Association of American Geographers*, 104(5), pp. 973–988.

Suhardiman, D., Giordano, M., Keovilignavong, O. and Sotoukee, T. (2015). Revealing the hidden effects of land grabbing through better understanding of farmers' strategies in dealing with land loss. *Land Use Policy*, 49, pp. 195–202.

Sutton, K. (2001). Agribusiness on a grand scale - FELDA's Sahabat complex in East Malaysia. *Singapore Journal of Tropical Geography*, 22(1), pp. 90–105.

Thu, P. M. (2012). Access to land and livelihoods in post-conflict Timor-Leste. *Australian Geographer*, 43(2), pp. 197–214.

Thulstrup, A. (2015). Livelihood resilience and adaptive capacity: Tracing changes in household access to capital in central Vietnam. *World Development*, 74, pp. 352–362.

Townsend, R. (2015). *Ending poverty and hunger by 2030*. Washington, DC: World Bank Group.

Turner, S. (2011). Making a living the Hmong way: An actor oriented livelihoods approach to everyday politics and resistance in Upland Vietnam. *Annals of the Association of American Geographers*, 102(2), pp. 403–422.

UNEP. (2011). *Oil palm plantations: Threats and opportunities for tropical ecosystems*, UNEP Global Environmental Alert Services (GEAS). Nairobi, United Nations Environment Programme.

United Nations. (2015). *Transforming our world: The 2030 agenda for sustainable development*. G. Assembly. New York: United Nations.

UNODC. (2014). *Southeast Asia opium survey*. Bangkok: UNODC.

Urano, M. (2014). Impacts of newly liberalised policies on customary land rights of forest-dwelling populations: A case study from East Kalimantan, Indonesia. *Asia Pacific Viewpoint*, 55(1), pp. 6–23.

Varkkey, H. (2012). Patronage politics as a driver of economic regionalisation: The Indonesian oil palm sector and transboundary haze. *Asia Pacific Viewpoint*, 53(3), pp. 314–329.

Wicke, B., Sikkema, R., Dornburg, V. and Faaij, A. (2011). Exploring land use changes and the role of palm oil production in Indonesia and Malaysia, Land Use Policy, 8, 193–206.

Woods, K. (2013). *The politics of the emerging agro-industrial complex in Asia's 'final frontier': The war on food sovereignty in Burma*. Food Sovereignty: A Critical Dialogue. International Conference, Yale University.

World Bank. (2015). *World development indicators*. Washington, DC: World Bank.

Ziegler, A. D., Fox, J. and Xu, J. (2009). The rubber juggernaut. *Science*, 324, pp. 1024–1025.

Ziv, G., Baran, E., Nam, S., Rodríguez-Iturbe, I. and Levin, S. A. (2012). Trading-off fish biodiversity, food security, and hydropower in the Mekong River Basin. *Proceedings of National Academy of Science*, 109(15), pp. 5609–5614.

Zoomers, A. (2010). Globalisation and the foreignisation of space: Seven processes driving the current global land grab. *Journal of Peasant Studies*, 37(2), pp. 429–447.

25
LABOR, SOCIAL SUSTAINABILITY AND THE UNDERLYING VULNERABILITIES OF WORK IN SOUTHEAST ASIA'S SEAFOOD VALUE CHAINS

Simon R. Bush, Melissa J. Marschke and Ben Belton

Introduction

The seafood industry, comprised of capture fisheries, aquaculture and their supporting value chains, plays a major role in the economy and society of Southeast Asia. Fish is the most important source of animal protein in many countries in the region (Belton and Thilsted 2014) and plays a central role in cuisine and culture in both inland and coastal areas. Expansion of the region's seafood sector has seen sustained increases in capture fisheries output and, more recently, the meteoric rise of aquaculture, with reported growth in the two sub-sectors averaging 2.9 percent and 9.7 percent per annum, respectively since 1990 (Figure 25.1). Fish make vital contributions to livelihoods and nutrition in vulnerable rural and coastal communities, but are also important to the urban middle class, whose growth – to include a projected 66 percent of the region's population by 2030 (Kharas 2010) – is likely to result in even higher levels of demand for seafood products (Hall et al. 2011). The region is also a major supplier of seafood exports, such as farmed shrimp and wild caught tuna, to the Global North.

The growth and significance of seafood production has given rise to a range of sustainability concerns in both the capture and aquaculture sub-sectors that are common across Southeast Asia. According to FAO statistics, landings from capture fisheries in Southeast Asia have been fished down to five to 30 percent of their unexploited levels, and in the Gulf of Thailand alone biomass is estimated to be only 8 percent of 1965 levels (Salayo et al. 2008). In addition, the aquaculture 'boom' has occurred across the region in a largely uncoordinated manner, leading to the loss of critical habitat and biodiversity, water pollution and problems with aquatic animal disease (e.g., Hall et al. 2011; Richards and Friess 2015). These highly visible environmental impacts have resulted in efforts to foster more sustainable supplies of seafood for both regional food security and export trade through private and public forms of regulation that have aimed at improving resource management and productivity. Consequently, social dimensions of sustainability, including issues such as well-being, human rights and gender equality, have historically

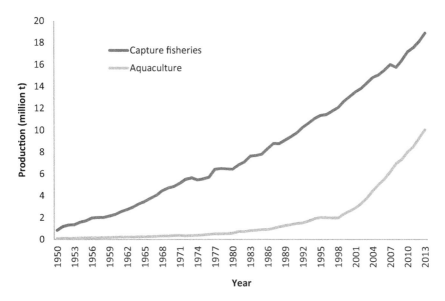

Figure 25.1 Capture fisheries and aquaculture output in Southeast Asia, 1950–2013
Source: Fish Stat J online database (FAO 2016)

been largely overlooked by activists, policy makers and academics alike, as has their fundamental relationship to the equitable use and stewardship of marine and coastal resources.

This unbalanced understanding of the relationship between social and environmental sustainability has recently begun to shift as labor conditions in Southeast Asia's seafood sector have come under increasing scrutiny, following the success of the Environmental Justice Foundation's "Seafood not Slavefood" campaign (see EJF 2015) and a series of high profile exposés by investigative journalists for the *Guardian*, *New York Times* and Associated Press (Table 25.1). The most striking of these has exposed conditions of 'slave labor' in Thai off-shore fisheries and shrimp peeling sheds. Although labor rights abuses in the sector were reported earlier (Derks 2010; Pearson et al. 2006), these recent stories have been highly influential in placing labor and the conditions under which it is employed at the center of debates over seafood sustainability. This scrutiny has led to a series of rapid responses by seafood suppliers, governments, retailers and NGOs in both exporting Southeast Asian nations and importing countries in the North. But while these responses have focused immediate attention on labor rights abuses, they have not necessarily dealt with the underlying causes of which these abuses are symptomatic – including, as argued by Rigg (2016), unequal levels of economic development across the region that push members of already vulnerable populations to migrate.

Understanding the underlying causes of unsustainable labor practices requires a multi-dimensional perspective on how production and supply chain level decisions influence both social and environmental sustainability in globally connected sectors like fisheries and aquaculture. In this chapter we adopt a value chain approach to understand how the agency of firms and workers are structured by both inter-firm power relations and 'extra-transactional' actors and institutions (Barrientos et al. 2011; Riisgaard 2009), and their implications for social and environmental sustainability in seafood production. Building on 'tiered' approaches to understanding internal and external influences over the structure and function of value chains (e.g., Ponte and Sturgeon

Table 25.1 Timeline of key media, NGO and academic coverage

Year	Coverage	Note
2010	Academic (c.f., A. Derks, 2010; Resurrección & Sajor 2010)	Detailed analysis of Khmer fishers in Thailand's off-shore sector and migrants working in Thailand's shrimp farms
2013	NGO (EJF starts "Seafood not Slavefood" campaign	A sustained advocacy campaign with original, detailed research into human trafficking in Thailand's off-shore fisheries
2014	Media (*Guardian* investigation)	Traced one supply chain to illustrate how inhumane labor was part of shrimp supply chains sold to the EU within its "Modern Day Slavery" series
2014	US TIP 2014 report	Thailand downgraded to Tier 3 status
2014	Creating of Project Issara and the "Sustainable Shrimp Task Force"	The business community begins to work together to better understand labor abuse in supply chains and begins providing funds to support workers
2015	Media (AP investigation)	Exposed human trafficking on Thai boats in off-shore waters in Indonesia
2015	EU Yellow Card	Thailand given six months to clean up Thailand's fisheries influencing a major policy reorientation (including labor)
2015	Media (*New York Times* series)	Explored shady recruitment practices to trick migrants into off-shore fisheries, as part of its "The Outlaw Ocean" series
2015	US TIP 2015 report	Thailand's Tier 3 status continues
2015	Verite report	Nestle admits to labor abuse within its pet food supply chain; coincides with several US lawsuits against major retailers for drawing on 'slave' labor in their supply chains
2016	Seafood Summit	At major Seafood advocacy event, retailers and the business community acknowledge that social issues in seafood need to be prioritized
2016	Associated Press	Wins Pulitzer Prize for Public Service (based on "Seafood from Slaves" investigation)

2013; Bush et al. 2014), we identify three interlinked 'sites' of labor exploitation that are key in shaping the social sustainability of the seafood industry (see Figure 25.2).

First, we focus, at the 'micro' scale, on the conduct of value chain actors (e.g., operators of fishing boats, fish farms and fish processing facilities), giving attention to the conditions under which labor is engaged and work is performed, as well as to the agency of workers themselves. Second, at the 'meso' level, we address the overall performance of seafood value chains in terms of inclusivity and equity, asking who is excluded or incorporated and how benefits and costs are distributed among actors. Third, we assess the 'macro' political economic and geographical context in which Southeast Asian seafood value chains are situated, and the ways in which this influences the conduct of actors in the chain and its performance as a whole.

Based on this three-tiered analysis we also address the range of market, state and non-state driven regulatory mechanisms that may be brought to bear in ensuring greater social sustainability at each of these interlinked sites. In doing so, we raise the 'social' out of the sustainability shadows and draw attention to the fact that seafood sustainability is as fundamentally dependent

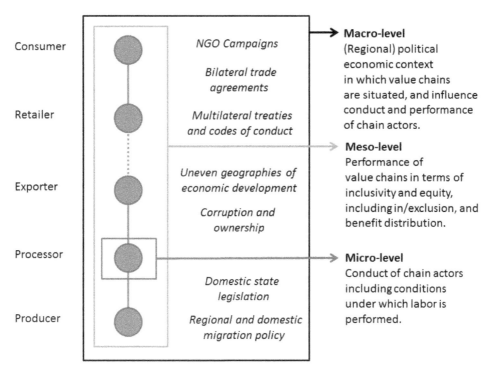

Figure 25.2 Social issue areas in the seafood employment chain

on the well-being of the people who produce it, as it is on the ecological viability of fisheries and farming systems.

Labor and value chain conduct

The multi-sited nature of the seafood industry means that abuses of labor are possible at multiple value chain nodes (e.g., boats/farms, harbors/collection points/factories and the transport that connects them), falling under the control of a variety of interlinked but often separate actors. This focus on nodes corresponds to the 'micro' level of analysis identified in Figure 25.2. The seafood sub-sectors of capture fisheries and aquaculture are also interlinked, with marine fish harvested by unfree labor on fishing vessels used as an input, in the form of feed, into farmed shrimp and fish value chains with their own labor problems. We use the term 'unfree' to refer to labor subject to varying degrees of coercion or restricted freedom, from either economic and state actors, that act to prevent voluntary exit from employment (see Strauss 2012). These interlinkages have been highlighted in the media, with off-shore fishing vessels that abused and held workers captive, selling low value, non-target fish to feed-mills that subsequently supplied major aquaculture producers including, most notably, the Thai conglomerate Charoen Pokphand (CP) which accounts for 10 percent of Thailand's shrimp exports, with customers including major northern retailers such as Wal-Mart, Costco and Carrefour (Marschke and Vandergeest 2016). These nodes have thus become sites of concern for regulators because of the lack of transparency around both who is engaged in labor and the conditions under which it is performed.

Despite the wealth that is generated from fisheries across Southeast Asia and the large number of people working in the sector (an incomplete estimate of employment in capture fisheries, aquaculture and related value chains in five countries in the region suggests that they employ over 21 million people – Table 25.2), labor practices often remain poorly regulated, if at all. The limited enforcement of what little regulation exists reflects, in part, the difficulty in monitoring value chain nodes rendered 'invisible' by the multiplicity of sites in which it occurs, the physical distance of off-shore fishing activity and the political distance created by elite ownership and control of the industry (Butcher 2004; Lebel et al. 2002).

This situation is compounded by the uneven geography of development in Southeast Asia. As seen in a number of sectors, relatively more affluent countries such as Thailand and Malaysia deploy labor in seafood value chains originating from neighboring countries with generally lower levels of economic opportunity like Cambodia, Laos and Myanmar. Work performed by these migrants is often less visible and less regulated than the formal or semi-formal work performed by citizens of receiving economies (Ananta and Arifin 2004). Indeed, as De Genova (2002) argues, and as we go on to describe, the active disempowerment of such international migrants within the region can be seen as systemic; an implicit goal of national policies designed to both regulate and discipline flows of migrant workers.

Thus, while international buyers, certification bodies and environmental NGOs alike have often assumed that national legislation or international labor policy is sufficient to ensure compliance with labor standards, the opposite has often proved the case. This is not to suggest that the issue of labor is absent from fisheries-related policies: Southeast Asian countries including Thailand and Vietnam have an extensive set of national labor-related policies linked to fisheries, including labor codes, anti-trafficking policy and safety-at-sea policy that have been updated and revised in recent years. However, these existing instruments are poorly enforced, largely because they defer to international normative frameworks that are vaguely written and provide

Table 25.2 Employment in capture fisheries in selected Southeast Asian countries

Item	Marine capture	Freshwater capture	Aquaculture	Total
		Cambodia		
Fishing/farming	39,534	597,002	No data	636,536
Other jobs	37,879	962,121	No data	1,000,000
Total	77,413	1,559,123	No data	1,636,536
		Indonesia		
Fishing/farming	1,841,909	555,000	2,797,005	5,193,914
Other jobs	784,000	382,000	1,215,258	2,381,258
Total	2,625,909	937,000	4,012,263	7,575,172
		Philippines		
Fishing/farming	1,388,173	226,195	605,336	2,219,704
Other jobs	No data	No data	222,026	222,026
Total	1,388,173	226,195	827,362	2,441,730
		Thailand		
Fishing/farming	168,140	3,131,355	217,883	3,517,378
Other jobs	No data	No data	145,190	145,190
Total	168,140	3,131,355	363,073	3,662,568
		Vietnam		
Fishing/farming	607,100	2,834,238	1,744,900	5,186,238
Other jobs	No data	No data	900,056	900,056
Total	607,100	2,834,238	2,644,956	6,086,294
Grand total	**4,866,735**	**8,546,732**	**7,847,654**	**21,261,121**

Source: World Bank 2008; Phillips et al. in press

little to no guidance on implementation or enforcement. For example, the Food and Agriculture Organization (FAO) 1995 voluntary Code of Conduct for Responsible Fisheries specifies that states should protect the rights of fishers and fish workers as a basis for addressing labor and employment conditions, but provides no further detail on the kinds of arrangements necessary to provide this protection. Furthermore, frameworks that do exist, such as the 2007 International Labor Organization (ILO) Work in Fishing Convention 188 that focuses specifically on fish worker rights (ILO 2007) have yet to be signed by any Southeast Asian country.

Private third party certification has the potential to serve as an alternative but complementary tool for regulating conditions of employment in the seafood industry, for states and the industry alike. However, the main wild caught fisheries certification scheme, the Marine Stewardship Council (MSC), has excluded social standards from its program, reflecting the environmental bias in seafood sustainability initiatives (Marschke and Vandergeest 2016). Another initiative, the Friend of the Sea's Sustainable Seafood Program does integrate social dimensions (Blackmore et al. 2015; FOS 2013), including basic labor standards, but is not regarded as credible as the MSC because of its weaker auditing requirements (Auld 2014). Aquaculture certifiers have been more rigorous overall in creating standards for a range of social issues, focused predominantly on labor, but also extending to community conflict. In many instances, these private social standards make reference to ILO or state legislation, for example in regards to labor conditions and conflict resolution over land resources, and, in doing so, may support the enforcement of national and international public rule making.

Is there a role for industry and the private sector in nudging the social agenda forward? For actors in supply chains serving northern markets there may be potential to, if nothing else, uphold buyer and retailer reputations in the EU or North America. For example, a group of seafood producers, major northern-based retailers, government units and NGOs (including Oxfam, EJF, WWF and others) set up the "Shrimp Sustainable Supply Chain Task Force" in July 2014 in Thailand. The task force has undertaken a series of actions with respect to traceability and vessel monitoring, Fisheries Improvement Projects (FIPs) and lobbying for legal changes to fisheries laws in Thailand and beyond (Marschke and Vandergeest 2016). In another example of private intervention Thai Union, a large seafood conglomerate, has agreed to pay broker fees for migrant workers to ensure they do not become indebted (Tang et al. 2016). The challenge in both cases is that while such efforts may ensure that some strands of the supply chain are free from 'slave labor,' they may also create 'sustainability enclaves' (Whitington 2012); improving the working conditions of certain sets of workers, whilst other parts of the sector remain unregulated.

In Thailand, the Royal Ordinance on Fisheries (Government of Thailand 2015) has, for the first time, (in a move linked to international policy pressure, which we discuss later) brought Thailand's fisheries policy in line with international standards and includes provisions for preventing all forms of forced labor, and directives related to labor inspection. Multiple implementing agencies with varied mandates have been involved in Thailand's fisheries reform, i.e., the Department of Fisheries (fish), the Ministry of Labor (off-shore workers) and the Ministry of Human Development and Social Security (human trafficking), illustrating the complexity of governing fish-sea-labor interactions. For migrant workers, however, emerging policy does impose restrictions. While a series of 'One Stop' fish worker migrant registration centers were set up throughout Thailand's coastal provinces, allowing workers to register (or be registered by employers) for two three-month periods, migrant workers remain effectively tied to particular employers: they cannot leave an employer unless they prove an employer has abused them, and the reporting requirements are cumbersome (ILO 2015). As Marschke and Vandergeest (2016) point out, this only serves to drive some migrants further underground. Instead they argue that allowing migrant workers to find work legally without restrictions could better enable them to express their agency in avoiding the worst employers.

Notwithstanding these challenges, the move among Thailand and other Southeast Asian countries to update their fisheries-focused policies in acknowledgement of the severity of labor issues represents a move in the right direction. Policy is being re-written to account for social conditions: ensuring a continuing commitment to implementation once the media spotlight has moved on will be key if this momentum is to be maintained.

Value chain performance and equity

Corresponding to meso-level analysis in Figure 25.2, we now turn to the wider performance of value chains in terms of the equitable distribution of benefits among chain actors. This has been widely conceptualized in relation to trade and labor in terms of achieving what is commonly termed 'inclusive business' (Márquez et al. 2010; Mendoza and Thelen 2008; Vorley et al. 2009). Inclusion can take many forms, but refers in general to the improved 'terms of incorporation' (Hickey and Toit 2013) of an individual or community into a given sector and/or value chain, at either (primary) production or (secondary) processing stages. The assumption is that if the conditions under which people operate can be regulated, then it is possible to improve the terms of their inclusion into value chains, by enabling access as well as enhancing the capability for economic and social empowerment (Mendoza and Thelen 2008). For the seafood sector, efforts to promote 'inclusion' have focused on a range of direct and indirect interventions aimed at increasing market access and/or ensuring higher prices for smallholder aquaculture farmers and fishers.

Research has suggested that both fisheries and aquaculture certification can improve conditions for fishers by opening up access to export markets (e.g., Wessels 2002; Oosterveer and Spaargaren 2011). However, these claims are often based on the assumption that markets offer 'premium' prices for eco-labeled product. While it appears that some products from a subset of MSC certified fisheries receive a premium in certain European and North American markets (e.g., Roheim et al. 2011), there is little evidence that they are passed back down the chain to fishers or fish farmers – either large or small. Instead most evidence shows that any economic benefits are absorbed by other chain actors, and most small-scale fishers and smallholder farmers are unable to comply with eco-standards because of the costs involved and limited capacity to upgrade production practices (Bush et al. 2013; Jonell et al. 2013; Marschke and Wilkings 2014). Alternative assessment methodologies and collective group certification models have been introduced by eco-certifiers to increase participation (see, for example, Ha et al. 2013), but these models and methods are designed predominantly with compliance in mind, rather than improved terms of incorporation or more equitable distribution of the resultant benefits.

The most direct intervention aimed at improved benefit sharing in seafood chains is the Fair Trade (USA) standard for fisheries developed in 2013. The primary goal of the Fair Trade standard is to support 'social sustainability' by redressing power imbalances in international trade (Nicholls and Opal 2005). By being certified against this standard, more durable and equitable trade relations are established, with requirements for developing country producers to receive higher economic returns. A key reason why Fair Trade certification of fisheries has not occurred earlier has been the ambiguous role played by traders, who often exploit small-scale fishers through interlocked credit transactions, whilst simultaneously performing important social welfare functions (Bush and Oosterveer 2007; Ruddle 2011). The one and only small-scale coastal community fishery to have successfully applied for Fair Trade certification has managed to rearrange these local relations to the benefit of fishers. But as Bailey et al. (2016) argue, the long-term success of the Fair Trade model is dependent on strong working relationships between local intermediaries and processors. Furthermore, based on experience in other Fair Trade certified sectors (e.g., Raynolds 2012), it is not yet clear whether the immediate benefits of higher dock-side prices and investment in community infrastructure will lead to the empowerment of these communities over the long-term.

Although framing social issues in the seafood sector in terms of inclusive business opens up space for imagining alternative forms of social and economic development in Southeast Asia, it is unlikely that such approaches, including private regulation through certification, will lead to widespread or fundamental changes in the way the market value of seafood is distributed. Given the underlying selectivity of eco-certification toward more capital-intensive fisheries and aquaculture systems (e.g., Belton et al. 2010; Pérez-Ramírez et al. 2012), it is unlikely that widespread market access for less powerful players will be the long-term result. More directed approaches to improved benefit sharing such as Fair Trade hold greater promise in this regard, but widespread upscaling will be required if significant change is to take place. Furthermore, while Fair Trade is designed for community level interventions, it remains unclear whether such models can also be applied to improving catch share programs on industrial vessels where some of the worst forms of labor rights abuses occur. It also remains to be seen whether Fair Trade will also move to include improved benefit sharing further down the chain at secondary and tertiary processing plants, which is currently not the case.

In the long-term, different forms of collective action may prove more promising than market-led approaches in improving inequity in the seafood sector. Cooperatives and clusters for both fishers and fish farmers have been developed in different parts of the region such as in the Philippines, Vietnam and Indonesia, with such goals in mind (e.g., Baticados 2004; Padiyar et al. 2012). Yet, many have folded without external state or donor support, and few have moved beyond technical aspirations to deliver real economic returns. Furthermore, as Ha et al. (2013) argue, the high level capabilities required to effectively participate in collective organizations can mean that more vulnerable members benefit less, or not at all.

Industry associations are also common across the region. While inclusive of a wide range of fishers, many of these associations are designed to lobby for industry interests rather than deal with social or environmental sustainability issues. Given the often politically enmeshed ownership structures of industrial fisheries and aquaculture in the region, there is also no guarantee that members of such associations are not implicated in abuses of labor. Nevertheless, certain forms of collective action may still offer a promising means of addressing value chain performance with respect to both the conduct of labor and equity of benefit distribution.

Addressing underlying vulnerabilities

Focusing on the conduct of actors at specific nodes in seafood value chains with respect to labor, and on the performance of seafood value chains as a whole with respect to equity and inclusiveness, provides a means of identifying entry points and options for addressing the vulnerabilities of those already working in the seafood industry. Such an approach does not, however, necessarily confront the underlying reasons why people are vulnerable to exploitation in the first place. Addressing these issues also requires attention to the macro political economy (see Figure 25.2) that ultimately creates or reinforces many of the conditions under which exploitation occurs in Southeast Asia.

It has been argued that human exploitation in seafood production is related to the environmental risk profile of the sector. In these accounts, declining catches due to overexploited fish stocks and margins squeezed by rising costs render abuses of labor a 'rational choice' for fishing vessel owners (Couper et al. 2015). More subtly, the inherent uncertainty and temporal variability involved in marine fishing may also lead boat operators to seek to balance low demand for labor during slack periods with sufficient supply at peak times, while simultaneously minimizing costs and maximizing extraction of the surplus value of labor, leading to a variety of forms of unfree labor (Belton et al. in press). In shrimp aquaculture, the high risk and unpredictability of disease might lead farmers and processors to make similar decisions regarding labor in order to manage unexpected fluctuations in levels of production. But while such accounts are intuitively appealing, they also risk being overly deterministic in positioning exploitation of labor as a direct outcome of the over-exploitation of natural resources and susceptibility of farmed organisms to disease and other environmental shocks. As shown by Stringer et al. (2016) in their study of Korean vessels operating in New Zealand waters, abuses of seafood labor also occur in nations with superficially strong regulatory frameworks, suggesting that these tendencies are compounded by the limited ability, or will, of nation states to govern on the high seas.

The contexts in which social mobility and migration occur are also important determinants of vulnerability to exploitation. Poverty or lack of opportunity are major drivers for men and women to leave their places of origin in order to seek better prospects, both domestically and in other more prosperous states within the region. Domestic migration can be related to seasonal movements of labor involving large numbers of people from impoverished rural areas. In some cases these movements are historical, as seen in the Cambodian inland fisheries around the Tonle Sap (Bene and Friend 2011). In other cases, seafood-related migration flows are in response to uneven national economic development. In Myanmar, for example, internal migration for work in anchovy fisheries in Rakhine State (Okamoto 2010) and fish ponds around Yangon (Belton et al. 2015) offer the prospect of incomes that, while relatively modest, are higher and/or more dependable than would be available to those in remote rural areas, with broadly positive outcomes for most of those involved.

More significant, however, are flows of international migrant labor, which offer much greater potential rewards than would be possible from internal migration, but often also involve a greater range of attendant risks and vulnerabilities. The most well-known of these, in the context of the scandals outlined above, are from Myanmar and Cambodia to Thailand, but flows are almost certainly far more complex and involve a greater range of countries and sectors than is widely recognized (as noted by Cheung 2012 on the plight of the Rohingya). Whilst possessing certain peculiarities with respect to labor conditions, such as the off-shore and 'out-of-sight' nature of fishing, Southeast Asian seafood value chains are by no means exceptional. As widely noted in migration studies within the region (e.g., Kaur 2014), the agriculture, construction and hospitality sectors (to name but a few) also rely on large volumes of cheap unskilled or low-skilled migrant labor to maintain their competitiveness.

Migration and the exploitation of migrants play out regionally in a context of uneven economic development and power disparities. Once mobile, migrants are vulnerable to exploitation by numerous and varied actors, including labor smugglers and brokers, officials seeking personal gain and employers profiting from cheap labor in destination countries. However, migrants are not just simply victims of predatory institutions and individuals, but are also active agents making choices calculated to maximize welfare. As Derks (2010) argues, Cambodian migrants working on Thai fishing vessels are capable of acting with a considerable degree of agency in seeking to optimize employment outcomes. These migrants, she points out, are not therefore bound to their employers under conditions that can be equated to 'slavery.' Instead they negotiate their way within a space of structural violence, characterized by regulations and networks of power relations that work to 'immobilize' and render them as easily disciplined and disposable bodies. The illegal or semi-legal status of some migrants, or bonds within a variety of debt-labor relations, may subject them to everyday intimidation by agents of the state via arrest, detention and the threat of deportation or physical violence, with little or no means of redress.

As documented for women migrants working on Thai shrimp farms (Resurrección and Sajor 2010), ambiguous labor status can also create space to simultaneously capitalize upon labor opportunities and meet familial and reproductive expectations. However, as observed in Myanmar, arrangements such as these in which 'husband and wife' couples are hired jointly as a single unit to tend fish or shrimp ponds also implicitly undervalue women's labor (both economically and symbolically), as compared to that of individual male workers (Belton et al. 2015). The division of labor within seafood value chains is also highly gendered, with men overwhelmingly performing physically demanding work on fishing vessels at sea, while on-shore processing activities tend to be constructed as a female domain. Therefore, even among exploited populations, wider inequalities along gender lines are reproduced.

Public and private attempts to regulate the conduct of seafood value chains with respect to labor and their performance with respect to equity and benefit sharing face the challenge of addressing these underlying conditions and resulting vulnerabilities and complexities. The most significant policy changes to date appear to have been those in response to pressure from the European Union's Illegal, Unreported, and Unregulated (IUU) policy and the US State Department annual Trafficking in Persons (TIP) report. Importing states are also pushing for greater oversight in global value chains through, for example, the *California Transparency in Supply Chains Act 2010* and the United Kingdom's *Transparency in Supply Chains Modern Slavery Act 2015*. Much of the pressure for change has come from retailers, some of whom are facing legal action from consumer groups in the United States for knowingly supplying 'slave-seafood.' Contrary to what is often assumed, private initiatives in this case, such as Project Issara, a Thai NGO supported by major international brands and retailers to provide multi-lingual victim support (see www.projectissara.org, accessed 17 July 2016), have followed rather than advocated regulatory change. The extent to which these initiatives can address vulnerabilities embedded within wider political economies of poverty, migration and uneven economic development remains to be seen, however – especially considering, as documented by EJF (2015), the entrenched corruption that exists in regional governments.

Beyond international regulatory approaches, regional governments and the wider development community also need to engage in reducing the causes underlying longer-term, persistent vulnerabilities across Southeast Asia's seafood sector. Fully acknowledging the vital role of migration across the Southeast Asian region and ensuring better regulation occurs to enable migrant workers to improve their working conditions is key. NGOs such as the Migrant Worker's Rights Network (a Thai-based organization of migrant workers from Myanmar), and national research institutions such as Thailand's Chulalongkorn Center for Migration Studies play important roles in this regard in advocating for worker rights. However, migrant workers also need to be able to use official channels to reduce the risks of unfree, abusive labor conditions, and to be able to organize in support of fairer working conditions via trade unions or other mechanisms.

At the same time, long-term commitments to work on more inclusive growth and poverty alleviation are required in order to raise living standards in home countries. Fisheries management within the region also needs to be dramatically improved by, for instance, ensuring a combination of improved coastal and sea tenure and the regulation and enforcement of state-led conservation and management measures to ensure these resources provide more secure livelihood opportunities. Drawing from the Sustainable Development Goals (SDG), for example, SDG two (zero hunger), SDG five (gender equality) and SDG fourteen (life below the water) could provide useful targets for reducing vulnerabilities within Southeast Asia's seafood sector. Achieving these goals will require serious commitments to action long after the dramatic media exposés have subsided.

Conclusions

The profile of 'social issues' has recently become more visible than ever before in debates around seafood sustainability after long being side-lined in discussions and regulation. One striking effect of this historical omission is that remarkably little is presently known about labor and social welfare in a sector with huge regional and global reach and importance. Building on the work of Ponte and Sturgeon (2013), we have used a three-tiered value chain analysis-based approach to identify the confluence of node-specific and wider contextual conditions that influence social sustainability in the Southeast Asian seafood industry. This relational approach both highlights complexity and identifies places and processes that can be made more visible, legible and ultimately better governed.

Focusing on labor as a keystone of social sustainability is in one sense reductionist. However, as illustrated in Figure 25.2, directing attention to the conditions governing the conduct and performance of seafood value chains with respect to labor, and linking these to the wider political economic context in which they are embedded, makes it possible to simultaneously inform interlinked debates around social sustainability and broader regional development. Although focusing on these three socially and spatially nested areas of concern reveals the interconnectedness of social issues within the industry in a systematic manner, the approach also highlights the limitations of narrow sectoral approaches in dealing with complex issues such as unfree labor. We thus conclude that while the industry is well placed to self-regulate in certain areas, the links between conditions under which labor is engaged and wider patterns of migration and structural inequality mean that searching for social sustainability solutions entails looking well beyond the immediate production of seafood.

The chapter also emphasizes the importance of understanding the wider political economy of decision-making around both the environmental and social sustainability of fisheries and aquaculture production. In summary, it appears that the weak governance institutions in Southeast Asia that enable people to be made unfree reflect similar governance weaknesses that enable the overexploitation of fishery resources. This in turn points to the need to move beyond managerial approaches to seafood sustainability that represent and justify policy interventions in terms of social and environmental production choices, and pay greater attention to higher tier drivers inequality and vulnerability (cf. Bush and Marschke 2016). Such an integrated approach emphasizes the need to view the sustainability of the Southeast Asian seafood sector not only in terms of environmental performance, but as fundamental to ensuring human well-being, gender equality and human rights.

Finally, we argue that further research is needed to better understand the conduct and performance of value chains with respect to labor and inclusion in globally integrated industries such as seafood in Southeast Asia. Explanations that emphasize proximate causes such as the remote nature of fishing as major factors in abuses of labor fail to account for more structural determinants of exploitation that are evident throughout the seafood supply chain in nodes such as land-based farms and processing factories. Focusing on the underlying drivers of unfree labor relations requires looking beyond the specificities of seafood production, to understand the role played by both state and industry actors in perpetuating the vulnerability of migrant and non-migrant workers to exploitation in low wage, low skilled employment in multiple sectors throughout the region. Such analysis is likely to reveal that the social and political processes involved in (re)producing these vulnerabilities are generic in nature, driven by a complex mix of uneven development, regional economic integration and environmental degradation.

References

Ananta, A. and Arifin, E. N. (2004). *International migration in Southeast Asia*. Singapore: Institute of Southeast Asian Studies.

Auld, G. (2014). *Constructing private governance: The rise and evolution of forest, coffee, and fisheries certification*. New Haven, CT: Yale University Press.

Bailey, M., Bush, S., Oosterveer, P. and Larastiti, L. (2016). Fishers, fair trade, and finding middle ground. *Fisheries Research*, 182, pp. 59–68.

Barrientos, S., Mayer, F., Pickles, J. and Posthuma, A. (2011). Decent work in global production networks: Framing the policy debate. *International Labour Review*, 150, pp. 297–317.

Baticados, D. B. (2004). Fishing cooperatives' participation in managing nearshore resources: The case in Capiz, central Philippines. *Fisheries Research*, 67, pp. 81–91.

Belton, B., Hossain, M. A. R., Thilsted, S. H. (in press). Labour, identity and wellbeing in Bangladesh's dried fish value chains. In: D. Johnson, ed., *Social wellbeing and the values of small-scale fisheries*. Dordrecht: Springer, xx–xx.

Belton, B., Hein, A., Htoo, K., Kham, L. S., Nischan, U., Reardon, T. and Boughton, D. (2015). *Aquaculture in transition: Value chain transformation, fish and food security in Myanmar*. International Development Working Paper 139. Yangon: USAID.

Belton, B., Murray, F., Young, J., Telfer, T. and Little, D. C. (2010). Passing the panda standard: A TAD off the mark? *Ambio*, 39, pp. 2–13.

Belton, B., Thilsted, S. H. (2014). Fisheries in transition: Food and nutrition security implications for the global South. *Global Food Security*, 3, pp. 59–66.

Bene, C. and Friend, R. M. (2011). Poverty in small-scale fisheries: Old issue, new analysis. *Progress in Development Studies*, 11, pp. 119–144.

Blackmore, E., Norbury, H., Mohammed Yassin, E., Cavicchi Bartolini, S. and Wakeford, R. (2015). *What's the catch? Lessons from and prospects for Marine Stewardship Council certification in developing countries*. London: IIED.

Bush, S. R., Belton, B., Hall, D., Vandergeest, P., Murray, F. J., Ponte, S., Oosterveer, P., Islam, M. S., Mol, A. P. J., Hatanaka, M., Kruijssen, F., Ha, T. T. T., Little, D. C. and Kusumawati, R. (2013). Certify sustainable aquaculture? *Science*, 341, pp. 9–10.

Bush, S. R. and Marschke, M. J. (2016). Social and political ecology of fisheries and aquaculture in Southeast Asia. In: P. Hirsch, ed., *Handbook of the environment in Southeast Asia*. London: Routledge, 224–238.

Bush, S., Oosterveer, P., Bailey, M. and Mol, A. P. J. (2014). Sustainability governance of chains and networks: A review and future outlook. *Journal of Cleaner Production*, 107, pp. 8–19.

Bush, S. R. and Oosterveer, P. (2007). The missing link: Intersecting governance and trade in the space of place and the space of flows. *Sociologia Ruralis*, 47, pp. 384–399.

Butcher, J. G. (2004). *The closing of the frontier: A history of the marine fisheries of Southeast Asia c. 1850–2000*. Singapore: Institute of Southeast Asian Studies.

Cheung, S. (2012). Migration control and the solutions impasse in South and Southeast Asia: Implications from the Rohingya experience. *Journal of Refugee Studies*, 25, pp. 50–70.

Couper, A., Smith, H. D. and Ciceri, B. (2015). *Fishers and plunderers: Theft, slavery and violence at sea*. London: Pluto Press.

De Genova, N. P. (2002). Migrant "illegality" and deportability in everyday life. *Annual review of Anthropology*, 31(1), pp. 419–447.

Derks, A. (2010). Migrant labour and the politics of immobilisation: Cambodian fishermen in Thailand. *Asian Journal of Social Science*, 38, pp. 915–932.

EJF. (2015). *Pirates and slaves: How overfishing in Thailand fuels human trafficking and the plundering of our oceans*. London: Environmental Justice Foundation.

FAO. (2016). *FAOSTAT-statistical database*, 2016. Rome: Food and Agricultural Organisation of the United Nations.
FOS. (2013). *Friend of the Sea Standard, FOS – wild – generic sustainable fishing requirements*. Friend of the Sea. Available at: http://www.friendofthesea.org/public/page/fos%20wild%20gssi%2030%2009%2016%20final.pdf. Accessed 19 July 2017.
Government of Thailand. (2015). *The royal ordinance on fisheries B.E. 2558 (2015): Thailand's strictest law dealing with IUU fishing and human trafficking issues*. Bangkok: Government of Thailand.
Ha, T. T. T., Bush, S. R. and van Dijk, H. (2013). The cluster panacea? Questioning the role of cooperative shrimp aquaculture in Vietnam. *Aquaculture*, 388–391, pp. 89–98.
Hall, S. J., Delaporte, A., Phillips, M. J., Beveridge, M. and O'Keefe, M. (2011). *Blue frontiers: managing the environmental costs of aquaculture*. Penang: WorldFish.
Hickey, S. and Toit, A. (2013). Adverse incorporation, social exclusion, and chronic poverty. In: A. Shepherd and J. Brunt, eds., *Chronic poverty: Concepts, causes and policy*. London: Palgrave Macmillan, 134–159.
ILO. (2007). *C188 work in fishing convention*. Geneva: International Labour Organization.
ILO. (2015). *Review of the effectiveness of the MOUs in managing labour migration between Thailand and neighbouring countries*. Bangkok: International Labour Organisation.
Jonell, M., Phillips, M., Rönnbäck, P. and Troell, M. (2013). Eco-certification of farmed seafood: Will it make a difference? *Ambio*, 42, pp. 659–674.
Kaur, A. (2014). Managing labour migration in Malaysia: Guest worker programs and the regularisation of irregular labour migrants as a policy instrument. *Asian Studies Review*, 38, pp. 345–366.
Kharas, H. (2010). *The emerging middle class in developing countries*. Paris: OECD Development Centre.
Lebel, L., Tri, N. H., Saengnoree, A., Pasong, S., Buatama, U. and Thoa, L. K. (2002). Industrial transformation and shrimp aquaculture in Thailand and Vietnam: Pathways to ecological, social, and economic sustainability? *Ambio*, 31, pp. 311–323.
Márquez, P., Reficco, E. and Berger, G. (2010). Introduction: A fresh look at markets and the poor. In: P. Márquez, E. Reficco and G. Berger, eds., *Socially inclusive business – Engaging the poor through market initiatives in iberoamerica*. Cambridge: Harvard University Press, 1–25.
Marschke, M. and Vandergeest, P. (2016). Slavery scandals: Unpacking labour challenges and policy responses within the off-shore fisheries sector. *Marine Policy*, 68, pp. 39–46.
Marschke, M. and Wilkings, A. (2014). *Is certification a viable option for small producer fish farmers in the global south? Insights from Vietnam*. Mar. Policy 50, Part A, 197–206.
Mendoza, R. U. and Thelen, N. (2008). Innovations to make markets more inclusive for the poor. *Development Policy Review*, 26, pp. 427–458.
Nicholls, A. and Opal, C. (2005). *Fair trade: Market-driven ethical consumption*. London: Sage.
Okamoto, I. (2010). The movement of Rural Labour: A case study based on Rakhine State. In: N. Cheesman, M. Skimore and T. Wilson, eds., *Ruling Maynmar: From cyclone Nargis to national elections*. Singapore: Institute of Southeast Asian Studies, 168–197.
Oosterveer, P. and Spaargaren, G. (2011). Organising consumer involvement in the greening of global food flows: The role of environmental NGOs in the case of marine fish. *Environmental Politics*, 20, pp. 97–114.
Padiyar, P. A., Phillips, M. J., Ravikumar, B., Wahju, S., Muhammad, T., Currie, D. J., Coco, K. and Subasinghe, R. P. (2012). Improving aquaculture in post-tsunami Aceh, Indonesia: Experiences and lessons in better management and farmer organizations. *Aquaculture Research*, 43, pp. 1787–1803.
Pearson, E., Punpuing, S., Jampaklay, A., Kittisuksathit, S. and Prohmmo, A. (2006). *Underpaid, overworked and overlooked: The realities of young migrant workers in Thailand*. Bangkok: International Labour Organisation.
Pérez-Ramírez, M., Phillips, B., Lluch-Belda, D. and Lluch-Cota, S. (2012). Perspectives for implementing fisheries certification in developing countries. *Marine Policy*, 36, pp. 297–302.
Phillips, M., Subasinghe, R. N. T., Kassam, L., Chan., C.Y. (in press). *Aquaculture big numbers: FAO fisheries and aquaculture technical paper*. Rome: Food and Agriculture Organisation of the United Nations.
Ponte, S. and Sturgeon, T. (2013). Explaining governance in global value chains: A modular theory-building effort. *Review of International Political Economy*, 21, pp. 1–29.
Raynolds, L. T. (2012). Fair trade flowers: Global certification, environmental sustainability, and labor standards. *Rural Sociology*, 77, pp. 493–519.
Resurrección, B. P. and Sajor, E. E. (2010). "Not a Real Worker": Gendering migrants in Thailand's Shrimp farms. *International Migration*, 48, pp. 102–131.
Richards, D. R. and Friess, D. A. (2015). Rates and drivers of mangrove deforestation in Southeast Asia, 2000–2012. *Proceedings of the National Academy of Sciences*, 113(2), pp. 344–349.
Rigg, J. (2016). *Challenging Southeast Asian development: The shadows of success*. New York: Routledge.

Riisgaard, L. (2009). Global value chains, labor organization and private social standards: Lessons from East African cut flower industries. *World Development*, 37, pp. 326–340.

Roheim, C. A., Asche, F. and Insignares, J. (2011). The elusive price premium for ecolabelled products: Evidence from seafood in the UK market. *Agricultural Economics*, 62, pp. 655–668.

Ruddle, K. (2011). "Informal" credit systems in fishing communities: Issues and examples from Vietnam. *Human Organization*, 70, pp. 224–232.

Salayo, N., Garces, L., Pido, M., Viswanathan, K., Pomeroy, R., Ahmed, M., Siason, I., Seng, K. and Masae, A. (2008). Managing excess capacity in small-scale fisheries: Perspectives from stakeholders in three Southeast Asian countries. *Marine Policy*, 32, pp. 692–700.

Strauss, K. (2012). Coerced, forced and unfree labour: Geographies of exploitation in contemporary labour markets. *Geography Compass*, 6, pp. 137–148.

Stringer, C., Whittaker, D. and Simmons, G. (2016). New Zealand's turbulent waters: The use of forced labour in the fishing industry. *Global Networks*, 16, pp. 3–24.

Tang, L., Llangco, M.O.S. and Zhao, Z. (2016). Transformations and continuities of issues related to Chinese participation in the global seafarers' labour market. *Maritime Policy & Management*, 43(3), pp. 344–355.

Vorley, B., Lundy, M. and MacGregor, J. (2009). Business models that are inclusive of small farmers. In: C. A. Da Silva, D. Baker, A. W. Shepherd, C. Jenane and S. Miranda-da-Cruz, eds., *Agro-industries for development*. Wallingford: CABI for FAO and UNIDO, 186–222.

Wessels, C. R. (2002). The economics of information: Markets for seafood attributes. *Marine Resource Economics*, 17, pp. 153–162.

Whitington, J. (2012). The institutional condition of contested hydropower: The Theun Hinboun – International rivers collaboration. *Forum for Development Studies*, 39, pp. 231–256.

World Bank. (2008). Small-scale capture fisheries: A global overview with emphasis on developing countries. PROFISH series. Washington DC: World Bank. Available at: http://documents.worldbank.org/curated/en/878431468326711572/Small-scale-capture-fisheries-a-global-overview-with-emphasis-on-developing-countries.

26
OIL PALM CULTIVATION AS A DEVELOPMENT VEHICLE

Exploring the trade-offs for smallholders in East Malaysia

Fadzilah Majid Cooke, Adnan A. Hezri, Reza Azmi, Ryan Morent Mukit, Paul D. Jensen and Pauline Deutz

Introduction

In humid tropical countries oil palm is seen as the most profitable form of rural land use (Sayer et al. 2012). As such, it has come to dominate the economies of several Southeast Asian countries, notably Indonesia and Malaysia. The benefits of oil palm as a boom crop have prompted some to refer to it as 'green gold' for its promise of increasing state revenue and poverty alleviation (Meijaard and Sheil 2013). However, focusing on East Malaysia, the rapid and large-scale conversion of lands suitable for agriculture, often with state support, has opened up conflicts resulting from encroachment into lands claimed under customary rights as well as concerns over the long-term social, economic and ecological sustainability (Majid Cooke 2013; Cramb and Sujang 2016). Nonetheless, some local communities with landownership rights are participating in the production of oil palm by choice, participating largely as smallholders. Notably, smallholders make a significant contribution to the global supply of palm oil, accounting for some 40 percent of supply (RSPO 2016a). Thus, some profits from palm oil production are going to small farmers and directly contributing to the economic development of rural communities. This development is in accordance with the Malaysian government's vision of oil palm as a source of poverty alleviation. Engaging with a market economy, however, brings disadvantages as well as benefits to small farmers. These disadvantages have been expressed as trade-offs: for example in terms of farmers' loss of subsistence or complimentary food production in exchange for the expectation of an increased income from a commodity crop (Agarwala et al. 2014).

Significantly, smallholders, as defined by the Roundtable on Sustainable Palm Oil (RSPO) as those cultivating 50 hectares of land, but often much less, are not a homogeneous group. There are numerous cultural and linguistic groups across East Malaysia. There are also variations in security of land tenure, which has become a prominent discussion with respect to oil palm production given infringements on indigenous land and challenges to prove ownership held under traditional (largely unwritten) systems of ownership and access rights (Majid Cooke 2002; Majid Cooke et al. 2011; Cramb and Sujang 2011; Cramb and Sujang 2016). Other smallholder groups consider oil palm cultivation as a means to create new rights in their resettled sites, having lost their original lands to infrastructure developments. Although estimates of smallholder incomes have been done elsewhere (Majid Cooke et al. 2006 and Cramb and Sujang 2016), less attention has been paid to the drivers for the decision made by independent smallholders to switch to oil palm. However, the aspiration and perceived needs of smallholders are drivers in decision-making. Also important is the role of agency, namely, an individual's ability to make reasoned livelihood choices, whether out of perceived necessity or otherwise.

Over the last decade there has been a significant change to the international palm oil market that provides smallholders with at least the appearance of a further choice. Wider Western attitudes to the environmental impact of oil palm cultivation have helped create a demand for Certified Sustainable Palm Oil (CSPO), initially under the auspices of the RSPO. The stipulations of the RSPO include requirements intended to both support and improve the performance of smallholders. Following the choice 'to grow or not to grow' oil palm, smallholders have the additional choice, at least in theory, of being certified and eligible for any related price premium. Non-governmental organizations (NGOs), as well as palm oil processors, users and others in the supply chain, have been actively involved in communicating these choices and consequent operational conditions to smallholders, in addition to promoting sustainability initiatives outside of the certification schemes (Potter 2015).

This chapter examines the decision to cultivate oil palm under the different conditions that independent smallholders experience in the East Malaysian oil palm frontier. The context that shapes smallholders' understanding of their own participation in the production chain is analyzed using three distinct case studies. Drawing on multiple academic disciplines and a perspective from practice, this transdisciplinary paper poses the questions: How do smallholders perceive the trade-offs involved in growing oil palm? To what extent do certification schemes and NGO participation help the smallholders offset that trade-off? And, how does land-tenure influence available options?

Understanding the context of smallholders' decision to grow oil palm

Although entitled to choose how their land is used, smallholder choice is constrained by personal and wider circumstance. Circumstances that influence the choice whether to cultivate oil palm or not can be understood in relation to well-being. As a concept well-being can, among other things, be understood in terms of an individual or group's health, their safety or simply their relative prosperity. Additionally, it can be considered in the context of individual and local community views and action in terms of their relational position within the local, national and global context (Agarwala et al. 2014). Context, in this instance, therefore refers to relational aspects of change between the individual and society, social and ecological factors, micro and macro spaces as well as power differentials within and across households and groups that affect the distribution of entitlements (Sen 1985; Gough and McGregor 2007). Here we consider the perception of smallholders in exercising their power with respect to crop choice, and also the limitations and outcomes of using that power.

Smallholders demonstrate agency in the way in which trade-offs are calculated and the benefits and costs of different contexts and options are weighted. For example, poor local villagers may be prepared to suffer from environmental degradation arising from large-scale infrastructure or agricultural development in return for developmental benefits, like increased income or access to healthcare and education. Hence, the practical implication of the literature on trade-offs is that farmers demonstrate agency and actively calculate and assess potential costs and benefits. Yet this agency varies greatly according to individuals' circumstances.

Consequently, the different conditions under which smallholders produce oil palm needs to be understood in relation to the different political ecology contexts of opportunities and costs that shape both challenges and trade-offs (Adams and Hutton 2007). The two significant challenges faced by smallholders that require consideration are land tenure insecurity (Majid Cooke 2013), and economic benefits arising through certification schemes that have been extended to smallholders by the RSPO. In principle, indigenous smallholders can exercise the power of choice by working toward certification as RSPO principles, despite difficulties in implementing them (discussed later), are a step ahead of government efforts in Southeast Asia in recognizing indigenous rights (Appalasamy 2013).

Smallholder production in Sabah and Sarawak in context

Independent smallholders, typically, are individual households growing oil palm who receive limited or no subsidy and are free to sell their fresh fruit bunches (FFBs) to traders or directly to mills (Nagiah and Reza 2012; Brandi et al. 2015). Independent smallholders have been found to be more efficient financially and are able to participate in the production in a more effective way than assisted smallholders (Majid Cooke et al. 2011). In the area of certification, however, smallholders that are linked to a company certification scheme and contractually bound to a given mill have a greater capacity to adopt required policies and production methods because of the formal assistance provided by a scheme (Brandi et al. 2015).

Independent farmers face greater challenges in some respects than assisted ones, such as insecure land tenure (experienced by assisted smallholders to a lesser extent). In East Malaysia most of the independent smallholders are indigenous peoples (culturally and linguistically distinct from the majority Malay population in Peninsular Malaysia and overwhelmingly rural-based) who tend to grow oil palm on lands claimed under customary rights. Under the Torrens system embedded in the Sabah Land Ordinance 1930 (SLO 1930), lands claimed under customary rights are recognized by the state only when such lands are titled, creating a dilemma for indigenous communities whose traditional access rights to land are broader based and more inclusive. Because of the long, bureaucratic and often uncertain process for getting individual Native Titles, much of the land claimed under customary rights are untitled (Majid Cooke et al. 2011).

Notably, absence of conflict over land to be included in certification schemes is an important aspect of the certification principles and criteria of the RSPO (RSPO 2013). However, for smallholders, conflicts often occur due to many factors, especially overlaps of lands claimed under customary rights with officially drawn territories (forest reserves, state parks) or lands awarded under license to oil palm or timber companies (Majid Cooke 2013). Thus, whilst the RSPO requirement for documented and uncontested land ownership is designed to protect mostly corporate members from accusations of benefiting from the confiscation of indigenous lands, it raises the barriers to membership for some indigenous people. Furthermore, this situation can be compounded by the RSPO being unable to rapidly deal with extreme cases of land conflict when they arise. As an example, despite attempts by the state to mediate, the land dispute between IOI Group – the second largest oil palm producer in Malaysia – and the indigenous (Kayan and Kenyah communities) of Long Teran Kananin Sarawak continued for many years before IOI's RSPO certification was suspended in 2016 (Potter 2015; RSPO 2016b). Such a long running saga is clearly not in the best interests of promoting CSPO, or the well-being of smallholders largely or wholly reliant on incomes from oil palm, certified or otherwise.

Smallholders and sustainable palm oil certification

The RSPO was formally established in 2004 in response to concerns about the environmental (primarily conservation and biodiversity) risks posed by the rapidly growing demand for palm oil. It was a collaborative effort between a range of stakeholders, including: NGOs, oil refiners and processors such as AAK UK, the Malaysian government, as represented by the Malaysian Palm Oil Association (MPOA), and the consumer goods manufacturer, Unilever. RSPO currently has a multi-stakeholder membership of 2,678 organizations based in numerous countries comprising growers, processors, traders, retailers, banks and NGOs (RSPO 2016c). The RSPO certification standard covers eight 'Principles' and 43 'Criteria,' ranging from commitments to transparency and the use of best cultivation practices, through to conservation of natural resources and responsible development (see RSPO 2013). As of April 2016, RSPO covered 3.66 million ha of plantations, 345 mills, 66 growers and 3,199 supply chain facilities producing 13.7 million tonnes of CSPO, or 21 percent of the annual global supply of palm oil (RSPO 2016d).

Aiming to promote a global sustainable palm oil standard, the RSPO was envisaged as a body that encouraged company and smallholder compliance through market and supply chain demands. One of the most pressing yet delicate issues surrounding RSPO, however, is how to design a certification system that takes into account the nuanced economic and social problems faced by smallholders. Indeed, though there are approximately 3 million oil palm smallholders in the world, only 5.6 percent are certified to RSPO standards (see RSPO 2016d). This calls into question the effectiveness of the RSPO as a potential supporter of smallholder rights, whether because few smallholders have been reached by RSPO or because the requirements of participation involve conditions that some smallholders cannot or will not meet (discussed later).

To date, divisions have appeared amongst oil palm stakeholders around the priorities and standards of certification. The RSPO offshoot, Palm Oil Innovation Group, calls for more stringent certification policies, particularly in relation to environmental standards. Some companies, notably large Southeast Asia-based growers, think existing standards have already gone too far. In 2015 the Malaysian Palm Oil Board launched its own certification scheme in the form of Malaysian Sustainable Palm Oil (MSPO). MSPO was designed to be more sympathetic and accessible to growers than the RSPO standards. However, this implies that MSPO standards are more lax than those of RSPO, and given also that civil society actors were not consulted in the development of MSPO standards, MSPO has suffered a credibility problem among some buyers (Potter 2015). It may not be deemed acceptable by the international community. RSPO and other multi-stakeholder initiatives, such as the International Sustainability and Carbon Certification scheme, thus remain the most widely accepted CSPO schemes.

Regardless of the certification scheme, it should be noted that, despite the putative benefits, for example to company credibility, from the use of certified palm oil, currently the supply of CSPO outpaces demand. Thus, only half the current global supplies of CSPO are sold as such, the rest is sold as conventional uncertified stock. This lack of demand creates concern for the attractiveness of instigating and enduring the certification process. Furthermore, this certification adoption is exasperated by the price premium for CSPO being small, even where a market for CSPO exists. The 2015 average price premium per tonne of CSPO had plunged to US$1.68 with a year low of 28 US cents; by comparison, the 2008 price per tonne was US$40–50 (see GreenPalm 2016).

Herein, smallholders' decision-making and influences in relation to the adoption of certification schemes, and the outcomes of working to RSPO standards, are examined.

Independent smallholders and certification in lower Kinabatangan, Sabah

Largely being an industry initiative, the RSPO has generated questions about its ability to make certification more inclusive, particularly for meeting the needs of smallholders who, even with small premiums, may be financially unable to implement its principles and practices (Nesadurai 2013; Brandi et al. 2015). In this instance, these authors (ibid) argued that RSPO relies on the support it directly or indirectly enjoys from NGOs operating in the civil sphere who influence companies along the supply chain. An important question then for this chapter is: How have indigenous groups (working with NGOs), largely in the civil sphere, exercised power to support a more inclusive and beneficial process of production for smallholders through certification? This question is approached, below, through a case study, combining academic observations and knowledge formed from within an NGO through direct experience of implementing RSPO principles.

Described here is the work of Wild Asia, a social enterprise NGO, engaged in promoting more inclusionary production practices for smallholders in Malaysia using RSPO guidelines. The statement provided by Wild Asia (Box 26.1) focuses on the Lower Kinabatangan area. The presented example highlights how certification could be used to implement best practice cultivation in cases where smallholder land tenure within villages is relatively secure.

Box 26.1 Certification in the Kinabatangan, Sabah

The Wild Asia Group Scheme or WAGS has been operational in the Kinabatangan region since January 2013. This management scheme was pioneered by Wild Asia to explore methods and approaches to group certification, to provide support as well as to create incentives and benefits for independent small producers. As the scheme is modeled to meet the requirements of the RSPO it offers a pathway for producers to move from being "traceable"[1] to certification. The Kinabatangan scheme was first initiated through a partnership with Nestle (which was already working with Kinabatangan smallholders) and another NGO, Solidaridad (a social NGO). The scheme now covers 6 villages in the region and includes 115 individual small producers with a total area of palm oil of 392.7 ha.

Tenure wise, 20% of the group members have Native Title to their land. Yet, a majority of the group members (78%) have cultivated land which has been subject to an application for alienation (Land Application) under customary land provisions of the Sabah Land Ordinance 1930. So far, no producers surveyed have been excluded from the group due to inability to their demonstrate legitimate land claims or existing land disputes. Moreover, almost all the producers say that they had minimal outside assistance to develop their farms (land clearing and planting materials), though some farmers have received Government assistance for replanting.

There are a number of lessons from directly running such a scheme. Firstly, many of the villages have been used to participating in Government programmes (housing, education) where they receive something, no matter how small (aid, materials). But in the WAGS programme all that was offered was advice, though efforts were also made to help farmers gain a better price for their FFB or gain priority access to a buyer during peak crop (when there is a risk of being excluded and crops are turned away). These additional benefits are only possible if a local buyer (dealer or miller) is found who is willing to provide these benefits. The strategy has been to link the certified producers to RSPO-certified mills. This helps create the opportunity to link up, through a trade or physical chain, all the way to global palm buyers.

Despite not being able to offer anything immediately tangible, participation in the programme has been growing. Villagers' enthusiasm for group certification is captured in the growing number of members each year: 36 farmers in 2014, 115 in 2015, and 150 (projected) in 2016.[2] The focus from 2016 onwards is to identify and create model farms to be able to demonstrate how good management practices can improve yields and lower costs. A lesson that has been learnt is that it could be more effective to focus on a few and demonstrate success,

> *rather than trying to immediately engage the many. With an emphasis on group sharing and data analysis of the records maintained by the farmer, we could naturally begin to influence a wider circle of people. Ultimately, we need to be able to demonstrate that self-improvement is possible and that gains in yields (or lower costs) are indeed a more sustainable practice (not only for enhancing net income from oil palm, but for applying the same methods to other production activities).*
>
> Wild Asia, 24 February 2016

Independent smallholder oil palm under extreme conditions, Bakun, Sarawak

The chapter turns now to a different context in which oil palm is produced, a setting grappling with additional insecurities associated with forced resettlement.[3] A total of 15 longhouses from the upper Balui, which were home to an estimated 9,161 individuals, including some semi-nomads, were resettled into sedentary settlements at Sungei Asap for the construction of the Bakun dam. The cost of resettlement was funded by the federal government, yet the actual implementation of resettlement was undertaken by the state government (interview: Sarawak Hidro, 29 June 2015).

Experience with resettlement reveals a major social concern regarding access to land. The provision of 3 acres of land to make up for lands drowned by the reservoir was considered distinctly inadequate by the resettled villages, as captured in the following quotes:

> "In our old place, our land was large and they replaced it with only 3 acres. In one family there are so many siblings and the land is not enough for one family or for family expansion."
>
> "The 3 acres were used up during the first year we moved here with pepper, cocoa and other crops. Now, we want to plant rubber and oil palm, but the plot is not enough, it is already full. If we want to plant outside the three acres, they will prohibit us."
>
> (Quotes from Focus Group Discussion (FGD), Uma Badeng, 19 April 2015)

Consequently, resettled villagers began occupying new land around the resettlement sites to grow oil palm with about 80 percent of residents of the Kenyah longhouse at Uma Bakah resorting to occupying company land surrounding their longhouses – claiming their rights over the land by right of occupation as is customary and as acknowledged in the Sarawak Land Code 1958 (SLC) Section 5 (2) (a) to (e). Smallholders' rationale for this action was:

> Whoever works the land first, will have the rights as long as we plant that area. . . . Whoever wanted to argue our right, we will bring them (the companies) to the Land and Survey Office. That's why those who don't want to or who don't have the courage to fight for the state land do not have enough land.
>
> (Quotes from FGD, Uma Badeng, 16 April 2015)

Coincidentally, this land and the land not inundated at the Bakun dam site is also targeted by private oil palm and logging companies, which typically have better access to state land use permits than resettled villagers. These multiple interests in the land, held by the government, the local communities and the plantation and logging companies, create conflicts over land use, particularly between private companies and communities. Hence, it can be said that smallholders at Bakun are re-occupying land for 'strategic agriculture' reasons, similar to those in the Baram region of Sarawak more than a decade ago (Majid Cooke 2002).

Similar to Sabah's SLO 1930, recognition of customary rights under SLC 1958, as interpreted by Sarawak state government, is narrow. Only occupied lands that have titles are acknowledged as being owned through customary rights, otherwise, they remain 'state land' (Majid Cooke 2013). This is in contradiction to the broader forms of access available to indigenous people under traditional systems of entitlement.

Farmers and oil palm companies at Bakun are set on a long-term path of conflict. Without due state care at the time of resettlement, people were forced to solve their problems themselves in the best way they knew how, namely using SLC 1958 (i.e., land rights established by way of prior occupation). In line with the spirit of volunteerism in RSPO, respect for such land rights has to come from the companies themselves, which can be problematic because their openness in part is contingent on their supply chain position.

As such, in the belief that 'bigger is better,' the struggle for rights could be a long process if companies are not receptive and continue to get state support because of their propensity for large-scale development projects (Sovacool and Bulan 2013). Such an orientation in economic development tends to work in favor of large-scale plantations. For this reason, some NGOs recognize that the process of respecting customary rights has to start with directly working with local communities to raise awareness. In sum, smallholder cultivation of oil palm on contested lands at Bakun is an example of 'independence' fought for under precarious conditions, not of the smallholders' making.

Independent smallholders and certification in central Sarawak

A certification scheme in Sarawak, initiated by an oil palm company, NGOs and local longhouse communities, has sought to translate RSPO principles into practice.[4] Initiated in 2010, an RSPO Smallholder Group Scheme (SGS) was established between farmers from longhouses A, B and C, and an oil palm company owning and operating a nearby mill. The SGS was expected to promote a 'win-win' situation between smallholders and the oil palm company in terms of economic, environmental and social benefits. The scheme built on an existing relationship established in 1981 when the company first entered the area and mapped the border between the longhouse territory, or *'pemakai menua,'* and company land in order to initiate a rattan plantation. In 1996, the company started planting oil palm and began giving the smallholders free seedlings in 2003. By late 2015, much of the longhouse lands, including their *temuda* (fallow lands) and their forest reserves (*pulau galau*), had been planted to oil palm.

An evaluation of the motivation for growing oil palm confirmed similar drivers for engaging in mono-crop agriculture as at Bukit Garam in the Kinabatangan, namely the desire for more cash, albeit over production of subsistence crops (see Majid Cooke et al. 2006 and Cramb and Sujang 2016 for similar observations in Sabah and elsewhere in Sarawak). More specifically, perceived benefits of oil palm farming are seen to be: escaping poverty and having the income required to finance new desires, especially children's education. The below quote captures sentiments expressed by many at the study site:

> It is true that we can get rice for free (from planting it). But other things, we cannot get for free. Nowadays, money is important. My children (who grow oil palm) take care of me. They buy many things for me. They are the ones who make my life easier

now. In the old days when I'm the one who took care of my children, we suffered from poverty. But now, my children take care of me, I feel the suffering has become less.

(Farmer 6, December 2015)

Such sentiments and thoughts around certification premiums helped to motivate smallholders toward joining the SGS. However, based on interviews, the local mill is not guaranteed a premium for certified oil and the smallholders in any case can supply no more than 7 percent of total CSPO that the SGS sponsoring mill requires. This places the mill in a difficult situation, given that because smallholders understood certification would bring a price premium for their FFB (Study Notes, November 2015). Thus, given fluctuating certified palm oil prices, the initial enthusiasm for being certified through the SGS somewhat faded. A respondent commenting on the certification scheme said:

We don't really think it is something special. Now they [the oil palm company] buy our FFB at a low price and sometimes the price can become lower than before. A few years back, the price was more than RM 800. But now it is between RM 380 to RM 400.

(Farmer, 25 November 2015)

Being members of the company's certification SGS, however, the farmers have benefits besides a price premium, such as buying fertilizer and agrochemicals from the company at a discounted rate. They can also borrow the company's farm machinery and rent – at cost – company transport for moving their FFB to the mill. Being a member of the group also affords smallholders training in important aspects of field operations and management, including accounting, safe chemical handling procedures and better cultivation practices that should lead to higher FFB yields. Auditing is undertaken to ensure that smallholders abide by the RSPO 'Principles and Criteria,' such as those concerning the best practices, safe chemical use and environmental conservation (e.g., RSPO 2013). Thus, although the smallholders question the benefits of scheme membership for short-term financial reasons, there are immediate and longer term wider economic and social well-being benefits.

Despite formal training and the informal presence of learning networks amongst related smallholders that have facilitated independent farm developments, smallholders struggle to fully meet the RSPO principles. With low levels of formal education, most smallholders are unable to manage the paperwork required to comply with auditing and some still do not wear the necessary protective equipment for applying agrichemicals, due to the cost of such equipment. During the field study (21 December 2015), one farmer commented: "RSPO wants us to do filing. But I am unable to do it." Another stated: "When applying the herbicide, they want us to wear masks, gloves, safety shoes and apron. We don't really follow that . . . We don't have much money to buy it." Indeed, fully observing all the RSPO principles is a costly process, with another farmer observing: "We are farmers, we are not rich. If your yield is high, then you get more money, if your yield is poor, then you won't have much." Thus, money saving is paramount to a smallholder when preparing and managing their land.

Notably, RSPO allows its principles to be interpreted and applied locally. This means there is freedom to adapt the implementation strategy according to conditions on the ground and, in the case of smallholders, perhaps find ways of saving money. For instance, when subjected to national interpretation, RSPO Principal 4, on 'appropriate and best practices,' implies that smallholders should be allowed to clear their '*temuda*' using traditional slash and burn practices for areas of

2 hectares or less. Indeed, some Malaysian NGOs hold this view since, from the perspective of smallholders, the slash and burn method of land clearance requires little effort and thus saves smallholders money (interviews Kota Kinabalu, May 2016). However, for such a method to become viable policy, it would have to be sufficiently 'sold' to the wider palm oil industry whose representatives are major players in the RSPO Board and the development of their principles. Industry interests understandably may be keen to escape from being scrutinized over the oil palm industry's role in forest fires and Southeast Asia's notorious haze (Varkkey 2013).

New smallholder desires and trade-offs

Once oil palms are bearing fruit, smallholders' cultivation of other cash crops (such as cocoa, rubber or pepper) wanes as do their growing of rice, a staple for the majority of smallholder households, so that eventually, oil palm mono-cropping dominates (Majid Cooke et al. 2006; Cramb and Sujang 2016). Becoming largely or entirely dependent on one crop involves a significant trade-off and change in lifestyle for smallholders. For example, the oil palm industry is subject to fluctuations in world commodity prices,[5] which in turn are heavily dictated by a global oversupply of palm oil. Any notable dip in palm oil prices is intrinsically felt by smallholders who are working at the very end of the industry supply chain. Rather than spreading the risks of cash crop farming across several crops and supplementing production with food crops, livelihood and aspired lifestyles are dependent on one crop, its ongoing demand and economic value. Such a level of risk is, generally, new to smallholders but is intrinsically linked to the trade-offs they make for a perceived better future.

With long-time land use change and much loss of forest foods in addition to the distant location of or absence of markets, it should be noted that food security for many indigenous villages and farmers in East Malaysia has been sporadic. Thus, it must be acknowledged that there are at least some shades of grey in the essentially black and white scenarios that are associated with trade-offs. Nevertheless, in summarizing smallholder rational and motivational trade-offs, with mass education and modern infrastructure being relatively new to many parts of East Malaysia, the expectation of meeting nascent desires for education, access to healthcare, communication networks, transport infrastructure and vehicles drives villages to adopt and persist with oil palm and its associated developmental promises. This persistence, however, means trading off existing environmental integrity of lands and exchanging subsistence and complimentary crop farming with cash crop income and other development benefits. One rationale for this accepted trade-off is summed up in the hope for a better future through education, portrayed below:

> *Bila ada pelajaran, anak saya boleh pergi kemana mana, dan kerja dimana sahaja di dunia. Saya mahu dia tahu itu.* (With education, my daughter can go anywhere in the world. I want her to know that.)
>
> (A smallholder quoted in: Majid Cooke et al. 2006, 48)

Discussion and conclusion

The chapter has described the drivers for oil palm cultivation under different contexts experienced by independent smallholders in East Malaysia. The key driver for smallholder oil palm production is economic, namely desire for a life out of poverty, a potential long-term future with economic and social sustainability. An additional strategic driver is that of securing long-term rights to land.

The different paths taken by smallholders across the case studies occur within a mosaic of competing perceptions, conditions and entitlements. Entitlements go beyond the physical (natural resource access) to include a mixture of pre-existing socio-cultural systems of values, norms, practices and power relations that affect both economic security and land tenure. Local level dynamics and interaction with diverse state and corporate oil palm actors result in physical as well as socio-cultural change and trade-offs between subsistence food production and focusing on cash crops.

The path taken by some smallholders in the Lower Kinabatangan River in Eastern Sabah is different from that adopted by those in Sarawak. Some disenchantment with oil palm and certification has been felt by the group in central Sarawak. However, in the Lower Kinabatangan area, with sufficient NGO support, the number of indigenous smallholder groups choosing to work toward certification is increasing. These groups work toward RSPO certification principles and criteria because, despite difficulties implementing them, they are still a step ahead of government efforts of observing indigenous rights. However, because of the RSPO principle of promoting harmony and avoiding conflict, RSPO cannot intervene in lands that are inscribed with conflicting claims as at the Bakun dam in Sarawak. When conflicts between indigenous groups and plantation companies are not resolved by state processes, RSPO certification can nevertheless be removed as happened in the IOI Group case.

RSPO sees smallholders as a group to be protected but also a group that is subject to standards, standards that can be expensive and difficult to meet without company or NGO support. Thus, trade-off for smallholders do not necessarily equate with agency to choose. Environmental protection or the protection of indigenous rights capacity building for effective smallholder negotiation is key to improving wider agency outcomes. NGOs engaged in implementing RSPOs principles on the ground are attempting to provide this capacity.

The chapter has shown that smallholders can be dependent on and influenced by mills; and mills (be they certified or not) are implicated in market processes taking place at multiple sites and levels, involving industry, state and non-state actors at local, national and international levels. The cumulative effect exerts pressure on smallholders to conform to standards that in some cases they are not able or willing to meet. Under such conditions, RSPO certification, it could be argued, is an experiment at engaging with the complexity of structural positions. From the perspective of smallholders, this can lead to an unpredictability of results. It could be more useful therefore to see the RSPO standard not as a blueprint per se, but as a set of guidelines or ambitions for producers to actively explore solutions and to keep the principles of dialogue, openness and best management practices alive for both smallholders and business. Certification is then a form of validation that producers are engaged in implementing. One thing the RSPO does provide is a set of principles and guidelines that, among others, enhances the need for more inclusive partnerships, the respect of native rights and for enhancing diversity within what otherwise would be a mono-crop landscape.

In each of the different contexts in which oil palm is grown, smallholder cultivation of the crop tends to be accompanied by a decrease in subsistence agriculture. While agricultural diversification for agro-ecological and production security purposes has been the modus operandi for small farmers throughout history, the drive toward intensification (specialization) under oil palm is overwhelming, resulting in smallholders looking at their own food and economic security in a new way. Many smallholders in the case studies perceived that economic security can ensure food security because with increased income, rice can be bought. Additionally, degradation of the environment that provided many of a smallholder's needs, is weighed against the perceived promise of oil palm being able to meet family needs. Though smallholder groups cultivating oil palm have indeed benefitted from steady employment and a regular income (Cramb and Sujang 2016), the long-term social and economic sustainability of such trade-offs can be questioned on several grounds. Monoculture brings vulnerability to total crop failure due, for example, to climatic and biological causes and, economically, there is ongoing risk of price fluctuations and even market failure.

Choices made by smallholders on trade-offs relating to land, food production, income and local ecological integrity are contextual and largely made based on available information and how they are positioned within the local, national and global context. Long-term consideration about the impacts of socio-ecological change on smallholder well-being may or may not be included in smallholder equations, thus creating a space and challenge for future practice and prescription.

Notes

1 'Traceability,' or knowing where your FFB comes from and how it was grown, is a technical challenge for refiners and mills buying from third-party millers/growers. Not knowing the source of FFB exposes stakeholders to the risk of purchasing FFB produced using practices that violate the principles of RSPO. To meet traceability standards, a system must be in place to ensure all FFB has been produced to RSPO standards.
2 In the neighboring Beluran district where Wild Asia also operates, there were 42 program members in 2013, 173 farmers in 2014 and 201 members in 2015.
3 We extend thanks to Dr. Frauke Urban who provided us with the opportunity to work on the UK Economic and Social Research Council (ESRC) grant: "China goes global: A comparative study of Chinese hydropower dams in Africa and Asia" (ES/10320X/1). We also thank Dr. Urban's co-researcher, Dr. Giuseppina Siciliano, for her valuable contribution.
4 Study fieldwork was conducted between November and December 2015: Eight in-depth interviews and five FGDs were held in three longhouses in the central Sarawak interior. Location of research site, longhouses names and respondents are withheld to observe conditions attached to the undertaking of the project. Interviews were also conducted with representatives of the mill. This case study was funded by a Newton Institutional Links. Grant 172702808 from the British Council to the University of Hull and Universiti Malaysia Sabah.
5 Crude palm oil prices have fluctuated between 1800 RM and 2800 RM per tonne between April 2015 and April 2016 (see: www.mpob.gov.my. Accessed 16 April 2016.

References

Adams, W. and Hutton, J. (2007). People parks and poverty: Political ecology and biodiversity. *Conservation and Society*, 5(2), pp. 147–183.
Agarwala, M., Atkinson, G., Fry, B., Homewood, K., Mourato, S., Rowcliffe, J., Wallace, G. and Gullard, E. J. (2014). Assessing the relationship between human well-being and ecosystem services: A review of framework. *Conservation and Society*, 12(4), pp. 437–447.
Appalasamy, S. (2013). RSPO 2nd Annual surveillance assessment (ASA02). Public summary report. BSI Group Singapore. Available online www.bsigroup.com/LocalFiles/en-ID/RSPO%20Public%20Summary%20Reports/2013/ASA2_Sime%20Darby%20SOU2%20RSPO%20Public%20Summary%20Report_Oct%202013.pdf accessed 24/7/2017.
Brandi, C., Cabani, T., Hosang, C., Schirmbeck, S., Westermann, L. and Wiese, H. (2015). Palm oil: Challenges for smallholder certification under RSPO. *Journal of Environment and Development*, 24(3), pp. 292–314.
Cramb, R. and Sujang, P. (2016). Oil palm smallholders and state policies in Sarawak. In: R. Cramb and J. F. McCarthy, eds., *The oil palm complex, smallholders, agribusiness and the state in Indonesia and Malaysia*. Singapore: National University Pres,. pp. 247-282.
Cramb, R. and Sujang, P. S. (2011). 'Shifting Ground' renegotiating land rights and livelihoods in Sarawak, Malaysia. *Asia Pacific Viewpoint*, 52, pp. 136–147.

Gough, I. and McGregor, J. A. (2007). *Wellbeing in developing countries, from theory to research*. Cambridge: Cambridge University Press.

GreenPalm. (2016). *Market overview* [Online]. Available at: greenpalm.org/the-market/market-overview. Accessed 31 May 2016.

Majid Cooke, F. (2002). Vulnerability, control and oil palm in Sarawak: Globalisation and a new era? *Development and Change*, 33(2), pp. 189–211.

Majid Cooke, F. (2013). Constructing rights, indigenous peoples at the national inquiry into customary rights to land in Sabah in June 2012. *Sojourn: Special Focus on Borneo*, 28(3) pp. 516–541.

Majid Cooke, F., Dimbab, N. and Norhafizah, S. (2006). *Learning by doing, social transformation of smallholder oil palm economies of Sabah and Sarawak*, Malaysia, UNESCO Report.

Majid Cooke, F., Toh, S. and Vaz, J. (2011). *Community-Investor business models: Lessons from the oil palm sector in East Malaysia*. London, Rome and Kota Kinabalu: IIED/IFAD/FAO/Universiti Malaysia Sabah.

Meijaard, E. and Sheil, D. (2013). Oil palm plantations in the context of biodiversity conservation. In: S. A. Levin, ed., *Encyclopedia of biodiversity*. 2nd ed., Elsevier Science Publishers, Netherlands. 5, 600–612.

Nagiah, C. and Reza, A. (2012). A review of smallholder oil palm production: Challenges and opportunities for enhancing sustainability a Malaysian perspective. *Journal of Oil Palm and Environment*, 3, pp. 114–120.

Nesadurai, H. (2013). Food security, the palm oil-land conflict nexus, and sustainability: A governance role for a private multi-stakeholder regime like the RSPO? *The Pacific Review*, 26(5), pp. 505–529.

Potter, L. (2015). *Managing oil palm landscapes: A seven country survey of the modern palm oil industry in Southeast Asia, Latin America and West Africa*. Bogor, Indonesia: Centre for International Forestry Research.

RSPO. (2013). *Principles and criteria for the production of sustainable palm oil*. Kuala Lumpur: Malaysia, Roundtable on Sustainable Palm oil.

RSPO. (2016a). *Roundtable on sustainable palm oil: Smallholders* [online]. Available at: www.rspo.org/members/smallholders. Accessed 12 July 2016.

RSPO. (2016b). *Roundtable on sustainable palm oil: News and events* [online]. Available at: www.rspo.org/news-and-events/announcements/notice-to-rspo-members-on-the-suspension-of-ioi-groups-certification. Accessed 14 July 2016.

RSPO. (2016c). *Roundtable on sustainable palm oil: Members* [online]. Available at: www.rspo.org/mambers/page/?. Accessed 18 July 2016.

RSPO. (2016d). *Roundtable on sustainable palm oil: Impacts* [online]. Available at: www.rspo.org/about/impacts. Accessed 17 April 2016.

Sayer, J., Ghazoul, J., Nelson, P. and Klintuni Boedhihartono, A. (2012). Oil palm expansion transforms tropical landscapes and livelihoods. *Global Food Security*, 1(2), pp. 114–119.

Sen, A. (1985). Wellbeing agency and freedom, the Dewey lectures 1984. *The Journal of Philosophy*, 82(4), pp. 16–221.

Sovacool, B. and Bulan, L. C. (2013). They'll be damned: The sustainability implications of the Sarawak Corridor of Renewable Energy (SCORE) in Malaysia. *Sustainability Science*, 8, pp. 121–133.

Varkkey, H. (2013). Patronage, politics, plantation fires and transboundary haze. *Environmental Hazards*, 12(3–4), pp. 200–217.

27
DISASTERS AND DEVELOPMENT IN SOUTHEAST ASIA

Toward equitable resilience and sustainability

Frank Thomalla, Michael Boyland and Emma Calgaro

Introduction

Disasters, occurring as a result of complex interactions between human action and natural hazards (see Turner et al. 2003; Wisner et al. 2004), are a growing threat to the development and prosperity of nations, societies and communities across Southeast Asia (Gupta 2010).[1] Disaster *risks* are being created within development; developmental processes largely determine the exposure of people, resources and assets to hazards, susceptibility to suffering harm and capacity to cope with and adapt to new conditions (see UNISDR 2015). This currently 'negative' two-way relationship between *disasters and development* needs to be transformed to a condition where development reduces risk and disasters provide opportunities for enhanced development and resilience. This chapter provides a review of the historical trends and impact of disasters and the status of disaster risk in Southeast Asia. It then discusses the importance of reducing that risk by tackling the underlying drivers of exposure and vulnerability generated inside development as well as pursuing equitable outcomes in resilience processes. Significant risk reduction in the region could be achieved by transforming developmental processes that have historically created risk toward processes driven by equity, resilience and sustainability.

Disasters in Southeast Asia

Southeast Asia is one of the most disaster-prone regions of the world. Almost all types of natural hazard-based disasters occur there; meteorological (i.e., heat waves and tropical storms – also known as cyclones and typhoons), hydrological (i.e., floods), climatological (i.e., droughts and wildfires) and geophysical hazards (i.e., earthquakes, tsunamis and volcanic eruptions) (Gupta 2010).[2] The coastlines of the Philippines and Vietnam are particularly exposed to tropical storms that form in the northwestern Pacific Ocean and track westwards, and the coastline of Myanmar is exposed to tropical storms forming in the North Indian Ocean and tracking eastwards (Figure 27.1). As part of the 'Ring of Fire' the Philippines and Indonesia are exposed to frequent earthquakes due to the location of the boundaries of the Indo-Australian, Eurasian and Philippine tectonic plates.

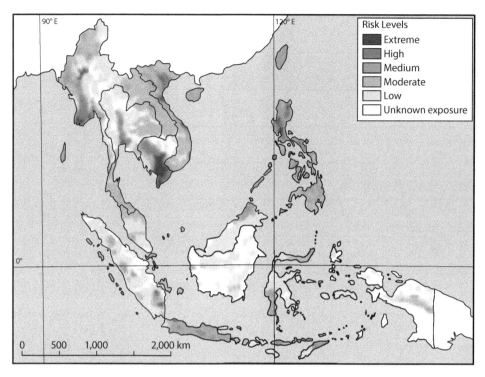

Figure 27.1 Mortality risk distribution of selected hydro-meteorological hazards (tropical cyclone, flood, rain-triggered landslide) in Southeast Asia

Source: UNESCAP and UNISDR 2012

The occurrence of disasters in Southeast Asia has been rising over the past half a century. Although in part due to improvements in disaster reporting and data collection, this trend is predominantly due to a marked increase in the number of tropical storms, floods, droughts and extreme temperature events (Figure 27.2). The Intergovernmental Panel on Climate Change (IPCC) (2012) reports that climate change is contributing to increases in the frequency and intensity of such climatological and hydro-meteorological hazard-based disasters. The occurrence of geophysical hazard-based disasters (i.e., earthquakes, tsunamis and volcanic eruptions) has only increased marginally during the same period. According to the International Disaster Database (EM-DAT) at the Centre for Research on the Epidemiology of Disasters (CRED), over 1,400 disasters occurred in Southeast Asia between 1970 and 2016, but the occurrence of these events is heavily concentrated in four nations: Philippines (527 disasters; representing 37 percent of the total); Indonesia (402 disasters; 28 percent); Vietnam (189 disasters; 13 percent); and Thailand (125 disasters; 9 percent) (CRED and Guha-Sapir 2017).[3] Together, these four nations account for 87 percent of all disaster events during this time period. Singapore (0), Brunei Darussalam (1) and Timor-Leste (9) have suffered very few natural hazard-based disasters in the same period. Specific multi-hazard hotspots in Southeast Asia include: the Ayeyarwady (Irrawaddy) Delta in Myanmar; the Chao Phraya Delta in Thailand; the Mekong Delta in Cambodia and Vietnam; the eastern coastline of Vietnam up to the Red River Delta;

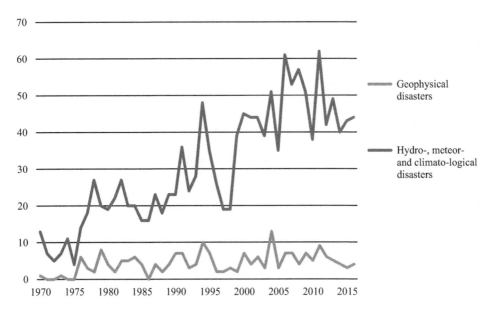

Figure 27.2 Total number of disasters [(i) hydro- meteor- and climate-logical disasters, and (ii) geophysical disasters] that have occurred in Southeast Asia, 1970–2015

Source: Data from CRED and Guha-Sapir 2017

Manila and other pockets across the Philippines; and many of the populated Indonesian islands (UNESCAP 2015).

Despite the uneven spatial distribution of events, the risk of disaster – particularly due to extreme climate change and variability-related hazard events – is high across much of the region. The 2016 World Risk Index (Garschagen et al. 2016) confirms Southeast Asia's status as a 'hotspot' region for natural hazard risk (see Table 27.1). The World Risk Index assesses the risk of becoming the victim of a disaster (natural hazard-based disasters) and ranks the severity of disaster risk for 171 nations. In their risk determination, Garschagen et al. (2016) adopt the widely accepted formula for determining risk (see Turner et al. 2003; Wisner et al. 2004; Cardona et al. 2012), where risk is equal to exposure of a population to natural hazards multiplied by vulnerability, and in turn, where vulnerability is a function of susceptibility, coping capacities and adaptive capacities (or lack thereof). In Southeast Asia, exposure is a particularly pronounced driver of risk (Garschagen et al. 2016).

Five Southeast Asian nations are listed in the World Risk Index's top 20 countries with the Philippines ranked as the third most at-risk country, with very high exposure and high vulnerability, susceptibility and a lack of coping capacities. Urbanization is seen as an important factor in disaster risk across Southeast Asia and the globe, as urban populations continue to grow rapidly (Birkmann et al. 2014). UNESCAP (2015) estimates that, in 2014, 17 Southeast Asian cities with a collective population of 46 million were at extreme risk from multiple hazards and projects the number of those at extreme risk to increase to 66 million people by 2030. These data point to high risk in the region but also show striking contrasts between nations. Further, significant at-risk populations are concentrated in exposed urban areas that, as we will discuss later in this chapter, are growing more rapidly than rural areas due to urbanization and focused economic development activities.

Table 27.1 The 2016 World Risk Index for Southeast Asian nations (adapted from Garschagen et al. 2016)

Rank	Nation	Risk Index (%)	Exposure (%)	Vulner-ability (%)	Suscept-ibility (%)	Lack of coping capacities (%)	Lack of adaptive capacities (%)
3	Philippines	26.70	52.46	50.90	31.83	80.92	39.96
7	Brunei Dar.	17.00	41.10	41.36	17.40	63.17	43.53
9	Cambodia	16.58	27.65	59.96	37.55	86.84	55.49
12	Timor-Leste	15.69	25.73	60.98	49.93	81.39	51.61
18	Vietnam	12.53	25.35	49.43	24.95	76.67	46.67
36	Indonesia	10.24	19.36	52.87	30.09	79.49	49.04
42	Myanmar	8.90	14.87	59.86	35.63	87.00	56.93
86	Malaysia	6.39	14.60	43.76	19.02	67.52	44.73
89	Thailand	6.19	13.70	45.22	19.34	75.53	40.79
100	Lao PDR	5.59	9.55	58.51	37.41	84.37	53.76
159	Singapore	2.27	7.82	28.99	14.24	49.44	23.28

Key

Very high
High
Medium
Low
Very low

Southeast Asia's history of devastating disasters has resulted in a major loss of lives and economic damages across the region. Between 1970 and 2016, over 425,000 people from Southeast Asia have perished from disasters (CRED and Guha-Sapir 2017). Most of these deaths were caused by a handful of major disasters (Table 27.2). For instance, the 2004 Indian Ocean tsunami event resulted in approximately 175,000 deaths alone in Indonesia, Thailand, Malaysia and Myanmar. Yet, mortality rates are only part of the loss equation. Disasters result in both direct and indirect costs; they undermine human security and well-being, and cause damage and losses to ecosystems and ecosystem services, property, infrastructure, livelihoods, economies, and places of cultural and recreational significance (UNEP 2015; IPCC 2012). The risk of displacement from disasters, an impact that is largely a consequence of people's vulnerability, is increasing in the region at a rate faster than population growth (Lavell and Ginnetti 2014).

Direct and insured economic losses and damages from weather and climate-related hazards have increased substantially in recent decades (Figure 27.3). The most economically damaging regional events in the last 45 years have occurred in Thailand and the Philippines. The prolonged flood event in Central Thailand and the Bangkok Metropolitan Area in 2011 caused over US$45 billion worth of damage (World Bank 2012a). Typhoon Haiyan resulted in US$10 billion worth of damage in the Philippines in November 2013. Although relative risk – measured as a proportion of population or GDP – remains stable (UNEP 2011), it is worth noting that the proportion of economic losses and damages resulting from disasters in the region that are uninsured is often significant and unaccounted for in official records.

The disaster trends discussed above reflect just one side of the disaster-development relationship. Over the past decades as both disaster magnitude and frequency has increased, Southeast Asia has developed tremendously in terms of both social and economic development. However, as wealth has grown, more socially and economically valuable assets, infrastructure and services are now at risk. As we explore in the next section of this chapter, population growth has been mirrored with growth in economies and wealth that have by and large concentrated people and development in sprawling urban areas that are often located in hazard-exposed places (UNEP 2016). The complex relationships between disaster impacts and development processes are unpacked through the lens of disaster risk drivers at play in the region.

Table 27.2 Major disasters affecting Southeast Asia in terms of loss of life and economic damage, 1970–2016

Year	Hazard (disaster name)	Affected Southeast Asian nations	Total deaths	Economic damage (million US$)
2004	Earthquake and Tsunami (Indian Ocean Earthquake and Tsunami)[1]	Indonesia Thailand Malaysia Myanmar	165,818 8,345 80 71 (174,314)	4,519 1,000 500 500 (6,519)
2006	Earthquake	Indonesia	6,592	3,155
2008	Tropical Storm (Cyclone Nargis)	Myanmar	138,366	4,000
2011	Flood	Thailand	877	40,371
2013	Tropical Storm (Typhoon Haiyan/Yolanda)	Philippines	7,415	10,136

1 This disaster also affected countries from outside of Southeast Asia, but the data displayed in Table 27.2 pertain only to Southeast Asian nations.

Source: Data from CRED and Guha-Sapir 2017

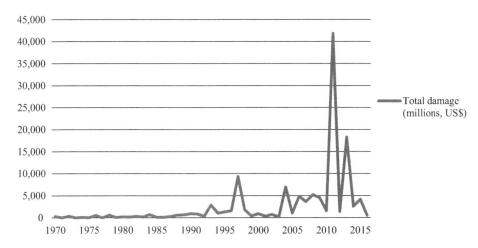

Figure 27.3 The total economic impact (damages) of disasters in Southeast Asia, 1970–2015
Source: Data from CRED and Guha-Sapir 2017

Development and disaster risk in Southeast Asia

Disaster risk varies greatly across Southeast Asia due to the geographical, environmental, climatological, social, economic and political diversity of the region. Increased regional vulnerability and risk levels to natural hazards are due to the compounding effect of multiple and interconnected drivers that include: high levels of poverty and inequality, demographic change and settlement patterns, increasing population density in growing urban areas, unplanned and rapid urbanization, unsustainable uses of natural resources, weak institutional arrangements, and non-risk-informed policies (UN General Assembly 2015a). For instance, the relationships between demographics, land use and environmental degradation have been shown to work together to perpetuate and even create vulnerability to disasters (Zou and Thomalla 2010). The potential for high levels of urban risk to natural hazards is a particular concern (ADW and UNU-EHS 2014). Climate change adds another dimension to the risk equation. Disaster risk related to meteorological, hydrological and climatological hazards is increasing due to climate change and development patterns (IPCC 2012). This section examines the main drivers of vulnerability and risk in Southeast Asia and how they work together to create and perpetuate disaster risk over time.

Poverty and inequality

Poverty remains a significant socio-economic development challenge in Southeast Asia despite a long-term decrease of poverty rates in Vietnam (World Bank 2012b), Thailand (UNDP 2014a), Indonesia (Statistics Indonesia 2014) and the Philippines (UNDP 2009). The UNDP estimate that there are 38 million people in the region living in multidimensional poverty[4](UNDP 2014b). The true total is thought to be significantly higher due to the exclusion of Myanmar in these calculations due to a lack of data (UNDP 2014b); it is estimated that a quarter of the population – 13 million people – may be below the income poverty line (UNDP 2014c).

The implementation of various economic growth and development policies across the region has greatly contributed to reductions in poverty levels in several countries, but not necessarily to reducing levels of inequality (Oxford Policy Management 2003; Rigg and Oven 2015;

Martinez 2017). An analysis of the national income distribution in the region – measured by the Gini Index[5] – indicates concerns regarding wealth inequality within several nations. Of the 142 nations for which data are available, the least developed countries (LDCs) Cambodia, Lao PDR and Timor-Leste all rank globally in the bottom third (UNDP 2014b). Generally in the region, rates of improved equality and human rights are still lagging behind poverty reduction rates, as there is strong evidence to suggest that economic growth has exacerbated inequalities within the region (UN-HABITAT 2010). Evidence of the correlation between levels of poverty, inequality and disaster risk highlights the importance of more transformative efforts to reduce poverty and inequality for reducing disaster risk and vulnerability (Cardona et al. 2012; UNISDR 2015).

Demographic change and urbanization

The majority of countries in Southeast Asia are developing countries with higher than global average population growth rates (Gupta 2010). Cambodia, Laos, Myanmar and Timor-Leste are considered Least Developed Countries (LDCs). Cambodia, Lao PDR and Timor-Leste have the highest annual population growth rates in the region (UNDP 2014b). Between 1992 and 2010, the region's total population grew by 28.3 percent (UNEP 2014). In 2013, the population of ASEAN was close to 620 million. Nearly half of these people (286 million) are living in urban areas. The rise in urban population growth is set to continue; the ASEAN national average urban population growth per year (2.6 percent) is double that of the national average (1.3 percent) (UNDP 2014b). Growing populations, often concentrated in higher-risk urban areas, are increasing the actual and proportional number of people exposed to hazards, thus increasing the potential risk of disaster losses in the region continuing to rise.

Much of this rapid and unplanned urbanization has occurred in high-risk locations, which increases disaster risk levels (UNESCAP and UNISDR 2012). Of the estimated 46 million people at extreme risk within the multi-hazard hotspots identified in Section 2 (UNESCAP 2015), a large proportion reside in Bangkok, Thailand, Manila, Philippines, and Can Tho and Ho Chi Minh City, Vietnam. An estimated 75 percent of the region's population reside within 100 kilometers of the coast, which increases their exposure and vulnerability to tropical cyclones, floods, storm surges and climate change impacts such as sea-level rise (UNEP 2014). Those living in urban slums are particularly vulnerable to natural hazards due to extremely low coping capacity – i.e., lack of capital, assets, social safety nets, access to basic infrastructure and services, food and water insecurity, and insurance. Of a total urban population of 286 million, 31 percent, or almost 89 million people, are living in urban slums in Southeast Asia (UN-HABITAT 2010). However, as a result of the prioritization of urban poverty reduction initiatives in several countries, the proportion of urban populations living in slums and the rate of urban slum population growth has been declining in recent years (Figure 27.4), particularly in Thailand (UN-HABITAT 2006). Cambodia and Lao PDR have reduced slum numbers and populations by providing more adequate housing for those living in slums, yet the percentage of total urban populations living in slums in these two countries is over 75 percent, as of 2014 (UNDP 2014b). Across Southeast Asia, and particularly in Indonesia and the Philippines where slum dwellers total close to 50 million, there still exists significant socio-political barriers to lifting people out of slum conditions, such as the costs and challenges associated with relocation and a certain lack of political will and recognition of slum populations in general. As absolute slum numbers continue to grow, so too do people's vulnerability to disasters as slums persist in the most exposed areas of cities and countries (UNESCAP and UNISDR 2012).

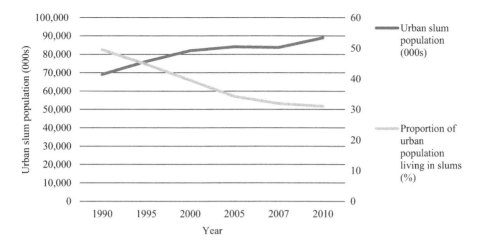

Figure 27.4 Trends showing urban slum population numbers and the proportion of urban populations living in slums in Southeast Asia, 1990–2010

Source: Data from UN-HABITAT 2010

Unsustainable natural resources and ecosystems use

The intense and unsustainable use of natural resources and degradation of vital ecosystems, especially in catchment and coastal areas, is a significant disaster risk driver in Southeast Asia. The pressure on natural resources is increasing due to an intensification and expansion of agriculture in several countries and high-risk areas, an increase in the material and resource intensity in the region, and rapidly rising resource consumption rates (UNEP 2014). Land degradation and increased water scarcity, caused by over-farming, can intensify the impacts of floods and droughts; increased surface runoff leads to flash flooding whilst the withdrawal of excessive amounts of freshwater and groundwater for crop irrigation can increase drought sensitivity levels (UNISDR 2007) and can exacerbate conflict within and between communities and localities in catchments (see Miller this volume).

Deforestation also heightens the risk of floods and landslides. Tree coverage reduces floods and landslides by limiting surface runoff and stabilizing slopes during periods of heavy rains. Between 1990 and 2000 an average of 2.4 million hectares of forest were lost annually in Southeast Asia with the rate slowing to 1.8 million hectares per year between 2000 and 2010 (UNEP 2014). With the exception of the highly urbanized state of Singapore, the average percentage change per country is 5 percent, with Cambodia the worst offender having reduced its forest cover by at least 13.7 percent since 2000 (Yale University 2014). Deforestation caused by illegal mining and logging activities and the subsequent increase in flood and landslide risk is a concern in the rural provinces of central and northern Luzon in the Philippines. The isolation of the affected communities in Northwest Luzon further compounds their risk levels. In the event of a disaster, roads can become impassable and disaster responders are unable to reach affected areas in a timely manner (IRIN 2010). The underlying model of economic development over environmental protection is driving unsustainable practices and resource use in the pursuit of economic development and poverty alleviation. This is a key trade-off in development decision-making: Prioritize immediate and rapid growth and therefore sacrifice sustainability, or vice versa?

Governance structures and processes

Governance systems, power structures and the strength of institutions that reinforce dominant power structures are critical in influencing vulnerability and resilience levels (Adger 2006; Bankoff 2003). This includes both formal (e.g., government and their institutions and the policies and laws that support and uphold these power and governance systems) and informal governance systems (e.g., culturally embedded power systems such as patron-client relationships in Thailand or the caste system in India and the ideologies, religious beliefs and social norms that reinforce these systems). Together, these multi-scaled governance structures and processes regulate access and entitlements to resources that are needed to effectively respond to risk (Adger 2006). By conferring access to some, while restricting entitlements and influence for others, they also influence (directly or indirectly, explicitly or inexplicitly) many of the other key drivers of risk and vulnerability in Southeast Asia, i.e., demographic trends, patterns of poverty, inequality and marginalization, and broader development pathways and patterns (Adger 1999; Pelling 2003; Wisner et al. 2004). These patterns of resource access and developmental trends are also shaped by wider national and international systems and the ideologies that underpin them (Cannon 1994). For example, capitalism and neoliberalism have created a system of specialized global inequality; this has produced 'ghettos of capitalism,' places that have been excluded from cycles of cumulative development, widened the gap between rich and poor and influenced broader patterns of vulnerability and resilience to risk (Bankoff 2003; Pelling 2003; Rigg and Oven 2015).

The link between development and increased risk and disaster events is a long-recognized issue with no easy answers (Bankoff 2003; Pelling 2003). Disasters are a function of development; a consequence of unsolved cumulative developmental problems that place certain people at greater risk to each other and their environment (Bankoff 2003). Natural hazards are merely the trigger event that destabilizes the already vulnerable coupled human-environment system (Birkmann 2006; Comfort et al. 1999; IRDR 2011; Pelling 2003). The distribution of international disaster aid has also been criticized for perpetuating the cycle of underdevelopment and disasters by reproducing pre-disaster conditions of dependence and underdevelopment (Pelling 2003). Disaster outcomes in Indonesia and Myanmar following the 2004 tsunami and the 2008 Cyclone Nargis demonstrate the strong influence that power in governance systems have on vulnerability levels. Political instability in Aceh, Indonesia and Myanmar at the time the disasters struck is cited as a reason for major delays in the delivery of aid and overseas assistance as well as longer-term recovery trajectories in both countries. Yet, such pre-existing political conditions also drove the underlying vulnerability of those affected, thus exacerbating the impact in two disaster-development systems: pre- and post-disaster (Fan 2013).

Yet disaster outcomes are not always negative; they can also become catalysts for change (Bankoff 2003; Calgaro et al. 2014). They create space for political, economic and social adjustments triggering adaptations in human behavior, modification to governance structures and, in some instances, can spur political resistance, regime changes and the downfall of social systems (Fagan 2009; Davis 2001). This was evident in the actions of two Thai tourism communities that were affected by the 2004 tsunami (Calgaro et al. 2014). Under the leadership of the dominant landowner families on the island of Phi Phi Don, the community collectively resisted and defeated governmental plans to redevelop the island into an exclusive resort enclave. In the worst affected resort town of Khao Lak, small businesses realized the power of having one united voice. Consequently, they formed a new tourism representative group that directly petitioned the Thai prime minister for much needed assistance. The process of disaster diplomacy – that takes place before and after disasters – can also facilitate wider positive changes like peace efforts, if supporting pre-existing conditions are already in place (Kelman 2012). For example, long-standing political conflicts between the Indonesian government and militants in Aceh were resolved amidst the international humanitarian response to the 2004 tsunami. Despite some setbacks, the negotiated peace deal signed on 15 August 2005 is still in effect (Gaillard et al. 2008).

Given the influence that governance and decision-making processes have on vulnerability levels (particularly those related to social welfare, economic growth, natural resources and development), it is imperative that good governance – processes and structures that open space for positive and inclusive decision-making among multiple stakeholders – be actively included as one of the cornerstones of any disaster risk reduction (DRR) action plan. For example, the role of customary institutions grounded in local contexts, of which there are many types across the region, should be better recognized, and their inherent resilience captured for wider risk reduction efforts (Boyland et al. 2017).

Climate change and variability

Climate change is also increasing disaster risk. Climate change is causing changes in the frequency, intensity, timing and spatial coverage of extreme events (IPCC 2012). According to the IPCC (2012), climate change will result in an increase in the frequency of heat waves, heavy precipitation, rising wind speed of tropical cyclones and increasing intensity of droughts. Increases in events related to heavy precipitation and coastal high water, the latter due to rising sea levels, are also predicted with high levels of confidence (IPCC 2012). Given the increasing occurrence of hydro-meteorological hazards in Southeast Asia (Figure 27.1), for instance those associated with annual monsoon and typhoon seasons, many nations can expect increasing hazard frequency and intensity to lead to greater disaster impacts if future climate change is not considered across both DRR and adaptation strategies as well as through more risk-informed policies in key development sectors.

Southeast Asia is expected to be one of the worst affected regions from the impacts of climate change (Gupta 2010; UNEP 2014). In 2011, six of ten most vulnerable countries to climate change worldwide were in Asia and the Pacific (UNEP 2014), with three of the ten countries most at risk worldwide in Southeast Asia. Sea level rise is expected to be a major threat. By 2050 sea level rise predictions suggest that more than 20 million people are at risk in Indonesia, nearly 15 million people are at risk in the Philippines, and nearly 10 million are at risk in Vietnam (UNEP 2014). Myanmar, Malaysia and Thailand are also expected to be highly exposed to sea level rise (ibid).

The section has summarized evidence that many of the conceptual and global underlying disaster risk factors, for instance as identified in the *Sendai Framework for DRR: 2015–2030* (Sendai Framework), are driving high rates of exposure and vulnerability in Southeast Asia (UN General Assembly 2015a). Many of these drivers and trends are the result of complex and integrated development processes both at and across scales within the region. Efforts to reduce disaster risk, particularly in a context of a changing climate, need to address these underlying factors systematically and by focusing on the ways in which development is creating and exacerbating risk. Until development is transformed to be more equitable, resilient and sustainable, disaster risks will continue to increase in Southeast Asia (UNISDR 2015).

Disaster risk reduction in Southeast Asia

A review of DRR initiatives and their outcomes over the last decade by UNISDR indicate that DRR activities still tend to be largely reactive rather than proactive (see UNISDR 2009b; UNISDR 2011a; UNISDR 2013a; UNISDR 2015; UNISDR 2011b; UNISDR 2011c;

UNISDR 2013b). To move forward, efforts will need to shift from a focus on short-term recovery to more long-term resilience building. However, much work has been done to strengthen DRR and development governance, in particular establishing governance structures and processes needed to support the implementation of the Hyogo Framework for Action 2005–2015 (HFA) and the Sendai Framework (UNISDR 2013b; UN General Assembly 2015a). Policy and systemic responses to the range of disaster risks differ between and within nations, as does the interpretation and implementation of DRR or disaster risk management (DRM) practices. Efforts urgently needed to unify responses and systematically reduce disaster risks globally have been guided by the HFA and now the Sendai Framework, but much more needs to be done to achieve transformative change. The main advances in DRR in meeting the five HFA Priorities for Action and the challenges that continue to inhibit advancement are outlined below.

Ensuring that DRR is a national and local priority with a strong institutional basis for implementation (HFA Priority 1)

ASEAN countries reported substantial progress in: (i) strengthening the institutions responsible for DRR; (ii) making DRR a national priority for development; (iii) developing legal instruments, strategies and documents toward a more proactive approach in disaster management; and (iv) establishing national and subnational bodies that implement DRR across sectors and levels of government (ASEAN 2013).

Six key challenges remain (ASEAN 2013). First, because the shift from disaster response to DRR is relatively new it is not yet well understood. This inhibits the dissemination and socialization of DRR policies. Second, the roles and responsibilities of national and local agencies require greater definition. Third, enhanced coordination of disaster-related laws and regulations among different sectoral agencies and levels of government is needed, particularly at the lower levels of government. Yet technical capacity at the local level remains limited due to a high turnover of government personnel. A lack of capacity at the local level also constrains the mainstreaming of DRR into local development processes and functions. The fifth challenge is related to financial constraints. While countries tend to have a national fund for disaster response and emergency relief, accessing these resources can be difficult if there are no concrete mechanisms and policies for the utilization of the fund and a poor understanding of the financial policies among the decision-makers. Finally, greater engagement of different sectors and stakeholders in DRR processes is required.

The identification, assessment and monitoring of disaster risks and enhanced early warning (HFA Priority 2)

To improve understanding of trans-boundary disaster risks at the regional level, Southeast Asia has established disaster information systems, including sub-regional hazard, risk and vulnerability information through the ASEAN Disaster Monitoring and Response System (DMRS) (UNISDR 2013b). However, these systems do not explicitly seek to holistically address both disaster and climate risks. Data scarcity, the lack of data sharing and weak coordination amongst key scientific and research agencies have been reported as reasons for this (UNISDR 2013b). These challenges prevent high-level commitment to implementing trans-boundary assessments (UNISDR 2013b).

At the national level, most countries have created initiatives to identify, assess and monitor disaster risks and to enhance early warning systems (EWS). However, not all countries have developed nationwide multi-hazard risk assessment systems; most risk assessments are hazard- and location-specific (ASEAN 2013). At the national level, most countries in the region lack the capacity to conduct comprehensive needs and impacts assessments after a major disaster without immediate assistance from the international community. Despite the growing appreciation of risk analysis as a basis for DRR across the region, challenges exist. Improving local capacities would enable more detailed risk analyses. However, improving knowledge and information sharing of disaster risks caused by different types of environmental hazards alone is not enough – more effort is needed to translate this knowledge into actionable methods, tools and guidelines for reducing risk and building resilience. Established EWS have varying levels of sophistication, but each system is linked with other systems or agencies in neighboring countries to address transboundary risks. However, the translation of risk and early warning information into risk-reducing actions is limited and in-country capacity is low and concentrated within specific stakeholders and agencies. There is a need to further enhance capacities at the local level in order to have more detailed and multi- and hazard-specific risk analysis and action at this level (ASEAN 2013).

Using knowledge, innovation and education to build a culture of safety and resilience at all levels (HFA Priority 3)

Countries have strengthened their disaster information systems by enhancing risk assessment models and tools (ASEAN 2013). Most have increased public awareness of DRR through various communication channels and the media, as well as pursuing strategies to promote a culture of resilience through the mainstreaming of DRR in the education sector, particularly school curricula. The major outstanding challenge is reaching people in remote areas on time. Dissemination is hampered by limited communication networks and internet connectivity outside urban areas. Furthermore, many people do not seek risk information or they interpret and act upon it differently for cultural reasons, such as world views, beliefs and values (Thomalla et al. 2015). Coordination challenges also exist between institutions. Risk information is not effectively conveyed from national offices to the district/city level, which prevents dissemination to communities at risk. Similarly, the flow of information from the community level to the district/city and national levels also needs to be improved. Inter-sectoral collaboration and coordination remains weak at all levels, posing challenges to resource sharing and the enforcement of policies, resulting in duplication and fragmentation. Greater efforts are needed to attract the participation of the private sector and more community members for public awareness campaigns to mobilize sufficient resources and to build a culture of disaster resilience. This is particularly important for highly vulnerable or hard to reach groups, like people with disabilities, who are routinely marginalized in everyday life in many parts of Southeast Asia and are wholly missing from mainstream DRR thinking and practice (Gartrell and Hoban 2013; UNESCAP 2012).

Reduce the underlying drivers of vulnerability (HFA Priority 4)

Most countries have initiated DRR mainstreaming in relevant sectors (ASEAN 2013). However, Southeast Asian countries have only just started to holistically address the underlying causes of risk, specifically those relating to climate change, through the development of regional DRR strategies. ASEAN member states are increasingly approaching underlying risks through regional guidelines for mainstreaming DRR into national development planning processes and by advocating for improved cross-sectoral coordination (UNISDR 2013b). Strengthening such regional cooperation would also enhance national and sub-national efforts of countries in the region. However, many gaps remain.

Underlying drivers of vulnerability to risk, such as those discussed in the previous section, continue to undermine DRR mainstreaming successes despite efforts to address them. The enforcement of strategies and relevant laws and regulations remains weak in many countries due to a lack of technical DRR capacity. This inhibits both preparedness and response, particularly at sub-national levels where disaster management agencies may lack the required understanding of regulatory and policy processes (UNISDR 2013b). Reducing vulnerability and exposure levels is difficult for national and local governments responsible for regulating the use of natural and land resources, overseeing the location and design of public infrastructures, and protecting the natural environment. And whilst there is some progress in providing better relief and rehabilitation after a disaster has occurred, regional plans for post-disaster recovery are still absent (UNISDR 2013b).

Strengthening of disaster preparedness for effective response at all levels (HFA Priority 5)

Effective risk reduction demands greater synchronization of disaster- and climate-related laws governing different sectoral agencies, at multiple levels of government up from the local level to the sub-regional and regional levels. Acknowledging this, Southeast Asia's governments are determined to improve their DRR capabilities and to increase their efforts in coordinating international humanitarian responses. Across the region, there are many examples of strengthened institutions, policies and practices as countries step up national and sub-national efforts to better link DRR, climate change adaptation and development (ASEAN 2013). For example, Djalante et al. (2012) examined Indonesia's efforts to improve DRR by implementing the HFA Priorities for Action. Improvements had been made to early warning systems, databases and risk knowledge due to advances in space-based technologies that link high-tech and local low-tech solutions and space- and ground-based systems, and a greater understanding of loss and damage systems. These are examples of practical measures that are being implemented to predict, monitor and communicate disaster risks to improve overall understanding and response.

At the regional level, and in support of the HFA initiative, some important cooperation and coordinated action on DRM in Southeast Asia occurs under the guidance of the Association of Southeast Asian Nations (ASEAN). In 1976, DRM was identified as one of eight ASEAN priorities and DRR has featured in ASEAN efforts ever since. ASEAN, through its various mechanisms and frameworks, is striving to build capacity of member states to manage disasters and enhance collaboration and partnership in the region, and has set a good example for other regions to learn from.

The current coordinating body, the ASEAN Committee on Disaster Management (ACDM), consists of heads of national agencies responsible for disaster management in the ASEAN Member Countries and has overall responsibility for coordinating and implementing the regional activities on disaster management (UNISDR 2011a). In pursuit of building disaster resilience among member nations, ACDM developed an ASEAN Regional Programme on Disaster Management (ARPDM) to support the development of a framework for action on disaster management and to provide a platform for cooperation and dialogue between regional and international actors for the period of 2004 to 2010. When the 2004 Indian Ocean Tsunami struck, affecting four of the ASEAN member states, work on establishing the ASEAN Regional Disaster Management Framework was already underway. The disaster influenced many discussions and negotiations among ASEAN leaders on the necessity of strengthening emergency relief, rehabilitation, reconstruction and prevention. This resulted in the formulation of the ASEAN Agreement on Disaster Management and Emergency Response (AADMER) that came into force in 2009 (UNESCAP and UNISDR 2010).

AADMER aims to improve ASEAN's capacity for effective and efficient disaster management by putting in place supportive policies, systems, plans, procedures, mechanisms and institutional and legal frameworks, at both the regional and national levels. It also provides support for risk reduction measures that link with climate change adaptation. AADMER has assisted member states in enhancing commitment toward regional collaboration in mainstreaming DRR into national development policies by making this commitment legally binding (Oxfam International 2014).

The ASEAN Agreement is currently the only HFA-related legally binding agreement in the world. This regional role played by ASEAN is particularly crucial considering that many hazards are regional or trans-boundary in scope, so a coordinated regional response is likely to be more effective and efficient. However, according to Oxfam International (2014), ASEAN must continue to work toward improving regional resources and the exchange of information, knowledge, expertise, funds and other resources in order to further support the implementation of DRR and adaptation policies and actions in ASEAN member states, as well as increase members' capacity to implement policies by enhancing partnerships between various actors.

How the post-2015 development agenda will be translated into action in Southeast Asia to ensure that investments in all development sectors consider DRR to achieve sustainable development remains to be seen. Throughout the Asia Pacific region, there is still a long way to go in integrating DRR with development planning and programming in all development sectors and across all levels (UNESCAP 2015). Efforts to strengthen resilience to disaster and climate-related risks must not only address increasing vulnerability and exposure levels but also ensure development decisions that are more inclusive and equitable by co-creating actionable knowledge that supports at-risk communities to shape their own development.

Transforming development and disaster risk for equitable resilience

As discussed throughout this chapter, development decision-making frequently compounds and exacerbates vulnerability levels as well as creating new risks (Schipper et al. 2016). For example, current unsustainable development pathways that have caused environmental degradation and increased greenhouse gas emissions are increasing the risks posed by natural hazards. These issues are evident in Southeast Asia. Development in the region has led to a decline in poverty, but inequality persists and furthermore climate change is a compounding factor in the threat against historic development achievements. Disaster impacts can also interfere with development pathways; the destruction of major development investments can set back development gains by years or even decades at an immense social and economic cost. In light of these issues, UNISDR (2009a) stress that DRR can counter these undesirable outcomes by contributing to the achievement of sustainable development, through reduced losses and improved development practices. Exactly how disaster risk relates to development, and what specific DRR and development practices are required to reduce vulnerability and build resilience, therefore needs to be better understood.

Launched in March 2015, the Sendai Framework (UN General Assembly 2015a) was the first major agreement of the post-2015 development agenda. The Sendai Framework reaffirms that development is crucial to reducing disaster risk as poverty, weak institutions, poor infrastructure and other development-driven factors are root causes of vulnerability. It also considers DRR as an integral part of social and economic development and essential for sustainable development and sets out targets for the post-2015 development agenda. These include a substantial reduction in mortality, in the numbers of people affected by disasters, economic losses and damage to critical infrastructure. The United Nations Sustainable Development Goals (SDGs) launched in September 2015 (UN General Assembly 2015b) also reaffirm the urgent need to reduce the risk of disasters and includes 25 targets relating to DRR in ten of the 17 SDGs.

Accepting that underlying disaster risks are generated inside development (UNISDR 2015), the question of *how* to transform development (and, in turn, disaster risk) comes to the fore. Aims to increase resilience without first asking the questions of 'for whom and to what purpose' may uphold systems and processes that fortify embedded states of inequality in terms of resource access and patterns of vulnerability and undesirable risk (see Lebel et al. 2006). Increasing *equitable* resilience presents as a worthier goal. Equitable resilience would mean addressing underlying failures in development and DRR, challenging the conditions and processes that perpetuate or sustain them, and opening possibilities for whole-scale transformation (Thomalla et al. 2016).

To significantly reduce disaster risk, we need to transform the way 'typical' DRR is done by critically unpacking the 'locked-in' relationship between development and disaster risk. Taking a systems approach, Thomalla et al. (2016) identify three interconnected pathways for unlocking this complex relationship, namely:

1 Addressing development-disaster risk trade-offs in decision-making;
2 Pursuing equitable resilience to reduce inequity in risk reduction and development; and
3 Achieving principles of inclusion, collaboration, social learning and system innovation through adaptive DRR governance.

Addressing development-disaster risk trade-offs in decision-making

Both disasters and resources are outcomes of human-environment interactions. It is therefore crucial to understand these links and to foster closer collaboration between organizations in both the DRR and development sectors. Linking risk reduction and resource management is a first step toward addressing trade-offs to minimize disaster losses while maximizing benefits (i.e., efficient and sustainable use of resources). Yet in recent decades, DRR research has drifted away from such a balanced view, focusing mainly on risk management and emergency response. While the need to mainstream DRR into development has been widely emphasized, resource exploration and use remain the dominant concerns and rationales in development decision-making.

Key trade-offs center on the distribution of power; the inequity of decision-making processes and outcomes; weighing up current and future demands in development; and finally weighing up a range of different risks, both disaster and non-disaster, that arise from one or another development decision, as well as the different probabilities of those risks. These trade-offs capture both the risk rationale (i.e., how risks are conceived and perceived, and how they are weighed against one another and prioritized) and the processes through which development and risk trade-offs are framed, deliberated and negotiated.

Pursuing equitable resilience to reduce inequity in risk reduction and development

Governments increasingly recognize the need to build the resilience of societies and ecological systems to a range of risks. However, such efforts rarely address the question of *whose* resilience should be built and why, and how they address the root causes of vulnerability. Consequently, efforts to build resilience often focus on technological solutions such as the construction of flood retention walls or drainage canals, which tend to protect primarily the wealthy, foreign investments and large businesses.

Addressing the inequity of development and DRR processes is a central pillar of how to transform these processes. Four ways in which more equitable and resilient development practice can reduce disaster risk are, (i) recognizing subjectivities, i.e., how social contexts, relations of power and categorizations are used to assign social and economic entitlements; (ii) ensuring inclusion and representation as well as agency, as opposed to exclusionary processes that disenfranchise certain groups; (iii) promoting transformation of the system(s) when it is no longer delivering well-being or reducing risks for a majority of people; and (iv) by working across geographical and temporal scales and levels of governance.

Achieving principles of inclusion, collaboration, social learning and system innovation through adaptive DRR governance

'Adaptive governance' refers to social, institutional, economic and ecological aspects of governance that help to build social-ecological resilience (see Folke et al. 2005). While adaptive governance has been extensively theorized in relation to natural resource management (e.g., Chaffin et al. 2014), there has been less analysis of how it can secure the integration of disaster risk and development. There is therefore a pressing need to understand what adaptive governance might look like in the context of transforming the relationship between development and disaster risk, and what institutions and processes would be needed – at the global, regional, national and sub-national scales – to achieve such transformations. Munene et al. (2016) argue that the presence of attributes of adaptive governance in the Sendai Framework creates the potential to serve as a catalyst for transformative change in DRR. These attributes are polycentric and multi-layered institutions; inclusion and collaboration; self-organization and networks; and social learning and system innovation (Djalante et al. 2011). However, there are considerable challenges, including, for example, power relations, hegemony and measuring resilience, that must be overcome for adaptive governance to achieve such a transformative potential. Finding solutions to these challenges will depend on political will and the presence and actions of engaged citizens and empowered social movements; the protection of advocacy groups and whistle-blowers; opportunities for collaborative learning; and the sustained combined effort of all DRR stakeholders.

This chapter has shown that there has been considerable progress in creating and strengthening DRR in Southeast Asia under the HFA and Sendai Framework. Moving forward into the era of the agenda for sustainable development, the Sendai Framework presents an opportunity for transformative change by enabling an adaptive governance approach to DRR. Yet, to reduce risks now and in the future, more needs to be done across governance levels in Southeast Asia to tackle the trade-offs inherent in development decisions that drive exposure and vulnerability levels, and to pursue equitable outcomes in resilience processes. Current and projected rates of disaster risk in the region call for a new research-for-policy agenda that enables and institutionalizes transformations for equitable resilience and sustainable development for all.

Notes

1 In this chapter, the region of Southeast Asia refers to the countries of the Association of Southeast Asian Nations (ASEAN): Brunei Darussalam, Cambodia, Indonesia, Laos, Malaysia, Myanmar, Philippines, Singapore, Thailand and Vietnam; as well as Timor-Leste.

2 Biological and technological hazard-group disasters, such as epidemics, are not included for analysis in this chapter.
3 A disaster is included in the EM-DAT database when either ten (10) or more people are reported as killed, one hundred (100) or more people are reported as affected, a state of emergency is declared, or there is a call for international assistance (CRED, 2016).
4 Multidimensional poverty is defined by the UNDP in the 2014 Human Development Report as poverty that is made up of several factors that constitute poor people's experience of deprivation – such as poor health, lack of education, inadequate living standard, lack of income, disempowerment and poor quality of work (UNDP, 2014b). The figure of 38 million excludes Brunei Darussalam, Malaysia, Myanmar and Singapore because national data are not available for these countries (UNDP, 2014b).
5 The Gini coefficient, or Gini index, is a statistical measure of income distribution of a country's residents and is commonly used to show income inequality.

References

Adger, N. (1999). Social vulnerability to climate change and extremes in coastal Vietnam. *World Development*, 27, pp. 249–269. doi:10.1016/S0305-750X(98)00136-3

Adger, W. N. (2006). Vulnerability. *Global Environmental Change*, 16, pp. 268–281. doi:10.1016/j.gloenvcha.2006.02.006

ADW, UNU-EHS. (2014). *World Risk Report – The city as a risk area*. Bonn, Germany: Alliance Development Works (ADW) and United Nations University – Institute for Environment and Human Security (UNU-EHS).

ADW, UNU-EHS. (2015). *World risk report 2015 – Focus: Food security*. Berlin and Germany: Bundnis Entwicklung Hilft (Alliance Development Works – ADW) and United Nations University – Institute for Environment and Human Security (UNU-EHS).

ASEAN. (2013). *ASEAN Regional progress report on the implementation of the Hyogo framework for action (2011–2013)*. Jakarta, Indonesia: The Association of Southeast Asian Nations (ASEAN).

Bankoff, G. (2003). Vulnerability as a measure of change in society. *International Journals of Mass Emergencies Disasters*, 21, pp. 5–30.

Birkmann, J. (2006). Measuring vulnerability to promote disaster-resilient societies: Conceptual frameworks and definitions. In: J. Birkmann, ed., *Measuring vulnerability to natural hazards: Towards disaster resilient societies*. Tokyo: United Nations University Press, 9–54.

Birkmann, J., Garschagen, M., Mucke, P., Schauder, A., Seibert, T., Welle, T., Rhyner, J., Kohler, S., Loster, T., Reinhard, D. and Matuschke, I. (2014). *World risk report 2014*. Bündnis Entwicklung Hilft and UNU-EHS.

Boyland, M., Nugroho, A. and Thomalla, F. (2017). The role of the Panglima Laot customary institution in the 2004 Indian Ocean tsunami recovery in Aceh. In: R. Djalante, M. Garschagen, F. Thomalla and R. Shaw, eds., *Disaster risk reduction in Indonesia*. Cham, Switzerland: Springer International Publishing, 357–376. doi:10.1007/978-3-319-54466-3_14

Calgaro, E., Dominey-Howes, D. and Lloyd, K. (2014). Application of the destination sustainability framework to explore the drivers of vulnerability and resilience in Thailand following the 2004 Indian Ocean Tsunami. *Journal of Sustainable Tourism*, 22, pp. 361–383. doi:10.1080/09669582.2013.826231

Cannon, T. (1994). Vulnerability analysis and the explanation of "Natural" Disasters. In: A. Varley, ed., *Disasters, development and environment*. Chichester, New York, Brisbane, Toronto and Singapore: John Wiley & Sons, 13–30.

Cardona, O. D., Van Aalst, M. K., Birkmann, J., Fordham, M., McGregor, G., Perez, R., Pulwarty, R. S., Schipper, E. L. F. and Sinh, B. T. (2012). Chapter 2: Determinants of risk: Exposure and vulnerability. In: H. Decamps and M. Keim, eds., *Managing the risks of extreme events and disasters to advance climate change adaptation. A Special Report of Working Groups I and II of the Intergovernmental Panel on Climate Change (IPCC)*. Cambridge and New York: International Panel on Climate Change (IPCC), 65–108.

Chaffin, B.C., Gosnell, H. and Cosens, B.A. (2014). A decade of adaptive governance scholarship: Synthesis and future directions. *Ecology and Society*, 19. doi:10.5751/ES-06824-190356

Comfort, L., Wisner, B., Cutter, S., Pulwarty, R., Hewitt, K., Oliver-Smith, A., Wiener, J., Fordham, M., Peacock, W. and Krimgold, F. (1999). Reframing disaster policy: The global evolution of vulnerable communities. *Environmental Hazards*, 1, pp. 39–44. doi:10.3763/ehaz.1999.0105

CRED. (2016). *EM-DAT: The international disaster database* [WWW Document]. Available at: www.emdat.be. Accessed 19 January 2017.

CRED and Guha-Sapir, D. (2017). EM-DAT: The Emergency Events Database - Université catholique de Louvain (UCL) [WWW Document]. Available at: http://www.emdat.be. Accessed 19 January 2017.

Davis, M. (2001). *Late victorian Holocausts: El Niño Famines and the making of the third world*. London and New York: Verso.

Djalante, R., Holley, C. and Thomalla, F. (2011). Adaptive governance and managing resilience to natural hazards. *International Journal of Disaster Risk Science*, 2, pp. 1–14. doi:10.1007/s13753-011-0015-6

Djalante, R., Thomalla, F., Sinapoy, M. S. and Carnegie, M. (2012). Building resilience to natural hazards in Indonesia: Progress and challenges in implementing the Hyogo Framework for Action. *Natural Hazards*, 62, pp. 779–803. doi:10.1007/s11069-012-0106-8

Fagan, B. (2009). *Floods, famines and emperors: El Niño and the fate of civilizations*. New York: Basic Books.

Fan, L. (2013). *Disaster as opportunity? Building back better in Aceh, Myanmar and Haiti*. London: Overseas Development Institute.

Folke, C., Hahn, T., Olsson, P. and Norberg, J. (2005). Adaptive governance of social-ecological systems. *Annual Review of Environment and Resources*, 30, 441–473. doi:10.1146/annurev.energy.30.050504.144511

Gaillard, J-C., Clavé, E. and Kelman, I. (2008). Wave of peace? Tsunami disaster diplomacy in Aceh, Indonesia. *Geoforum*, 39, pp. 511–526. doi:10.1016/j.geoforum.2007.10.010

Garschagen, M., Hagenlocher, M., Comes, M., Dubbert, M., Sabelfeld, R., Lee, Y. J., Grunewald, L., Lanzendörfer, M., Mucke, P., Neuschäfer, O., Pott, S., Post, J., Schramm, S., Schumann-Bölsche, D., Vandemeulebroecke, B., Welle, T. and Birkmann, J. (2016). *World risk report 2016*. Bündnis Entwicklung Hilft and UNU-EHS.

Gartrell, A. and Hoban, E. (2013). Structural vulnerability, disability, and access to nongovernmental organization services in rural Cambodia. *Journal of Social Work in Disability & Rehabilitation*, 12, pp. 194–212. doi:10.1080/1536710X.2013.810100

Gupta, S. (2010). *Synthesis report on ASEAN countries disaster risks assessment – ASEAN disater risk management initiative*. World Bank, UNISDR, ASEAN and GFDRR. Available at: https://www.unisdr.org/we/inform/publications/18872

IPCC. (2012). *Managing the risks of extreme events and disasters to advance climate change adaptation (No. A Special Report of the Intergovernmental Panel on Climate Change Working Groups I and II (Field, C. B., V. Barros, T. F. Stocker, D. Qin, D.J. Dokken, K. L. Ebi, M. D. Mastrandrea, K. J. Mach, G-K. Plattner, S. K. Allen, M. Tignor, and P.M. Midgley))*. Cambridge and New York: Cambridge University Press.

IRDR. (2011). *The FORIN project: Forensic investigations of disasters*. Beijing: IRDR.

IRIN. (2010). *Philippines: Landslide risk increasing* [WWW Document]. IRIN. Available at: http://www.irinnews.org/report/90967/philippines-landslide-risk-increasing. Accessed 23 January 2017.

Kelman, I. (2012). *Disaster diplomacy: How disasters affect peace and conflict*. Abingdon, UK: Routledge.

Lavell, C. and Ginnetti, J. (2014). *The risk of disaster-induced displacement in Southeast Asia and China (Technical Paper)*. Norwegian Refugee Council (NRC) Internal Displacement Monitoring Centre (IDMC), Geneva, Switzerland.

Lebel, L., Anderies, J., Campbell, B., Folke, C., Hatfield-Dodds, S., Hughes, T. and Wilson, J. (2006). Governance and the capacity to manage resilience in regional social-ecological systems. *Ecology and Society*, 11. doi:10.5751/ES-01606-110119

Martinez, A. (2017). *Does lower inequality mean less poverty?* [WWW Document]. Available at: https://blogs.adb.org/blog/does-lower-inequality-mean-less-poverty. Accessed 23 January 2017.

Munene, M. B., Gerger Swartling, Å. and Thomalla, F. (2016). *The Sendai framework: A catalyst for the transformation of disaster risk reduction through adaptive governance? (Discussion Brief)*. Stockholm Environment Institute, Stockholm, Sweden.

Oxfam International. (2014). *Can't afford to wait: Why disaster risk reduction and climate change adaptation plans in Asia are still failing millions of people, Oxfam Briefing Note*. Oxford: Oxfam International.

Oxford Policy Management. (2003). *Trends in growth and poverty in Asia: An economic background paper for ASREP*. Oxford: Oxford Policy Management.

Pelling, M. (2003). *Natural disasters and development in a globalizing world*. Hove, UK: Psychology Press.

Rigg, J. and Oven, K. (2015). Building liberal resilience? A critical review from developing rural Asia. *Global Environmental Change*, 32, pp. 175–186. doi:10.1016/j.gloenvcha.2015.03.007

Schipper, E. L. F., Thomalla, F., Vulturius, G., Davis, M. and Johnson, K. (2016). Linking disaster risk reduction, climate change and development. *International Journal of Disaster Resilience Built Environment*, 7, pp. 216–228. Available at: http://dx.doi.org/10.1108/IJDRBE-03-2015-0014

Statistics Indonesia. (2014). *Statistical yearbook of Indonesia 2014*. Badan Pusat Statistik (BPS) – Statistics Indonesia, Jakarta, Indonesia.

Thomalla, F., Boyland, M., Tuhkanen, H., Ensor, J., Swartling, Å. G. and Salamanca, A. (2016). *Transforming the relationship between development and disaster risk: Insights from a year of research (Research Synthesis Brief)*, Stockholm Environment Institute, Stockholm, Sweden.

Thomalla, F., Smith, R., Schipper, E.L.F. (2015). Cultural aspects of risk to environmental changes and hazards: A review of perspectives. In: Companion, M., ed., *Disaster's impact on livelihood and cultural survival: Losses, opportunities, and mitigation*. Boca Raton, FL: CRC Press, 3–18.

Turner, B. L., Kasperson, R. E., Matson, P., McCarthy, J. J., Corell, R., Christensen, L., Eckley, N., Kasperson, J. X., Luers, A., Martello, M. L., Polsky, C., Pulsipher, A. and Schiller, A. (2003). A framework for vulnerability analysis in sustainability science. *PNAS*, 100, pp. 8074–8079.

UN General Assembly. (2015a). *Sendai framework for disaster risk reduction 2015–2030 (No. A/CONF.224/CRP.1)*. UN, Sendai, Japan.

UN General Assembly. (2015b). *Transforming our world: The 2030 agenda for sustainable development (Resolution adopted by the General Assembly on 25 September 2015 No. A/RES/70/1)*.

UNDP. (2009). *Philippine human development report 2008/2009*. United Nations Development Programme (UNDP), Manila, Philippines.

UNDP. (2014a). *Advancing human development through the ASEAN community – Thailand human development report 2014*. United Nations Development Programme (UNDP), Bangkok, Thailand.

UNDP. (2014b). *Human development report 2014 – Sustaining human progress: Reducing vulnerabilities and building resilience*. New York: United Nations Development Programme.

UNDP. (2014c). *About Myanmar* [WWW Document]. UNDP Myanmar. Available at: www.mm.undp.org/content/myanmar/en/home/countryinfo.html. Accessed 23 January 2017.

UNEP. (2011). *UNEP year book 2011: Emerging issues in our global environment*. United Nations Environment Programme, Nairobi, Kenya.

UNEP. (2014). *Keeping track of our changing environment in Asia and the Pacific*. Bangkok, Thailand: United Nations Environment Programme.

UNEP. (2015). *UNEP annual report 2014*. Nairobi, Kenya: United Nations Environment Programme.

UNEP. (2016). *Global environment outlook GEO-6 regional assessment for Asia and the Pacific*. United Nations Environment Programme (UNEP).

UNESCAP. (2012). *Disability, livelihood and poverty in Asia and the Pacific: An executive summary of research findings*. United Nations Economic Commission for Asia Pacific, Bangkok.

UNESCAP. (2015). *Disasters without borders: Regional resilience for sustainable development (Asia Pacific Disaster Report 2015)*. Bangkok, Thailand: United Nations Economic Commission for Asia Pacific.

UNESCAP and UNISDR. (2010). *The Asia Pacific disaster report, 2010: Protecting development gains – Reducing disaster vulnerability and building resilience in Asia Pacific*. Economic and Social Commission for Asia and the Pacific (ESCAP), The United Nations Office for Disaster Reduction (ISDR), Bangkok, Thailand.

UNESCAP and UNISDR. (2012). *The Asia-Pacific disaster report 2012: Reducing vulnerability and exposure to disasters*. United Nations Economic and Social Commission for Asia and the Pacific (UNESCAP); United Nations International Stategy for Disaster Reduction (UNISDR), Bangkok, Thailand.

UN-HABITAT. (2006). *State of the world's cities 2006/07*. Nairobi, Kenya and London: Earthscan.

UN-HABITAT. (2010). *The state of Asian cities 2010/11*. UN Human Settlements Programme (UN-HABITAT), Fukuoka, Japan.

UNISDR. (2007). *Drought, desertification and water scarcity*. Geneva, Switzerland: United Nations International Strategy for Disaster Reduction (UNISDR).

UNISDR. (2009a). *UNISDR terminology on disaster risk reduction*. Geneva, Switzerland: UNISDR.

UNISDR. (2009b). *2009 global assessment report on disaster risk reduction: Risk and poverty in a changing climate: Invest today for a safer tomorrow*. Geneva, Switzerland: United Nations Office for Disaster Reduction (UNISDR).

UNISDR. (2011a). *Global assessment report on disaster risk reduction: Revealing risk redefining development*. United Nations Office for Disaster Reduction (UNISDR), Geneva, Switzerland.

UNISDR. (2011b). *Hyogo framework for action 2005–2015 mid-term review*. Geneva, Switzerland: United Nations Office for Disaster Risk Reduction (UNISDR).

UNISDR. (2011c). *HFA progress in Asia Pacific regional synthesis report 2009–2011*. Bangkok, Thailand: UNISDR Regional Office for Asia and Pacific.

UNISDR. (2013a). *From shared risk to shared value: The business case for disaster risk reduction (No. ISBN 978-92-1-132038-1)*. Global Assessment Report on Disaster Risk Reduction. United Nations Office for Disaster Reduction (UNISDR), Geneva, Switzerland.

UNISDR. (2013b). *The hyogo framework for action in Asia and the Pacific 2011–2013, progress review of HFA 2011–2013*. UNISDR.
UNISDR. (2015). *2015 Global assessment report on disaster risk reduction. Making development sustainable: The future of disaster risk management*. United Nations Office for Disaster Reduction (UNISDR).
Wisner, B., Blaikie, P., Cannon, T. and Davis, I. (2004). *At risk: Natural hazards, people's vulnerability and disasters*. London: Routledge.
World Bank. (2012a). *Thai flood 2011: Rapid assessment for resilient recovery and reconstruction planning (Vol. 2 of 2): Final Report (English), Thai flood 2011: Rapid assessment for resilient recovery and reconstruction planning*. World Bank, Bangkok, Thailand.
World Bank. (2012b). *Well Begun, not yet done: Vietnam's remarkable progress on poverty reduction and the emerging challenges*. World Bank, Hanoi, Vietnam.
Yale University. (2014). *Environmental performance index*. New Haven, CT: Yale University Press.
Zou, L. and Thomalla, F. (2010). Social vulnerability to coastal hazards in Southeast Asia: A synthesis of research insights. In: C.T. Hoanh, B. Szuster, S. Cam, A. Ismail and A. Noble, eds., *Tropical Deltas and Coastal Zones: Food Production, Communities and Environment at the Land-Water Interface, Comprehensive Assessment of Water Management in Agriculture Series*. Wallingford, Oxfordshire, UK and Cambridge, MA: CABI, 367–383.

28
UPSCALED CLIMATE CHANGE MITIGATION EFFORTS
The role of regional cooperation in Southeast Asia

Noim Uddin and Johan Nylander

Introduction

Sustainable development is dependent upon the successful attainment of climate change mitigation targets, with failure to meet these targets threatening existing and future development gains. The need for every region[1] to contribute to reaching the target of limiting global temperature rise to 2^0C or below provides a strong imperative for action (IPCC 2014a). As a global environmental and social development challenge, mitigation of climate change requires a global solution, yet regions through the coordinated actions of countries could effectively play an important complementary role to such efforts as regional similarities and differences can enable more efficient and appropriate mitigation measures to be implemented (IPCC 2014b). The regional dimension of development is recognized as being critical for an effective and coordinated response to an ever growing number of development challenges (UN 2011). However, as global agreement to address climate change has proved difficult to achieve, regional cooperation[2] may provide useful opportunities to further accomplish global mitigation objectives, at least partially (IPCC 2014a). While the United Nations Framework Convention on Climate Change (UN FCCC) as a global climate agreement with its own targets and mechanisms has been necessary, it has been hampered by divergent views on its design and powers both in the past and for future implementation. The Paris Agreement is a significant achievement in terms of global efforts, building to a large extent on the voluntary plans and actions of countries at the national level (UN 2016). Developing countries that are particularly vulnerable to climate change could initiate vital action at a regional level, contributing to the protection of the collective good in the absence of a strong top-down regime (Ostrom 2010) on climate change mitigation. However, country level plans in developing countries in Asia are struggling and a regionally coordinated approach to taking strong, ambitious and rapid collective action on climate change is still lacking (Anbumozhi 2015). The submission of national plans (Intended Nationally Determined Contributions or INDCs) in advance of the UNFCCC Conference of Parties (COP) 21 Paris climate talks reinforced the role of nation states as key actors in realizing climate change targets. However, it is still only in the European Union (EU)[3] that there exists a strong regional approach to future climate change action in the form of the EU-wide INDCs (EU 2015). Regions can thus play an important role in delivering a coordinated approach toward global climate change response objectives. In the context of Southeast Asia, existing cooperation on energy, technology and other trans-boundary issues could potentially link with enhancing greenhouse gas (GHG) mitigation in order to support wider development goals.

Regional cooperation ranges from forums for economic cooperation and free trade among countries to deeper integration on economic, social, cultural and strategic matters as in the case of the EU. Similar to the EU, one of the Association of Southeast Asian Nations' (ASEAN's) objectives has been to accelerate economic growth along with social progress and cultural development for the region. As such, economic development remains as the prime motivator for ASEAN nations at the present (Chang 2014). Although ASEAN and the EU both share this central objective of economic growth and development, the EU has widened its mandate considerably to encompass a range of social, political and environmental issues, including climate change. In fact, the EU has been a central force in regional cooperation on climate change mitigation having established a carbon trading scheme since 2005 (IPCC 2014a). The implementation of a regional renewable energy policy in the EU, so crucial to achieving mitigation targets, was also very much dependent upon the national policies of EU member states.

In this chapter, the Southeast Asian region is examined as a case study of regional cooperation on climate change mitigation as a central element of sustainable development. ASEAN[4] was established as a regional organization by the Bangkok Declaration of 1967; originally composed of five countries – Indonesia, Malaysia, the Philippines, Singapore and Thailand – it was later enlarged with the inclusion of Brunei Darussalam in 1984, (Kreinin 1992), Vietnam in 1995, and Lao PDR and Myanmar on 1997 (ASEAN 2016) with objectives to accelerate economic growth and development through cooperation (Sovacool 2009). While there are several other regional fora across Southeast Asia, for example, APEC (Asia Pacific Economic Cooperation), BIMSTEC (Bay of Bengal Initiative for Multi-Sectoral Technical and Economic Cooperation), and the GMS (Greater Mekong Sub-region) program of the Asian Development Bank, this chapter primarily focuses on ASEAN because of the unique nature of cooperation across Southeast Asian nations and the range of issues it addresses (ASEAN 2016).

The specific question explored in this chapter is whether regional cooperation, including the application of upscaled market-based approaches to mitigation, can drive GHG mitigation efforts further. Upscaling means addressing mitigation in one or more sectors at a national or regional level, in contrast to project-by-project mitigation that was common under earlier climate change mitigation schemes, such as the Clean Development Mechanism (CDM). The value of upscaling lies in its potential to accomplish transformational change (Mersmann et al. 2014).

The chapter first discusses ASEAN's energy, GHG emissions and development paradigm and then discusses different approaches for assessing regional cooperation in Section 3. Section 4 provides an assessment of regional cooperation focusing on examples that provide opportunities for GHG mitigation from around the world before reflecting on potential links with Southeast Asia. Regional cooperation in Southeast Asia is then analyzed in Section 5 with a number of possible regional cooperation approaches considered, including the possibility for upscaled regional cooperation on trans-boundary issues, as discussed in Section 6. Section 7 provides concluding remarks.

Energy, GHGs emissions and the development paradigm – the case of ASEAN

Most of the member countries of ASEAN have a strong reliance on natural resources and long coastlines together with a high proportion of the population living within the Low Elevation Coastal Zone. This makes the Southeast Asian region highly vulnerable to the effects of climate change. These development challenges are compounded by a high dependency on fossil fuels and associated environmental concerns including air pollution and other environmental emissions (Karki et al. 2005). Hence, economic and environmental policies and practices in general have to be enhanced in order to achieve longer-term sustainable development in the region. GHG emissions associated with fossil fuel use are the focus here because reducing fossil fuel dependence is a major development concern in regards to reaching the 2^0C target. This does not imply that sustainable development should be reduced to a singular focus on a low-emission development model; however, a transition to low-emission development is the focus here due to the need to address climate change mitigation as part of ensuring future sustainable development.

ASEAN introduced a number of economic cooperation programs in the mid-1970s with a specific focus on energy that have been expanded and enhanced in recent years. For example, the ASEAN Council of Petroleum (ASCOPE) was formed in 1975, just after the first oil shock in 1973, in order to navigate energy challenges and mitigation measures in response to the oversupply or undersupply of petroleum (Karki et al. 2005). As a follow-up to this agreement, the ASEAN 1998 Summit adopted policies related to "Energy security and sustainability of energy supply" and also established a policy framework and implementation strategy for trans-ASEAN energy networks, including the ASEAN Power Grid and the Trans-ASEAN Gas Pipeline projects (Sovacool 2009; Yu 2003). Currently, the 4th ASEAN Plan of Action on Energy Cooperation (APAEC) covering the period during 2016 to 2020 (IEA 2015) is being prepared.

As one of the most dynamic economic regions in the world, ASEAN's economy varies according to the diversity of its energy resources and socio-economic factors. ASEAN's fast growing energy demand is driven by its economic and demographic growth as well as increasing urbanization. The demand for energy is projected to grow by 80 percent by 2040 from the current level in the region (IEA 2015). At the same time ASEAN governments need to provide access to the 120 million people in the region who still lack round-the-clock electricity (IEA 2015).

Meeting ASEAN's energy needs with unprecedented increases in fossil fuel use and associated GHG emissions would be detrimental to the climate, having flow-on effects on vulnerable communities and economic sectors through climate change impacts. Although Southeast Asia is not the main global CO_2 producer, its emissions of CO_2 will become significant if no or weak mitigation measures were to be implemented (Lee et al. 2013). According to International Energy Agency (IEA) estimates, CO_2 emissions will rise to 2.4 Gt in 2040 from 1.2 Gt in 2013 (IEA 2015). If the projected reliance on fossil fuels is realized, ASEAN is poised to become a significant global contributor to GHG emissions as the use of coal in electricity generation would rise to 50 percent by 2040 from 32 percent in 2013 (IEA 2015). Hence, mitigation of GHG emissions across the ASEAN region is an urgent and important task in the foreseeable future. Alternative energy sources and effective policy measures (for example, phasing out or reducing subsidies to fossil fuels,[5] demand side management and efficiency measures in existing energy systems) could complement electricity generation switching from fossil fuel systems to systems with a larger share of renewable energy. ASEAN member countries are also looking at market-based solutions in order to accomplish mitigation results at a larger scale (WB 2014).

Integrating climate policies into broader development policies facilitates the transition of developing economies toward a low-carbon development paradigm (Anbumozhi 2015). Climate change has been considered a core issue during recent ASEAN Summits, with member country leaders approaching climate change mitigation through the adoption of declarations or statements, such as the ASEAN Joint Statement on Climate Change 2014 (ASEAN 2014). Recognizing the challenging relationship between climate change policy on the one hand and equity and development on the other (Bosello et al. 2003), the countries have adopted these declarations or statements regularly (Lee et al. 2013). In general, climate policy has both positive and negative effects on income distribution and growth, and affects different groups in society in different ways, and therefore has implications on equity and development. At the same time, equity and development affects a country's decision to sign a climate treaty and therefore have implications on the effectiveness and efficiency of climate policy (Bosello et al. 2003). For example, according to the Intergovernmental Panel on Climate Change Fifth Assessment Report, mitigation policy could devalue fossil fuel assets and reduce revenues for fossil fuel exporters (IPCC 2014a). This could indeed affect fossil-fuel export-dependent countries' development objectives. While this is relevant to Gulf Cooperation Council Region countries, some ASEAN countries that depend on fossil-fuel export may also have similar experience, for example, Indonesia – as a major coal exporter in the ASEAN (IEA 2015). The pricing of GHG emissions, directly via carbon taxes or through cap-and-trade systems as part of any comprehensive strategy (OECD 2008), could be useful in the context of ASEAN. However, such an approach may fail to deliver affordable energy to the poorest in the ASEAN region since taxes may increase the price of electricity, as fossil-fuel subsidies amounted to US$36 billion in 2014 in the ASEAN (IEA 2015). However, any implementation of common actions across the region, such as GHG mitigation programs or projects in connection with development cooperation, appears to be challenging due to disparities among ASEAN member countries in terms of level of development, availability of resources and economic status, and not least, the weak structure of ASEAN itself (Areethamsirikul 2008).

Assessing regional cooperation approaches

A case study approach has been adopted for this study as it is appropriate for obtaining an in-depth understanding of a particular research context with a view to providing analytical generalizations (Yin 2003), focusing in particular on contemporary development issues in Southeast Asia.

The analytical methods used for this research include: collection of information from various sources and desktop assessment of literature and communication with key agencies. Besides information on regional cooperation on climate-relevant and climate-specific programs that have strong links with the mitigation of GHG emissions, information on regional cooperation on trans-boundary issues has also been gathered from government and donor agency reports.

This approach was employed in order to allow a systematic analytical assessment of available evidence regarding regional cooperation. To ensure the authenticity of gathered information, cross-comparison was carried out among different sources and the most reliable and up-to-date information has been incorporated. Uddin and Taplin (2015) have previously reviewed a number of regional cooperation contexts across the globe and assessed whether or not regional cooperation offers synergetic benefits, such as accelerated energy access (Uddin and Taplin 2015). However, this study explores another important development dimension, namely the links between regional cooperation in Southeast Asian countries and upscaling GHG mitigation in the region.

The region as a frame for action on climate policy and GHG mitigation has not as yet been given a lot of attention in the literature related to development, energy and climate change. Regional cooperation for example in the ASEAN could accelerate development through regional peace and stability (Kusuma 1998). It has however been used as a unit for analysis in political ecology (Bryant and Bailey 1997). From a development perspective, regions are now seen as playing a crucial role in constituting economic globalization as they are an important force in economic development (Sean 2010). As such, efforts at a regional level on climate change mitigation can shape economic development in positive ways at multiple scales.

Other types of regional cooperation, for example the G7, G20 or newly established Like Minded Group of Developing Countries (LMDC),[6] are not discussed in this chapter as the chapter mainly focuses on regional cooperation linked to ASEAN and GHG mitigation. However, links are made where relevant. For instance, it is worth mentioning that whereas Malaysia has been very active in the LMDC, advocating the responsibility of the industrialized countries to act on climate change, the Philippines has recently left this group. The varying support of ASEAN countries for a negotiation group such as LMDC shows that different approaches to climate policy exist across the region, presenting a particular challenge to regional cooperation efforts in this area.

Regional cooperation – examples from other regions and links with Southeast Asia

Regional economic cooperation that is relevant to climate change mitigation may be broadly classified into two main categories: climate-specific initiatives and climate-relevant initiatives (IPCC 2014a). The following section discusses regional cooperation initiatives in energy, trans-boundary management, technology and GHG mitigation, and provides an assessment of regional cooperation initiatives on climate change mitigation that are of relevance to Southeast Asia. However, the likely mitigation impact of most regional cooperation agreements is found to be limited, as also identified by Grunewald et al. (2013) due to their informal and mostly voluntary nature.

Regional cooperation – energy

The EU Regional Cooperation on Renewable Energy (EC 2009) has had mixed success with implementation, with the share of renewables accounting for 13 percent in 2011 compared to 8.5 percent in 2005. With binding national targets, growth in renewable energy has increased but needs to average 6.3 percent to meet the overall 2020 target (20 percent) (IPCC 2014a). As analyzed by Uddin et al. (2006), implementation of renewable energy policy in the EU was very much dependent upon the national policies of EU member countries (Uddin et al. 2006). A relatively successful form of regional cooperation linked with energy is the well-established Nord Pool – the common market for electricity in Scandinavia[7] that has allowed cross nation sharing of wind energy (Kopsakangas-Savolainen and Svento 2013). However, the regional setting of Scandinavia bears significant differences with the wider EU setting, particularly in regard to the institutional settings under EU and cooperation objectives. This experience has bearing for ASEAN member countries in regards to the advancement of renewables and cross-border energy trade. The Scandinavian Nord Pool experience also demonstrates that regional collaboration cannot be easily transferred to larger regional settings where significant differences in institutions and the extent of deregulation exist, as is the case when comparing the Nordic countries with the whole of the EU.

In a developing countries context, the African region has a number of power pools. However, the scale of trade within these power pools has been small, leading to continual inefficiencies in the distribution of electricity generation across the African continent and less integration with renewables – except for large hydropower (Eberhard et al. 2011). Examples from Africa could signal that ASEAN is on the right track in its attempt to collaborate with regional energy policy (ASEAN 2015).

Compared to the above examples of regional cooperation on energy from other regions, ASEAN's initiative on energy cooperation, the ASEAN Plan of Action for Energy Cooperation

(APAEC) 2016–2025, aims to enhance energy security cooperation and initiate steps toward connectivity and integration in ASEAN (ASEAN 2015).

Regional cooperation – transboundary issues

While regional cooperation on gas supply and transmission systems offers opportunities for switching from emission-intensive fossil fuels to use of gas for electricity generation, such regional gas grids are limited to certain geographical contexts, for example in East Asia and Eastern Europe. While the Trans-ASEAN Gas Pipeline has been presented as one of the core programs under ASEAN in the Plan of Action for Energy Cooperation, future gas networks will include Liquid Natural Gas (LNG). To date a number of bilateral gas interconnection network projects have already been completed and, although not aiming to support a transition to low-carbon development, a number of future regional programs seek to enhance petroleum security measures in ASEAN through an emergency petroleum sharing scheme in response to adverse supply and demand situations (ASEAN 2015).

Regional cooperation on hydropower is already well advanced in Southeast Asia and could potentially play an important role in electricity trade in Southeast Asia via joint development of transboundary river systems. Although hydropower is a less GHG-intensive form of energy, such development needs to comply with stringent requirements regarding environmental, social and economic sustainability (IPCC 2011). In regard to transboundary river systems and hydropower development, sustainable hydropower development in Mekong River Region by the Mekong River Commission (Mekong River Commission member countries include: Lao PDR, Cambodia, Thailand and Vietnam) and through the ADB funded GMS program could enhance energy access to local communities and industrial development (MRC 2010) and open power trade within ASEAN region. Moreover, it could encourage poorer ASEAN countries – for example, Lao PDR and Cambodia – to access cheaper energy (Shi 2016). However, such large-scale hydropower development should ensure mitigation of food or water security aspects.

Regional cooperation – technology focused

Technology-focused regional cooperation initiatives, supportive of GHG mitigation, include clean energy, low-carbon development and climate resilient growth as foci. One example of this form of cooperation is the Energy and Climate Partnership of the Americas (ECPA), which focuses on power sector integration, advancement of renewables, promotion of sustainable energy innovation and climate change adaptation (ECPA 2014). An example of a non-traditional[8] form of regional cooperation includes the Asia Pacific Partnership on Clean Development and Climate (APP)[9], which initially focused on clean technologies in power generation and other emission-intensive industries (Taplin and McGee 2010), and a more recent initiative, the Clean Energy Ministerial (CEM), which aims to leverage clean energy technology experience from different regions (CEM 2014).

Another example of technology focused cooperation is Sustainable Energy Technology at Work (SETatWork), which aimed to transfer EU technologies and Clean Development Mechanism (CDM) know-how (Uddin et al. 2015)[10] to some Southeast Asian countries (EC 2011). The Carbon Sequestration Leadership Forum (CSLF)[11] also aimed to develop and deploy carbon capture, use and storage technologies. South-South cooperation has been fostered via the IBSA (India, Brazil and South Africa) Trust Fund implemented to assist LDCs with, for example, solar energy programs for rural electrification (UNDP IBSA 2013).

While many regional cooperation initiatives are focused on technology cooperation, the GHG mitigation aspect is often lost, however, most of them have potential to link with GHG mitigation. Technology transfer under regional cooperation appears to be in both existing regional settings or in settings where member countries are able to cooperate bilaterally or multilaterally. Inter-regional cooperation on technology transfer could potentially enhance cooperation in support of climate change mitigation between industrialized and developing countries.

Regional cooperation – GHG mitigation

The European Union (EU) Emission Trading Scheme (ETS) is by far the largest emission trading system in the world, covering over 12,000 installations belonging to over 4,000 companies and initially over 2Gt of annual CO_2 emissions from across 28 EU member states as well as Norway. The CDM under the Kyoto Protocol Mechanism is tied to the EU ETS and emerged as one of the main climate finance mechanisms that leveraged investments in GHG mitigation projects in developing countries, including Southeast Asian countries. Among the over 7,705[12] registered projects, most projects were implemented in East Asia, South Asia, Southeast Asia, Latin America and the Caribbean (URC 2014). Compared to other regions, Sub-Saharan Africa has very few registered CDM projects. In contrast, CDM program of activities (PoA) or pCDM[13] that allow bundling an unlimited number of small size CDM projects differs markedly, as Sub-Saharan Africa's share is ten times higher than the number of CDM projects registered in the same region, while East Asia and South Asia shares are only a third compared to the total number registered CDM projects in the regions. The reason for this more balanced distribution of PoAs is the higher attractiveness for project proponents of small-scale projects in low-income contexts (Hayashi et al. 2010). Carbon credit buyers are concentrated mainly in Western Europe and there was initially a strong demand from the EU ETS for credit generated through CDM projects, but during the last few years, the market has been driven mainly by voluntary climate compensation.

Besides being linked as an offset program to the regional EU ETS, CDM has not been implemented as a regional program in participating countries, however, when CDM Programs of Activities are implemented in several countries it appears to indicate that regionally coordinated initiatives could play a wider role in mitigation of GHGs. Other programs such as the Western Climate Initiative (WCI)[14] and Regional Greenhouse Gas Initiative (RGGI)[15] have not been assessed as they fall outside the regional definition adopted in this chapter.

Regional cooperation initiatives in Southeast Asia

This section focuses on regional cooperation in Southeast Asia. In the area of regional energy cooperation initiatives in ASEAN these have mainly been motivated by concerns about security of energy supply (Kuik et al. 2011), energy access (Bazilian et al. 2012), rising fossil fuel imports and rapidly growing GHG emissions and air pollution (IEA 2010a; Cabalu et al. 2010; IEA 2010b; UNESCAP 2008). Table 28.1 provides current regional cooperation across Southeast Asia, with ASEAN initiatives being most prominent.

In terms of intra-regional cooperation, ASEAN countries have an active agenda on many energy policy fronts. The ASEAN Centre for Energy (ACE) is an intra-regional body under ASEAN that facilitates and coordinates the work of specialist organizations, including the Forum of Heads of ASEAN Power Utilities/Authorities (HAPUA), the ASEAN Council on Petroleum (ASCOPE), the ASEAN Forum on Coal (AFOC), the Energy Efficiency and Conservation Sub-sector Network (EE&C-SSN) and the New and Renewable Sources of Energy

Table 28.1 Regional cooperation in the Southeast Asian region

Regional cooperation	Type of agreement	Goals/Strategies	Comments
ASEAN Centre for Energy (ACE)	Energy	Catalyst for economic growth and activities on energy	ACE is an intergovernmental body that facilitates and coordinates the work of specialist energy organizations
ASEAN+3	Energy	Energy security, oil market and natural gas, renewable energy and energy conservation	ASEAN and Japan, China and Korea, focus on regional cooperation on energy security
ASEAN Petroleum Security Agreement	Energy	Petroleum sharing schemes	Petroleum supply security in ASEAN
ASEAN Council of Petroleum	Energy	Offshore oil and gas infrastructures	Guidelines on decommissioning of oil and gas infrastructures
ASEAN Power Grid	Electricity	Electricity supply security	Enhance electricity supply and distribution
ASEAN Energy Market Integration	Energy	Electricity access, spill-over effects	Proposed under ASEAN Energy Cooperation
Trans-ASEAN Gas Pipeline	Energy	Energy resources security	Exploitation energy resources
APAEC	Energy	Energy cooperation	Boost in renewable-based electricity generation
COGEN	Technology focused	Technology dissemination	A number of pilot cogeneration technologies in ASEAN
APEC-EWG	Technology focused	Energy cooperation	Maximize energy sector's contribution in economic growth
AEEC-Japan	Technology focused	Energy technology cooperation	Efficient energy solutions, smart grids
APCTT	Technology focused	Develop, transfer and adaptation of technology	UN led cooperation among ESCAP countries
BIMSTC	Trans-boundary	Enabling environment for economic development	International organization involving countries from ASEAN
Mekong River Commission	Trans-boundary	To enhance sustainable use of water resources, including for hydropower	Opportunity for GHG mitigation
ADB GMS Regional Power Trade	Trans-boundary	Environmental sustainability of future power plants	Regional power coordination center

Source: Compiled by authors

Notes: COGEN (Cogeneration) – EC-ASEAN Cogeneration Program;

Mekong River Commission – member countries are Cambodia, Lao PDR, Thailand and Vietnam with dialogue partners China and Myanmar;

GMS (Greater Mekong Sub-region) – member countries are Cambodia, China (Yunnan and Guangxi Zhuang Autonomous Region), Lao PDR, Myanmar, Thailand and Vietnam;

ESCAP (Economic and Social Commissions for Asia and the Pacific) – member countries are from Asia and the Pacific regions;

BIMSTC (Bay of Bengal Initiative for MultiSectoral Technical and Economic Cooperation) – member countries are Sri Lanka, Bangladesh, India, Myanmar, Thailand, Bhutan and Nepal;

APCTT (Asian and Pacific Centre for Transfer of Technology of UN ESCAP) – member countries are from Asia and the Pacific regions.

Subsector Network (NRE-SSN). One example of transforming relevant regional policies into action during the implementation of the APAEC (ASEAN Plan to Action for Energy Cooperation) 2004–2009 was the achievement of the 10 percent target to increase renewable energy-based capacities for electricity generation (ASEAN 2010; IEA 2010a; Sovacool 2009). In addition to efforts to establish the ASEAN Power Grid and Trans-ASEAN Gas Pipeline, the ASEAN Energy Market Integration (AEMI) initiative takes off from existing efforts toward greater ASEAN Energy Cooperation, however AEMI is an example of stronger regional cooperation as it involves integrating markets (Navarro et al. 2013).

A technology-focused example of regional cooperation is the Energy Working Group of the APEC (Asia Pacific Economic Cooperation) forum, which seeks to maximize the energy sector's contribution to the APEC region's economic and social well-being (APEC 2014). Of the 21 member economies seven ASEAN countries (Brunei Darussalam, Indonesia, Malaysia, Philippines, Singapore, Thailand and Vietnam) are members of APEC. Hence, regional cooperation under APEC also offers an opportunity to address GHG mitigation.

Programs and initiatives under technology-based regional institutes, for example Asia Energy Efficiency and Conservation Collaboration Centre (AECC) under Energy Conservation Centre of Japan (ECCJ), the Asian and Pacific Centre for Technology Transfer (APCTT) under UNESCAP, could potentially be linked with GHG mitigation – especially in the energy and industrial sector.

Regional cooperation initiatives in Southeast Asia – potential for GHG mitigation

As mentioned above there are a number of existing forms of cooperation across the region that could potentially be linked to stronger GHG mitigation action. For example, regional cooperation that directly focuses on energy interconnection through the ASEAN Power Grid and Trans-ASEAN Gas Pipeline (Shi and Mallik 2013) could potentially contribute to GHG mitigation. In other contexts, for example, the Nordic grid integration optimized efficiency of electricity production and resulted in reduced emissions of GHGs (Finnsson 2016). Moreover, the interconnectors between Norway/Sweden and Germany and between Norway/Denmark and UK have shown a strong influence in terms of decreased CO_2 emissions (Fingrid 2014). This is because interconnections assisted in replacing fossil fuel-based electricity production with renewables but also added flexibility to the system by optimization of the system generation dispatch during peak/off-peak periods and intermittent wind periods. A related example of policy collaboration is the Green Certificates program between Sweden and Norway that started during 2003 and since 2012 resulted in a common market for Sweden and Norway called the el-certificate market (Wolfgang et al. 2015). In this regard AEMI could play an important role in explicitly addressing the GHG mitigation opportunities accompanying an integrated energy market. The ASEAN Economic Community (AEC) provides an institutional framework for the integration of energy markets, which are currently fragmented (Shi 2016); however, progress in ASEAN energy market integration remains slow due to a lack of will. Additionally some ASEAN member countries are reluctant to initiate energy market integration given their early stage of state building as integration is perceived as potentially undermining nation building efforts.

In the context of energy security in realizing the potential of GHG mitigation, the ASEAN+3 CDM Cooperation Program has been named as an ASEAN+3 Cooperation on Mitigation Program by ASEAN+3 Ministers of Energy. Despite the concern with mitigation demonstrated in these initiatives, other regional cooperation initiatives involving ASEAN member countries (for example, regional cooperation on higher education, communicable disease and pandemic preparedness, competition policy and security cooperation) have shown limited concern with GHG mitigation. In other words, GHG mitigation is likely only to be successful in regional cooperation initiatives on energy if this is an explicit focus. Upscaling potentially provides opportunities to strengthen such efforts.

Upscaling of GHG mitigation objectives can take place as both bottom-up and top-down approaches linking with existing and relevant regional cooperation in the region. For instance, it would be possible to have multinational CDM-programs (as they are present in Africa and the Middle East). The idea of replication could also work in the context of ASEAN if barriers, such as a lack of adequate policy frameworks, institutional settings, markets, finance, technological development, human resources and slow diffusion rates of new technologies, are mitigated appropriately (Uddin and Taplin 2009). This is a type of cooperation that could be driven by companies, project developers and communities in the region and under the ASEAN Cooperation on Energy framework. The second way is more top-down, where ASEAN member countries decide to collaborate on policy or energy generation. The ASEAN+3 Cooperation on Mitigation Program could strengthen partnerships to learn from CDM in each member country and upscale CDM Programs and experiences across the region. Yet despite these possibilities, the lack of infrastructure and policy framework continues to act as barriers to regional integration in the region, especially the prevailing fossil fuel subsidies and strong focus on national energy security rather than on regional energy security taken by many countries.

The direction that many developing countries, including those in Southeast Asia, take to meet their energy needs will have profound impacts on climate change and the economy (Lehmann et al. 2014). This warrants that any instrument that facilitates both economic development and mitigation of GHGs has to be designed carefully. Any negative impact due to GHG mitigation policy, for instance on equity, needs to be taken into consideration during the design of any mitigation policy framework (IPCC 2014b). While a number of instruments have been tested in several ASEAN member countries, for example, in Thailand and in Vietnam (Uddin et al 2010; Uddin et al. 2009), upscaling of these instruments across the ASEAN region through regional cooperation could facilitate sustainable development beyond mitigation of GHGs. One such pathway could potentially be result-based finance linked with global carbon governance (Nylander 2015).

The framework for carbon finance and result-based finance under the Paris Agreement is still to be developed. Currently, a generally advocated concept is carbon pricing, which includes both taxes on emissions as well as trading and rewarding emission reductions through issuance of carbon credits. However, this frame while widely adopted, e.g., through emissions trading implemented from California to China, Europe, Kazakhstan to Korea, is yet to be operationally reflected in the Paris Agreement. The Paris Agreement has established 'co-operative approaches' that may open for linking of emissions trading schemes or possibly providing a window for bilateral collaboration such as Japan's Joint Crediting Mechanism, and a 'mechanism for mitigation and sustainable development' that could play a similar role as the CDM.

Over 180 countries, including many in Southeast Asia, have now submitted their Intended Nationally Determined Contributions (INDCs) to the UNFCCC, where many countries refer positively to the use of international carbon markets. The bottom-up approach of the INDC scheme allows countries, following the first submission of their contributions, to also examine other options including bilateral and multilateral collaborations to extend emission reductions further. For this process to be successful, having a facilitating structure and tools that suit different purposes in different places – including market-based approaches – could prove to be very useful. While the tools available post-2020 are not yet known in detail, upscaling of existing regional cooperation initiatives such as ASEAN+3 Cooperation Mitigation and several other initiatives such as elements of the carbon pricing readiness demonstrated by Indonesia, Thailand and Vietnam (WB 2014) could form building blocks under a future ASEAN-market-based framework. Several countries in the region are already planning emissions trading schemes (e.g., Thailand and Vietnam) and the future linking of these schemes could enhance cost-efficiency.

The bottom-up approach could also imply using the Nationally Appropriate Mitigation Action (NAMA) concept. NAMAs could be something of a hybrid of national and regional approaches as suggested by the World Bank (Benitez 2013), where international finance (public or private result-based finance) helps drive integration through assisting the development of NAMAs that are similar but country specific. In other words, there will be several tools available for ASEAN countries if they wish to pursue regional collaboration with regard to emission reduction objectives, tools that could provide for cost efficient reductions in the region. Market approaches as such, and the Paris Agreement's approaches for collaboration with regard to the implementation of the national plans (INDCs), could open a window of opportunity for addressing GHG mitigation in regional energy collaboration settings. These approaches may also provide other potential development benefits in relation to energy security and socio-economic development.

Concluding discussion

In order to realize GHG mitigation for sustainable development, regional cooperation efforts in Southeast Asia require harmonizing technical and regulatory potential as well as consideration of the diversity of national development interests and strategies. Creating regional markets could provide the foundation for regional action on climate change – building on the fact that the momentum for regional cooperation is evolving through the creation of a single market (via ASEAN) as regions play a crucial role in constituting economic globalization. However, regional efforts should be seen as complementary rather than an alternative to global action on climate change mitigation. Due to the increasing interdependence amongst Southeast Asian nations in relation to energy, market-based approaches to mitigation seem most appropriate for an ASEAN-wide approach reflecting on the success of the EU ETS in supporting renewables and the mitigation of GHGs. In the context of ASEAN, regional cooperation associated with a CDM-like program could in fact facilitate environmental integrity and assist with ensuring sustainability requirements in addition to achieving the GHG mitigation outcomes associated with future climate change policy, programs and projects.

While many Southeast Asian countries have set national climate policy targets (under their INDCs), existing cooperation in ASEAN could be strengthened to better address future climate change challenges, through, for example, the setting of an ASEAN climate policy target thus supporting a stronger focus on action. One potential pathway could be realizing a regional NAMA based on cooperation within ASEAN. European regional cooperation directly linked with energy (e.g., the Nord Pool) has been successful, but other forms of regional cooperation based on climate specific and climate relevant aspects could assist substantial mitigation opportunities if linked with appropriately designed market mechanisms and regulatory instruments. ASEAN's experience to date with earlier energy cooperation could potentially help to facilitate development of such a mechanism as a starting point.

In conclusion, regional cooperation can provide a linkage between global and national/sub-national action on climate change mitigation and can also complement national and global action. In essence, regional cooperation toward climate change mitigation could reinforce the *ASEAN Way* to a new level and could essentially drive the sustainable development agenda of the ASEAN Economic Community.

Acknowledgements

The authors acknowledge financial support from the Swedish Energy Agency. The editors of the *Handbook of Southeast Asian Development* are gratefully acknowledged. The authors acknowledge comments from Professor Ros Taplin, University of New South Wales, Sydney on an earlier version of this chapter. The usual disclaimer applies.

Notes

1 Regions can be understood as either supra- or sub-national spatial units; here region is understood at the supra-national level. Regions can be differentiated by level of development, geographical settings and common interests (e.g., economy, trade, etc.). Our understanding of region is based on geographical setting but also includes consideration of regional cooperation based on common interests.
2 Regional integration is often connected with regional cooperation; however, this chapter focuses only on regional cooperation.
3 The European Union currently has 28 member states; Britain voted to exit the EU in 2016.
4 Association of Southeast Asian Nations (ASEAN) member states include: Brunei Darussalam, Cambodia, Indonesia, Lao PDR, Malaysia, Myanmar, Philippines, Singapore, Thailand and Vietnam.
5 Current members of Friends of Fossil Fuel Subsidy Reform are Costa Rica, Denmark, Ethiopia, Finland, New Zealand, Norway, Sweden and Switzerland. http://fffsr.org/about/
6 G7 – Group of 7 countries, member countries are: Canada, France, Germany, Italy, United Kingdom and United States; G20 – Group of 20 countries, member countries are: Australia, Canada, Saudi Arabia, United States, India, Russia, South Africa, Turkey, Argentina, Brazil, Mexico, France, Germany, Italy, United Kingdom, China, Indonesia, Japan, South Korea and the European Union. LMDC – Like Minded Group of Developing Countries who organize themselves as a negotiation block in international climate negotiations.
7 Covering Denmark, Sweden, Norway and Finland.
8 Nontraditional regional cooperation refers to cooperation that includes countries from different regions and based on any common interest.
9 APP member states include: Australia, Canada, China, India, Japan, Korea and the United States.
10 CDM is one of the three flexible mechanisms (CDM, Joint Implementation and International Emission Trading) under the Kyoto Protocol (Article 12) to the United Nations Framework Convention on Climate Change (UNFCCC). The two principal objectives of CDM projects are: assisting Annex I countries in meeting their GHG emissions reduction targets cost-effectively and promotion of sustainable development in non-Annex I countries according to Article 12(2) of the Kyoto Protocol.
11 The CSLF is currently comprised of 25 members, including 24 countries and the European Commission. CSLF member countries represent over 3.5 billion people on six continents, or approximately 60% of the world's population. Collectively, CSLF member countries comprise 80% of the world's total anthropogenic CO_2 emissions.
12 As of 10 April 2016.
13 Under a program of activities (PoA) it is possible to register the coordinated implementation of a policy, measure or goal that leads to emission reduction. Once a PoA is registered, an unlimited number of component project activities (CPAs) can be added without undergoing the complete CDM project cycle. Compared to regular CDM project activities, this programmatic approach has many benefits, particularly for less developed countries or regions. www.cdm.unfccc.int
14 Provincial cap-and-trade systems, seven western US states and four Canadian provinces.
15 RGGI is the first market-based regulatory program in the United States to reduce greenhouse gas emissions. RGGI is a cooperative effort among the states of Connecticut, Delaware, Maine, Maryland, Massachusetts, New Hampshire, New York, Rhode Island and Vermont to cap and reduce CO_2 emissions from the power sector.

References

Anbumozhi, V. (2015). *Low carbon green growth in Asia: What is the scope for regional cooperation?* Jakarta: ERIA.
APEC. (2014). *APEC energy working group* [online]. Asia Pacific Economic Cooperation. Available at: www.ewg.apec.org/about.html. Accessed 2 September 2014.

Areethamsirikul, S. (2008). *The impact of ASEAN enlargement of economic integration: Successes and impediments under ASEAN political institution*. Madison: University of Wisconsin.

ASEAN. (2010). *ASEAN plan of action for energy cooperation (APAEC) 2010–2015*. Jakarta: ASEAN Centre for Energy.

ASEAN. (2014). *ASEAN joint statement on climate change 2014*. Jakarta: ASEAN.

ASEAN. (2015). *ASEAN plan of action for energy cooperation (APAEC) 2016–2025*. Jakarta: ASEAN Center for Energy.

ASEAN. (2016). *ASEAN – About ASEAN* [online]. Available at: www.asean.org/asean/about-asean/overview/. Accessed 4 April 2016.

Bazilian, M., Nussbaumer, P., Eibs-Singer, C., Brew-Hammond, A., Modi, V., Sovacool, B. and Ramana, V. (2012). Improving access to modern energy services: Insights from case studies. *The Electricity Journal*, 25, pp. 93–114.

Benitez, P. (2013). *A regional NAMA framework for Solar PV in the Caribbean*. Washington, DC: The World Bank.

Bosello, F., Carraro, C. and Buchner, B. (2003). Equity, development and climate change control. *The Journal of European Economic Association*, 1, pp. 601–611.

Bryant, R. L. and Bailey, S. (1997). *Third world political ecology*. London: Routledge.

Cabalu, H., Cristina, A. and Manuhutu, C. (2010). The role of regional cooperation in energy security: The case of the ASEAN+3. *International Journal of Global Energy Issues*, 33, pp. 56–72.

CEM. (2014). *Clean Energy Ministerial* [online]. CEM. Available at: www.cleanenergyministerial.org/. Accessed 2 September 2014.

Chang, F. K. (2014). Economic and security interests in Southeast Asia. *Orbis*, 58(3), pp. 378–391.

Eberhard, A., Rosnes, O., Shkaratan, M. and Vennemo, H. (2011). Africa's power infrastructure: Investment, integration, efficiency. In: V. Foster and C. Briceno-Germendia, eds., *Directions in development*. Washington, DC: World Bank.

EC. (2009). *Directive 2009/28/EC of the european parliament and of the council of 23 April 2009 on the promotion of the use of energy from renewable sources*. Brussels: EC.

EC. (2011). *SET at work* [online]. Available at: http://setatwork.eu/. Accessed 2 September 2014.

ECPA. (2014). *Energy and climate partnership of the Americas* [online]. ECPA. Available at: http://ecpamericas.org/About-ECPA.aspx. Accessed 2 September 2014.

EU. (2015). *Submission by Latvia and the european comission on behalf of the european union and its member states EU*. Brussels: EU.

Fingrid. (2014). *Nordic grid development plan 2014*. Helsinki: Fingrid.

Finnsson, P. T. (2016). Decarbonised energy systems – Nordic countries are where the IEA wants the world to be in 2040. *Green Growth Web Magazine, Nordic Way*, Norden.

Grunewald, N., Butzlaff, I. and Klassen, S. (2013). *Regional agreements to address climate change: Scope, promise, funding and impacts*. Gottingen: University of Gottingen.

Hayashi, D., Muller, N., Feige, S. and Michaelowa, A. (2010). *Towards a more standardised approach to baselines and additionality under the CDM*. Zurich: Perspectives.

IEA. (2010a). *Deploying renewables in Southeast Asia trends and potentials*. Paris: IEA.

IEA. (2010b). *Energy poverty: How to make modern energy access universal*. Paris: IEA.

IEA. (2015). *Southeast Asia energy outlook 2015*. Paris: OECD.

IPCC. (2011). *Renewable energy sources and climate change Mitigation*. Cambridge: Cambridge University Press.

IPCC. (2014a). *Mitigation of climate change*. New York: Cambridge University Press.

IPCC. (2014b). *Mitigation of climate change – Summary for policy makers*. New York: Cambridge University Press.

Karki, S. K., Mann, M. D. and Salehfar, H. (2005). Energy and environment in the ASEAN: Challenges and opportunities. *Energy Policy*, 33, pp. 499–509.

Kopsakangas-Savolainen, M. and Svento, R. (2013). Promotion of market access for renewable energy in the nordic power markets. *Environmental and Resource Economics*, 54, pp. 549–569.

Kreinin, M. E. (1992). Effects of economic integration in industrial countries on ASEAN and the Asian NIEs. *World Development*, 20, pp. 1345–1366.

Kuik, O. J., Lima, M. B. and Gupta, J. (2011). Energy security in a developing world. *Wiley Interdisciplinary Reviews: Climate Change*, 2, pp. 627–634.

Kusuma, S. (1998). Thirty years of ASEAN: Achievements through political cooperation. *The Pacific Review*, 11, pp. 183–194.

Lee, Z. H., Sethupathi, S., Lee, K. T. and Bhatia, S. (2013). An overview on global warming in Southeast Asia: CO2 emission status, efforts done, and barriers. *Renewable and Sustainable Energy Reviews*, 28, pp. 71–81.

Lehmann, A., Uddin, N. and Nylander, J. (2014). *Designing new market based mechansim*. Uppsala: Climate Policy and Market Advisory.

Mersmann, F., Olsen, K., Wehnert, T. and Boodoo, Z. (2014). *From theory to practice: Understanding transformational change in NAMAs*. Bonn: UNEP DTU.

MRC. (2010). *Initiative on sustainable hydropower (ISH) 2011–2015 document*. Phnom Penh, Cambodia: Mekong River Commission.

Navarro, A., Tri Sambodo, M. and Todoc, J. (2013). Energy market integration and energy poverty in ASEAN. *Philippine Institute for Development Studies*, 50, pp. 1–23.

Nylander, J. (2015). *Carbon trading in a Paris agreement Stockholm: FORES – Forum for reforms, entrepreneurship and sustainability*. Stockholm: FORES.

OECD. (2008). *Climate change mitigation: What do we do?* Paris: OECD.

Ostrom, E. (2010). Polycentric systems for coping with collective action and global environmental change. *Global Environmental Change*, 20, pp. 550–557.

Sean, O. R. (2010). Globalisation and regional development. In: A. Pike and J. Tomaney, eds., *Handbook of local and regional development*. London: Routledge, 17–29.

Shi, X. (2016). The future of ASEAN energy mix: A SWOT analysis. *Renewable and Sustainable Energy Reviews*, 53, pp. 672–680.

Shi, X. and Mallik, C. (2013). Assessment of ASEAN energy cooperation within the ASEAN economic community. *ERIA Discussion Paper Series*, ERIA.

Sovacool, B. (2009). Energy policy and cooperation in Southeast Asia: The history, challenges, and implications of the trans-ASEAN gas pipeline (TAGP) network. *Energy Policy*, 37, pp. 2356–2367.

Taplin, R. and Mcgee, J. (2010). The Asia-Pacific partnership: Implementation challenges and interplay with Kyoto. *Wiley Interdisciplinary Reviews: Climate Change*, 1, pp. 16–22.

Uddin, N., Blommerde, M., Taplin, R. and Laurance, D. (2015). Sustainable development outcomes of coal mine methane clean development mechanism Projects in China. *Renewable and Sustainable Energy Reviews*, 45, pp. 1–9.

Uddin, N. and Taplin, R. (2015). Regional cooperation in widening energy access and also mitigating climate change: Current programs and future potential. *Global Environmental Change*, 35, pp. 1–8.

Uddin, S. N. and Taplin, R. and Yu, X. (2010). Towards sustainable energy future – Exploring current barriers and potential solution in Thailand. *Environment, Development and Sustainability*, 12(1), pp. 63–87.

Uddin, S. N. and Taplin, R. (2009). Trends in renewable energy strategy development and the role of CDM in Bangladesh. *Energy Policy*, 37, pp. 281–289.

Uddin, S. N., Taplin, R. and Yu, X. (2006). Advancement of renewables in Bangladesh and Thailand: Policy intervention and institutional setting. *Natural Resources Forum*, 30, pp. 177–187.

Uddin, S. N., Taplin, R. and Yu, X. (2009). Sustainable energy future for Vietnam: Evolution and implementation of effective strategies. *International Journal of Environmental Studies*, 66, pp. 83–100.

UN. (2011). *The regional dimension of development and the UN Systems*. New York: United Nations.

UN. (2016). *Paris agreement*. New York, United Nations.

UNDP IBSA. (2013). Overview of Project Portfolio. New York: UNDP.

UNESCAP. (2008). *Energy security and sustainable development in Asia and the Pacific*. New York: UN.

URC. (2014). *CDM pipeline*. Frederiksborgvej: UNEP Risø Centre.

WB. (2014). *Putting a price on carbon – Carbon pricing readiness*. Washington, DC: World Bank.

Wolfgang, O., Jaehnert, S. and Mo, B. (2015). Methodology for forecasting in the Swedish – Norwegian market for el-certificates. *Energy*, 88, pp. 322–333.

Yin, R. (2003). *Case study research: Design and methods*. 3rd ed. Thousand Oaks, CA: Sage.

Yu, X. (2003). Regional cooperation and energy development in the Greater Mekong Sub-region. *Energy Policy*, 31, pp. 1221–1234.

29
CAN PAYMENTS FOR ECOSYSTEM SERVICES (PES) CONTRIBUTE TO SUSTAINABLE DEVELOPMENT IN SOUTHEAST ASIA?

Andreas Neef and Chapika Sangkapitux

Introduction

The degradation of Southeast Asian ecosystems is progressing at a rapid pace. Large forest areas have disappeared or are threatened by illegal logging, mining and conversion into agro-industrial plantations. River basins are increasingly polluted by agrochemicals, industrial pollutants and sediments, and mountain watersheds have been degraded by unsustainable management practices. The major reason identified by economic theory is that these ecosystems generally belong to the category of 'common-pool resources' (CPRs), where there is intense rivalry of consumption, while it is costly to exclude individuals from using the resources for their own benefit or to prevent them from reaping the benefits from their protection (Ahlheim and Neef 2006). While the most common approach in Southeast Asia has been for governments to impose a command-and-control regime of environmental governance (e.g., Neef et al. 2003; Forsyth and Walker 2008), the environmental economics literature suggests a number of different policy measures to address such problems and encourage environment-friendly action. A conventional environmental policy tool is the taxation of activities that are considered environmentally harmful. Yet, given the high number of potential users of a typical CPR, such as a rainforest, a tropical wetland or a river basin, administering such a tax – including its enforcement and monitoring – would be difficult and the transaction costs would be prohibitively high (Ahlheim and Neef 2006). Economists have therefore suggested alternative policy instruments that emphasize economic incentives (Coase 1960; Engel et al. 2008). One approach is to encourage biodiversity conservation and environmentally benign agricultural practices by rewarding activities that preserve CPRs and their associated 'ecosystem services' instead of taxing their consumption (Ahlheim and Neef 2006). Such ecosystem services include, for instance, the supply of wild food, fodder, medicinal plants and energy (provisioning services), carbon sequestration, climate regulation and water purification (regulating services), and spiritual, historical and recreational values and benefits (cultural services). Payments for Ecosystem Services (PES) hold the promise of being more effective in halting environmental degradation in ecologically fragile areas than conventional command-and-control approaches that have largely failed in fostering resource conservation, particularly in the context of Southeast Asia (Neef and Thomas 2009). PES have been most prominently defined as "a (1) voluntary transaction where (2) a well-defined ES (or corresponding land use) is (3) being 'bought' by a (minimum one) ES buyer (4) from a (minimum one) ES provider (5) if and only if ES provision is secured (conditionality)" (Wunder 2008, 280). Local and global demand for enhancing ecosystem services is steadily rising against the background of climate change and increased global ecological vulnerability. Hence, the idea of providing public support for conservation programs and for direct payments to land managers maintaining or restoring such services is slowly gaining currency among national policy-makers and in the international donor community (Rosales 2003; Tomich et al. 2004; Engel et al. 2008; Wunder et al. 2008a; Neef and Thomas 2009; Gómez-Baggethun et al. 2010). Yet the rationale for Payments for Ecosystem Services has been the subject of some controversy since the late 1990s. While the proponents count on new impulses for global biodiversity protection and environmental conservation and even expect positive effects in terms of poverty alleviation and sustainable 'green' development, the critics of such market-based concepts and instruments fear the transformation of biodiversity and ecosystem services into marketable commodities, associated with the enclosure and commodification of nature and its biological diversity in the sense of a neoliberal form of "Green Capitalism" (e.g., McAfee 1999; McGregor et al. 2014). Critical scholars and human rights NGOs have also expressed concerns that PES may restrict poor and landless rural people's access to natural resources that are crucial for sustaining their livelihoods (e.g., McElwee et al. 2014).

The chapter contributes to this debate through providing a comparative perspective of experiences with PES schemes in three Southeast Asian countries. Section 2 describes the various ways in which PES schemes have been classified. Section 3 presents a theoretical-conceptual framework for analyzing processes, relationships and institutional environments related to PES mechanisms and then discusses various PES pilot projects in Vietnam and Indonesia in the light of the framework. A more in-depth case study of a particular watershed in Northern Thailand is provided in Section 4, followed by a summative conclusion in Section 5.

Classification of Payments for Ecosystem Services (PES)

Bulte et al. (2008) divide Payments for Ecosystem Services schemes into three broad categories according to their function. 'Payments for pollution control' serve as a complement or alternative to the 'polluter-pays' principle. For instance, upstream farmers may agree to reduce the use of agro-chemicals and – in return – receive payments for the provision of improved water quality from downstream beneficiaries, such as residents and/or drinking water companies. 'Payments for the conservation of natural resources and ecosystems' may include payments by international donors for the conservation of pristine forest areas in Southeast Asia that provide habitats for endangered wildlife species and contribute to global climate regulation through sequestering carbon. An example of the third category, 'Payments to generate environmental amenities,' would be when a national government pays upland farmers for setting agricultural land aside to plant trees for stabilizing the soil to prevent flooding of downstream areas (Neef and Thomas 2009).

Another way of classifying PES is to distinguish them according to the spatial-administrative level at which payments are transferred. In a sub-national PES scheme upstream farmers in one watershed may be directly rewarded by downstream residents in the same watershed to switch to environment-friendly practices (cf. Sangkapitux et al. 2009 for the case of Thailand). In a national PES system, resource managers are paid for protecting existing forests or planting new trees (cf. McElwee et al. 2014 and Pham et al. 2014 for the case of Vietnam). In an international PES arrangement, international donors such as USAID or the World Bank may provide funds for a national government to ensure that rainforests rich in biodiversity are protected from encroachment by illegal squatters. A much debated PES approach at the international level is the so-called 'Reducing Emissions from Forest Degradation and Deforestation (REDD+),' which revolves around the idea of paying national governments or local communities for not clearing forestland (e.g., Angelsen (ed.), 2008; see also McGregor and Thomas for the case of Indonesia in chapter 30, this volume).

Wunder et al. (2008a) have classified PES schemes according to the provenance of the payments. In user-financed programs the service buyers are the actual service users, e.g., a hydro-electric power company that benefits from reduced deforestation around a dam reservoir. In government-financed programs the service buyer is a third party, typically a government entity that may or may not receive funding from an international donor. Whereas user-financed programs are fully voluntary for both providers and buyers, government-financed programs tend to be voluntary on the providers' side only. As the case studies discussed in sections 3 and 4 will show, the distinction between user- and government-financed programs is not always clear-cut. Hybrid forms are more the norm than the exception, with strong government interference on one side of a continuum and weak interference from government actors on the other (cf. McElwee et al. 2014).

Experiences and lessons learned from PES pilot projects in Southeast Asia

Theoretical-conceptual framework

Neef and Thomas (2009) proposed the following theoretical-conceptual framework on the basis of three sets of prerequisites for functioning PES schemes in the context of Southeast Asian countries:

- **Identification of the PES market:** There are three basic components of the PES market that determine whether a PES scheme can be developed: (1) the ecosystem service(s) involved in the scheme need to be specified; (2) the potential providers of the ecosystem services need to be identified and express their willingness to participate voluntarily in the PES scheme; and (3) the potential buyers of the ecosystem services need to be aware of the PES concept and willing to pay for the ecosystem services provided.
- **PES processes and relationships:** Once the PES market components have been identified, key processes and relationships need to be established in a deliberative process that involves all relevant stakeholders: (1) all actors need to agree upon the types and magnitude of the rewards to be provided; (2) the rules for deciding under which conditions rewards will be provided or denied need to be defined, with particular emphasis on transparency and conditionality; and (3) trust needs to be established between the buyers and providers of ecosystem services, preferably through face-to-face dialogue and with support from credible intermediaries.
- **Institutional environment of PES:** PES markets and associated processes and relationships can only function well under a conducive institutional environment. This requires (1) reliable, experienced and well-respected intermediary organizations that facilitate the interaction between ES providers and buyers; (2) a supportive legal and regulatory framework, preferably at national level; and (3) well-defined and officially recognized property rights of the ES providers, including customary rights to communally managed resources.

Table 29.1 Major prerequisites for PES schemes in the Southeast Asian context

Prerequisites	Components		
Identification of the PES Market	Specification of ecosystem services (ES) involved	Identification of potential ES providers	Awareness of potential ES buyers
PES Processes and Relationships	Definition of types, forms and levels of rewards	Transparency and conditionality	Trust between ES buyers and providers
Institutional Environment of PES	Credible intermediaries to facilitate the PES scheme	Supportive legal and regulatory framework	Recognition of ES providers' resource rights

Source: Based on theoretical framework developed by Neef and Thomas (2009)

Payments for ecosystem services in Vietnam

Vietnam is currently the only Southeast Asian country that has incorporated PES into its national legislation and mainstreamed the concept as part of their natural resource governance strategy, which warrants a closer look at the implementation of the late socialist country's nationwide PES scheme and its impact on resource management practices and rural livelihoods.

Identification of the PES market in Vietnam

Over the past decade, the PES concept has gained increasing currency in Vietnam. In the mid-2000s, a number of small pilot projects were initiated by international donors, such as Winrock International (USA), the International Development Research Centre (Canada) and the Swedish International Development Cooperation Agency (e.g., The and Ngoc 2006).

Two major pilot projects were set up in the Central Highlands' Lam Dong Province and in north-western mountainous Son La Province in 2008 under the prime minister's Decision No. 380 QD-TTG entitled "On the Pilot Policy on Forest Environment Service Charge" (McElwee et al. 2014). Under these pilots, the Ministry of Agriculture and Rural Development started to charge fees from hydroelectric power companies, irrigation projects, water supply works and industrial estates for forest environmental services on a pilot basis. As these pilots were considered successful after a brief trial period of only two years, a new national policy was passed in 2010 under the title "On the Policy for Payment for Forest Environmental Services" (Decision 99 ND-CP). The Decree introduced mandatory PES schemes and required 'beneficiaries/buyers' of ecosystem services to pay fees either directly to 'providers/sellers' or indirectly into a Forest Protection and Development Fund to be set up at the provincial level (McElwee et al. 2014). Thereby, the government hopes to drastically reduce the use of central government funds for forest conservation efforts (ibid.).

Meanwhile, private sector demand for ecosystem services remains low. As a consequence, the interest in PES schemes has been mainly concentrated in the international donor community. Some 13 smaller-scale PES and PES-style projects were recorded in individual provinces in 2014 that involved donor-funded transfers rather than truly user-funded PES (Pham et al. 2010; McElwee et al. 2014). A major barrier for developing voluntary markets for ecosystem services in Vietnam has been the lack of understanding of the concept by the private sector as potential service buyer and communities as ecosystem service providers alike (Pham et al. 2008; Neef et al. 2013). Studies by McElwee et al. (2014) in the major pilot sites in Son La and Lam Dong found that only about one-third of the interviewed PES recipients had actually heard of PES and very few of them were aware that it was a program supported by payments from environmental service users.

PES processes and relationships in Vietnam

As discussed above, voluntariness is one of the principal features of PES markets. Yet, in the case of Vietnam's national PES program, ecosystem service buyers and sellers are "identified by law and must take part in the program" and "[t]he level of PES payments is administratively set by the Vietnamese government through legislation" (Pham et al. 2014, 889), hence both participation and monetary transactions are mandatory and the payments are not determined by a market mechanism as prescribed by PES theorists. Some authors have argued that through mainstreaming PES schemes the Vietnamese government has actually tightened its control over natural resource use and increased state power by using "PES as a potential tool for internal territorialization" (Suhardiman et al. 2013, 66).

Another issue of concern in the Vietnamese context is conditionality. McElwee et al.'s study in Son La and Lam Dong found that "many participants treated PES as a government entitlement fund and not a conditional environmental fund" (2014, 434) and were therefore not inclined to change their land use systems. Pham et al. (2014) also stated that although enhancement of forest quality, soil erosion control and water regulation are targeted outcomes of PES in Vietnam, there are currently no specific requirements for monitoring these ecosystem services. This raises the question whether the payments can be an effective incentive mechanism for environmental protection.

A further challenge to the viability and effectiveness of PES schemes in Vietnam is that transaction costs are very high, because landholdings are often very small and fragmented, the terrain tends to be rugged and communities are often remote, which makes monitoring of compliance and regular transfers of payments to individual land managers an onerous and complicated task (The and Ngoc 2006; Neef et al. 2013; Pham et al. 2014). McElwee et al. (2014) found that in the government-piloted site in Son La Province 52,000 forest owners, of which 45,000 were individual households, have contracted directly with the PES provincial fund, driving up transaction costs. Group contracts were introduced as an alternative, but studies by Pham et al. (2014) in various communes of Son La Province have shown that such contracts are often complex and raise questions of fair distribution of benefits in a local context where village elites and powerful groups dominate the decision-making process. In addition, due to their negative experience with the cooperative period in the 1970s and 1980s many farmers are reluctant to work together toward a common goal, thus reducing the level of collective action that is needed for ecosystem protection at the community level (Neef et al. 2013).

Another major impediment to sustainable PES schemes in Vietnam is the relatively high opportunity costs (Phuc et al. 2012). Son La Province, for instance, is one of the major corn-producing areas of Vietnam, and many farmers have become more affluent through intensive corn cultivation on erosion-prone hillsides. These farmers are not likely to shift to more environment-friendly forms of agriculture, given the relatively low payments from PES schemes and the insecurity of their regular and long-term disbursement.

Institutional environment of PES in Vietnam

Forests in Vietnam were primarily managed by State Forest Enterprises (SFEs) from the 1950s until the 1980s. The shift from a centrally planned to a more market-driven economy under the *doi moi* (renovation) policy reduced government control over the country's forests. Under the

1993 Land Law rural households could obtain forestland certificates in return for signing forest protection contracts. The government also instigated a number of reforestation campaigns to reverse deforestation, such as Program 327 (1993–1998) and its successor Program 661 (1999–2010), popularly called the Five-Million-Hectare Program (Phuc et al. 2012). Pilot PES schemes in Vietnam's Son La and Lam Dong provinces were initially backed by Decision No 380, which provided general guidance for provincial administrations on how to collect fees from ES service 'buyers,' such as hydroelectric power companies, water companies and tourism businesses (Pham et al. 2008). Decision 99 – described earlier – formalized the country-wide introduction of mandatory PES schemes at provincial level.

The formal land allocation and registration system in Vietnam could – in principle – provide a relatively sound basis for the establishment of PES schemes (Neef et al. 2013). Yet, all agricultural and forest land formally belongs to the Vietnamese government, and there are often overlapping land claims, e.g., between formal control rights of forestland and informal customary rights (Pham et al. 2008). In a case study on an international PES pilot project in Ba Vi National Park, Phuc et al. (2012) found that various stakeholders established contrasting claims and rights to land, thus inciting conflicts between park officials and villagers. The authors concluded that the majority of the households were excluded from the benefits of the PES scheme (ibid). In the past, frequent re-allocation of land by the government in many rural areas also led to tenure insecurity and thus reduced the motivation of individual farmers to invest in improved ecosystem management (Neef et al. 2013). The Land Law of 2013 extended the allocation of land use certificates for cropland to 50 years, but it may take several years to establish a sufficiently high level of trust in the new legislation among small farmers, which would then possibly increase their motivation to make long-term investments in more environment-friendly practices.

In sum, the institutional environment in Vietnam does not seem to be very conducive to the establishment of market-based PES systems in their classical definition. The mandatory participation of 'buyers' and 'sellers' in most existing schemes, the involuntary nature of the monetary transactions, the lack of monitoring and enforcement of ecosystem service provision and the high transaction and opportunity costs raise doubts whether PES can become an efficient, effective and sustainable mechanism to enhance the provision of ecosystem services while at the same time improving the livelihoods of Vietnam's rural population.

Payments for ecosystem services in Indonesia

With its large territory, vast tracts of remaining, albeit dwindling rainforest – the world's third largest – and recent decentralization processes in natural resource management, Indonesia has become one the most popular target countries for project-based PES schemes.

Identification of the PES market in Indonesia

The PES concept has become popularized in Indonesia in recent years through several pilot projects (Munawir and Vermeulen 2007; Leimona et al. 2009; van Noordwijk and Leimona 2010). In most existing PES schemes the environmental services requested by downstream buyers are (1) provision of reliable water flows and reduced sedimentation, e.g., in the case of hydroelectric facilities (Leimona et al. 2015) and (2) improvement of water quality, e.g., in the case of local water suppliers (Munawir and Vermeulen 2007). Environmental services are mostly provided through tree planting (e.g., Huang and Upadhyaya 2007; Munawir and Vermeulen 2007) or community-based forest conservation or agroforestry schemes (e.g., Leimona et al. 2015). Yet the causality between tree planting and water supply remains a hotly debated issue often marked by myths and misconceptions, such as the simple 'more trees = more water' equation, which has been refuted by several studies (e.g., Calder 2004).

Flexibility in land use decisions under PES schemes appears to be much greater in Indonesia than in Vietnam (Huang and Upadhyaya 2007; Munawir and Vermeulen 2007). Farmers in the Brantas catchment, East Java, for instance, have dismissed civil engineering measures, such as terracing, as too costly and have chosen high-income tree crops instead (Munawir and Vermeulen 2007, cited in Neef et al. 2013). In other areas, jungle rubber, fruit-based agroforestry systems or organic coffee plantations have been established under PES schemes (van Noordwijk and Leimona 2010; Leimona et al. 2015).

Environmental service buyers are more diverse in Indonesia than in other Southeast Asian countries. Private drinking water companies, hydroelectric power companies, local governments and the central government are all involved in various types of PES schemes across the country (Munawir and Vermeulen 2007; Leimona et al. 2009; van Noordwijk and Leimona 2010).

Because Indonesia is a country that – despite widespread deforestation, e.g., for oil palm plantations – is still endowed with a large area of pristine forests, expectations of benefitting from REDD+ and other international carbon market arrangements are also high, although the experience with voluntary carbon trading projects has been mixed (see McGregor, chapter 30, this volume).

PES processes and relationships in Indonesia

Most PES schemes in Indonesia allocate rewards in cash and in kind to groups rather than to individual farmers in order to reduce transaction costs that otherwise would be prohibitively high (Arifin 2006; Huang and Upadhyaya 2007). Yet, there are not enough data available to determine whether PES programs induce lower transaction costs than traditional conservation approaches (Wunder et al. 2008a). It is also difficult to quantify in-kind benefits – such as improved tenure security, development of trust between the government and land managers or enhanced negotiation skills – that accrue to ecosystem service providers.

Appropriate monitoring mechanisms can enhance the viability of PES schemes, but this has often been difficult in Indonesian pilot projects. Hard evidence of delivery of watershed services has proved elusive and reported impacts of PES-induced land use changes are often based more on local perceptions than on measurements, as Porras et al. (2008) have noted (cited in Neef and Thomas 2009). Leimona et al. (2015, 19) found that the preconditions for the Coasean conceptualization of PES as a market-led incentive mechanism have hardly been met in their case studies in Indonesia and argue that it would be more accurate to describe the existing schemes as "co-investment in environmental stewardship as opposed to a strict and prescriptive PES definition." Pirard et al. (2014) suggest that there is a continuum between large-scale government programs and purely market-based PES. In their study of multi-stakeholder PES partnerships in two Indonesian sites, they found that despite the polycentric governance of these PES schemes one actor was able to dominate the decision-making process.

Lack of trust and power differentials between service providers and buyers were major constraints in other Indonesian PES schemes. In a pilot project in the Brantas catchment, for instance, the district government did not trust in local people's engineering skills for soil and water conservation and argued that the necessary standards could only be met by professional contractors (Munawir and Vermeulen 2007, cited in Neef and Thomas 2009). The authors also found that sellers of environmental services were not treated as equals in PES negotiations, had little influence on decision-making and were concerned that trees planted under PES schemes would ultimately be claimed by the government (Munawir and Vermeulen 2007).

An important question is which groups could benefit the most from PES schemes in the complex ethnic and socio-economic setting of Indonesia. Seeberg-Elverfeldt et al. (2009) found in their study on the potential of carbon payments for supporting less intensive, cacao-based agroforestry systems in Sulawesi that poorer, indigenous farmers were more likely to benefit from well-targeted payments than more affluent, migrant farmers. They also concluded that such payments could have positive effects both in terms of fighting deforestation and alleviating rural poverty. Yet, to date, there has been very little empirical evidence that PES as a stand-alone policy has made a substantial contribution to alleviating rural poverty in Indonesia (Leimona et al. 2015).

Institutional environment of PES in Indonesia

Among Southeast Asian countries, Indonesia has one of the most advanced and conducive legal and regulatory framework for PES schemes. Although policy-makers have not yet come up with a specific framework for PES in Indonesia, "several laws and policies can be interpreted as providing basic rules and incentives" (Munawir and Vermeulen 2007, 15), such as the Water Resources Act, the Environmental Management Act and a national decentralization program. In 1999, the forestry law was substantially revised, with more provision for local management and with enhanced recognition of customary rights. A decision of the Constitutional Court in 2012 provisioned that the Indonesian government was constitutionally required to recognize and respect the customary (*adat*) rights of traditional forest communities. Yet, many ambiguities remain, particularly in areas where the economic interests of the state and large-scale agribusiness investors have undermined local communities' rights, for example in areas suitable for oil palm, rubber or timber plantations (e.g., Li 2002; Gellert 2015).

Intermediaries play a pivotal role in the success of PES mechanisms, particularly in cases where providers of ES are not familiar with formal contracts, are not formally registered as a group and cannot operate bank accounts (as in a case reported by Munawir and Vermeulen 2007 and cited in Neef and Thomas 2009) or where corporate buyers are not used to negotiating directly with farmers. The majority of intermediaries in Indonesian pilot PES projects have been local NGOs and international donors and organizations (Suyanto et al. 2005; Huang and Upadhyaya 2007; Pirard et al. 2014). Evidence from these case studies suggests that intermediaries can both reduce and increase transactions costs, depending on the experience of mediating agencies, scale of interventions and local contexts (Neef and Thomas 2009).

Several pilot case studies in Indonesia suggest that improving tenure security can provide additional incentives for reliable ES provision (Munawir and Vermeulen 2007; Huang and Upadhyaya 2007; Leimona et al. 2009; Leimona et al. 2015). In the case of weakly defined property rights in Kalimantan, PES were found to induce more secure property rights by raising the value of natural resources to local people (Wunder et al. 2008a). A pilot project in Sumberjaya has helped farmer groups to get access to conditional community forestry permits that subsequently covered around 70 percent of the protection forest in the area (Suyanto 2007). In other settings, however, donors that had been approached for funding PES schemes eventually withdrew their support because of uncertain land tenure and the perceived risk of PES money exacerbating inter-village conflicts over land (Wunder et al. 2008b).

In sum, the Indonesian case studies suggest that the configurations and outcomes of PES schemes can vary substantially and are contingent upon how economic, social, political and institutional factors intersect in a particular local context.

From hypothetical PES markets to sustainable PES schemes? A case study of Mae Sa watershed in Northern Thailand

The following case study is based on our long-term research in the Mae Sa watershed, spanning from 2000–2015. While our initial studies have laid the theoretical and conceptual foundations of PES schemes, we have also looked at the practical implementation of a number of pilot PES projects in the watershed.

Thailand's PES policy background

Natural resource governance in Thailand has mostly followed a command-and-control approach from the second half of the twentieth century onwards. The Thai government continues to regard the state as the major, if not sole protector of forest resources, particularly in ecologically sensitive upland watersheds and protected areas where neither individual nor communal rights to resource management are recognized under the official legal framework (Walker and Forsyth 2008; Neef and Thomas 2009).

At the national level, the concept of PES was mentioned for the first time in the 11th National Economic and Social Development Plan (2012–2016). The establishment in 2007 of a specialized government agency, the Biodiversity Based Economic Development Office (BEDO), with the responsibility of implementing PES programs in various parts of the country signals that the government wants to place more emphasis on economic instruments and market-based mechanisms to maintain and improve ecosystem services and biodiversity in the Kingdom. However, most PES pilot projects that were in operation in various parts of Thailand at the time of writing (September 2015) had been implemented under the leadership of the Department of National Parks, Wildlife and Plant Conservation (DNP), an agency that is better known for its uncompromising 'fortress approach' to forest and nature conservation than for the establishment of multi-stakeholder partnerships that are an essential component of viable PES schemes.

Background of the study area

The Mae Sa watershed is located in Mae Rim district, Chiang Mai Province in northern Thailand, about 35 km northwest of Chiang Mai city (Figure 29.1). It is inhabited by northern Thai people (*khon muang*) and Hmong, an ethnic minority group that migrated to this area from Southwest China several decades ago. While the northern Thai communities are found both in the upstream and downstream parts of the watershed, the Hmong villages are all situated in upstream hillsides. Other major stakeholders in the area comprise tourist-related companies (such as resorts, restaurants and elephant farms), a communally managed drinking water company, a public tap water provider (Mae Sa Waterworks) and non-farming communities.

Some of the areas occupied by the Hmong communities overlap with the Suthep-Pui National Park, hence farmers have no secure land rights, but are allowed to practice agriculture on those plots that they already used before the establishment of the park in 1980s. While none of the Hmong farmers have official land titles, more than 50 percent of the northern Thai farmers in the watershed have some form of legally recognized land documents. Both groups engage in highly intensive, commercial cultivation of fruits, vegetables and cut flowers with detrimental impacts on the environment, expressed in high rates of soil erosion, agro-chemical pollution and unsustainable groundwater extraction. Researchers from the Uplands Program – a Thai-German collaborative research program – worked with local communities from 2000–2014 to find alternative land use practices that would be less harmful to the environment.

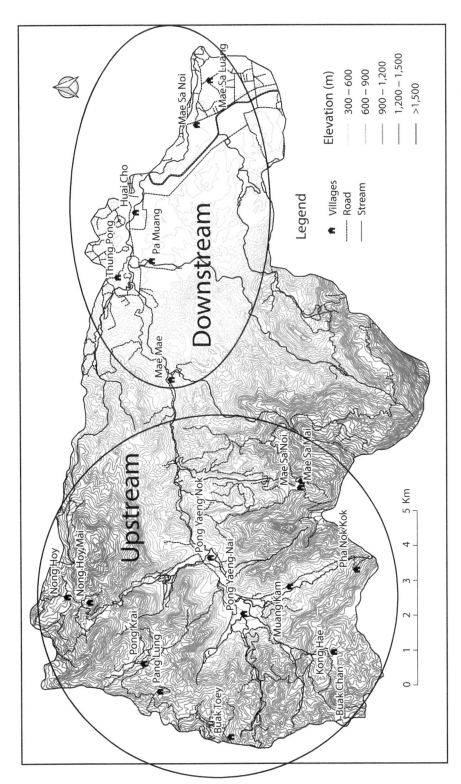

Figure 29.1 Map of Mae Sa watershed, Chiang Mai Province, Northern Thailand

Studies on hypothetical PES markets

An empirical study conducted by a research team under the leadership of the second author of this chapter and based on choice experiments found that upstream Thai and Hmong farmers in the Mae Sa watershed would be willing to adopt more sustainable and environment-friendly farming practices (for instance, by planting vegetative strips against soil erosion and by adopting less harmful pest-control measures and water-saving irrigation technologies) if adequate financial compensation were provided (Sangkapitux et al. 2009; Neef 2012). Interestingly, one of the findings of the study was that the poorer farmers in the upstream communities were more likely to accept regular payments from a PES scheme as this would provide secure and regular economic benefits, while the more affluent farmers did not see a reason why they should change their practices, which they deemed successful and profitable. Poor upstream farmers expected that the establishment of such schemes would enhance their tenure security in this protected watershed area (Sangkapitux et al. 2009; Neef 2012). In downstream farming communities and among tap water users, there was a clearly stated willingness to contribute financially for obtaining better water resources, although the overall amount proposed was only a fraction of what would be needed to adequately compensate upstream farmers for switching to environment-friendly farming practices (Sangkapitux et al. 2009; Punyawadee et al. 2010). Subsequently, our research team organized a meeting in 2010 with all major stakeholders in the Mae Sa watershed to gauge the opportunities and constraints of developing a viable PES scheme. The model proposed for the PES scheme in Mae Sa watershed is depicted in Figure 29.2.

Results from the stakeholder meeting suggested that other stakeholders such as the private sector engaging in tourism, industrial activities and government enterprises like Mae Rim Water Works were also in the position to pay for acquiring better water resources and ecosystem services. This was seen as a positive sign for the prospects of developing a market for environmental services in this watershed under broad stakeholder participation.

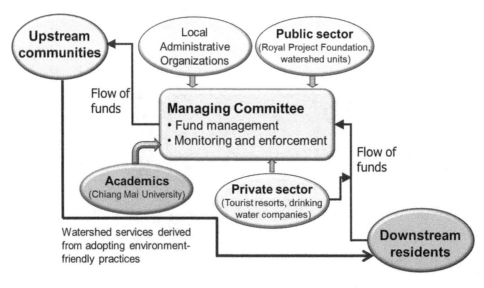

Figure 29.2 Proposed PES Model for Mae Sa watershed

However, barriers to put the PES into practice – as mentioned in the stakeholder meeting – were the following:

1) PES are based on locally driven and decentralized systems of managing natural resources, while the existing national policy approach continues to rely on a centralized command-and-control system. Hence, changes in laws and regulations, such as land titling, definition of resource ownership and water and forest resource management, would be needed as a basis for negotiation between service providers and beneficiaries of environmental services.
2) Downstream water users showed a tendency to believe that it is the government's duty to compensate the upstream famers for their shift toward agricultural conservation practices. This group also expressed relatively strong mistrust that upstream communities would indeed change their farming practices if they received compensation.
3) Environmental services in the form of an improvement in quality and quantity of water resources were difficult to define, underscoring the need for independent monitoring to help provide further evidence. A workable monitoring and enforcement system needs to be established to help facilitate the effective implementation of PES.

Practical implementation of PES schemes

Hmong communities in northern Thailand tend to have a reputation for being notorious shifting cultivators and forest destroyers. Yet, in several Hmong villages in the Mae Sa watershed, a range of traditional and contemporary forest protection strategies can be found. Villagers in Mae Sa Mai and Mae Sa Noi hold an annual ceremony to protect a large area of forest around the tallest tree (*ntoo xeeb*), which is believed to be the residence of the guardian spirit of the community (Neef 2012). Several community members founded a Natural Resource and Environment Conservation Club in the early 1990s, initially to engage in forest fire prevention activities. In 1997, it linked up with the Forest Restoration Research Unit (FORRU), an international NGO based at Chiang Mai University's Faculty of Science, and established a community-based tree nursery to produce tree seedlings from local species for reforestation purposes on abandoned land. A number of organizations (e.g., the World Wildlife Fund Thailand, the PATT Foundation and various private companies) sponsored a range of forest restoration activities. One example is a local news magazine – *Citylife Chiang Mai* – which provided funds for the annual planting of 0.56 hectares of forest over several years to offset the company's annual carbon output (Neef 2012).

Since 2012, the USAID-funded **L**owering **E**missions from **A**sia's **F**orests (LEAF) project – implemented by Winrock Foundation – has sought to establish a viable PES scheme in the wider Mae Sa watershed area. According to one of the project managers, initial attempts were unsuccessful, as it was difficult to identify private sector actors that were willing to engage in a long-term commitment as 'buyers' of ecosystem services. Major progress was made in the course of 2015 with the conclusion of a contract with the AURA mineral water company, a subsidiary of Tipco Foods Public Company Limited, that operates in the Mae Sa watershed. Using groundwater from the watershed, this company relies directly on the surrounding ecosystems and the communities as potential 'service' providers. The company agreed to provide regular payments to a local northern Thai community – Ban Pong Khrai – over an initial period of two years to restore 1.6 ha of forests in a sub-catchment area related to the groundwater supply of the company. The community has a long-standing history of collaboration with the Royal Forest Department (RFD), which needed to give permission on using the land for restoration.

The local subdistrict (*tambon*) administrative organization (TAO) has provided logistic support for implementing the PES scheme, while the DNP and the Royal Project Foundation has given technical support to the community. FORRU, the organization that had previously supported small-scale forest restoration schemes in neighboring Hmong communities, has provided ecological and technical knowledge through restoration training and regular monitoring of the site. Only native species have been used for the reforestation scheme, and the initially planted 3,340 tree seedlings were bought from a nursery of the Hmong community of Ban Mae Sa Mai, thus providing a one-off financial benefit to another village in the watershed. All relevant stakeholder groups are represented in the AURA PES Committee. The role of the committee is to approve the action plans submitted by Ban Pong Khrai community, recommend the evaluation criteria for planned activities, approve the self-reports submitted by the community, oversee the payment process and resolve disputes and disagreements among the various actors (S. Soonthornnawaphat, personal communication).

While the AURA PES scheme does not closely resemble the PES model that was earlier proposed by Thai researchers for the Mae Sa Watershed (cf. Figure 29.2) – as it does not include downstream residents and involves only one private actor as an ES buyer and a single upstream community as an ES provider – the scheme may be a first step toward a more comprehensive PES program encompassing all major stakeholders in the watershed. Yet it still needs to stand the test whether it is sustainable beyond the two-year PES contract and viable without continued support from external donors and intermediary organizations.

Summary and conclusions

Payments for Ecosystem Services (PES) have been promoted in several Southeast Asian countries as alternatives to traditional command-and-control approaches to environmental conservation. Drawing on the theoretical-conceptual framework developed by Neef and Thomas (2009), this chapter has provided a review of experiences and lessons learned from various PES schemes in Vietnam and Indonesia, where recent policy changes have made the institutional environment more conducive to such approaches. Yet, securing long-term commitment of corporate 'buyers' of environmental services has proven difficult, making such schemes overly reliant on donor and/or national government funds and on intermediary roles played by academics and external experts. Voluntariness and conditionality – two important principles of the PES approach – are particularly difficult issues in late socialist, semi-authoritarian regimes, such as Vietnam. There are also well-founded concerns that government-implemented PES schemes may enhance state control of forestland rather than support community- or household-based conservation based on economic incentives.

Most PES schemes in Thailand have been tested at the pilot project and small catchment level because a comprehensive legal framework at the national level is still lacking and government officials remain reluctant to devolve decision-making power over forest conservation and use to local communities. Our long-term research on the theoretical foundations and practical implementation of PES approaches in Thailand suggests that small-scale user-financed schemes may be more efficient than large-scale government-financed ones, as they can be better targeted, monitored and tailored to particular local needs and conditions. Such small-scale schemes are also more conducive to face-to-face social exchange between providers and 'buyers' of ecosystem services, which is oftentimes an important prerequisite for building trust and converging expectations among the two parties.

However, in a context where local communities traditionally do not have a strong voice in managing forests and other natural resources – which is the norm rather than the exception in most Southeast Asian countries – PES schemes have yet to provide evidence that they can help strengthen customary land rights and increase local control over natural resources.

In terms of their contribution to alleviating rural poverty, most studies confirm that PES schemes tend to provide only supplemental income to low-income ecosystem service providers and in many cases do not even match the opportunity costs of alternatives, often environmentally damaging land uses. Therefore, even in those rare cases where PES schemes can be successfully implemented without unsustainable external support, they should only be regarded as a small component of a more comprehensive strategy toward environmental conservation and rural development.

Experiences from case studies in the three countries suggest that in their current manifestation and scale of implementation, PES schemes in Southeast Asia are neither a panacea for solving pressing environmental problems (as originally suggested by PES theorists and proponents) nor do they represent a major shift toward the commodification of nature by corporate actors (as feared by critical scholars and activists). Yet they need to be seen as evolving and contested policy spaces in which government entities, private actors, NGOs, local communities and individual land managers constantly renegotiate the degree of control over natural resources and the environment.

References

Ahlheim, M. and Neef, A. (2006). Payments for environmental services, tenure security and environmental valuation: Concepts and policies towards a better environment. *Quarterly Journal of International Agriculture*, 45(4), pp. 303–318.

Angelsen, A. (ed.) (2008). *Moving Ahead with REDD: Issues, options and implications*. Bogor, Indonesia: Center for International Forestry Research (CIFOR).

Arifin, B. (2006). Transaction cost analysis of upstream-downstream relations in watershed services: Lessons from community-based forestry management in Sumatra, Indonesia. *Quarterly Journal of International Agriculture*, 45(4), pp. 361–377.

Bulte, E. H., Lipper, L., Stringer, R. and Zilberman, D. (2008). Payments for ecosystem services and poverty reduction: Concepts, issues, and empirical perspectives. *Environment and Development Economics*, 13, pp. 245–254.

Calder, I. R. (2004). Forests and water – Closing the gap between public and science perceptions. *Water Science and Technology*, 49(7), pp. 39–53.

Coase, R. H. (1960). The problem of social cost. *Journal of Law and Economics*, 3(1), pp. 1–44.

Engel, S., Pagiola, S. and Wunder, S. (2008). Designing payments for environmental services in theory and practice: An overview of the issues. *Ecological Economics*, 65, pp. 663–674.

Forsyth, T. and Walker, A. (2008). *Forest guardians, forest destroyers: The politics of environmental knowledge in Northern Thailand*. Seattle, WA: University of Washington Press.

Gellert, P. K. (2015). Palm oil expansion in Indonesia: Land grabbing as accumulation by dispossession. In: States and Citizens: Accommodation, Facilitation and Resistance to Globalization. *Current Perspectives in Social Theory*, 34, pp. 65–99.

Gómez-Baggethun, E., de Groot, R., Lomas, P. L. and Montes, C. (2010). The history of ecosystem services in economic theory and practice: From early notions to markets and payment schemes. *Ecological Economics*, 69, pp. 1209–1218.

Huang, M. and Upadhyaya, S. K. (2007). *Watershed-based payment for environmental services in Asia*. Working Paper No. 06–07 [online] Available at: www.oired.vt.edu/sanremcrsp/wp-content/uploads/2013/11/Sept.2007.PESAsia.pdf

Leimona, B., Joshi, L. and van Noordwijk, M. (2009). Can rewards for environmental services benefit the poor? Lessons from Asia. *International Journal of the Commons*, 3(1), pp. 82–107.

Leimona, B., van Noordwijk, M., de Groot, R. and Leemans, R. (2015). Fairly efficient, efficiently fair: Lessons from designing and testing payment schemes for ecosystem services in Asia. *Ecosystem Services*, 12, pp. 16–28.

Li, T. M. (2002). Engaging simplifications: Community-based resource management, market processes and state agendas in upland Southeast Asia. *World Development*, 30(2), pp. 265–283.

McAfee, K. (1999). Selling nature to save it? Biodiversity and green developmentalism. *Environment and Planning D: Society and Space*, 17, pp. 133–154.

McElwee, P., Nghiem, T., Le, H., Vu, H. and Tran, N. (2014). Payments for environmental services and contested neoliberalisation in developing countries: A case study from Vietnam. *Journal of Rural Studies*, 36, pp. 423–440.

McGregor, A., Weaver, S., Challies, E., Howson, P., Astuti, R. and Haalboom, B. (2014). Practical critique: Bridging the gap between critical and practice-oriented REDD+ research communities. *Asia-Pacific Viewpoint*, 55(3), pp. 277–291.

Munawir and Vermeulen, S. (2007). *Fair deals for watershed services in Indonesia*. London: International Institute for Environment and Development (IIED).

Neef, A. (2012). Fostering incentive-based policies and partnerships for integrated watershed management in the Southeast Asian uplands. *Southeast Asian Studies*, 1(2), pp. 247–271.

Neef, A., Ekasingh, B., Friederichsen, R., Becu, N., Lippe, M., Sangkapitux, C., Frör, O., Punyawadee, V., Schad, I., Williams, P. M., Schreinemachers, P., Neubert, D., Heidhues, F. Cadisch, G., Nguyen The Dang, Gypmantasiri, P. and Hoffmann, V. (2013). Participatory approaches to research and development in the Southeast Asian uplands: Potential and challenges. In: H. L. Fröhlich, P. Schreinemachers, G. Clemens and K. Stahr, eds., *Sustainable land use and rural development in Southeast Asia: Innovations and policies for mountainous areas*. Berlin, Heidelberg, New York, Dordrecht and London: Springer, 321–365.

Neef, A., Onchan, T. and Schwarzmeier, R. (2003). Access to natural resources in Mainland Southeast Asia and implications for sustaining rural livelihoods – The case of Thailand. *Quarterly Journal of International Agriculture*, 42(3), pp. 329–350.

Neef, A. and Thomas, D. (2009). Rewarding the upland poor for saving the commons? Evidence from Southeast Asia. *International Journal of the Commons*, 3(1), pp. 1–15.

Pham, T. T., Campbell, B. M., Garnett, S., Aslin, H. and Hoang, H. H. (2010). Importance and impacts of intermediary boundary organizations in facilitating payment for environmental services in Vietnam. *Environmental Conservation*, 37(1), pp. 64–72.

Pham, T. T., Hoang, M. H. and Campbell, B. M. (2008). Pro-poor payments for environmental services: Challenges for the government and administrative agencies in Vietnam. *Public Administration and Development*, 28, pp. 363–373.

Pham, T. T., Moeliono, M., Brockhaus, M., Le, D. N., Wong, G. Y. and Le, T. M. (2014). Local preferences and strategies for effective, efficient, and equitable distribution of PES revenues in Vietnam: Lessons for REDD+. *Human Ecology*, 42, pp. 885–899.

Phuc, X. T., Dressler, W. H., Mahanty, S., Pham, T. T. and Zingerli, C. (2012). The prospects for Payment for Environmental Services (PES) in Vietnam: A look at three payment schemes. *Human Ecology*, 40, pp. 237–249.

Pirard, R., de Buren, G. and Lapeyre, R. (2014). Do PES improve the governance of forest restoration? *Forests*, 5, pp. 404–424.

Porras, I., Grieg-Gran, M. and Neves, N. (2008). All that glitters: A review of payments for watershed services in developing countries. *Natural Resources Issues 11*. London: International Institute for Environment and Development (IIED).

Punyawadee, V., Sangkapitux, C., Kornsurin, J., Pimpaud, N. and Sonwit, N. (2010). Assessment of implicit prices for water resource improvement of tap water users in the downstream of the Mae Sa watershed, Chiang Mai Province. *Thammasat Economics Journal*, 28(4), pp. 1–28 (in Thai).

Rosales, R. M. P. (2003). *Developing pro-poor markets for environmental services in the Philippines*. London: International Institute for Environment and Development (IIED).

Sangkapitux, C., Neef, A., Polkongkaew, W., Pramoon, N., Nongkiti, S. and Nanthasen, K. (2009). Willingness of upstream and downstream resource managers to engage in compensation schemes for environmental services. *International Journal of the Commons*, 3(1), pp. 41–63.

Seeberg-Elverfeldt, C., Schwarze, S. and Zeller, M. (2009). Payments for environmental services – Carbon finance options for smallholders' agroforestry in Indonesia. *International Journal of the Commons*, 3(1), pp. 108–130.

Suhardiman, D., Wichelns, D., Lestrelin, G. and Hoanh, C. T. (2013). Payments for ecosystem services in Vietnam: Market-based incentives or state control of resources? *Ecosystem Services*, 6, pp. 64–71.

Suyanto, S. (2007). *Conditional land tenure: A pathway to healthy landscapes and enhanced livelihoods*. RUPES Sumberjaya Brief No. 1. Bogor, Indonesia: World Agroforestry Center (ICRAF).

Suyanto, S., Beria, L., Permana, R. P. and Chandler, F. J. C. (2005). *Review of the development of the environmental services market in Indonesia*. Working Paper 71. Bogor, Indonesia: World Agroforestry Center (ICRAF).

The, B. D. and Ngoc, H. B. (2006). *Payments for environmental services in Vietnam: Assessing an economic approach to sustainable forest management*. EEPSEA Research Report 2006-RR3. Economy and Environment Program for Southeast Asia, Singapore.

Tomich, T. P., Thomas, D. E. and van Noordwijk, M. (2004). Environmental services and land use change in Southeast Asia: from recognition to regulation or reward? *Agriculture, Ecosystems and Environment*, 104(1), pp. 229–244.

Van Noordwijk, M. and Leimona, B. (2010). Principles for fairness and efficiency in enhancing environmental services in Asia: Payments, compensation, or co-investment? *Ecology and Society*, 15(4), p. 17. [online] Available at: www.ecologyandsociety.org/vol15/iss4/art17/

Wunder, S. (2008). Payments for environmental services and the poor: Concepts and preliminary evidence. *Environment and Development Economics*, 13, pp. 279–297.

Wunder, S., Campbell, B., Frost, P. G. H., Sayer, J. A., Iwan, R. and Wollenberg, L. (2008b). When donors get cold feet: the community conservation concession in Setulang (Kalimantan, Indonesia) that never happened. *Ecology and Society*, 13(1), pp. 12, [online] Available at: www.ecologyandsociety.org/vol13/iss1/art12/

Wunder, S., Engel, S. and Pagiola, S. (2008a). Taking stock: A comparative analysis of payments for environmental services programs in developed and developing countries. *Ecological Economics*, 65, pp. 834–852.

30
FOREST-LED DEVELOPMENT? A MORE-THAN-HUMAN APPROACH TO FORESTS IN SOUTHEAST ASIAN DEVELOPMENT

Andrew McGregor and Amanda Thomas

Introduction

Southeast Asian forests are amongst the most diverse and extensive in the world (see Figure 30.1). Many of the species and ecologies inhabiting Southeast Asian forests are endemic to the region, contributing to the unique cultures and societies of those evolving with them. In this chapter we explore these forest-society relationships in the context of development. We do so by drawing from recent work that troubles the subject-object binary that has paralyzed so much work on human-nature relations to consider forests as development actors in themselves. This diverts from more common approaches in which forests generally provide a backdrop to development debates, or, at best, are considered in terms of how development impacts forests – usually in terms of forest loss and the associated impacts on human and animal communities. While the impact of development upon forests is a theme within this chapter, we focus more explicitly upon the influence of forests upon development. As such we see forests not as inert objects within development processes but dynamic assemblages of interacting things, whose materiality influences the types of development that can and does occur. This is not a retreat to some sort of restrictive environmental determinism but instead an approach that acknowledges that development is never a purely human endeavor – instead it derives from complex negotiations involving diverse assemblages of human and non-human actors.

Our approach draws from ongoing efforts within the social sciences and environmental humanities to decenter human agency (Latour 1993, 2004; Haraway 2008; Bennett 2010). A rich literature has evolved that emphasizes the interconnections between humans and their environments, arguing that humans are neither exceptional in their capacity for agency, nor separable from the contexts in which they are embedded. Instead attention is devoted to the relations that enable particular actions to take place, where agency is derived from particular assemblages of human and non-human actors extending across time and space. These more relational approaches, sometimes referred to as more-than-human, aim to break down human-nature dualisms to emphasize that humans do not evolve or act in isolation, or according to detached human rationalities, but instead respond to, are dependent upon and enabled by material and embodied relations with other people, species and things. In this chapter we are interested in forests as one of these things. This chapter explores what kinds of politics and ways of life are enabled by forest-human assemblages (Kohn 2013), as well as the politics and ways of life that form in the wake of the removal of forests from these assemblages. Much of the research that has sought to decenter human agency has focused on animals, and particularly ethical entanglements with them. There has been less focus on plants because they are "easy to take for granted" (Head and Atchison 2009, 237) and the ethical connections can seem more distant. Plants, particularly trees, are an important focus of this chapter, however we engage with forests as assemblages of plants, humans, (non-human) animals and other non-human things (such as fire, soil and carbon) that enable or constrain development pathways. As such forest-led development refers to development shaped by this assemblage of entangled human and non-human things.

With some notable exceptions (e.g., Gibson-Graham 2011; Hill 2015; Weisser 2015) development studies have yet to fully engage with more-than-human theories. Yet the work is central to rethinking development. The challenge is to approach development, in the same way Durbeck et al (2015, 120) outline for the Anthropocene, "not from the normative viewpoint of exclusive human agency (human exceptionalism) but rather from the perspective that we are a species living in conjunction with our co-species and interacting with – not just impacting or controlling – the weather, the water flows, the landscape." We see value within such approaches as a means to rethink development pathways in the context of widespread environmental degradation and associated poverty and marginalization. We also believe development researchers

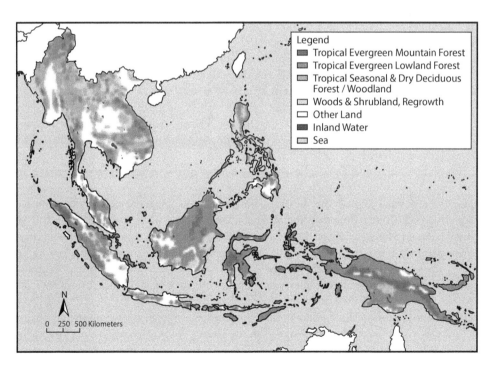

Figure 30.1 Regional extent of tropical forest in Southeast Asia (including Papua New Guinea) derived from Spot Vegetation 1 km data 2000

Source: Stibig et al. 2014

have much to contribute to more-than-human debates, by learning from societies in which the Western hyper-separation between humans and nature is not as acute, providing conduits to amplify Indigenous and non-Western theories of the nonhuman. As such more-than-human development research has the potential to expand the political and institutional spaces for Indigenous voices and amodern ontologies, providing alternatives to the socioecological violence inherent in human-nature dualisms.

In what follows we initially provide a brief regional overview of the status of forests in Southeast Asia. We then analyze three common forest-development pathways from a more-than-human perspective, emphasizing the agency of forests in enabling these directions. We conclude with some thoughts about the value of more-than-human approaches in supporting more ethical and context-specific development approaches.

Forests in Southeast Asia

Forests are contested spaces in Southeast Asia. Conflicts emerge from the diverse human and non-human stakeholders seeking to extract a wide range of experiential and material benefits. At the core of many of these debates are differences over what a forest is, what its primary role should be, who should have the rights to interact with it and who benefits from its use. Official forest definitions have their roots in the colonial period when forest science, developed in Germany, was imported to the Southeast Asian context (Vandergeest and Peluso 2006; Potter 2003). The aim of scientific knowledge systems was "to make legible the extractive potential of trees" (Wong et al. 2007, 646). Classification systems were developed that identified different forest types based on characteristics such as climate, vegetation, slope and elevation – with some identified as more appropriate for timber extraction than others. As such the term forests has become a term that encompasses a wide diversity of ecological systems with the most common category being tropical rainforests but also including categories like tropical dry forests, mountain forests, mangrove forests, teak forests, freshwater and peat forests. Other forest classification systems focused on how they should be used – as production, conservation or plantation forests, for example. Largely absent from these classifications are the people living within or alongside them – instead forests are depicted as natural spaces for expert management.

The purposeful construction of forests as human-less spaces benefited colonial empires and post-colonial states who generally claimed authority over forested spaces in order to build forest industries and expand state territory. Processes of mapping and classification bolstered state claims to land stretching from the mountain forests of northern Myanmar through to the coastal mangrove forests of Papua. In the post-World War II period colonial powers and newly independent regimes were often violent in their ambitions to feed a resource-hungry Europe and secure power through forests (Bryant 2014). The socio-ecological histories, livelihoods, knowledges and materialities of millions of people whose lives were intimately entwined with forests were effectively erased or disrupted through colonial processes. The classification system developed by Pwo Karen, for instance, includes a range of categories incorporating physical features (such as soil, slope, plant species) with the history of human interactions (for instance, time since last cultivation) that determine whether and how land can be cultivated (Wong et al. 2007). In contrast, colonial practices of mapping and classification distinguished 'forests' from 'agriculture,' demarking what practices were suitable and where – often criminalizing those groups who continued to practice forms of agriculture within forests (Hecht 2014; Wong et al. 2007; Buergin 2003). As such governing bodies used seemingly objective scientific classification systems as political instruments for exerting territorial control in the name of development. Peluso and Vandergeest (2001) use the term 'political forests' to emphasize that Southeast Asian forests are not natural spaces, but outcomes of inter-related social, political and ecological processes.

The Food and Agriculture Organization (FAO) has become central to promoting and embedding scientific ways of knowing forests in the post-colonial period (Peluso and Vandergeest 2001; Vandergeest and Peluso 2006). It is the key organization for data gathering and policy leadership on food, agriculture and forests in the region. Yet despite their role as a central forest-knowledge-producer, data produced by the FAO is problematic, as definitions of forest shift across space, and the dynamic properties of forests themselves trouble easy measurement and classification (see Grainger 2008, 2014). Core to current debates are definitions of primary (implicitly and problematically implying pristine or 'untouched' forests), secondary (forest that has been degraded or cleared at some point) and plantation forests (are they 'forests' at all?). Despite these limitations some broad trajectories of change are clear from FAO data. Of all the world's tropical regions, Southeast Asia is experiencing the highest rate of deforestation. From 2000 to 2010 there was a total estimated decrease in forest area of almost 9,000,000 hectares per year, including 230,000 hectares of biodiverse primary forest (FAO 2011). Between 1990 and 2010 forest cover declined by 33 million hectares, an area larger than Vietnam, to a total of 214 million hectares (FAO 2011). Deforestation is concentrated in the forest-rich areas of Indonesia and Myanmar, whereas Vietnam and the Philippines have shown signs of an overall increase in forest cover (see Figure 30.2). Payn et al. (2015) caution, however, that forest plantations, rather than more opportunistic regrowth, drive forest growth. In Vietnam, for example, only 1 percent of total forest area is considered primary forest, while 25 percent is plantation forest (FAO 2011).

Deforestation is fueled by an array of factors, with the most obvious being agricultural expansion and timber harvesting. Road construction, the installation of transmission lines and dam building play a role in opening up forested areas to these activities. Stibig et al. (2014) identify the conversion of forests into cash crop plantations as being the main cause of deforestation

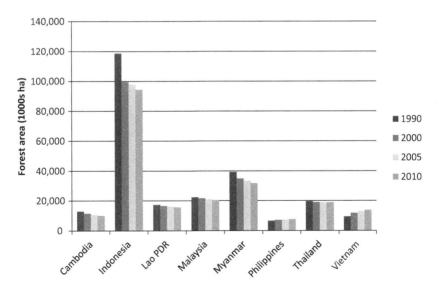

Figure 30.2 Change in forested area by country and year

Source: FAO (2011)

in Southeast Asia between 1990 and 2010. Oil palm plantations are the most important driver in insular Southeast Asia, with coffee, tea, rubber and commercial tree plantations also important, particularly in continental Southeast Asia (see Figure 30.3). Non-sustainable timber harvesting is prominent in heavily forested areas, particularly northern Myanmar, Kalimantan and West Papua. While the value of forest product exports (forestry, wood, pulp and paper and wooden furniture) has soared across the region (FAO 2014, see Figure 30.4), the proportion of total GDP from forestry has declined, most significantly in Cambodia and Malaysia (FAO 2014, see Figure 30.5). Despite the declining significance of forestry at the national and regional scale forest industries can provide important sources of income and employment at sub-national scales.

Efforts to curb deforestation are typically beset by problems related to governance, corruption, economies, land rights and inequality. The FAO (2011) argues that without strengthening state policies and bolstering liberal institutions in SEA countries, deforestation and the loss of forests will continue. However this ignores the benefits states receive from removing forests – much deforestation is state assisted and approved. Instead there are deeper problems relating to the ontological construction of human-forest relationships within development processes. In what follows, we focus on three types of relationships – based on swidden, extractive industries and conservation – to analyze the role of forests in shaping development pathways. We use the term forest-led development deliberately, to emphasize how the material characteristics of Southeast Asian forests enable these pathways, while also recognizing the broad assemblage of human and non-human actors involved. Our aim is to show that forests are important development actors in themselves, and consideration of such dynamism can contribute to more desirable development directions.

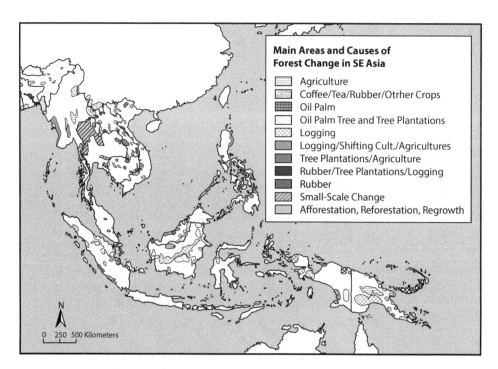

Figure 30.3 Regional pattern of main areas and causes of forest change in Southeast Asia
Source: Stibig et al. 2014

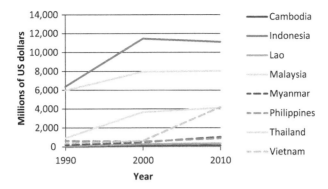

Figure 30.4 Value of forest products exports (in millions USD at 2011 prices and exchange rates)
Source: FAO (2014)

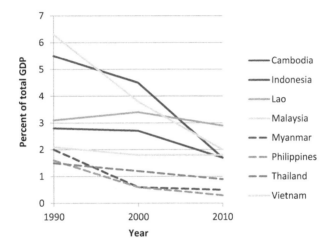

Figure 30.5 Contribution of the forestry sector to total GDP
Source: FAO (2014)

Density, diversity and swidden development

Southeast Asian forests host some of the most diverse biomes on the planet. A rich mix of trees and plants compete for sunlight, providing conditions for the flourishing of a wide variety of animals, insects and other creatures/species. This diversity has supported long established development pathways amongst minority groups in Southeast Asia in at least two ways. First, the sheer density of undergrowth typical of Southeast Asian forests, in combination with other geographical factors such as slope, terrain, elevation and remoteness from other settled areas, has generally obstructed the spread of colonial empires and slowed the subsequent expansion of national development initiatives. Quite simply, forests, while acting as a magnet for resource extraction in some cases (see next section), have made some communities, such as hill tribes of continental Southeast Asia, or the Dayak people of Borneo, much less accessible to mainstream development and nation-building efforts than others (Scott 1985). Second, forest ecologies have traditionally provided local communities with resources to enable subsistence-oriented livelihoods, bolstered by trade of forest products with outsiders. This has allowed them to retain relatively isolated lifestyles if so desired.

These long established development pathways reflect an intimate relationship between people and forests and the natural world, with forests shaping and reflecting everyday life. Forests provide food, medicines, materials as well as important social and spiritual spaces and experiences for communities that rely on them. As a consequence societies have co-evolved or 'become with' (Wright 2014) the fluctuating conditions of forests. The availability of forest products for human use shifts with time, reflecting "sharp seasonal fluctuations owing to the phenomenology of plants, migration patterns of animals and climatic conditions" (De Jong et al. 2003, 108). The trade of forest products such as sandalwood, rattan, minerals, forest rubber, spices and edible bird's nests have brought opportunities and conflicts to those living with forests, underpinning the growth of pre-colonial kingdoms, colonial trading empires and post-colonial states. Complex agro-ecological systems have evolved with forests, providing food and livelihoods for forest communities. The most important of these is swidden agriculture.

Swidden agriculture, sometimes referred to as shifting cultivation or slash-and-burn, has been practiced for centuries in Southeast Asia. It is impossible to get accurate figures on the numbers of swidden farmers however recent estimates suggest between 14 million and 34 million people are involved (Mertz 2009), a figure that has declined in the past half century due to changing political economies (Fox et al. 2009). Swidden agriculture blurs distinctions between forested land and agricultural land, and practices of cultivation and gathering. While there is a lot of variety within swidden agriculture, the general approach is to rotate fields rather than crops, distinguishing it from settled forms of agriculture. Typically farmers clear a section of forest to grow a diversity of crops over a short period of time (0–3 years) before leaving the area to fallow and regenerate over a much larger period of time (typically 5–20 years), and rotating to a different plot. The fallow regenerates as a type of secondary forest, sometimes accompanied by more permanent productive crops (such as rubber trees for example) in increasingly complex and diversified systems. However increasing land pressures have reduced the land under swidden agriculture, shortening fallow times and constraining ecological diversity (Wong et al. 2007). Livelihoods are maintained and bolstered by forest gardens, the collection of other forest products, and increasingly through trade, employment and migration of family members to non-swidden areas.

The swidden agriculture practiced in and alongside forests has played a key role in the societies and cultures of many ethnic minority cultures across the region. Particular knowledges, practices, social relations, identities and religious rituals and beliefs are bound up in swidden. For example, communities need intimate knowledge of local ecological systems and seasons to be successful. One study revealed over 1,000 useful forest products known and utilized by just two swidden communities in Sarawak (Christensen 2002). As Brookfield et al. (1995, 29) write about the forests of Borneo, "To its occupiers, the forest is a known set of resources to be worked over; much if not most of it has been exploited for centuries or millennia." There are similarly rich social relations, and religious practices bound up in clearance and harvesting. Locally significant rituals led Cramb et al. (2009) to refer to a "swidden centred religious world" (see also Lim et al. 2012). While often invisible to the untrained eye a complex mosaic of land use types may exist, including sacred sites, agroforestry gardens, forest gardens, active and fallow swidden zones, and closed and open forests (Lim et al. 2012). Forests have enabled these unique cultures and agricultural systems to evolve at some distance from the settled agriculture and urban centers of lowland societies.

The capacity of forests to regenerate and provide nourishing soils for swidden has underpinned this style of development over generations. However swidden lifestyles are, and have been for some time, experiencing rapid decline (Cramb et al. 2009). There are a variety of reasons for this but underpinning many of them is the view, often emanating from states within Southeast Asia, that swidden is inappropriate for modern nations. Swidden has generally been seen by governments and development agencies as a primitive form of agriculture that contributes to poverty and environmental degradation – while also potentially threatening the cohesiveness and reach of the state (for a discussion see Brookfield et al. 1995; Scott 1985). It doesn't fit the administrative divisions between increasingly professionalized forestry and agricultural departments, neither of which have been particularly keen to include it as a viable activity in their portfolios (Fox et al. 2009; Peluso and Vandergeest 2001). Nor does it fit modernist binaries between humans and nature, use and conservation, blurring the boundaries of what actually constitutes a forest. Hence development programs have rarely sought to assist communities in enhancing and building from swidden practices but, on the contrary, have attempted to shift people away from the practice altogether. Swidden communities in the highlands of mainland Southeast Asian, for example, have long been targets of government sponsored resettlement programs, often with international assistance (for instance, Baird and Shoemaker 2007).

State-led development has seen the extension of roads and transmission lines to increasingly remote areas, opening swidden forests to government oversight, market forces, commodification and a range of economic activities. With states benefiting little from swidden practices and remote communities having little political sway, old but often ineffective laws and policies that criminalized or restricted them – many swidden communities are legally considered to be squatters on government-owned land – are more likely to be implemented. There are examples where communities have been forcibly relocated to allow for plantation, timber or conservation initiatives – a violent form of primitive accumulation led by state and private sector actors (Hall 2012). Population and development policies have created further pressures, with Indonesia's Transmigrasi scheme and Malaysia's FELDA program, for example, shifting millions of people to what are considered underpopulated forested areas as a means of promoting development in those regions. This has expanded the spaces of settled agriculture, reducing fallowing times and requiring the intensification of swidden systems through agricultural inputs, leading to a reduction of the overall sustainability of the practice. In addition many farmers actively pursue new opportunities through settled agriculture or plantation and timber industries in order to access the associated development benefits of improved infrastructure and services that come with that (see Hall 2012). Development has also drawn young people away from these more traditional lifestyles to seek employment and livelihood opportunities elsewhere.

As a consequence the relationship between forests, swidden communities and state-led development has been an uneasy one. Some swidden populations have openly embraced the benefits development brings, such as health and education services, and electricity, and abandoned old lifestyles. For others the increasing reach of development industries in the form of timber harvesting, plantation economies and conservation initiatives has forced them from their lands. For those who remain, however, swidden continues to be part of increasingly complex livelihoods, albeit often covertly or without permission from the state, in which the benefits of development, in terms of education, goods and services co-exist with swidden lifestyles. In these places swidden may be practiced alongside timber harvesting, cash crops, wage labor, livestock rearing and seasonal migration. These diverse activities rely on forest and other non-human agencies, improving the resilience of communities by providing an alternative source of food and income, or what Lian (1993, 330) once termed a "basic survival kit," when formal economies fail (Cramb et al. 2009). In contrast, forcibly separating people from forests, as has happened too often, can increase vulnerabilities and contribute to further social and ecological degradation. Generating policies to allow for the continuance of swidden agriculture, as opposed to dispossession and its cessation, would seem good development practice, recognizing that forests actively contribute to the well-being of some of Southeast Asia's most marginalized people.

Wood, fire and extractive development

Wood is a valuable commodity. It is solid but moveable and shapeable – being used in a variety of things such as construction, furniture, ornaments, arts and utensils, and used as fuel for cooking and warmth through burning. These characteristics – its solidity, malleability, transportability and combustibility – and that it grows freely from seeds in soils to form trees has underpinned very different development pathways in Southeast Asia. Indeed tree trunks were part of the original attraction of Southeast Asia for colonial empires, providing valuable materials to feed the industrializing economies of Europe. Certain types of trees had valuable characteristics that attracted more attention and resources than others, leading to a concentration of timber activities in particular areas, profiting colonial expansion. The teak forests of Myanmar, for example, have been subjected to timber operations since the mid-nineteenth century, and were promoted as symbols of progress (Bryant 2014). However, European-derived forestry models have not adapted well to much more socioecologically diverse tropical contexts, most notably in their focus on managing timber extraction for national and international economies as opposed to forest products and swidden agriculture for local community economies (Potter 2003).

In most Southeast Asian countries the state claims forested land as its own, irrespective of de facto relationships on the ground, and assumes the authority to determine forest uses. These generally revolve around four functions: production (timber), conversion (to agriculture), protection (for human benefit such as preventing erosion or flooding) and conservation (for ecological reasons). Production and conversion forests have proved particularly lucrative with state-industry-community-forest assemblages forming to generate employment and income through timber extraction or via forest clearance to enable plantation and mining economies. The state provides the necessary permissions; the private sector provides the capital, equipment and expertise; local communities often provide labor; and forests provide timber and/or space that can be cleared for plantations and mining ventures. The focus of such arrangements is on profit, with the financial benefits concentrated amongst the owners and shareholders of extractive industries, as well as state actors through fees, taxes, corruption and other forms of rent. Other benefits trickle through to the local scale – such as employment in harvesting, transport, sawmills, and pulp and paper factories, plantations and mining sites – creating new skills, identities, networks, dependencies and opportunities. These benefits come at the expense of forests that lose their capacities to sustain the habitats, soils, atmosphere and biodiversity that are crucial to many human and non-human forest communities.

This style of development, which is the dominant approach to forests in the region, has attracted plenty of criticism. Production forests are meant to be managed sustainably with repeated harvests over 20- or 30-year cycles, yet this is rarely the case. Timber harvesting and the roads built to facilitate their operations can cause intense forms of forest degradation and contribute to agricultural expansion from which forests are unlikely to ever fully recover (FAO 2011). Conversion forests are completely removed in order to redirect land use to settled agriculture, with oil palm, rubber, cashew nut, coconut and sugar cane plantations, as well as shrimp ponds in coastal areas, being key drivers of forest conversion (FAO 2011). For plantation or mining concession holders forests are seen to inhibit, rather than enable, development. Only a small amount of the revenue generated by such activities finds its way to local communities – and many are adversely affected by it. Instead logging has financed military activities in unstable areas, such as Aceh's long independence struggle (Kingsbury and McCulloch 2006) and violent Khmer Rouge activities in Cambodia (Global Witness 1995), and boosted the power and wealth of political and business elites across the region. Illegal logging is common with organized crime gangs working with local communities and corrupt officials to extract and export wood from under-policed areas (see Telapak and the Environmental Investigation Agency 2005).

Forests have, in a sense, proved uncooperative to state-led development. Most significantly, logged over land is much more prone to fire than primary forest. The increase in timber harvesting has contributed to the intensity and magnitude of fire events in the region. As Schinidele et al (1989, 71, cited in Brookfield et al. 1995, 175) conclude in regards to the devastating Kalimantan fires of 1982/83 "it was not the drought which caused this huge fire, but the changed condition of the forest." In addition the woodiness of forests has encouraged concession holders and smallholders to clear land through slash-and-burn methods, which are much quicker and cheaper than cutting and clearing, but can lead to wildfires. Fires have devastated societies, ecologies and economies, not only in forested areas but much further afield as well. At the time of writing the 2015 Indonesian fires have killed 19 people, contributed to 500,000 cases of respiratory chest infections, are expected to contribute to 100,000 premature deaths, and will have a $50 billion impact on Indonesia's economy (Balch 2015). Huge smoke hazes spread from fires in Indonesia to Malaysia and Singapore almost every dry season, negatively impacting the health, well-being and economies of millions of people, accelerating carbon emissions and creating tensions in ASEAN (see Eaton and Radojevic 2001; Atkinson 2014). Forest fires are very much a part of this development pathway.

Forest removal also contributes to the degradation of watersheds, air and water quality, soil and biodiversity, whilst also contributing to local increases in heat and flash flooding. The death of several hundred villagers in deforestation-related mudslides in Thailand, for example, shocked the nation into declaring a nationwide logging ban in 1989 (see Hirsch and Lohmann 1989). In a similar way President Suharto's ambitious Mega Rice Project cleared peat forests in an attempt to create agricultural land but instead destroyed forest soils and productivity, eventually contributing to the abandonment of the multi-billion dollar project (Ken-ichi 2003). In a more local and immediate sense the biodiverse non-human communities that exist in forests may not cooperate in clearance, with infectious diseases more prominent in deforested or degraded areas (Gottdenker et al. 2014) as are human-animal conflicts (Tan et al. 2015).

This imbalance, where the beneficiaries of forest industries are concentrated amongst elites at national and international levels while the socio-ecological costs are experienced at much more local scales, has inspired a number of initiatives aimed at making timber harvesting more equitable and sustainable. One approach has been to grow plantation forests to take the pressure off natural forests. Plantation forests in Southeast Asia and southern China now extend over 7 million hectares (ACIAR 2014). However they come with their own set of problems including corporate ownership, exotic single use species, ecological degradation, forced displacement and land tenure conflicts (see Hirsch and Lohmann 1989). Other initiatives focus on certifying the sustainability of forest production through standards like that provided by the Forest Stewardship Council or through the European Union's Forest Law, Enforcement and Governance (FLEGT) project which seeks to reduce illegal logging. Finally there is community forestry in which the responsibility for timber harvesting and maintaining forest health is maintained by those most affected, people at the local scale. Despite being marginal to most forest planning, community forests are becoming increasingly recognized as a viable alternative to more centralized forest systems. They are diverse multifunctional systems embedded with shifting local economies, cultures and ecologies, usually run by farmers pursuing a form of agroforestry (Michon et al. 2007).

The capacity of forests to generate wood, then, is at the center of international assemblages oriented at profiting from Southeast Asia's forests. Wood enables particular forms of resource-based development that can be directed and financed at a distance, privileging international political economies of forest extraction over the place-based felt, embodied and reciprocal relationships that can emerge through local interactions. However, those focused on forest extraction do not adequately recognize forests as assemblages made up of dynamic interacting things, of which timber, as well as forest communities, tigers and soils are legitimate parts. The agency of human/non-human as well as non-human/non-human interactions within these assemblages – for example the agency of forests in preventing or contributing to mudslides or fires – become unsettled through such industries, creating new development risks. For the millions of swidden agriculturalists and other human and non-human communities reliant on forest ecologies, extractive industries can force radical changes in their relations with forests and associated development directions. While some welcome the opportunity to diversify their livelihoods through employment in forest and plantation industries, others resist, concerned their customary lands and livelihoods will be destroyed. However, organized resistance, as opposed to everyday resistance (Scott 1985), has been rare. Instead the materialities of forests themselves – in terms of things like combustibility, density and quality – have created the main obstacles for forest extraction. As extraction continues forests release carbon to the atmosphere, contributing to hotter, dryer climates and more combustible forests.

Carbon, ecology and preservationist development

A third form of forest-led development is being enabled through the function of Southeast Asian forests within global ecologies. Forests extract carbon dioxide from the atmosphere and turn it into biomass, thereby acting as global carbon sinks. Southeast Asian forests are also important biodiversity zones, housing four globally significant, but threatened, hotspots of biodiversity[1] (Sodhi et al. 2004). Such functions are increasingly being recognized through the discourse of 'ecosystem services,' attracting a different type of preservation-oriented development actor. While environmental groups have long valued and sought to protect Southeast Asian forests for their biodiversity values, interest in carbon has only really been of interest since the UNFCCC Bali Climate Change conference in 2007. The carbon sequestration capacities of forests are reorienting forest values and economies toward preservation.

Preservation is nothing new in Southeast Asia. At a local scale sections of forests have long been preserved for religious or ecological reasons and professional foresters reserved broad swathes of forest in the colonial period. Currently 18 percent of the region's forests, or 9 percent of total landmass, is officially under some sort of reserve status (FAO 2011). Protected areas differ in terms of the amount of human involvement in the reserve – from strict nature reserves that exclude people in order to preserve areas for their biological and scientific merit, to much more integrated reserves involving sustainable forest use (UNEP-WCMC 2014). Apart from obvious ecological benefits, reserves also provide opportunities for tourism development, particularly around National Parks or World Heritage Sites – of which Southeast Asia has 13 listed for their natural values. However reserves come with their own set of problems and can have similar impacts on people's livelihoods as extractive industries.

The most controversial approach to forest reserves is sometimes referred to as 'fortress conservation' in which people are barred from or limited in how they interact with forests through state-imposed policies involving 'fences and fines' (Fletcher 2010). As the state officially owns forest land in most countries, when a reserve is established they can criminalize those living within it or their livelihoods – which may involve hunting, harvesting and swidden. Buergin (2003), for example, has detailed how the state criminalized Karen people living on their homelands in northern Thailand when the Thung Sai Yaresuan Wildlife Sanctuary was created. In this case rice barns were burned down and people forcibly moved out of the reserve ostensibly to protect biodiversity. Such approaches – where the state uses conservation rationales to oppress ethnic groups and other marginalized people in order to promote their vision of development – are not uncommon (Fairhead et al. 2012). Other approaches seek to implement Integrated Conservation and Development programs (ICDP) that focus on generating alternative lifestyles for people previously reliant on protected areas as a means of reducing pressure on forests. However these projects often repeat the human-nature dualisms underpinning fortress conservation and have also experienced problems in implementation. State-community, national-local and human-nature tensions pervade such arrangements, with the consequences of failed projects experienced most acutely by those most reliant upon them.

Forest preservation efforts have recently received a boost as a result of concerns about the contribution of deforestation to global climate change. Forest loss and the associated degradation of peatlands is estimated to contribute 12–20 percent of global emissions (McGregor 2015). As a consequence, the UNFCCC have been piloting an instrument to prevent forest loss and to encourage the expansion of forests through a payment-based system known as Reducing Emission from Deforestation and forest Degradation (REDD+). As Indonesia has the region's largest amount of forest land and the fastest rate of deforestation, it has become a leading site for REDD+ development, attracting over US$2 billion worth of public and private sector financing commitments (grants, loans and investments) at the time of writing (REDDX n.d.). The program seeks to develop governance, measurement and reporting systems that will allow countries to record and verify improvements in overall carbon stock, which can then be traded through carbon markets. It provides a means for recognising and paying for the global climate services of Southeast Asian forests and compensating for the opportunity costs of protecting, as opposed to exploiting, forests.

REDD+ is a controversial program in which market forces based on the so-called green economy are expected to step in where the state has failed in preventing forest loss (McGregor 2010). Small REDD+ projects oriented at voluntary carbon markets and involving diverse combinations of public, private and community sector actors have sprung up across Southeast Asia however there is little evidence to suggest that carbon emissions have declined in any significant way, and few projects are actively generating carbon credits. The barriers are many (e.g., Howson and Kindon 2015; Lounela 2015), not least the vested interests of influential groups across Southeast Asia in timber and plantation economies. However resistance has also been expressed at a more local scale as communities fear a further round of exclusion and marginalization under the REDD+ umbrella (Eilenberg 2015). Nevertheless there have been some hopeful signs that REDD+ can lead to improvements in the visibility and claims of forest communities and slow or prevent the type of land clearance associated with forest industries (Astuti and McGregor 2015; McGregor et al. 2015). Only time will tell.

In Indonesia a majority of environmental and Indigenous NGOs at the national scale have become involved in REDD+ planning and development, strategically engaging in order to unsettle the ongoing degradation of socioecological systems associated with resource industries. There can be no doubt their engagement has allowed them to influence and improve forest governance in the country. However it has also provided a sort of tacit approval that market mechanisms have a role to play in preventing forest loss. This is a significant shift in forest politics in the country – one that has the potential to reap large rewards in attracting private finance for conservation but also comes with large risks regarding the profit-focused agenda of private capital (Dixon and Challies 2015). Development in these areas becomes increasingly linked to the movement of carbon through the atmosphere, the absorption capacity of different forest types, fires and alternative carbon management technologies. However it also provides new opportunities for people to oppose resource industries and pursue new development pathways. What is clear, however, is that REDD+ will not work unless it is geared to the interests and aspirations of those living with forests, rather than the more distant interests of carbon investors (McGregor 2015).

More-than-human development

In the three pathways discussed above we have emphasized the agency of forests, positioning them as core to the diverse human/non-human development assemblages that have formed. It is the density of forests, their regenerative abilities, their contributions to soils and watersheds, the wood they produce, their diversity, habitats, combustibility and carbon storage abilities that are shaping, and are shaped by, development. Southeast Asian identities, societies, cultures, empires, professions, economies, trading networks, nations, politics and power relationships have emerged from and continue to reflect the agencies of forests – the region would look radically different without them. We end this chapter not by promoting one particular pathway over another, but instead by considering how this more dynamic view of forests challenges development thinking, and what new principles might be generated for pursuing equitable outcomes. When forest agencies and assemblages are acknowledged development is reoriented toward working with entanglements of humans and non-humans in ways that resist the hyper-separation of humans and nature that has contributed to so much degradation of people and places around the world. The anthropocentrism of human development is replaced by a more caring relational ethic and sense of responsibility, oriented around enhancing the intimate more-than-human connections that form in place.

If, as we have argued, forests possess a form of agency in shaping development, can we think about development in more-than-human ways? Our analysis suggests that development occurs not through independent human decision-making but through human relationships with other agentic things, where agency is a form of creativity and the "capacity to make something new appear or occur" (Bennett 2010, 31). Southeast Asian forests produce and are produced by rattan, rain, tigers, timber, livelihoods, identities and empires, while stabilizing landscapes, fertilizing soils and preventing or enabling fires. In doing so, forests trouble simplistic depictions of human development that rest solely on human agency and construct humans as the central locale for ethical commitments. Instead development can be thought of as assemblages of humans and non-humans arranged for particular ends. When forests disappear so too do those development pathways, and the societies, cultures, ecosystems and economies bound up within them.

The unfortunate fixation on the human within development theory and practice blinds us to the central role of non-humans in contributing to equitable and sustainable futures. Is it possible to move beyond the human and – to borrow from Anna Tsing (2012) – to think of development as an 'interspecies relationship'? Can we be more worldly and generous in thinking about development beyond the human, in which forests, trees, animals and the people that depend on and constitute them are promoted to the foreground rather than the background of development debates? Such an approach is likely to be more inclusive, equitable and ethical, being sensitive to spatial differences and building from existing interconnections rather than replacing them with more destructive development models. A more-than-human approach to swidden farming, for example, would seek to build from and strengthen existing practices in ways that enhance all the human and non-human agents that are situated within the complex place-based relationships that constitute life. Rather than focus on humans (in development), or nature (in conservation), there is an opportunity to focus on 'becoming with' (Wright 2014) and recognize how humans are small parts of much greater networks that co-evolve (Thomas 2015). As the shadow of the Anthropocene sweeps across the planet, development oriented toward living with forests, of developing, recognizing and respecting diverse interspecies relationships, rather than a narrowly-focused human development that benefits some humans over others, or particular charismatic non-humans, would seem an appropriate pursuit (McGregor 2017).

While such speculation can be criticized for being disconnected from the broader political ecologies governing development and forests, highlighting interspecies relationships is hardly radical or unimportant to those who are interacting with forests on a daily basis. A more-than-human approach can provide a framework to explore the diverse ways in which people's lives are deeply entwined and co-dependent on what we now know as forests, and to promote a relational ethics and politics based on respect and responsibility generated within these intimate relationships. It can make Indigenous knowledges and marginalized people and species more visible and encourage place-specific development interventions oriented at becoming with, rather than becoming without, forests. Reorienting development toward improving interspecies relationships by recognizing and working with non-human agencies can provide a lens to assess the strengths and weaknesses of current development pathways and provide rationales for assembling new ones.

Note

1 The hotspots are Indo-Burma (parts of Bhutan, Nepal, eastern India, southern China through to Vietnam and southern Thailand), Sundaland (Malaysia through to Bali, including Borneo), Wallacea (eastern Indonesia from Lombok and Sulawesi to Malaka but ending before West Papua) and the Philippines.

References

Astuti, R. and McGregor, A. (2015). Governing carbon, transforming forest politics: A case study of Indonesia's REDD+ Task Force. *Asia Pacific Viewpoint*, 56(1), pp. 21–36.

Atkinson, C. L. (2014). Deforestation and transboundary haze in Indonesia: Path dependence and elite influences. *Environment and Urbanization ASIA*, 5(2), pp. 253–267.

Australian Centre for International Agricultural Research (ACIAR). (2014). *Sustainable plantation forestry in South-East Asia*. ACIAR Technical Reports No. 84. Canberra: Australian Centre for International Agricultural Research.

Baird, I. and Shoemaker, B. (2007). Unsettling experiences: Internal resettlement and international aid agencies in Laos. *Development and Change*, 38(5), pp. 865–888.

Balch, O. (2015). *Indonesia's forest fires: Everything you need to know*. Available at: www.theguardian.com/sustainable-business/2015/nov/11/indonesia-forest-fires-explained-haze-palm-oil-timber-burning. Accessed 23 November 2015.

Bennett, J. (2010). *Vibrant matter: A political ecology of things*. Durham, NC and London: Duke University Press.

Brookfield, H., Potter, L. and Byron, Y. (1995). *In place of the forest: Environmental and socio-economic transformation in Borneo and the Eastern Malay Peninsula*. Tokyo and New York: United Nations University Press.

Bryant, R. L. (2014). The fate of the branded forest: Science, violence, and seduction in the world of teak. In: S. B. Hecht, K. D. Morrison and C. Padoch, eds., *The social lives of forests: Past, present, and future of woodland resurgence*. Chicago and London: University of Chicago Press, 220–230.

Buergin, R. (2003). Shifting frames for local people and forests in a global heritage: The Thung Yai Naresuan Wildlife Sanctuary in the context of Thailand's globalization and modernization. *Geoforum*, 34(3), pp. 375–393.

Christensen, H. (2002). *Ethnobotany of the Iban and the Kelabit*. Sarawak Forest Department, NEPCon Denmark, Denmark: University of Aarhus.
Cramb, R. A., Pierce Colfer, C. J., Dressler, W., Laungaramsri, P., Trang Le, Q., Mulyoutami, E., Peluso, N. and Wadley, R. L. (2009). Swidden transformations and rural livelihoods in Southeast Asia. *Human Ecology*, 37(3), pp. 323–346.
De Jong, W., Tuck-Po, L. and Ken-ichi, A. (2003). The political ecology of tropical forests in Southeast Asia: Historical roots of modern problems. In: L. Tuck-Po, W. de Jong and K. Abe, eds., *The political ecology of tropical forests in Southeast Asia: Historical perspective*. Kyoto: Kyoto University Press, 1–28.
Dixon, R. and Challies, E. (2015). Making REDD+ pay: Shifting rationales and tactics of private finance and the governance of avoided deforestation in Indonesia. *Asia Pacific Viewpoint*, 56(1), pp. 6–20.
Durbeck, G., Schaumann, C. and Sullivan, H. (2015). Human and non-human agencies in the Anthropocene. *Ecozone*, 6(1), pp. 118–136.
Eaton, P. and Radojevic, M. (eds.) (2001). *Forest fires and regional haze in Southeast Asia*. Huntington, NY: Nova Science Publishers.
Eilenberg, M. (2015). Shades of green and REDD: Local and global contestations over the value of forest versus plantation development on the Indonesian forest frontier. *Asia Pacific Viewpoint*, 56(1), pp. 48–61.
Fairhead, J., Leach, M. and Scoones, I. (2012). Green grabbing: A new appropriation of nature? *Journal of Peasant Studies*, 39(2), pp. 237–261.
Fletcher, R. (2010). Neoliberal environmentality: Towards a poststructuralist political ecology. *Conservation and Society*, 8(3), pp. 171–181.
Food and Agriculture Organisation (FAO). (2011). *Southeast Asian forests and forestry to 2020: Subregional report of the Second Asia-Pacific forestry sector outlook study*. Bangkok: FAO.
Food and Agriculture Organisation (FAO). (2014). *Contribution of the forestry sector to national economies, 1990–2011*, eds., A. Lebedys and Y. Li. Forest Finance Working Paper FSFM/ACC/09. Rome: FAO.
Fox, J., Fujita, Y., Ngidang, D., Peluso, N., Potter, L., Sakuntaladewi, N., Sturgeon, J. and Thomas, D. (2009). Policies, political-economy, and swidden in Southeast Asia. *Human Ecology*, 37(3), pp. 305–322.
Gibson-Graham, J. K. (2011). A feminist project of belonging for the Anthropocene. *Gender, Place and Culture*, 18(1), pp. 1–21.
Global Witness. (1995). *Forests, famine and war: The key to Cambodia's future*. Available at: www.globalwitness.org. Accessed 21 July 2016.
Gottdenker, N. L., Streicker, D. G., Faust, C. L. and Carroll, C. R. (2014). Anthropogenic land use change and infectious diseases: A review of the evidence. *EcoHealth*, 11(4), pp. 619–632.
Grainger, A. (2008). Difficulties in tracking the long-term global trend in tropical forest area. *PNAS*, 105(2), pp. 818–823.
Grainger, A. (2014). Pan-tropical perspectives on forest resurgence. In: S. B. Hecht, K. D. Morrison and C. Padoch, eds., *The social lives of forests: Past, present, and future of woodland resurgence*. Chicago and London: University of Chicago Press, 84–96.
Hall, D. (2012). Rethinking primitive accumulation: Theoretical tensions and rural Southeast Asian complexities. *Antipode*, 44(4), pp. 1188–1208.
Haraway, D. (2008). *When species meet*. Minneapolis and London: University of Minnesota Press.
Head, L. and Atchison, J. (2009). Cultural ecology: Emerging human-plant geographies. *Progress in Human Geography*, 33(2), pp. 236–245.
Hecht, S. B. (2014). The social lives of forest transitions and successions: Theories of forest resurgence. In: S. B. Hecht, K. D. Morrison and C. Padoch, eds., *The social lives of forests: Past, present, and future of woodland resurgence*. Chicago and London: University of Chicago Press, 97–113.
Hill, A. (2015). Moving from "matters of fact" to "matters of concern" in order to grow economic food futures in the Anthropocene. *Agriculture and Human Values*, 32(3), pp. 551–563.
Hirsch, P. and Lohmann, L. (1989). Contemporary politics of environment in Thailand. *Asian Survey*, 29(4), pp. 439–451.
Howson, P. and Kindon, S. (2015). Analysing access to the local REDD+ benefits of Sungai Lamandau, Central Kalimantan, Indonesia. *Asia Pacific Viewpoint*, 56(1), pp. 96–110.
Ken-ichi, A. (2003). Peat swamp forest development in Indonesia and the political ecology of tropical forests in Southeast Asia. In: L. Tuck-Po, W. de Jong and K. Abe, eds., *The political ecology of tropical forests in Southeast Asia: Historical perspectives*. Kyoto: Kyoto University Press, 133–151.
Kingsbury, D. and McCulloch, L. (2006). Military business in Aceh. In: A. Reid, ed., *Verandah of violence: The background to the Aceh problem*. Singapore: Singapore University Press, 199–224.
Kohn, E. (2013). *How forests think: Toward an anthropology beyond the human*. Berkeley and Los Angeles: University of California Press.

Latour, B. (1993). *We have never been modern*. Cambridge, MA: Harvard University Press.
Latour, B. (2004). *Politics of nature: How to bring the sciences into democracy*. Translated by C. Porter. Cambridge, MA, London and England: Harvard University Press.
Lian, F. J. (1993). On threatened peoples. In: H. Brookfield and Y. Byron, eds., *South-East Asia's environmental future: The search for sustainability*. Tokyo and New York: United Nations University Press, 322–337.
Lim, H. F., Luohui, L., Camacho, L. D., Combalicer, E. A. and Singh, S. K. K. (2012). Southeast Asia. In: J. A. Parrotta and R. L. Trosper, eds., *Traditional forest-related knowledge: Sustaining communities, ecosystems and biocultural diversity*. The Netherlands: Springer, 357–394.
Lounela, A. (2015). Climate change disputes and justice in Central Kalimantan, Indonesia. *Asia Pacific Viewpoint*, 56(1), pp. 62–78.
McGregor, A. (2010). Green and REDD? Towards a political ecology of deforestation in Indonesia. *Human Geography*, 3(2), pp. 21–34.
McGregor, A. (2015). Policy: REDD+ in Asia Pacific. *Nature Climate Change*, 5, pp. 623–624.
McGregor, A. (in press). Critical development studies in the Anthropocene. *Geographical Research*.
McGregor, A., Challies, E., Howson, P., Astuti, R., Dixon, R., Haalboom, B., Gavin, M., Tacconi, L. and Afiff, S. (2015). Beyond carbon, more than forest? REDD+ governmentality in Indonesia. *Environment and Planning A*, 47(1), pp. 138–155.
Mertz, O. (2009). Trends in shifting cultivation and the REDD mechanism. *Current Opinion in Environmental Sustainability*, 1, pp. 156–160.
Michon, G., Foresta, H., Levang, P. and Verdeaux, F. (2007). Domestic forests: A new paradigm for integrating local communities' forestry into tropical forest science. *Ecology and Society*, 12(2). Available at: www.ecologyandsociety.org/vol12/iss2/art1
Payn, T., Carnus, J-M., Freer-Smith, P., Kimberley, M., Kollert, W., Lio, S., Orazio, C., Rodriguez, L., Neves Silva, L. and Wingfield, M. J. (2015). Changes in planted forests and future global implications. *Forest Ecology and Management*, 352, pp. 57–67.
Peluso, N. and Vandergeest, P. (2001). Genealogies of the political forest and customary rights in Indonesia, Malaysia, and Thailand. *The Journal of Asian Studies*, 60(3), pp. 761–812.
Potter, L. (2003). Forests versus agriculture: Colonial forest services, environmental ideas and the regulation of land-use change in Southeast Asia. In: L. Tuck-Po, W. de Jong and K. Abe, eds., *The political ecology of tropical forests in Southeast Asia: Historical perspectives*. Kyoto: Kyoto University Press, 29–71.
REDDX. (n.d.). *Indonesia*. Available at: http://reddx.forest-trends.org/country/indonesia/overview. Accessed 27 October 2015.
Scott, J. C. (1985). *Weapons of the weak: Everyday forms of peasant resistance*. New Haven, CT: Yale University Press.
Sodhi, N. S., Koh, L. P., Brook, B. W. and Ng, P. K. L. (2004). Southeast Asian biodiversity: An impending disaster. *Trends in Ecology and Evolution*, 19(12), pp. 654–660.
Stibig, H-J., Achard, F., Carboni, S., Rasi, R. and Miettinen, J. (2014). Change in tropical forest cover of Southeast Asia from 1990 to 2010. *Biogeosciences*, 11, pp. 247–258.
Tan, C. K. W., O'Dempsey, T., Macdonald, D. W. and Linkie, M. (2015). Managing present day large-carnivores in 'island habitats': Lessons in memoriam learned from human-tiger interactions in Singapore. *Biodiversity and Conservation*, 24(12), pp. 3109–3124.
Telapak and the Environmental Investigation Agency. (2005). *The last frontier: Illegal logging in Papua and China's massive timber theft*. Available at: www.eia-international.org/wp-content/uploads/The-Last-Frontier.pdf. Accessed 20 November 2015.
Thomas, A. C. (2015). Indigenous more-than-humanisms: Relational ethics with the Hurunui River in Aotearoa New Zealand. *Social and Cultural Geography*, 16(8), pp. 974–990.
Tsing, A. (2012). Unruly edges: Mushrooms as companion species. *Environmental Humanities*, 1, pp. 141–154.
United Nationals Environment Programme-World Conservation Monitoring Centre (UNEP-WCMC). (2014). *IUCN protected area management categories*. Available at: www.biodiversitya-z.org/content/iucn-protected-area-management-categories. Accessed 27 October 2015.
Vandergeest, P. and Peluso, N. (2006). Empires of forestry: Professional forestry and state power in Southeast Asia, part 2. *Environment and History*, 12(4), pp. 359–393.
Weisser, F. (2015). Efficacious trees and the politics of forestation in Uganda. *Area*, 47(3), pp. 319–326.
Wong, T., Delang, C. O. and Schmidt-Vogt, D. (2007). What is a forest? Competing meanings and the politics of forest classification in Thung Yai Naresuan Wildlife Sanctuary, Thailand. *Geoforum*, 38, pp. 643–654.
Wright, S. (2014). More-than-human, emergent belongings: A weak theory approach. *Progress in Human Geography*, 39(4), pp. 391–411.

INDEX

Page numbers in bold indicate tables; those in italics indicate figures.

AAK UK 333
Abebe, T. 218–219
Abu-Lughod, J. L. 56
accountability, authoritarianism and 32–34
adaptive DRR governance, principles of 357
ADB *see* Asian Development Bank (ADB)
ADB GMS Regional Power Trade **369**
AEEC-Japan **369**
Age of Ecology, The (Radkau) 53
agglomeration, economic gain from 121
aggregate trends in countryside **40**, 40–42
agrarian transition 232–233
agricultural development, land and 300–311; capitalist land relations, social/environmental implications of 309–310; changing role of 301–306, **302**, **303**, *303*, *304*; future of 310–311; introduction to 300–301; rural livelihoods, changing contexts of **306**, 306–308, *307*
agriculture and land, changing role of 301–306, *304*; Green Revolution technologies and 301–302; land allocation schemes and 305; land reforms and 301, **302**; land under cereal crop production **303**, *303*; oil palm demand and 304; property rights reforms and 305–306
Agriculture for Development (World Bank) 46, 48
Ahmad, Z. H. 114
Aiken, S. R. 60
Alston, P. 87
American War 17
Ansar, Z. 125
Anthropocene 53
APEC (Asia Pacific Economic Cooperation) Energy Working Group **369**, 370

Appadurai, A. 6, 7
Aquino, C. 71
Area Studies 7
Aritenang, A. 10
Arora-Jonsson, S. 266
Art of Not Being Governed, The (Scott) 59
ASEAN *see* Association of Southeast Asian Nations (ASEAN)
ASEAN Agreement on Disaster Management and Emergency Response (AADMER) 354–355
ASEAN Association of Radiologists 116
ASEAN Centre for Energy (ACE) 368, **369**
ASEAN Chamber of Commerce and Industry 113
ASEAN-China Free Trade Agreement 142
ASEAN Civil Society Conference 116, 117–118
ASEAN Committee of Permanent Representatives 115–116
ASEAN Committee on Disaster Management (ACDM) 354
ASEAN Cosmetics Association 116
ASEAN Council of Petroleum (ASCOPE) 364, 368, **369**
ASEAN Declaration of Commitment: Getting to Zero New Infections, Zero Discrimination, Zero AIDS-Related Deaths 192
ASEAN Declaration on the Protection and Promotion of the Rights of Migrant Workers 192
ASEAN Disaster Monitoring and Response System (DMRS) 352
ASEAN Economic Community (AEC) 10, 66, 109; deprivation and 240–241; disability-based discrimination 241; dynamic growth and 239–240; energy market integration and

370–371; foreign direct investment and 122; inequality and 240–241; member countries 239; neoliberal reforms through 109, 111–113; regional development and 239
ASEAN Energy Market Integration (AEMI) **369**, 370
ASEAN Forum on Coal (AFOC) 368
ASEAN Framework Agreement on Services (AFAS) 188–189
ASEAN Free Trade Agreement (AFTA) 122
ASEAN Free Trade Area 239
ASEAN Institutes of Strategic and International Studies (ASEAN-ISIS) 113
ASEAN Investment Area (AIA) 122
ASEAN Joint Statement on Climate Change 364–365
ASEAN People's Assembly 116
ASEAN Petroleum Security Agreement **369**
ASEAN Plan of Action on Energy Cooperation (APAEC) 364, 366–367, **369**, 370
ASEAN+3 **369**; CDM Cooperation Program 370–371, 372
ASEAN Power Grid 364, **369**, 370
ASEAN reform, civil participation in 109–118; contesting regional governance 117–118; overview of 109; reconfiguring regional governance 110–113; widening/limiting regional governance 113–117
ASEAN Regional Programme on Disaster Management (ARPDM) 354
ASEAN Socio-Cultural Community (ASCC) Blueprint 187, 192
Asia Energy Efficiency and Conservation Collaboration Centre (AECC) 370
Asian and Pacific Centre for Transfer of Technology (APCTT) **369**, 370
Asian Development Bank (ADB) 10, 65, 69, 144; loans to Indonesia 72, 76; loans to Philippines 72; sovereign approvals for developing member countries **75**
Asian familialism 174
Asian Financial Crisis 31, 33; regional integration following 239
Asian Infrastructure Investment Bank (AIIB) 80, 142, 150
Asian Values 6
Asia Pacific Economic Cooperation (APEC) 363
Asia Pacific Partnership on Clean Development and Climate (APP) 367
Asis, M. M. B. 178
Associated Press 317
associational social capital 137
Association of Khmer Rouge Victims in Cambodia 101
Association of Southeast Asian Nations (ASEAN) 3, 6, 29, 65, 66; healthcare and (*see* healthcare entitlements); members of 16; motto 16;

regional sources of visitors to **155**; *see also individual ASEAN organizations*
Aung, M. M. 116
Aung San, S. K. 92–93
Austrian East India Company 56
authoritarianism under neoliberalism 32–34

Baccaro, L. 88
backpacker tourism 157–158
Badan Pusat Statistik (BPS) 125
Bailey, M. 323
Baird, I. G. 46
Bakun, Sarawak, oil palm cultivation by independent smallholders in 335–336
Baldassar, L. 181
Ball, J. 11
Barker, R. 294
Bay of Bengal Initiative for Multi-Sectoral Technical and Economic Cooperation (BIMSTEC) 363, **369**
Beach, The (Garland) 159
Beazley, H. 11, 217
behavioral social capital 137
Belanger, D. 176, 177
Belton, B. 11–12
Benjamin, G. 60
Better Factories Cambodia Project (ILO) 87–88, 92
Biodiversity Based Economic Development Office (BEDO) 384
biophilia, defined 54
Bodetabek (Bogor-Depok-Tangerang-Bekasi) 121; *see also* industrial economies in suburbs
Bogart, W. T. 125
Bornstein, E. 252
Bovinsiepen, J. 103
Boyland, M. 12
brain drain, migration and 206–207
Brautigam, D. 146, 147, 148
Bremen, J. 277
Brodjonegoro, B. P. S. 128
Brookfield, H. 398
Bruneau, M. 276
Brunei 4; development indicators **8**, 9; ethnic minorities/indigenous peoples in **225**; freshwater withdrawals *291*; Human Development Index (HDI) **213**; irrigation water withdrawals *293*; remittance flows **201**; renewable water resources *290*; tourist arrivals by **155**, 156; water, sanitation and health indicators **295**; World Risk Index for **345**
Brunei Darussalam *see* Brunei
Buergin, R. 403
Bulte, E. H. 377
Bumiputera classification 231–232
Burma *see* Myanmar
'Burmanization' policy 230

Index

Bush, S. R. 11–12
Bylander, M. 204

Cahill, A. 133
Calgaro, E. 12
California Transparency in Supply Chains Act 2010 325
Cambodia 4; ADB sovereign approvals for **75**; aggregate trends in **40**; agriculture percentage of GDP for *307*; capture fisheries employment in **320**; cereal crop yield in **303**, *304*; change in forested area *395*; child sex tourism 160; Chinese aid to 143, 145, 149; development indicators **8**, 9; disability, defined 242; disability-based discrimination in 241; disability statistics 242; ECCC in 100–101; ethnic minorities/indigenous peoples in **225**; as exporter of low-skilled labor migrants 244–245; forest products exports, value of *397*; forest sector contribution to total GDP *397*; freshwater withdrawals *291*; Human Development Index (HDI) **213**; international justice tribunals in 66; irrigation water withdrawals **293**; land reforms in **302**; land under cereal crop production for *303*; malnutrition in children 214; modernization and religious institutions in 253–254; mountain people label in 229; neoliberalism and 29–30, 33; neoliberalist reform in 243; post-conflict reconciliation in 96, 97, 98–99; Rectangular Strategy 243; remittance flows **201**, 204; renewable water resources **290**; reparations in 102–103; rural, urban and national poverty rates for **47**; rural population of **306**; total cumulative MDO allocations to **75**; tourism sector in 154, 155–156; tourist arrivals by **155**; victim associations in 101; voluntourism in 161–162; water, sanitation and health indicators **295**; World Bank commitments for **75**, 82; World Risk Index for **345**
Cannon, T. 49
capital for productive activities, remittances role in providing 203
Capital in the Twenty-First Century (Picketty) 57
capitalism: defined 56; human-nature relationships and 56–57
capitalist land relations, social/environmental implications of 309–310
capitalocentric framings of economic diversity 135–138
Caraway, T. 93
Carbon Sequestration Leadership Forum (CSLF) 367
Carnegie, M. 132
Carpenter, J. P. 137
Carrefour 319
Carroll, T. 10

Castells, M. 54
Castro, J. 179
CAVR *see* Comissão de Acolhimento, Verdade e Reconciliação (CAVR)
Central Intelligence Agency 230
Centre for Research on the Epidemiology of Disasters (CRED) 343
Certified Sustainable Palm Oil (CSPO) 331, 333
Chao Khao 229
Charoen Pokphand Food Public Company Ltd 308, 319
Chase, R. S. 138
Chea, S. 98
the Chicago Boys 71
Chicago School of Economics 28
child labor, global discourses on 215–216
child mortality rates (CMR) **213**, 214
children, care provisioning for left-behind 177–179
children/youth, economic development and 211–221; child labor, global discourses on 215–216; children in disaster situations example 217–219; dominant perceptions *vs.* local realities 215–220; education and 214–215; health and 214; hunger and 214; indicators of 212–215; Myanmar forced migrant children example 219–220; overview of 211–212; poverty and 213–214; socio-economic status and 213, **213**; street children example 216–217
children/youth well-being, indicators of 212–215; education 214–215; health 214; hunger 214; poverty 213–214; socio-economic status 213, **213**
Chile, neoliberalism and 29
China Africa Development Fund 147
China disability, defined 242
China ExIm Bank 143, 147
Chinese aid 142–150; China as developmental partner and 149–150; Eight Principles for Economic Aid and Technical Assistance 143; geographical distribution of **144**; 'Go Out' Policy and 143–144; governance and 148–149; history of, globally and in Southeast Asia 143–146; introduction to 142–143; loan types 146; and OECD aid, similarities/differences in 146–148, **148**; as project-based 143; timeline of **145**
Chinese Chamber of Commerce of Metals, Minerals and Chemical Importers and Exporters 150
Chinese State Council 146
Choeung Ek 102–103
Choi, H. 204
Chulalongkorn Center for Migration Studies, Thailand 325
Citylife Chiang Mai (magazine) 387
Civilizations (Fernandez-Armesto) 55

Clark, J. 122
Clarke, G. 230
class, rural livelihoods and 44–47, **45**
Clean Development Mechanism (CDM) 363, 367
Clean Energy Ministerial (CEM) 367
climate change: development/disaster risk in Southeast Asia and 351; environmental degradation and 262; feminist political ecology (FPE) of 266–268; mitigation, regional cooperation and 362–372; rural livelihoods and 48–49
Coe, N. M. 239
Coffey, W. J. 125
COGEN (Cogeneration) **369**
Cohen, D. 100
Cold War, postwar politics and 17
Colloque Walter Lippmann 28
colonialism, human-nature relationships and 57–58
Comissão de Acolhimento, Verdade e Reconciliação (CAVR) 97, 101–102
Commission for Reception, Truth and Reconciliation 97
Commission on Filipinos Overseas 199
common-pool resources (CPRs) 376
community economies in Southeast Asia 131–139; within capitalocentric framings of economic diversity 135–138; described 134; diverse economic practices and 132–134, *133*; informal economy sector 135–136; introduction to 131–132; new research directions for 138–139; patron-client relationships and 136–137; social capital and 137–138
Compulsory Migrant Health Insurance (CMHI) 192
concessional loans 146
conditional lending 71
Condominas, G. 59–60
Conklin, H. C. 59
Connell, J. 10, 11
Constable, N. 176, 205
Convention on the Rights of Persons with Disabilities (CRPD) 241–242
conversion forests 400–401
Costco 319
Cramb, R. A. 138, 398
Crutzen, P. 53
cultural change, tourism and 164–165
cultural heritage tourism 159
Cupples, J. 266

Da, W. W. 181
dagyaw system 132
Dahles, H. 135–136
dams 290–292
Dawei Special Economic Zone 262–263, 264
Dayley, R. 311

Decent Work Agenda (ILO) 87, 88, 93
Declaration of Fundamental Principles and Rights at Work (ILO) 86–87
Declaration on Social Justice for a Fair Globalization 88
"Declining Economic Importance of Agricultural Land, The" (Schultz) 300
decropping countryside 271–279; challenging quintessentially rural spaces 275–278; changing rural space 272–273; community and shared labor expectations 274–275; introduction to 271–272; Nong Khwai village example 272–275; property and privacy 273–274; studies of 276–278; trees and normative landscapes 274
Deep Ecology 54
Deep Green Theory 54
deep marketization 65, 80–82; described 70; development policy of **73–74**
deforestation 349, *395*, 395–396, *396*
De Genova, N. P. 320
democracy, neoliberalism and 30–32
Democratic People's Republic of Korea (DPRK), Chinese aid to 143
demographic change, development/disaster risk in Southeast Asia and 348, *349*
Dercon, S. 48
Derks, A. 324–325
development: altruistic goals of 15; failures of 14; pros and cons of 14–15
development, in Southeast Asia 7–9; approach to 7; disjunctures and 7; in Eastern Indonesia 22–23; indicators **8**; introduction to 14–15; NICs and 19–20; in northern Thailand 18–19; in Philippines 20–22; politics of 17–18; regional analyses, sensitivity and 7, 9; region of 15–17
Development Assistance Committee (DAC), OECD 142; official development assistance (ODA) 142
development-disaster risk trade-offs, addressing 356
development institutions/economies in Southeast Asia: ASEAN reform, civil participation in 109–118; community economies 131–139; industrial economies in suburbs 120–129; International Labour Organization (ILO) 85–93; introduction to 65–67; multilateral development organisations (MDOs) 69–83; non-OECD aid 142–150 (*see also* Chinese aid); post-conflict reconciliation 96–105; tourism industry and 153–166; *see also individual institutions*
development policy tensions 232–234; agrarian transition and 232–233; environmental destruction and 233–234; resettlement policies and 233; tourism and 232
development shaping religion 252–254

Diamond, J. 53, 58
di Caprio, L. 159
Dick, H. W. 120
disability inclusion, globalization and 238–246; ASEAN Economic Community and 239–241; Cambodia as exporter of low-skilled labor migrants 244–245; disability-based discrimination, ASEAN region 241; institutionalized discrimination 241–243; labor migration and 243–244; neoliberalist reform in Cambodia and 243; overview of 238; regionalism and 239
disaster management, children and 217–219
disaster risk 342–357; development and 347–351; formula for determining 344; introduction to 342; reduction 351–355; in Southeast Asia 342–346; transforming development and, for equitable resilience 355–357
disaster risk, development and 347–351; adaptive governance and 357; climate change and 351; ecosystem degradation and 349; equitable resilience to reduce inequity in 356–357; governance structures/processes and 350–351; inequality and 347–348; poverty and 347–348; trade-offs in decision-making, addressing 356; transforming, for equitable resilience 355–357; unsustainable natural resource use and 349; urbanization and 348, *349*
disaster risk, feminist political ecology (FPE) of 266–268
disaster risk management (DRM) 352
disaster risk reduction (DRR) 217, 351–355; drivers of vulnerability, reducing 353–354; early warning systems 352–353; initiatives 351–352; institutional basis for implementation of 352; knowledge, innovation and education 353; overview of 351–352; preparedness, strengthening of 354–355
disasters, in Southeast Asia 342–346; deaths caused by 346, **346**; economic losses/damages from 346, **346**, *347*; mortality risk distribution of hydro-meteorological *343*; occurrences of 343–344, *344*; types of 342–343; World Risk Index and 344
diverse economic practices 132–134; ethical negotiations in 134; framing of *133*
Djalante, R. 354
Documentation Centre of Cambodia (DC-Cam) 102
donkey friends 157
Douglass, M. 174
drinking water supply 294–296, **295**
Duara, P. 61
Ducanes, G. 202–203
Dudgeon, D. 291
Durbeck, G. 393
Dutch East India Trading Company 56

early warning systems (EWS) 352–353
East Asian Miracle 29
East Timor: ethnic minorities/indigenous peoples in **225**; unity amid diversity in 231–232
Eat, Pray, Love (Gilbert) 159
ECCC *see* Extraordinary Chambers in the Courts of Cambodia (ECCC)
economic development, people and: children/youth and 211–221; decropping countryside and 271–279; disability inclusion, globalization and 238–246; ethnic minorities/indigenous groups and 224–235; feminist political ecology and 261–269; gender politics of care, migration and 173–182; healthcare entitlements 186–194; introduction to 169–171; religion and 250–258; remittance flows, migration and 198–208; *see also individual headings*
economic localization 121
economic urbanization 121
ecosystem degradation, development/disaster risk in Southeast Asia and 349
ecotourism 158
Eight Principles for Economic Aid and Technical Assistance, Chinese aid 143
Elliott, L. 110
Elmhirst, R. 265
Elson, R. E. 57
Emmerson, D. 15
employment/job deconcentration 122
Energy and Climate Partnership of the Americas (ECPA) 367
Energy Conservation Centre of Japan (ECCJ) 370
Energy Efficiency and Conservation Sub-sector Network (EE&C-SSN) 368
energy policy, regional cooperation 366–367
English East India Trading Company 56
environment, development and: agricultural, land and 300–311; disaster risk 342–357; forest-led 392–405; introduction to 281–283; oil palm cultivation by independent smallholders 330–340; Payments for Ecosystems Services (PES) 376–389; regional cooperation, climate change mitigation and 362–372; seafood value chains 316–327; water 285–297; *see also individual headings*
environmental change, tourism and 164–165
environmental degradation 262
environmental destruction, development policy tensions and 233–234
Environmental Justice Foundation (EJF) 317, 325
environmental state in Southeast Asia *see* human-nature relationships
equitable resilience 356; pursuing, to reduce inequity in risk reduction/development 356–357
ethnicity in Southeast Asia, defining 224–225

ethnic minorities/indigenous groups, economic development and 224–235; agrarian transition and 232–233; in Burma and Philippines 229–230; environmental destruction and 233–234; mountain people in Thailand and Cambodia 229; overview of 224–225, **225–227**; in post-socialist Laos and Vietnam 228–229; resettlement policies and 233; social movements supporting 234; tensions associated with 232–234; unity amid diversity in Indonesia, East Timor and Malaysia 231–232
Eurasian Exchange 56
European Union (EU) Emission Trading Scheme (ETS) 368
extractive development, wood/fire and 400–402
Extraordinary Chambers in the Courts of Cambodia (ECCC) 97, 100, 101; Legal Documentation Centre (LDC) 102; public memorialization sites 102–103

Fair Trade standard for fisheries (USA) 322–323
faith-based organizations (FBOs) 251
Federation of ASEAN Public Relations Organizations 113
feminist political ecology (FPE) 261–269; core defining feature of 262; of disaster risk/climate change 266–268; emergence of 261; focus of 262; of large-scale development investments 262–266; overview of 261; view from 261–262
Feminist Political Ecology: Global Issues and Local Experiences (Rocheleau, Thomas-Slayter and Wangari) 262
Ferguson, J. 136
Fernandez-Armesto, F. 55
Ferrer, M. C. 230
Ferry, W. C. 125
film tourism 159
Firman, T. 123
Fisheries Improvement Projects (FIPs) 321
five-five-ten rule loans 146
Five-Million-Hectare Program 381
flying grandmothers 181
Food and Agriculture Organization (FAO) 233, 395, 396; Code of Conduct for Responsible Fisheries 321
Ford, M. 65
foreign brides 175–177
foreign direct investment (FDI): expansion and 122–123; transnational relocation of 120
foreign investment, labor relations and 91–92
Forest Law, Enforcement and Governance (FLEGT), EU 401
forest-led development 392–405; extractive development and 400–402; introduction to 392–394, *393*; more-than-human development and 404–405; preservationist development

and 402–404; in Southeast Asia 394–396, *395, 396–397*; swidden agriculture and 397–400
Forest Restoration Research Unit (FORRU) 387–388
Forest Stewardship Council 401
Frank, A. G. 55
Freedom of Association project (ILO) 90, 91
French East India Trading Company 56
Friedman, M. 28
Friend of the Sea's Sustainable Seafood Program 321
Fujita Lagerqvist, Y. 11
Further India 5

Gaia 53–54
Gaonkar, D. P. 235
Garland, A. 159
Garschagen, M. 344
Gartrell, A. 11
Geertz, C. 59, 135, 137
gender and generation: interplay of *43*; rural livelihoods and 42–44, **44**
gender issues in Indonesia 22–23
gender politics of care, migration and 173–182; introduction to 173–174; paid care labor, importing 174–175; provisioning for left-behind children 177–179; transnational care strategies among privileged 179–182; unpaid care labor, importing 175–177
Generalized System of Preferences (GSP) Renewal Act 87
geoparks 158
Gerakan Aceh Merdeka (GAM, Free Aceh Movement) 231
Gerard, K. 10
GHG emissions, Southeast Asian region energy needs and 363–365
GHG mitigation, regional cooperation 368, 370–372
Ghoshal, B. 114
Gibson, K. 10, 133, 134
Gibson-Graham, J. K. 21, 132, 133, 139, 203
Giddens, A. 54
Gilbert, E. 159
Gill, S. 32
Gillan, M. 65
global care chains 173–174; *see also* gender politics of care, migration and
global health governance (GHG) 188
globalization, regionalism and 239
global spatial hypergamy 176
Global Witness 150
GO-NGO forums 114, 115
good governance, Chinese aid and 148–149
Good Society, The (Lippmann) 28
'Go Out' Policy, China 143–144
Gourou, P. 278

governance structures/processes, development/ disaster risk in Southeast Asia and 350–351
Graham, E. 178
Greater Mekong Sub-region (GMS) 363; tourism sector in 154
Greater Mekong Sub-regional Economic Cooperation Program 144
Green, P. 180–181
Green Capitalism 283
Green Certificates program 370
green culture 54
Green Revolution: rice production and 44; rural poverty and 47–48; technologies, agricultural productivity and 301–302
Grunewald, N. 366
G7 366
G20 366
Guardian 317
Guillain, R. 125
Gunn, G. 56

Ha, T. T. T. 323
Hak, P. 11
Harmadi, S. H. B. 128
Harper, S. E. 179
Hart, K. 135
Harvey, D. 28, 266
Hayek, F. 28
Haynes, J. 252
Heads of ASEAN Power Utilities/Authorities (HAPUA) 368
healthcare entitlements 186–194; to citizens and non-citizens 189–191; global health governance perspective 188; international health governance perspective 187–188; for internationally mobile peoples 191–193; introduction to 186–187; from national to regional/global governance 187–189
Heindel, A. 100
Heine-Geldern, R. 60
Hewison, K. 113
Hill, A. 10, 134
hill tribes 229
Hirsch, P. 11, 276
Hochschild, A. R. 173
Holliday, I. 230
host country development, migrant labor and 205
Huang, S. 11, 178
Hudalah, D. 10, 123, 125
Hughes, C. 98–99, 105
Hughes, R. 10
Huijsmans, R. 206
Human Development Index (HDI) 213
human-nature relationships 53–61; capitalism and 56–57; colonialism and 57–58; future of 60–61; indigenous knowledge and 58–60; introduction to 53–55; nature valued/appraised 55–56

hungus 132
Hyndman, J. 266
Hyogo Framework for Action (HFA) 2005–2015 352–355; HFA Priority 1 352; HFA Priority 2 352–353; HFA Priority 3 353; HFA Priority 4 353–354; HFA Priority 5 354–355; *see also* disaster risk reduction (DRR)

Ibrahim, A. 110
IBSA (India, Brazil and South Africa) Trust Fund 367
ICOLD 291
Ieng, S. 98
IHH Healthcare 190
Illegal, Unreported, and Unregulated (IUU) policy (EU) 325
Imanishi, K. 53–54
IMF *see* International Monetary Fund (IMF)
inclusive business, seafood sector as 322
indigenous groups *see* ethnic minorities/ indigenous groups, economic development and
indigenous knowledge, human-nature relationships and 58–60
Indonesia 4; ADB sovereign approvals for **75**, 76; aggregate trends in **40**; agriculture percentage of GDP for *307*; average farm size trends in **45**; capture fisheries employment in **320**; cereal crop yield in **303**, *304*; change in forested area *395*; Child Protection Law 215–216; child sex tourism 160; Christian NGOs operating in 256–257; development indicators **8**, 9; ethnic minorities/indigenous peoples in **226**; forest products exports, value of *397*; forest sector contribution to total GDP *397*; freshwater withdrawals *291*; gender/ small-scale agriculture in Eastern 22–23; greying of farm labor force in **44**; Human Development Index (HDI) **213**; IMF reforms in 78–79; irrigation water withdrawals **293**; Islamic *pesantren* education system in 257; Komodo National Park 162; land reforms in **302**; land under cereal crop production for *303*; loss of life/economic damage due to disasters in **346**; malnutrition in children 214; manufacturing sector in (*see* industrial economies in suburbs); neoliberalism and 30, 31; Payments for Ecosystems Services (PES) in 381–383; reconfiguring regional governance in 110; relationship with MDOs 72, 76; religion shaping development in 254, 255–256; remittance flows **201**, 207; renewable water resources *290*; rural, urban and national poverty rates for **47**; rural population of **306**; Social Security Management Agency 189–190; total cumulative MDO allocations to **75**; tourism sector in 154; tourist arrivals by **155**; unity amid diversity in 231–232; water,

sanitation and health indicators **295**; World Bank commitments for **75**; World Risk Index for **345**
Indonesian Prosperous Labor Union 87
industrial agglomeration, defined 121
industrial development, spatial pattern of 125–128; deconcentration and 125–126; specialization and 126–128
industrial economies in suburbs 120–129; deconcentration and 125–126; expansion, FDI and 122–123; introduction to 120–121; Location Theory and 121–122; small and micro industries (SMIs) and 123–125, **124**; spatial patterns of 125–128, **126**; specialization and 126–128, **128**
inequality, development/disaster risk in Southeast Asia and 347–348
infant mortality rates (IMR) **213**, 214
informal economy sector 135–136
infrastructure networks 122
institutional environment of PES 378, **379**, 380–381, 383
institutionalized discrimination 241–243
Integrated Conservation and Development programs (ICDP) 403
Intended Nationally Determined Contributions (INDCs) 362, 371
Intergovernmental Panel on Climate Change (IPCC) 343, 351, 365
internal renewable water resources (IRWR) 289
International Bank for Reconstruction and Development (IBRD) 81
International Convention on the Protection of the Rights of All Migrant Workers and Members of their Families 192
International Disaster Database (EM-DAT) 343
International Drinking Water Supply and Sanitation Decade 294
International Energy Agency (IEA) 364
International Financial Institutions (IFIs) 32
international health governance (IHG) 187–188
International Labor Organization (ILO) 10, 65, 66, 85–93, 211; Better Factories Cambodia Project 87–88, 92; challenges to 86; Convention 182 against the Worst Forms of Child Labor 211, 215; Decent Work Agenda 87, 88, 93; Declaration of Fundamental Principles and Rights at Work 86–87; as development actor 86–89; Freedom of Association project 90, 91; International Program against Child Labor (IPEC) 215; introduction to 85; Mission to Timor-Leste 88–89; in Myanmar 89–92; purpose of 86; Work in Fishing Convention 188 321
International Monetary Fund (IMF) 10, 30, 65, 69, 78–79; criticisms of 31
International Organisation of Migration 244

International Program against Child Labor (IPEC) 211, 215
International Sustainability and Carbon Certification 333
IOI Group 332
IPCC Assessment Report 48
irrigation **293**, 293–294

Jakarta Metropolitan Area (JMA) 123–125, **124**, *127*; *see also* small and micro industry (SMI)
Jawa 5
Jayasuriya, K. 111
JMA *see* Jakarta Metropolitan Area (JMA)
Johnson, C. 70
Joint Crediting Mechanism, Japan 371
Joko, W. 189
Jones, L. 111
Journal of the Indian Archipelago and Eastern Asia, The 15

Kaing Guek, E. 100
Kaplan, R. D. 60
Kecamatan Development Program (KDP) 79
Kelly, P. 11
Kemp, J. 271, 277
Kent, L. 99, 105
Ketut 159
Keyes, C. 278
Khieu, S. 98, 100
Khmer Rouge Study Tour 102–103
Khoon, C. C. 11
Kinabatangan, Sabah, oil palm certification in 333–335
Kitano, N. 147
Klein, N. 29
Knodel, J. 177
knowledge transfer, migration and 204
Kolbert, E. 54
Koning, J. 42
Korean War 17
Kragelund, P. 143
Ksaem Ksan (victim association) 101, 104

Labonne, J. 138
labor: migration, disability and 243–244; regulation, as development strategy 91; seafood value chain conduct and 319–322
Lam, T. 180
Lambin, E. F. 305–306
land, rural livelihoods and 46–47
Land Ethics 54
land grabbing 233, 307–308
land tenure systems 57
Laos 4; ADB sovereign approvals for **75**; aggregate trends in **40**; agriculture percentage of GDP for *307*; cereal crop yield in **303**, *304*; change in forested area *395*; child sex tourism 160;

Chinese aid to 145; development indicators **8, 9**; disability-based discrimination in 241; ethnic minorities/indigenous peoples in **226**; forest products exports, value of *397*; forest sector contribution to total GDP *397*; freshwater withdrawals *291*; Human Development Index (HDI) **213**; irrigation water withdrawals **293**; land reforms in **302**; land under cereal crop production for *303*; post-socialist cultural preservation in 228–229; remittance flows **201**; renewable water resources *290*; rural population of **306**; total cumulative MDO allocations to **75**; tourism sector in 154, 155; tourist arrivals by **155**; water, sanitation and health indicators **295**; World Bank commitments for **75**; World Risk Index for **345**

large and medium industry (LMI) 120–121, 125–126

large-scale development investments: dispossession and 263; feminist political ecology (FPE) of 262–266; gendered workings in 264–266; as green and clean 263–264; land/water grabbing and 263

Latt, S. 205

Law, L. 10

Leach, E. R. 60

Least Developed Countries (LDCs) 348

left-behind children, care provisioning for 177–179

Legal Documentation Centre (LDC) 102

Leigh, C. H. 60

Leimona, B. 382

Li, T. M. 47

liberal familialism 175

Lie, M. L. S. 181

Like Minded Group of Developing Countries (LMDC) 366

Lindayati, R. 57

Liu, Z. 146

Local Value Chain Development (LVCD) program, World Vision 22–23

Location Theory 121–122

Lockard, C. 57

Logan, J. R. 15

long-distant mothering 178

Lost in Thailand (film) 159

Lowering Emissions from Asia's Forests (LEAF) project 387

Lukasiewicz, A. 203

LVCD *see* Local Value Chain Development (LVCD) program, World Vision

Lyne, I. 137

Mae Sa watershed PES scheme 384–388; hypothetical PES markets, studies on *386*, 386–387; implementation of 387–388; PES Model for *386*; study area 384, *385*; Thailand's PES policy 384

Majid Cooke, F. 11, 12

Malaysia 4; ADB sovereign approvals for **75**, 76; aggregate trends in **40**; agriculture percentage of GDP for *307*; cereal crop yield in *303, 304*; change in forested area *395*; development indicators **8, 9**; disability, defined 242; disability-based discrimination in 241; ethnic minorities/indigenous peoples in **226**; forest products exports, value of *397*; forest sector contribution to total GDP *397*; freshwater withdrawals *291*; greying of farm labor force in **44**; Human Development Index (HDI) **213**; irrigation water withdrawals **293**; land reforms in **302**; land under cereal crop production for *303*; loss of life/economic damage due to disasters in **346**; medical tourism in 161; modernization of Islam in 253, 256; remittance flows **201**; renewable water resources *290*; rural, urban and national poverty rates for **47**; rural population of **306**; smallholders in East (*see* oil palm cultivation by independent smallholders); sports tourism in 160; total cumulative MDO allocations to **76**; tourism sector in 154; tourist arrivals by **155**; unity amid diversity in 231–232; water, sanitation and health indicators **295**; World Bank commitments for **75**, 76; World Risk Index for **345**

Malaysian Palm Oil Association (MPOA) 333

Malaysian Sustainable Palm Oil (MSPO) 333

Marcos, F. 31, 230, 254

Marine Stewardship Council (MSC) 321

maritime/insular Southeast Asia 4

Marschke, M. 11–12, 321–322

Marshall, A. 128

Maupain, F. 86

McElwee, P. 379–380

McGregor, A. 12, 122

McKay, D. 44, 203

McKinnon, K. 9–10

McNamara, R. 72

MDO *see* multilateral development organisations (MDOs)

medical tourism 160–161

Mekong Delta, landscape engineering in 293–294

Mekong River Commission 367, **369**

Mekong River dams 292

Mele, V. 88

Mennonite Central Committee (MCC) 254

Merchant, C. 58

Meyfroidt, P. 305–306

Miettinen, J. 233

migrant domestic workers 174–175

migrant wives 175–177

Migrant Worker's Rights Network 325

migration: brain drain and 206–207; feminization of 174; gender politics of care and 173–182; knowledge transfer and 204; labor, disability inclusion and 243–244; national economic growth and 201–202; poverty reduction and 202–203; remittance flows and 198–208; welfare gains and 202–203
Millennium Development Goals (MDGs) 14, 211–212; for reduced child mortality (CMR) 214; for reduced infant mortality (IMR) 214; of reducing extreme hunger 214; water supply and sanitation improvements 294–295
Miller, F. 11
Miller, M. 217
Mission to Timor-Leste (ILO) 88–89
modernization, development and 251; in Cambodia 253–254; in Malaysia 253, 256; in Singapore 252–253
Mohamad, M. 110
Mohan, G. 149
Molle, F. 294
Montefrio, M. J. F. 204
Mont Pèlerin Society 28
moral economy 57
more-than-human development 404–405
Moro National Liberation Front (MNLF) 230
Mortimer, R. 44
Mountbatten, L. 15
Mujahid, G. 181
Muller-Mahn, D. 49
multilateral development organisations (MDOs) 65, 69–83; crisis of 1997–98 and 77–80; deep marketization 80–82; introduction to 69–70; post-Washington consensus 77–80; total historical cumulative allocations to Southeast Asian countries **75–76**; Washington consensus 70–77
Mussomeli, J. A. 104
Mutual Recognition Arrangements 112
Myanmar 4; ADB sovereign approvals for **75**; aggregate trends in **40**; agriculture percentage of GDP for *307*; cereal crop yield in *303*, *304*; change in forested area *395*; child sex tourism 160; Chinese aid to 144, 145, 149; development indicators **8**, **9**; disability-based discrimination in 241; ethnic conflict in 229–230; ethnic minorities/indigenous peoples in **225**; forced migrant children from 219–220; foreign investment and 91–92; forest products exports, value of *397*; forest sector contribution to total GDP *397*; freshwater withdrawals *291*; Human Development Index (HDI) **213**; internal migration for fishery work in 324; International Labour Organization (ILO) in 89–92; irrigation water withdrawals **293**; labor regulation work in 91; land reforms in **302**; land under cereal crop production for *303*; Land Use Certificate (LUC) titling 264; loss of life/economic damage due to disasters in **346**; malnutrition in children 214; poverty level in 214; remittance flows **201**; rural population of **306**; total cumulative MDO allocations to **76**, 82; tourist arrivals by **155**; water, sanitation and health indicators **295**; women's labor value in 325; World Bank commitments for **75**; World Risk Index for **345**
Myanmar Garment Manufacturing Association 92

Naim, M. 148
Naipaul, V. S. 257
Nanyang (South Seas) 5
Nartsupha, C. 275
National Bureau of Statistics 125
national economic growth, migration and 201–202
Natural Experiments in History (Diamond and Robinson) 53
natural hazard-based disasters 342
natural tourism 158
Ne, W. 230
Neef, A. 12, 378, 388
neoliberal development, social/environmental consequences of 282
neoliberal developmental policy, phases of 70, **73–74**
neoliberalism: accountability under 32–34; authoritarianism under 32–34; Cambodia and 29–30, 33; Chile and 29; critical scholars and 27; democracy and 30–32; history of 29–30; hybridity of 34–35; Indonesia and 30, 31; International Financial Institutions and 32; introduction to 27–28; roots of 28; term use of 27; World Bank and 30–31
neoliberal reforms 239
Nevins, J. 99
New and Renewable Sources of Energy Subsector Network (NRE-SSN) 368, 370
New Development Bank (NDB) 80
New Human era 53
newly industrializing countries (NICs), Asia's 19–20, 70
newly industrializing economies (NIEs), Asia's 70
New York Times 317
Nguyen, C. V. 206
Nguyen, H. 11
NICs *see* newly industrializing countries (NICs), Asia's
Nong Khwai village decropping 272–275; community and shared labor expectations 274–275; property and privacy 273–274; trees and normative landscapes 274
non-governmental organizations (NGOs) 234

Index

Nord Pool 366
North American Free Trade Agreement 87
Noun, C. 98, 100
Nurchayati 207
Nusa Tenggara Timur (NTT), eastern Indonesia, LVCD program in 22–23
Nylander, J. 12

Ochiai, E. 175
OECD *see* Organization of Economic Cooperation and Development (OECD)
official development assistance (ODA) 69; defined 146
oil palm cultivation by independent smallholders 330–340; in Bakun, Sarawak 335–336; benefits of 330; certification in central Sarawak and 336–338; certification in lower Kinabatangan, Sabah and 333–335; context of decision for 331–332; desires/trade-offs for new 338; drivers for 338; environmental impact of 331; introduction to 330–331; paths taken by 339; in Sabah and Sarawak 332; sustainable palm oil certification and 332–333
Ong, A. 176
Open Working Group (UN) 88
Organization of Economic Cooperation and Development (OECD) 142; Chinese aid and, similarities/differences in 146–148, **148**; Development Assistance Committee (DAC) 142; official development assistance, defined 146; *see also* Chinese aid
Ormond, M. 11
Osborne, M. 4
Other Official Financing (OOF) 147
Otsuka, K. 46
Overseas Contract Workers (OCWs) 20–22
Overseas Filipino Workers (OFWs) 200
Ovesen, J. 229
Oxfam International 355

Pacific Island Development Forum 6
Pacific Island Forum 6
Palmer, L. 105, 288
Palm Oil Innovation Group 333
Paris Agreement 362; carbon finance/result-based finance under 371
Parreñas, R. 174, 178, 201
partial citizenship concept 176
patron-client relationships 136–137
Payments for Ecosystems Services (PES) 376–389; classification of 377–378; defined 377; identification of, market 378, **379**, 379–380, 381–382; in Indonesia 381–383; institutional environment of 378, **379**, 380–381, 383; introduction to 376–377; Mae Sa watershed case study 384–388, *385*, *386*; processes and relationships 378, **379**, 380, 382–383; in Vietnam 379–381
Payn, T. 395
Peace Corps 161
Pelras, C. 136–137
Peluso, N. 395
PES *see* Payments for Ecosystems Services (PES)
PES market, dentification of 378, **379**, 379–380, 381–382
PES processes and relationships 378, **379**, 380, 382–383
Pfau, W. D. 202
Pham, T. T. 380
Philippines 4; ADB sovereign approvals for 72, **75**, 81; aggregate trends in **40**; agriculture percentage of GDP for *307*; anti-discrimination laws 242; average farm size trends in **45**; capture fisheries employment in **320**; cereal crop yield in **303**, *304*; change in forested area *395*; Comprehensive Agrarian Reform Program (CARP) 49; development indicators **8**, **9**; ethnic conflict in 229, 230; ethnic minorities/indigenous peoples in **226**; forest products exports, value of *397*; forest sector contribution to total GDP *397*; freshwater withdrawals *291*; greying of farm labor force in **44**; Human Development Index (HDI) **213**; irrigation water withdrawals **293**; land reforms in **302**; land under cereal crop production for *303*; loss of life/economic damage due to disasters in **346**; Overseas Contract Workers in 20–22; religion shaping development in 254; remittance flows 199, 200, **201**, 202–203, 204, 206–207; renewable water resources *290*; rural population of **306**; total cumulative MDO allocations to **76**; tourist arrivals by **155**; water, sanitation and health indicators **295**; World Bank commitments for **75**; World Bank projects in 72; World Risk Index for **345**
Pholsena, V. 229
Phongphit, S. 275
Phuc, X. T. 381
Picketty, T. 57
Pingol, A. 179
Pirard, R. 382
plantation forests 395
Pol, P. 98
Polanyi, M. 28
Policy Reform Support Loans (PRSLs) 79
political ecology, described 261
political economy 57
political economy of development approach 69
political forests 395
Pollock, S. 55
Ponte, S. 326
Popkin, S. 57

population deconcentration 122
Porras, I. 382
post-conflict reconciliation, justice processes/ discourses of 96–105; in Cambodia 96, 97, 98–99; introduction to 96–98; reparations measures 102–104; in Timor-Leste 96; tribunal justice as 99–100; victim participation and 101–102
post-Washington consensus (PWC) 65, 77–80; described 70; development policy of **73–74**; technical assistance (TA) projects 79
poverty, development/disaster risk in Southeast Asia and 347–348
poverty and wellbeing, in rural areas **47**, 47–48
poverty reduction, migration and 202–203
Poverty Reduction Strategy Papers (PRSPs) 77
Prabawa, T. S. 135–136
preservationist development, carbon/ecology and 402–404
primary forests 395
privileged migrant families, transnational care strategies among 179–182
production forests 400–401
Project Issara 325
prosperity, defined 20–21
Pūras, D. 190

Quynh Nguyen, H. 206

Radkau, J. 53
Ramos, F. 71
Ratha, D. 202
Reagan, R. 28
Reaganomics 29
Reducing Emissions from Forest Degradation and Deforestation (REDD+) 378, 403–404
Reformasi movement 110
regional cooperation, climate change mitigation and 362–372, **369**; ASEAN GHG emissions and 363–365; assessing, approaches 365–366; energy policy 366–367; examples from other regions 366–368; GHG mitigation and 368, 370–372; initiatives, in Southeast Asia 368–370, **369**; introduction to 362–363; technology-focused initiatives 367–368; transboundary issues 367; types of 363
regional cooperation initiatives, in Southeast Asia 368–370, **369**
Regional Cooperation on Renewable Energy, EU 366–367
regional development: defined 239; globalization and 239
regional governance, ASEAN reform and: contesting 117–118; reconfiguring 110–113; widening/limiting 113–117
Regional Greenhouse Gas Initiative (RGGI) 368

Reid, A. 55, 56
Reilly, J. 149
relations of disjuncture 7
religion, development and 250–258; development shaping religion 252–254; intersection of, outcomes at 256–257; introduction to 250–252; religion shaping development 254–256
religion shaping development 254–256; in Indonesia 254, 255–256; outcomes of 256–257; in Philippines 254; in Thailand 255
remittance flows, migration and 198–208; brain drain and 206–207; capital for productive activities and 203; family separation and 207–208; host country development and 205; as insurance against economic crisis/ natural disaster 203–204; mapping 199–201, **201**; national economic growth and 201–202; optimism for 201–205; overview of 198–199; pessimism for 205–208; poverty reduction/ welfare gains and 202–203; returning skilled workforce and 204; unequal access/ inequalities and 205–206; working conditions and 207
resettlement policies 233
Resurrección, B. P. 11
Reynolds, C. 59
Rice Christians 255
Rigg, J. 7, 10, 276, 278, 317
Riley, L. 218–219
Rimmer, P. J. 120
Ritchie, M. 276
Road to Serfdom, The (Hayek) 28
Roberts, J. 159
Robinson, J. 53
Rocheleau, D. E. 262
Rodan, G. 113
Röpke, W. 28
Rougier, L. 28
Roundtable on Sustainable Palm Oil (RSPO): Certified Sustainable Palm Oil 331, 333; establishment of 332–333; requirement for documented/uncontested land ownership 332; Smallholder Group Scheme (SGS) 336–337; smallholders defined by 330; stakeholders in 333
Royal Ordinance on Fisheries (Government of Thailand) 321
rural livelihoods, Southeast Asia 39–50; aggregate trends in countryside **40**, 40–42; class and inequality of 44–47, **45**; climate change and 48–49; future for 49–50; gender/generation and 42–44, *43*, **44**; introduction to 39; land and 46–47; land and changing contexts of **306**, 306–308, *307*; poverty and wellbeing in **47**, 47–48; smallholders and **45**, 45–46
Rüstow, A. 28

Sabah Land Ordinance 332
Sacred Nature 54
Saengtienchai, C. 177
Salamanca, A. 10
Sangkapitux, C. 12
San Miguel Corporation 308
Sarawak, independent smallholders and certification in 336–338
Sarawak Rainforest World Music Festival 159
Sattayanurak, A. 311
Savage, V. R. 10
Save the Children 217–218
Schultz, T. 300
Scott, J. 6, 57, 59
Scott, J. C. 44–45
"Seafood not Slavefood" (EJF campaign) 317
seafood value chains 316–327; components of 316; Fair Trade standard for fisheries and 322–323; gendered division of labor within 325; governing fish-sea-labor interactions 320–322; growth and significance of 316–317, *317*; introduction to 316–319; labor and 319–322; performance and equity 322–323; social/environmental sustainability coverage 317, **318**; social issue areas in *319*; vulnerabilities, addressing underlying 323–326
secondary forests 395
Seeberg-Elverfeldt, C. 383
Sendai Framework for DRR: 2015–2030 (Sendai Framework) 351, 352, 355–356
Setthakhid Muubaan Thai nai Adit (Nartsupha) 275
Severino, R. 110
sex tourism 160
Shearmur, R. G. 125
shifting cultivation 398
Shinawatra, T. 110
shock doctrine 29
"Shrimp Sustainable Supply Chain Task Force" 321
Sidaway, J. D. 7
Silvey, R. 207–208
Sime Darby Berhad 308
Simpson, A. 234
Singapore 4; development indicators **8**, **9**; ethnic minorities/indigenous peoples in **226**; freshwater withdrawals *291*; Human Development Index (HDI) **213**; irrigation water withdrawals **293**; modernization and world religions in 252–253; Newly Industrialized Country label in 20; remittance flows 199, **201**; renewable water resources *290*; total cumulative MDO allocations to **76**; tourist arrivals by **155**; water, sanitation and health indicators **295**; World Gourmet Summit 159; World Risk Index for **345**
single ocean trade 56
Sitt, A. 116

Sixth Extinction, The (Kolbert) 54
slash-and-burn 232, 398
small and micro industry (SMI) 120–121; dynamics and resilience of 123–125; employment in, in JMA **124**; number of, in core and suburbs of JMA **124**; spatial redistribution of 125–126, **126**; specialization of 126–128, **128**
small business, tourism and 163–164
Smallholder Group Scheme (SGS) 336–337
smallholders 309; defined 330; incomes 330–331; palm oil market and (*see* oil palm cultivation by independent smallholders); rural livelihoods and **45**, 45–46; sustainable palm oil certification and 332–333
Sobritchea, C. 178
social capital 137–138
Socialist Republic of Vietnam 4
social movements supporting ethnic minorities/indigenous groups 234
socio-economic status, children's well-being and 213, **213**
Solheim, W. 59
Solidarity for Asian People's Advocacy (SAPA) network 115
Somphone, S. 118
Southeast Asia: aggregate trends in countryside **40**, 40–42; characteristics of 16; development in 7–9, **8** (*see also* development, in Southeast Asia); development institutions in (*see* development institutions/economies in Southeast Asia); disasters in 342–346; diversity in 3; environmental state in (*see* human-nature relationships); future of 3; identity 6; introduction to 3–4; land tenure systems 57–58; map of 5; neoliberalism in (*see* neoliberalism); political/cartographic designations of 16–17; progress/modernization in 17–18; as region 4–7, 15–17; rural livelihoods in 39–50 (*see also* rural livelihoods, Southeast Asia); term use of 15
Southeast Asia Collective Defense Treaty 17
South-East Asia Command (SEAC) 15
Southeast Asia in the Age of Commerce 1450–1680 (Reid) 56
Southeast Asian forests *393*, 394–396, *397*; causes of forest change in *396*; classification systems 394–395; definitions of 395; deforestation of 395, 395–396, *396*
Southeast Asian Treaty Organization (SEATO) 17
Special Panels for Serious Crimes (SPSC) 97, 100
sports tourism 159–160
Springer, S. 10
Standing, G. 87
Stark, R. 56
Start and Improve Your Own Business (SIYB) 88
"State of the World's Children" (UNICEF) 212
Stibig, H-J. 395–396
Stiglitz, J. 77

street connected children 216–217
Stringer, C. 324
structural adjustment 71
Sturgeon, T. 326
Sustainable Development Goals (SDGs) 212, 220–221, 326, 356
Sustainable Energy Technology at Work (SETatWork) 367
sustainable palm oil certification, smallholders and 332–333
Suvarnadwipa (Goldland) 5
Swedish East India Company 56
swidden agriculture, density/diversity and 232, 397–400

Tan-Mullins, M. 10, 149
Taplin, R. 365
Teaching and Temples Tour of Cambodia 161–162
technology-focused initiatives, regional cooperation 367–368
Thai, H. C. 177
Thai-German Highland Development Program 19
Thailand 4; ADB sovereign approvals for **75**, 77; aggregate trends in **40**; agriculture percentage of GDP for *307*; average farm size trends in **45**; capture fisheries, employment in **320**; cereal crop yield in **303**, *304*; change in forested area *395*; child labor laws 216; Chulalongkorn Center for Migration Studies 325; Compulsory Migrant Health Insurance scheme (CMHI) 193; cultural heritage tourism 159; development indicators **8**, 9; disability, defined 242; disability-based discrimination in 241; ethnic minorities/indigenous peoples in **226**; forest products exports, value of *397*; forest sector contribution to total GDP *397*; freshwater withdrawals *291*; gender/generation interplay in *43*; greying of farm labor force in **44**; healthcare coverage 189, 193; Human Development Index (HDI) **213**; irrigation water withdrawals **293**; land reforms in **302**; land under cereal crop production for *303*; loss of life/economic damage due to disasters in **346**; medical tourism in 161; mountain people label in 229; National Ecotourism Policy 158; PES policy 384 (*see also* Mae Sa watershed PES scheme); post-war aid/development investment in 18–19; religion shaping development in 255; remittance flows 199, **201**; renewable water resources *290*; rice pledging scheme 49; Royal Ordinance on Fisheries 321; rural, urban and national poverty rates for **47**; rural population of **306**; "Shrimp Sustainable Supply Chain Task Force" in 321; total cumulative MDO allocations to **76**; tourism sector in 154, 155; tourist arrivals by **155**; water, sanitation and health indicators **295**; women migrants working on shrimp farms in 325; World Bank commitments for **75**, 76–77, 78; World Risk Index for **345**
Thanh Giang, L. 202
Thatcher, M. 28
Thatcherism 29
Thein, H. H. 65
Thein, S. 90, 92, 116
Thomalla, F. 12, 356
Thomas, A. 12, 405
Thomas, D. 378, 388
Thomas-Slayter, B. P. 262
Thompson, E. 277–278
Thung Sai Yaresuan Wildlife Sanctuary 403
Timor-Leste 4, 6; aggregate trends in **40**; agriculture percentage of GDP for *307*; CAVR report and 102; cereal crop yield in **303**, *304*; development indicators **8**, 9; freshwater withdrawals *291*; Human Development Index (HDI) **213**; independence 16; international justice tribunals in 66; irrigation water withdrawals **293**; land reforms in **302**; land under cereal crop production for *303*; Mission to Timor-Leste projects 88–89; post-conflict reconciliation in 96, 99–100; remittance flows **201**; renewable water resources *290*; reparations in 103–104; rural population of **306**; Special Panels for Serious Crimes in 100; victim associations in 101–102; water, sanitation and health indicators **295**; World Risk Index for **345**
Tipco Foods Public Company Limited 387
Tong, C. K. 245, 253
total actual renewable water resources (TARWR) 289
tourism, development policy tensions and 232
tourism industry 153–166; backpacker tourism and 157–158; culture and 159; employment and 162–163; environmental/cultural change and 164–165; film tourism 159; introduction to 153–156; medical tourism and 160–161; natural places/ecotourism and 158; niches in 156–162; regional sources of visitors to ASEAN **155**; sex tourism 160; small business and 163–164; sport/adventure and 159–160; tourist arrivals by country in Southeast Asia **155**; voluntourism and 161–162
To Xuan Phuc 278
Toyota, M. 180
Trafficking in Persons (TIP) report (US State Department) 325
Tran, G. L. 177
Trankell, I. B. 229
trans-ASEAN energy networks 364
Trans-ASEAN Gas Pipeline 364, 367, **369**, 370
transboundary issues, regional cooperation 367
transnational care strategies among privileged migrant families 179–182

transnational corporations (TNC) 122
Transparency in Supply Chains Modern Slavery Act 2015 (UK) 325
tribunal justice, post-conflict reconciliation and 99–100
trickle-down economics 28
Tsing, A. 6, 404
Tubtim, T. 11
Tuol Sleng Museum 102–103
Turner, S. 11, 22

UBS-INSEAD 255
Uddin, N. 12, 365
Uddin, S. N. 366
UNESCAP 344
unfree labor, on fishing vessels 319
UNICEF: *Progress for Children* report, 2015 14
Unilever 333
UNISDR 351–352, 355
United Nations: Convention on the Rights of Persons with Disabilities (CRPD) 241–242; Millennium Development Goals (MDGs) 211–212; Sustainable Development Goals (SDGs) 212, 230–231
United Nations Convention on the Rights of the Child (UNCRC) 211, 219–220
United Nations Development Program (UNDP) 213
United Nations Framework Convention on Climate Change (UN FCCC) 362, 371, 403
United Nations Transitional Authority in Cambodia (UNTAC) 30
Unlad Kabayan Migrant Services Foundation Inc. 21–22, 203
unsustainable natural resources, development/ disaster risk in Southeast Asia and 349
UNTAC (UN Transitional Authority in Cambodia) 98
urban deconcentration 121–122
urbanization, development/disaster risk in Southeast Asia and 348, *349*
US-Cambodia Bilateral Textile Trade Agreement 87–88

Vaddhanaphuti, C. 229
Value Chain Development 22
Vandergeest, P. 276, 321–322, 395
van der Geest, S. 181–182
Van Schendel, W. 6, 15
Verghis, S. 11
Vientiane Action Program 114
Vietnam: ADB sovereign approvals for **75**; aggregate trends in **40**; agriculture percentage of GDP for *307*; capture fisheries employment in **320**; cereal crop yield in **303**, *304*; change in forested area *395*; Chinese aid to 143; cultural heritage tourism 159; development indicators **8**, 9; ethnic minorities/indigenous peoples in **227**; forest products exports, value of *397*; forest sector contribution to total GDP *397*; freshwater withdrawals *291*; Human Development Index (HDI) **213**; irrigation water withdrawals **293**; land reforms in **302**; land transfer in 49; land under cereal crop production for *303*; Payments for Ecosystems Services (PES) in 379–381; post-socialist cultural preservation in 228–229; remittance flows **201**, 202, 203, 206; renewable water resources *290*; rural, urban and national poverty rates for **47**; rural population of **306**; total cumulative MDO allocations to **76**, 81; tourism sector in 154–155; tourist arrivals by **155**; water, sanitation and health indicators **295**; World Bank commitments for **75**; World Risk Index for **345**
Vietnam Household Living Standards Survey 202, 206
Vision 2020 114
voluntourism 161–162

Walker, A. 278
Wal-Mart 319
Wang, G. 55
Wang, H. 176
Wangari, E. 262
Washington consensus 65; described 70; development policy of **73–74**; influence of, in MDO activities 71; in Philippines, impact of 71–72; Williamson on 71
water, development and 285–297; dams and 290–292; dimensions of **286–287**; distribution changes in 289–292; drinking water supply and 294–296, **295**; historical shifts in 287–289; integrated approaches to 287; introduction to 285; irrigation and **293**, 293–294; issues, approaching 286–287; water cultures and 296–297; water scarcity and 289–290, *290*, *291*
water cultures, relations with water and 296–297
water distribution, material changes in 289–292
water scarcity 289–290, *290*, *291*
welfare gains, migration and 202–203
Western Climate Initiative (WCI) 368
White, B. 44
Wichterich, C. 266–267
Wild Asia 334–335
Wilding, R. 181
Williamson, J. 71
Wilson, E. O. 54
Winrock Foundation 387
Withayapak, C. 276
Wolfensohn, J. 77
Wolters, O. W. 16, 56
Woods, O. 11

Woolcock, M. 137–138
World Bank 10, 30, 65, 69, 214; *Agriculture for Development* report 46, 48; commitments for Southeast Asian developing member countries **75**; East Asian Miracle 29; National Community Driven Development Project 82; neoliberalism and 30–31; projects in Philippines 72
World Health Assembly (WHA) 192
World Health Organization (WHO) 189
World of Living Things, The (Imanishi) 53–54
World of Work Report (ILO) 87
World Risk Index 344, **345**
World Trade Organization 239
World Until Yesterday, The (Diamond) 58
World Vision 22
Wright, S. 132
Wunder, S. 378

Yang, D. 204
Yeoh, B. S. A. 11, 177, 178
Yongchaiyudh, C. 110

Zhou, E. 143
Zomia 6